Springer Complexity

Springer Complexity is an interdisciplinary program publishing the best research and academic-level teaching on both fundamental and applied aspects of complex systems—cutting across all traditional disciplines of the natural and life sciences, engineering, economics, medicine, neuroscience, social, and computer science.

Complex Systems are systems that comprise many interacting parts with the ability to generate a new quality of macroscopic collective behavior the manifestations of which are the spontaneous formation of distinctive temporal, spatial or functional structures. Models of such systems can be successfully mapped onto quite diverse "real-life" situations like the climate, the coherent emission of light from lasers, chemical reaction-diffusion systems, biological cellular networks, the dynamics of stock markets and of the internet, earthquake statistics and prediction, freeway traffic, the human brain, or the formation of opinions in social systems, to name just some of the popular applications.

Although their scope and methodologies overlap somewhat, one can distinguish the following main concepts and tools: self-organization, nonlinear dynamics, synergetics, turbulence, dynamical systems, catastrophes, instabilities, stochastic processes, chaos, graphs and networks, cellular automata, adaptive systems, genetic algorithms, and computational intelligence.

The three major book publication platforms of the Springer Complexity program are the monograph series "Understanding Complex Systems" focusing on the various applications of complexity, the "Springer Series in Synergetics", which is devoted to the quantitative theoretical and methodological foundations, and the "Springer Briefs in Complexity", which are concise and topical working reports, case studies, surveys, essays and lecture notes of relevance to the field. In addition to the books in these two core series, the program also incorporates individual titles ranging from textbooks to major reference works.

Indexed by SCOPUS, INSPEC, zbMATH, SCImago.

Series Editors

Understanding Complex Systems

Founding Editor: Scott Kelso

More information about this series at https://link.springer.com/bookseries/5394

Till D. Frank

COVID-19 Epidemiology and Virus Dynamics

Nonlinear Physics and Mathematical Modeling

Springer

Till D. Frank
Department of Psychological Sciences
University of Connecticut
Storrs, CT, USA

ISSN 1860-0832 ISSN 1860-0840 (electronic)
Understanding Complex Systems
ISBN 978-3-030-97177-9 ISBN 978-3-030-97178-6 (eBook)
https://doi.org/10.1007/978-3-030-97178-6

This Springer imprint is published by the registered company Springer Nature Switzerland AG
The registered company address is: Gewerbestrasse 11, 6330 Cham, Switzerland

To those who have died from COVID-19

Preface

Coronavirus disease 2019 (COVID-19) is a disease that was first reported in China in December 2019. The disease is caused by the SARS coronavirus 2 (SARS-CoV-2). Since December 2019, the disease turned into a deadly pandemic. At the time of writing this book, the pandemic is still ongoing.

From an epidemiological point of view, populations are composed of interacting subpopulations. Likewise, the virus dynamics in an individual is characterized by several components that evolve in time and interact with each other. In view of this key feature of virus infections in populations and individuals, the question arises what is the appropriate physics perspective to address the pandemic and the disease? In general, what is the appropriate physics perspective to address epidemics and virus infections in individuals?

This book provides answers to this question. An analysis of the subpopulation dynamics reveals that epidemics in populations arise from instabilities in which subpopulations organize themselves to a whole in a certain way. Likewise, a virus infection in a human body arises from an instability and is characterized again by a certain organization of the cell populations and viruses involved. In short, the book demonstrates that epidemics in populations as well as virus infections in individuals are characterized by instabilities and an emerging order. The appropriate theoretical framework to address such phenomena is the theory of nonlinear physics and synergetics.

The aim of this book is to study COVID-19 epidemics in countries and SARS-CoV-2 infections in individuals from the nonlinear physics perspective and to model explicitly COVID-19 data observed in countries and viral load data observed in COVID-19 patients. The first part of this book provides a short technical introduction into amplitude spaces given by eigenvalues, eigenvectors and amplitudes. Emphasis is given to the description of positive eigenvalues and their corresponding unstable eigenvectors and amplitudes. The triple given by a positive eigenvalue, an unstable eigenvector, and its amplitude captures within the amplitude space description the key dynamics that underlies any instability-induced phenomenon of interest. In the second part of the book, mathematical models of epidemiology are introduced such as the SIR and SEIR models. The nonlinear physics descriptions of these models are worked out in amplitude space and the positive eigenvalues and the unstable eigenvectors are identified. COVID-19 outbreaks in various countries around the globe are analyzed. It is shown that these outbreaks were characterized by positive eigenvalues and evolved in time along unstable eigenvectors. The unstable eigenvectors define the characteristic order in which these outbreaks took place and, using the terminology of synergetics, are referred to as order parameters. The impacts of measures implemented to stop the spread of COVID-19 are also discussed. In particular, it is shown how intervention measures can lead to bifurcations that stabilize the originally unstable disease-free states of populations. Stabilizing disease-free states causes COVID-19 epidemics to subside at least locally and temporarily until these states are destabilized again and new COVID-19 waves are triggered. In the third part of the book, the individual level is considered. Virus dynamics models such as the TIV and TIIV models are described and, again, amplitude space descriptions involving eigenvalues, unstable eigenvectors, and amplitudes are worked out. Viral load data from a sample of COVID-19 patients are analyzed. It is shown that the disease in the bodies of those patients emerged from instabilities characterized by positive eigenvalues and that the disease states of those patients evolved along patient-specific unstable eigenvectors. Subsequently, the disease dynamics branched off from the unstable eigenvectors. According to the TIV and TIIV models, the disease induced a stabilization of the original instabilities leading to a recovery of the patients.

This book is written for researchers, modelers, and graduate students in physics and medicine, epidemiology and virology, biology, applied mathematics, and computer sciences. This book identifies the relevant mechanisms behind past COVID-19 outbreaks and in doing so can help efforts to stop future COVID-19 outbreaks and other epidemic outbreaks. Likewise, this book points out the physics underlying SARS-CoV-2 infections in individuals and in doing so supports a physics perspective to address human immune reactions to SARS-CoV-2 infections and similar virus infections.

Storrs, CT, USA Till D. Frank
December 2021

Acknowledgments I wish to thank Dr. H. Niko and Mrs. J. Krishnamoorthi of Springer Publishers for their help in preparing this manuscript.

Contents

Chapter 1
Introduction

This introductory chapter reviews some tragic milestones of the COVID-19 pandemic. Subsequently, the chapter introduces some basic concepts of epidemic viral infections and virus dynamics. The key concepts of nonlinear physics relevant for understanding the spreading of viral infections in populations and viral infections in individuals are also presented. The chapter is closed with considerations on religion and physics in order to set up the general framework of this book.

1.1 COVID-19 Outbreaks and SARS-CoV-2 Infections and the Physics Behind Them

The coronavirus disease 2019 (COVID-19) is an infectious disease that is caused by the SARS coronavirus 2 (SARS-CoV-2) [1]. Patients suffering from COVID-19 were first reported in December 2019 in Wuhan city, China [2]. During 2020, the disease spread out over other Asian countries, subsequently hit Europe and, eventually, by January 2021, could be found in all 8 geographic regions of the world: Africa, Asia, the Caribbean, Central America, Europe, North America, Oceania and South America [3]. While some people infected by SARS-CoV-2 do not experience any symptoms or experience only mild symptoms, for other people COVID-19 is a deadly disease. After just one year, by January 2021, the worldwide death toll was about 2,000,000 people [4]. Half a year later, by July 2021, a total of about 4,000,000 deaths associated with COVID-19 were reported worldwide [5]. For comparison purposes, Croatia, a European country, has a population of 4,100,000 people and Oklahoma, a US state, has a population of 4,000,000 people. That is, COVID-19 claimed within one and a half years the lives of as many people as there are living in Croatia or in the US state of Oklahoma. Other countries and US states in the range of 3.5 to 5 million people

T. D. Frank, *COVID-19 Epidemiology and Virus Dynamics*,
Understanding Complex Systems, https://doi.org/10.1007/978-3-030-97178-6_1

Table 1.1 Worldwide COVID-19 associated deaths as of July 2021 by comparison with populations counts of some countries and US states (with population counts taken from Ref. [6] accessed in August 2021). The down-arrow ⇓ in the first column indicates that the number of COVID-19 deaths is increasing and, consequently, over time will shift down the rows of the table

COVID-19 deaths Jan. 2020–July 2021	Countries and their populations	US states and their populations
4,000,000 ⇓	Uruguay 3,500,000	
	Eritrea 3,600,000	Connecticut 3,600,000
	Croatia 4,100,000	Oklahoma 4,000,000
	Kuwait 4,300,000	Oregon 4,200,000
	Panama 4,400,000	Kentucky 4,500,000
	Mauritania 4,800,000	Louisiana 4,700,000
	Ireland 5,000,000	Alabama 5,000,000

are listed in Table 1.1. They may also be used to put the number of 4,000,000 people into perspective.

Research on COVID-19 is an interdisciplinary effort. Efforts to model the spread of the disease in populations and the virus dynamics in individuals are frequently based on the principles of nonlinear physics [7, 8] and bifurcation theory [9]. In this context, synergetics, a particular school of nonlinear physics and self-organization [10–14] provides a promising framework since it traditionally is concerned with pointing out the general principles that underly seemingly unrelated phenomena in a variety of disciplines ranging from physics to biology, chemistry, economics, and medicine [10–15]. This book presents a detailed discussion of the general nonlinear physics principle that underly the spread of COVID-19 in populations and the multiplication of the virus causing COVID-19 in the bodies of COVID-19 patients. In doing so, fundamental mechanisms that determine COVID-19 outbreaks on the population level and SARS-CoV-2 infections on the level of individuals are revealed. According to such a mechanistic, nonlinear physics point of view, COVID-19 outbreaks and infections can be seen in the context of similar well-studied phenomena in a plenitude of systems and disciplines. Identifying the relevant nonlinear physics principles of COVID-19 outbreaks in populations and SARS-CoV-2 infections in individuals provides researchers with a departure point and a sound basis to discuss details and peculiarities of COVID-19 on the level of populations and individuals.

1.2 Epidemic Viral Infections

Infectious diseases, in general, are diseases caused by certain particles (or agents) that enter humans or animals and cause certain diseases in their bodies [16, 17]. Such particles can be for example viruses, bacteria, and worms [16, 17]. This book is concerned with infectious diseases caused by viruses, that is, with virus infections and viral infectious diseases. Viruses can be transmitted between people in various ways, for example, by person-to-person transmissions or airborne transmissions. In a person-to-person transmission virus is transmitted from a person X to another person Y. The transmission may involve direct or indirect contact. The human immunodeficiency virus (HIV) causing AIDS is frequently transmitted by sexual contact, that is, direct contact [16–18]. Syphilis and gonorrhea are further examples of sexually transmitted diseases that involve direct contact. However, they are bacteria (i.e., not viral) infections. Influenza is a viral infection that is typically transmitted between people by indirect contact. An infected person X can create droplets in the air that contain the influenza virus when coughing, sneezing, or talking. If a person Y standing nearby inhales some of those droplets, the virus enters the body of that person and the person may become infected [17, 19]. This transmission mechanism via air and droplets is referred to as airborne transmission. Objects may have influenza viruses produced by X on them. There is a chance that when a person Y touches such objects and, subsequently, touches his or her mouth, nose, or eyes, some viruses are picked up and enter the body of Y [17, 19]. Just like the influenza virus, it is believed that SARS-CoV-2 is transmitted by respiratory droplets and surface contacts [20, 21].

If a virus spreads out in a population of a particular region or country, the occurrence of the disease is referred to as an epidemic. If the infectious disease spreads out across several countries or continents, then the disease emergence is referred to as a pandemic. In view of the worldwide presence of COVID-19, not surprisingly, the World Health Organization (WHO) declared in 2020 the spread of COVID-19 as a pandemic [22]. In this context, note however, that in this book the term COVID-19 epidemic will be used when referring to the spread of COVID-19 in a certain country (e.g., China) or region (e.g., the European countries), respectively.

Historically, several epidemics and pandemics of viral infectious diseases have claimed millions of lives. Table 1.2 lists three of them. During 1918 and 1919 the influenza virus triggered a pandemic. Conservative estimates state that about 20 million people died during this pandemic [17, 23]. AIDS was first reported in 1981. The WHO estimates that since the beginning of the HIV pandemic up to the year 2020 about 36 million people died from HIV-related causes [24]. Finally, as mentioned above, as of July 2021, about 4 million COVID-19 associated deaths have been reported.

Looking at annual data, while in many parts of the world vaccines for influenza are available, influenza is still claiming the lives of many people every year. It is

Table 1.2 Three infectious disease epidemics/pandemics and their death tolls (as of 2020/2021)

Years	Epidemic/pandemic	Deaths
1918–1919	Influenza	20 million
1981–2020	AIDS	36 million
Jan. 2020–July 2021	COVID-19	4 million

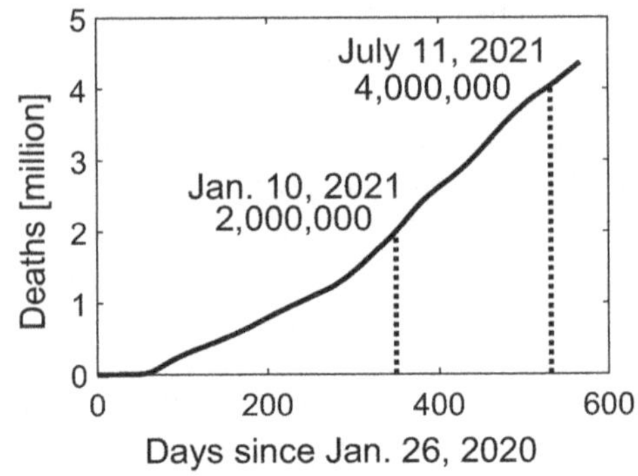

Fig. 1.1 Cumulative, worldwide COVID-19 deaths during the period from January 26, 2020 to August 15, 2021. Weekly data retrieved from Ref. [27] accessed on August 16, 2020

estimated that there are 300,000 to 600,000 influenza-related deaths worldwide every year [25, 26]. HIV-related deaths peaked around 2005 with about 2 million per year. The number dropped to about 700,000 deaths per year in the year 2020 [24]. As mentioned above, in the first year of the COVID-19 pandemic (more precisely by January 2021), COVID-19 had claimed 2 million lives. Since by July 2021 COVID-19 reached the tragic death toll of 4 million, it follows that the COVID-19 deaths per year for the year 2021 will be more than 2 million. Figure 1.1 shows a graph of the worldwide cumulative deaths associated with COVID-19 in the period from January 2020 to August 2021. The tragic milestones of 2 million and 4 million deaths are indicated as well.

The time course of epidemics has often been modeled with the help of differential equations [16–18]. Differential equations can describe the increase and decrease of the number of individuals belonging to a certain population. Two populations are of particular interest. The population of healthy individuals that do not infect others and the population of infectious individuals that infect others. The former are also called the susceptible individuals. Accordingly, the infectious individuals infect the susceptibles. In doing so, susceptibles become infectious individuals. Frequently, a third population is added and placed in-between the two aforementioned populations: the population of exposed individuals. Exposed individuals are those that have been infected by the infectious people [18]. However, they are not yet infectious. Only after a certain period, they become infectious and, consequently, make a transition from the population of exposed individuals to the population of infectious individuals. In the context of the three-population model, individuals make two transitions. They begin with as healthy individuals. Some of the healthy individuals become infected and make transitions to the population of exposed individuals. Subsequently, after a certain period, they become infectious themselves and make the transition to the class of infectious individuals. Note also that in the context of the three-population model, being infected does not mean the same as being infectious.

As we will see in the following section (Sect. 1.3), on the level of the virus dynamics within individuals there is a three-variable model that can be regarded as counterpart to the three-population model. In fact, the mathematical structures of both models are almost identical (Sect. 9.3.3). Therefore, according to those models, epidemic viral infections on the population level and the course of virus infections in individuals should exhibit similar characteristic properties.

1.3 Virus Dynamics

There are hundreds of different viruses causing diseases in humans, animals, and plants [23, 28]. Viruses are particles that contain deoxyribonucleic acid (DNA) or ribonucleic acid (RNA), enter certain cells, and are replicated by them [23, 29, 30]. That is, two key features of viruses are: they contain genetic material in terms of DNA or RNA and are produced by certain cells once they have entered them. The cells that are invaded and infected by a virus are often referred to as target cells [17, 25]. The humans, animals, or plants to which those target cells belong are referred to as hosts. The production of viruses by infected target cells may be described in terms of the following steps [23, 31]:

1. A virus enters a target cell.
2. The virus changes the biochemical reactions of the target cell.
3. As a result, the target cell produces the genetic material of the virus.
4. In addition, the target cell produces the remaining material that constitutes the virus.
5. The genetic material and the constituents are put together to form virus particles and the fully assembled particles are released from the cell.

As far as viral human infections are concerned, there is an important follow-up step: the viruses released from a virus-producing cell spreads around and enters other target cells. They, in turn, begin with the production of the virus and release viruses. This mechanism typically leads to an increase of the viral load in the body of an infected individual (i.e., the human host). In order to describe epidemic or pandemic viral infections, that is, the spread of a virus across a population in a given country or region and beyond, additional steps may be added such as that the virus enters the body of an individual in the first place and that it leaves the body of that individual, for example, by sneezing and talking or during sexual intercourse (see Sect. 1.2). In this context, the aforementioned five-step sequence may be modified like:

1. A virus enters the body of an individual.
2. The virus enters a target cell of the individual.
3. The virus changes the biochemical reactions of the target cell.
4. As a result, the target cell produces the genetic material of the virus.
5. In addition, the target cell produces the remaining material that constitutes the virus.

Table 1.3 Common entities of epidemics in populations and virus infections in individuals

Entities	Epidemics (population level)	Virus infections (individual level)
Healthy entities	Susceptible individuals	Target cells
Entities turning into or producing infectious entities	Exposed individuals	Infected, virus producing cells
Infectious entities	Infectious individuals	Viruses

6. The genetic material and the constituents are put together to form virus particles and the fully assembled particles are released from the cell.
7. The produced viruses enter further target cells (see step 2).
8. The individual transmits a part of the produced viruses to other individuals.

As far as human diseases are concerned, the influenza A virus causes seasonal influenza also called the flu [25]. The human immunodeficiency virus (HIV) causes the acquired immunodeficiency syndrome (AIDS) [16–18, 23]. The SARS coronavirus 2 (SARS-CoV-2) causes coronavirus disease 2019 (COVID-19) [1].

The time course of the virus concentration or viral load in an infected individual may be described with the help of a three-variable model given in terms of three coupled differential equations. The model involves the target cells, the infected, virus producing cells, and the virus [17, 23, 25]. Accordingly, viruses enter target cells and turn them into infected, virus producing cells. The infected, virus producing cells produce and release viruses that in turn infect other target cells. As mentioned in Sect. 1.2, there is an analogy between the three-population model for epidemics and the three-variable model for virus infections. Table 1.3 illustrates this analogy.

As will be shown in this book, the mathematical models are almost identical (see Chap. 9 and Sect. 9.3.3 in particular). The equations of both models exhibit the same structure. However, they differ with respect to the number of parameters. The virus dynamics model exhibits an additional parameter. The reason for this is that in the epidemiological model one exposed individual can only "produce" one infectious individual, namely, by turning into an infectious individual. In contrast, in the virus dynamics model, an infected, virus producing cell can produce many copies of the virus. Therefore, the virus dynamics model exhibits an additional parameter that describes how many viruses are produced (per time unit) by a single infected, virus producing cell. Overall, in view of the analogy between the two three-variable models and similar models in epidemiology, on the one hand, and models of virus dynamics, on the other hand, epidemics in populations and virus infections in individuals share common properties and exhibit similar phenomena. What can be seen in epidemics on the level of populations may be seen in individuals when looking at their disease progressions and vice versa (see also Sect. 1.5).

1.4 Instabilities

It has been argued above that the spread of infectious diseases in populations mimics what happens in the bodies of infected individuals and vice versa. However, there is a much broader perspective to address epidemics and virus dynamics: the perspective of nonlinear physics. An infectious disease spreads out in a population due to an instability [32, 33]. Likewise, a virus multiplies in the body of an individual due to an instability. Instabilities are very well-studied phenomena in nonlinear physics. Consequently, what happens during a pandemic such as the COVID-19 pandemic and what happens in a patient suffering from a viral infection such as a SARS-CoV-2 infection are not unpresented phenomena. From a nonlinear physics perspective, there is nothing fundamentally novel or special about infectious diseases in populations and individuals. Infectious diseases in population and individuals and, in particular, the COVID-19 pandemic and SARS-CoV-2 infections in COVID-19 patients, belong to the class of instability-related phenomena that have been extensively and thoroughly studied in nonlinear physics over more than 40 years [7, 8, 10–12].

A mechanical example of an instability is a marble placed on the top of a surface that is curved like a hill. If the marble is placed perfectly on the top, it remains in this position forever. However, a small perturbation of the marble makes that it rolls down the curved surface. Another mechanical example of an instability is the famous buckling instability of solid materials. If a straight, solid bar is pushed together (i.e., squeezed) a little bit, then the bar typically shortens a bit but remains its straight shape. If the squeezing forces become stronger and exceed a certain critical value, then the bar buckles. That is, the straight bar is deformed and a kink emerges. Heated fluids and gases are further examples of systems that exhibit instabilities [7, 10, 12, 34]. Many of those instabilities occur when fluids or gases are heated from one side such that there is a temperature difference across the layer or volume under consideration. A prominent example is the instability in the earth mantle. A part of the earth mantle is fluid. This part is not at rest. Due to temperature differences across the earth mantle the resting, homogeneous state is unstable. Rather, the fluid part of the earth mantle exhibits huge convection rolls in which fluid is streaming up and down like the cabins of a Ferris wheel in an amusement park [10]. Likewise, the air of the earth atmosphere does not show a resting, spatially homogeneous pattern. Such a pattern is unstable because the earth surface is hot relative to the outer space. Huge rotating cells composed of flowing air have emerged from that atmospheric instability and determine the spatio-temporal dynamic pattern of the earth atmosphere [10]. While the spatio-temporal dynamic fluid and gas patterns in the earth mantle and the earth atmosphere exists day after day, various short-lived patterns in the earth atmosphere emerge from instabilities as well. Roll-shaped clouds emerge under appropriate circumstances from instabilities of the clear sky air. Hurricanes emerge over the Atlantic Ocean during the fall months, when the temperature over the ocean is sufficiently high [10]. Moreover, lightening is an electrical instability of the air atmosphere.

In optical engineering, laser devices produce laser light when the ordinary light in their cavities becomes unstable [12]. In chemistry, spatially homogeneous states in which all chemicals are evenly distributed over space become unstable under appropriate conditions and colorful spiral patterns may emerge. Stripe and dot patterns on the skin of animals emerge from similar instabilities of the relevant biochemical systems [12, 35]. Human and animals make gait transitions when certain gaits become unstable [10]. For example, human walking becomes unstable at a certain speed such that individuals switch from walking to running. Likewise, horses switch from walk to trot at a certain speed at which the walking pattern becomes unstable. Brain activity patterns in humans emerge due to instabilities [10, 36]. Signaling pathways in cells can become unstable under certain circumstances such that the biochemistry in the cells switches from one biochemical reaction chain (or signaling pathway) to another chain (or pathway) [10]. Importantly, for all those instabilities the same principles of nonlinear physics apply. Moreover, all those instabilities are described by the same type of fundamental equations. These equations are called amplitude equations and will be discussed in general in Chap. 2. They will discussed in the context of epidemics of virus disease in Chaps. 4, 5, 6, and 8. They will be discussed in the context of viral infections in individuals in Chap. 10.

In order to illustrate the generality of the phenomenon that underlies epidemics and viral infections, Fig. 1.2 shows the increase of task-related brain activity in a certain area of a monkey brain due to an instability of the corresponding neural system (panel(a)) as discussed in Ref. [10] and the increase of COVID-19 cases in Italy during 2020 due to an instability of the state with zero COVID-19 infections (panel (b)). The respective data are shown by circles. In panel (b) the confirmed (i.e., diagnosed) COVID-19 cases are shown and modeled. In panels (a) and (b)

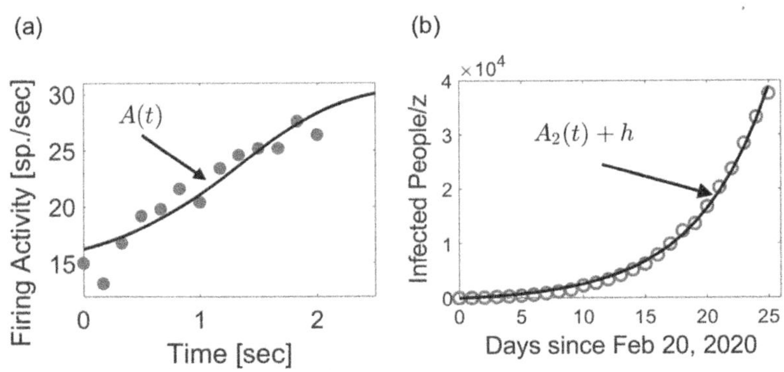

Fig. 1.2 Amplitude equation descriptions in neuroscience and epidemiology. Panel (**a**): Increasing firing activity of neurons (measured in spikes per seconds) in a particular region of the monkey brain (full gray circles) is described in terms of an amplitude $A(t)$ (black line) that evolves over time [10]. Panel (**b**): The number of confirmed individuals infected with COVID-19 in Italy during the first COVID-19 wave in 2020 (open gray circles) when rescaled by a scaling parameter z is described in terms of an amplitude $A_2(t)$ (black line) that evolves over time (see Sect. 4.5)

the model solutions of the relevant amplitude equations are shown by solid lines. Panel (b) will be discussed in Sect. 4.5. Figure 1.2 and the examples presented in the aforementioned paragraph should illustrate that whether we talk about a marble rolling down from an unstable position, lightening in the sky, stripe patterns on a zebra, a person switching from walking to a bus stop to running to a bus stop when a bus approaches, a task-related brain activity emerging in a human or animal brain, a SARS-CoV-2 infection increasing the viral load in a COVID-19 patient, or the COVID-19 pandemic emerging in the year 2020 in the human population of the planet Earth, then we talk always about the same fundamental phenomenon: the phenomenon of an instability.

1.5 Phase Transitions, Bifurcations, Unstable Eigenvectors, and Order Parameters

Instabilities are frequently induced by bifurcations [7, 10, 12]. Figure 1.3 illustrates three bifurcations: a bifurcation in a magnetic material (panel (a)), a bifurcation in a fluid heated from below (panel (b)), and a bifurcation in a population affected by a virus (panel (c)). As shown in panel (a), some type of magnetic materials do not show a

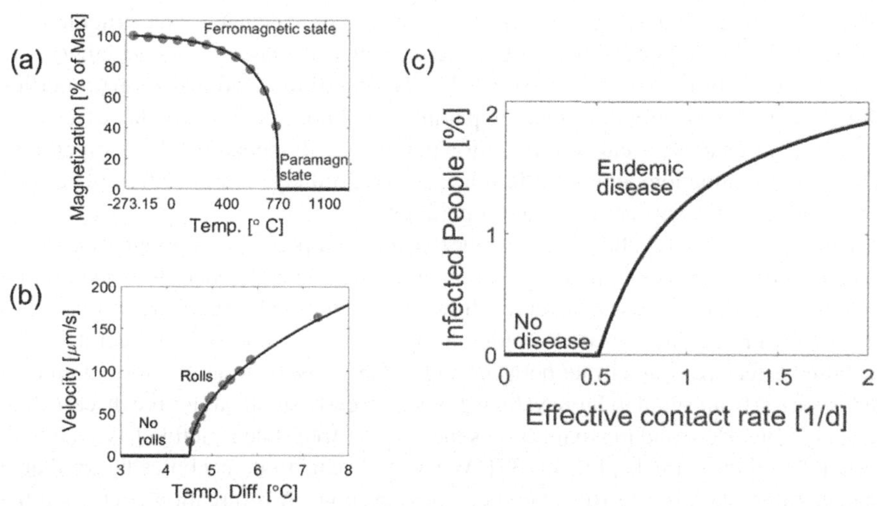

Fig. 1.3 Bifurcations in three systems. Panel (**a**): Bifurcation of a magnetic material from a non-magnetic to a magnetic state when temperature is lowered (circles describe data, while the solid line is the solution of theoretical model) [10]. Panel (**b**): Bifurcation of a fluid heated from below when the temperature difference is increased (circles describe data, while the solid line is the solution of an amplitude equation model) [10]. Panel (**c**): Bifurcation of the health or disease state of a population from a disease-free state to a state with an endemic fixed point (i.e., a state for which permanently a certain proportion of the population is infected)

magnetization at relatively high temperatures. At such temperatures their elementary magnets (i.e., spins) point in random directions. At sufficiently high temperatures, this kind of disordered state is stable. However, when the temperature is decreased below a critical value, the disordered state becomes unstable. The elementary magnets begin to point in the same direction. The material under consideration exhibits a non-zero magnetization that increases in magnitude when the temperature is lowered further (see panel (a)). Panel (a) shows experimental data as circles and the fit of a theoretical model as solid line [10]. The bifurcation from the non-magnetic to the magnetic state takes place at a temperature of about 770 °C.

Roll-shaped cloud-like patterns emerge in fluid and gas layers in laboratory experiments when the layers are heated from below and the temperature differences between the bottom and top of the layers exceed critical values [34]. That is, in those experiments increasing temperature differences make resting, homogeneous states unstable. In experimental studies such bifurcations can be revealed by measuring the roll velocity of the emerging rolls. Panel (b) shows measured roll velocities (as circles) as functions of the temperature differences in an experimental study on a fluid layer heated from below [10]. As can be seen in panel (b), in this experiment, rolls emerged at a critical temperature difference of about 4.5 °C. The stationary solution of a theoretical model given in terms of the cubic amplitude equation $\mathrm{d}A/\mathrm{d}t = \lambda A - cA^3$ is shown in panel (b) as well (solid line) and fits the data very well [10].

In general, when a state of a system becomes unstable due to a change of the conditions or a change of the structural properties of a system, we talk about a bifurcation. In particular, a change in the conditions or a change of the structure of a system such that the state of the system becomes unstable and a new state emerges is an example of a bifurcation [7, 9, 10, 12, 37]. The conditions and structural properties of a system are typically described by parameters. Those parameters that determine the stability of states are called bifurcation parameters. For example, the temperature of magnets and the temperature difference across fluid layers in the aforementioned laboratory experiments are bifurcation parameters.

Panel (c) of Fig. 1.3 shows the number of infected people in a population for an infectious disease that remains in the population over several generations (i.e., for an endemic infectious disease). Panel (c) shows the prediction for such circumstances as obtained from the three-variable epidemiological model addressed in Sect. 1.2. The so-called effective contact rate between individuals is shown on the horizontal axis and is varied from small to large values. For sufficiently small values the disease-free state is stable. Here and in what follows the disease-free state is defined as state with zero infected individuals [18, 32, 33]. When the contact rate becomes larger than a critical value, the disease-free state becomes unstable. A bifurcation occurs. In the population an endemic disease state emerges that describes that a certain proportion of the population under consideration is infected all the time. For the example shown in panel (c), the critical effective contact rate at which the bifurcation from a disease-free state to an endemic disease state takes place is 0.5 contacts per day. The notion of an effective contact rate will be explained in Chap. 3 and the graph shown in panel (c) of Fig. 1.3 will be derived in Sect. 3.5.2.

Figure 1.3 exemplifies bifurcations for three types of systems. As such bifurcations occur in all kind of systems [7, 10, 12]. Bifurcations take place when lightening emerges in a thunderstorm, hurricanes emerge over the oceans, lasers start to produce laser light, humans and animals make transitions from walking to running and so on. For laser devices the pumping current is a bifurcation parameter. When it exceeds a critical value, the ordinary light in the cavity becomes unstable [12]. For humans and animals locomotion speed is a bifurcation parameter. When locomotion speed exceeds certain critical values walking for humans and walking for horses become unstable [10]. Human and horses switch to run and trot, respectively.

Figure 1.4 illustrates that instabilities and bifurcations are related phenomena. In particular, as mentioned in Sect. 1.4, amplitude equations capture both bifurcation- and instability-phenomena. In fact, the theory of amplitude equations has been worked out in detail for bifurcations, as will be discussed in Chap. 2. Amplitude equations come with eigenvalues and eigenvectors that will be discussed in Chap. 2 as well. Among those eigenvectors there are special eigenvectors, the unstable eigenvectors, that determine how systems evolve close to instabilities. There is a general agreement across the various schools of nonlinear physics that unstable eigenvectors are at the heart of any instability-related phenomenon. Since epidemic outbreaks in populations and viral infections in individuals belong to the class of instability-phenomena, these entities of nonlinear physics, namely, unstable eigenvectors, have determined every epidemic outbreak and every individual viral infection that took place so far. In particular, unstable eigenvectors have determined every SARS-CoV-2 infection and every COVID-19 outbreak that took place so far and they will continue to do so at least in the near future.

Synergetics, which is a special school of nonlinear physics and self-organization, has the unique benefit to see beyond the scope of other schools. Synergetics relates bifurcations to phase transitions. Examples of phase transitions are transitions from water to ice and transitions of non-magnetic pieces of material to magnetic ones. Synergetics points out that bifurcations and phase transitions have some key properties in common as indicated in Fig. 1.4. A detailed discussion of this topic can be

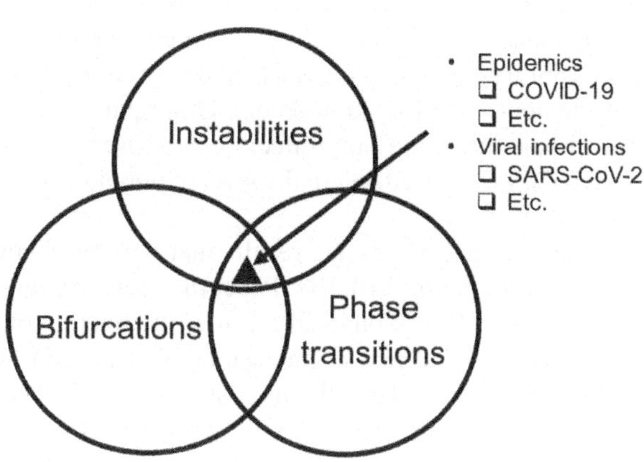

Fig. 1.4 Epidemics and viral dynamics are phenomena belonging to the broader classes of instability-, bifurcation-, and phase transition phenomena

found in the literature (e.g., see Refs. [10, 12]). At this stage only two comments will be made. The first comment is that phase transitions involve instabilities. For example, if water is put into a fridge, then it typically turns into ice at the freezing point, which is, zero degrees Celsius. However, under certain circumstances water can be cooled below the freezing point and remains in its liquid state. This so-called supercooled water is an unstable state. For example, throwing a dust particle into the supercooled water, it immediately turns into ice. Likewise, when experimental conditions are changed appropriately such that a non-magnetic piece of material becomes magnetic, then under the changed conditions the non-magnetic state still exists. However, it is unstable. The second comment is about terminology. Synergetics refers to the aforementioned key players of any instability, the unstable eigenvectors, as order parameters. The reason for this is that phase transitions (such as the transitions of non-magnetic materials to magnetic ones) feature order parameters that determine the emerging order. As mentioned above and as will be shown in detail in this book, populations exhibiting epidemic outbreaks such as COVID-19 outbreaks satisfy certain relationships that are determined by unstable eigenvectors. In other words, epidemic outbreaks, in general, and COVID-19 outbreaks, in particular, take place with certain organizations of their parts. They take place with a certain order. The unstable eigenvectors determine this kind of order. Following the terminology of synergetics, unstable eigenvectors are the parameters that determine the emerging order within populations at the beginning of epidemics and pandemics. They are order parameters. Likewise, the biological entities (e.g., the three entities mentioned in Sect. 1.3: target cells, infected cells, and virus concentration) that are changing in their numbers in the bodies of infected individuals satisfy certain relationships. Infectious diseases emerge in the body of infected individuals in some kind of order. This order is determined again by unstable eigenvectors. Again, following the terminology of synergetics, this means that viral infections in humans and, in particular, SARS-CoV-2 infections in COVID-19 patients are determined by certain order parameters.

Figure 1.4 points out that the three phenomena, phase transitions, bifurcations, and instabilities, have common properties. Viral infectious diseases on the level of populations and viral infections in individuals are placed in the middle of these phenomena. The spread of a virus in a population or in the body of an individual is an instability-phenomenon. Such epidemics and infections in individuals also feature key elements of bifurcation theory. Order parameters as used in the theory of phase transitions determine how infectious diseases (e.g., COVID-19) and viruses (e.g., SARS-CoV-2) spread out on the level of populations and in the bodies of individuals, respectively.

Finally, Fig. 1.5 presents results that will be obtained in later parts of this book (see Sections 5.8 and 10.2.1). Panel (a) shows the number of individuals diagnosed with COVID-19 in Wuhan city, China, as function of time (gray full circles) during the COVID-19 outbreak of the year 2020. The solution of an epidemiological model (solid black line) of the outbreak is shown as well. The reporting period shown on the horizontal axis is January 23 to February 11, 2020. The epidemiological model that was used to generate the solution shown in panel (a) is similar to the three-population

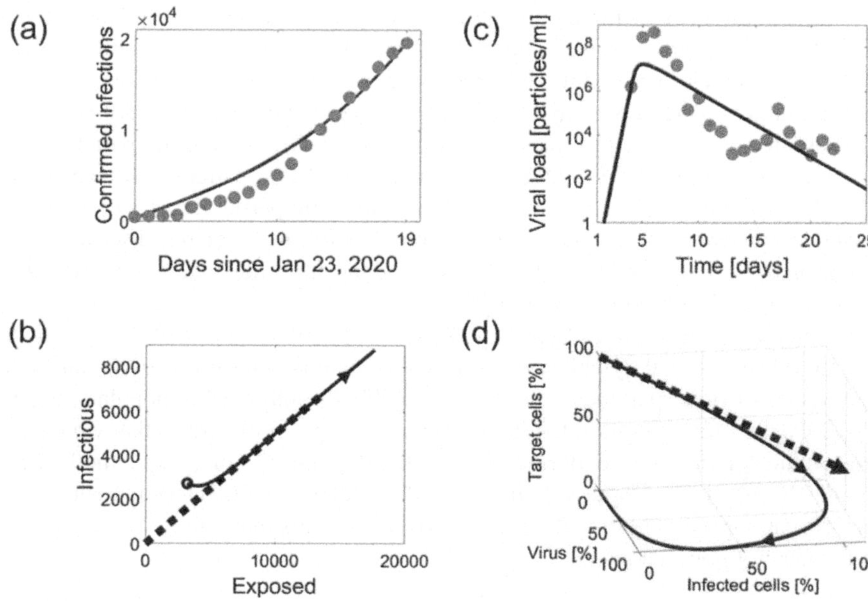

Fig. 1.5 COVID-19 outbreaks in populations and SARS-CoV-2 infections in individuals are instability-phenomena determined by unstable eigenvectors or order parameters. Panel (**a**) shows confirmed cumulative COVID-19 cases observed (gray circles) and modeled (solid line) for Wuhan city, China, as reported in Ref. [38] and will be discussed in Sect. 5.8. Panel (**b**) shows the underlying dynamics (solid line) in the state space of exposed (E) and infectious (I) individuals and the corresponding unstable eigenvector or order parameter of the unstable disease-free state (dotted black line). Panel (**c**) presents viral load observed (gray circles) and modeled (solid line) for a COVID-19 patient as reported in Ref. [39] and will be discussed in Sect. 10.2.1. Panel (**d**) presents the underlying dynamics (solid line) in the three-variable model space of target cells (T), infected cells (I), and viral load (V) and the corresponding unstable eigenvector or order parameter of the disease instability (dotted black line)

model discussed in Sect. 1.2 and involves exposed and infectious individuals. In panel (b) the number of infectious individuals is plotted versus the number of exposed individuals over time. In doing so, a so-called phase curve of the COVID-19 outbreak is obtained. In addition, panel (b) shows the unstable eigenvector or order parameter of the COVID-19 outbreak (dotted thick line). The circle shown in panel (b) indicates the initial disease state of the population of Wuhan city on January 23. Panel (b) demonstrates that after a short period, the population numbers of exposed and infected individuals approached the order parameter (unstable eigenvector) and, subsequently, the epidemic evolved along that vector. Overall, as can be seen in panel (b), the order parameter (unstable eigenvector) determined the way the number of exposed individuals increased relative to the number of infected individuals.

Panel (c) of Fig. 1.5 presents viral load data over time from a patient suffering from a mild SARS-CoV-2 infection (gray circles). Details about the patient data will be given in Sect. 10.2.1. On the horizontal axis, time is given in days after symptoms onset. The solution of the three-variable virus dynamics model introduced in Sect. 1.3 for that patient is shown as well (solid black line). Panel (d) shows the solutions for the three variables (target cells, infected target cells, and viral load) over time in the three-dimensional model space as a phase curve (solid thick line). All variables are presented as percentage values of their maximal values. The order parameter or unstable eigenvector of the SARS-CoV-2 infection of that patient is shown in the three-dimensional space as well (dotted thick line). Comparing the order parameter (unstable eigenvector) and the model solution, it follows that the disease in the patient under consideration evolved initially along the order parameter (unstable eigenvector) and, subsequently, branched off.

In summary, Fig. 1.5 illustrates that a key principle of nonlinear physics, namely, that the relevant initial dynamics of an unstable state is determined by its unstable eigenvector or order parameter, holds for COVID-19 outbreaks in populations and SARS-CoV-2 infections in individuals. Moreover, Fig. 1.5 illustrates that what happens on the population level during the COVID-19 pandemic that started in the year 2019/2020 may mimic what is going on in the bodies of COVID-19 patients. Vice versa, what happens in the bodies of COVID-19 patients may mimic what happens on the level of populations.

1.6 Religion and Physics

This book is the straightforward consequence of a previously published book by the same author [10] that is about the nonlinear physics perspective of the world, in general, and humans, in particular. In this context, the author would like to point out what physics, in general, can do and what physics cannot do. Physics, in general, is a particular perspective of this world. This perspective considers humans as special cases of entities that are part of this world together with many other entities that are also part of this world. All those entities satisfy the same physical principles. There is only one physics that holds for all and everything. There exist many alternative perspectives of this world and many perspectives of humans alternative to the physics perspective, in particular. For example, there are religious and spiritual perspectives. Physics describes the world within the physics perspective [10]. Physics does not address alternative perspectives. In particular, physics does not address mystic and religious concepts. That is, what physics cannot do is to address or explain such concepts. Mystic and religious concepts are outside of the scope of physics. However, physics does not negate such concepts as such. In other words, while mystic and religious concepts do not have a place in physics, this does not mean that they do not have a place at all. In a similar vein, while religious content is not part of physics, this does not mean that physics states that religious content is irrelevant at all. In summary, on the one hand, mystic and religious elements are not part of physics. On the other hand, physics does not make statements about the elements of religious and spiritual perspectives.

There are many mystic, religious-motivated concepts in modern medicine, biology, epidemiology, and virology. In view of what has been said before, it follows that physics does not explain such concepts. Not explaining such concepts does not mean that from a physics perspective there is something missing. In contrast, the concepts of physics, in general, and nonlinear physics, in particular, are completely sufficient to describe any aspect of the world including any phenomenon related to infectious diseases such as COVID-19.

The author as a physicist presents in the subsequent chapters of this book some key nonlinear physics aspects of infectious diseases, in general, and COVID-19, in particular. The author as a private person, not as a physicist, believes that religion is important. The author also believes that religious-motivated concepts in medicine, biology, epidemiology, and virology are important and would be even more important if researchers would address them as what they are: religious concepts. This book provides the reader with the nonlinear physics perspective of epidemics and viral infections such as COVID-19 epidemics and SARS-CoV-2 infections, respectively. In view of this perspective, the reader may discover various mystic and religious concepts that are currently used in medicine, biology, epidemiology, and virology to deal with COVID-19 and other infectious diseases (e.g., HIV/AIDS). These concepts are not part of the physics of COVID-19 or any other infectious disease. However, this does not mean that they are useless, "wrong", or irrelevant.

References

1. World Health Organization, Naming the coronavirus disease (COVID-19) and the virus that causes it, in *Naming the coronavirus disease (COVID-19) and the virus that causes it*, https://www.who.int/emergencies/diseases/novel-coronavirus-2019
2. World Health Organization, Novel coronavirus (2019-nCoV) Situation Report-1, 21 Jan. 2020, in *Novel Coronavirus (2019-nCoV) Situation Report-1, 21 Jan 2020*, https://www.who.int/emergencies/diseases/novel-coronavirus-2019/situation-reports (2020)
3. World Health Organization, COVID-19 Weekly Epidemiological Update-17, in *COVID-19 Weekly Epidemiological Update-17*, https://www.who.int/emergencies/diseases/novel-coronavirus-2019, 6 December 2020
4. World Health Organization, COVID-19 Weekly Epidemiological Update 23, 19 Jan 2021, in *COVID-19 Weekly Epidemiological Update 23, 19 Jan 2021*, https://www.who.int/emergencies/diseases/novel-coronavirus-2019/situation-reports (2021)
5. World Health Organization, COVID-19 Weekly Epidemiological Update 47, 6 July 2021, in *COVID-19 Weekly Epidemiological Update 47, 6 July 2021*, https://www.who.int/emergencies/diseases/novel-coronavirus-2019/situation-reports (2021)
6. United Nations Population Division, World prospect data, in *World prospect data*, https://population.un.org/wpp/ (2021)
7. G. Nicolis, *Introduction to Nonlinear Sciences* (Cambridge University Press, Cambridge, 1995)
8. Y. Oono, *The Nonlinear World: Conceptual Analysis and Phenomenology* (Springer, Berlin, 2013)
9. J. Guckenheimer, P. Holmes, *Nonlinear Oscillations, Dynamical Systems, and Bifurcations of Vector Fields* (Springer, Berlin, 1983)
10. T. Frank, *Determinism and Self-organization of Human Perception and Performance* (Springer, Berlin, 2019)

11. H. Haken, *Synergetics. An Introduction* (Springer, Berlin, 1977)
12. H. Haken, *Synergetics: Introduction and Advanced Topics* (Springer, Berlin, 2004)
13. H. Haken, J. Portugali, Information and selforganization: a unifying approach and applications. Entropy **18** (2016). Article 197
14. G. Wunner, A. Pelster, *Self-organization in Complex Systems: The Past, Present, and Future of Synergetics* (Springer, Berlin, 2016)
15. H. Haken, G. Schiepek, *Synergetik in der Psychologie* (Hogrefe, Gottingen, 2006). (in German)
16. H.W. Hethcote, The mathematics of infectious diseases. SIAM Rev. **42**, 599–653 (2000)
17. M. Martcheva, *An Introduction to Mathematical Epidemiology* (Springer, New York, 2015)
18. K. Rock, S. Brand, J. Moir, M.J. Keeling, Dynamics of infectious diseases. Rep. Prog. Phys. **77** (2014). Article 026602
19. Centers for Disease Control and Prevention, USA, Seasonal influenza: how flu spreads, in *Centers for Disease Control and Prevention, USA*, https://www.cdc.gov/flu/about, Accessed 2021
20. C.W.M. Ong, G.B. Migliori, M. Raviglione, et al., Epidemic and pandemic viral infections: impact on tuberculosis and the lung. Eur. Respir. J. **56** (2020). Article 2001727
21. World Health Organization, Newsroom: transmission of SARS-CoV-2: implications for infection prevention precautions, in *Newsroom: Transmision of SARS-CoV-2: Implications for Infection Prevention Precautions*, https://www.who.int/news-room/commentaries, 9, 2020
22. World Health Organization, Coronavirus disease 2019 (COVID-19) Situation Report-51, in *Coronavirus Disease 2019 (COVID-19) Situation Report-51*, https://www.who.int/emergencies/diseases/novel-coronavirus-2019, 11, 2020
23. M.A. Nowak, R.M. May, *Viral Dynamics: Mathematical Principles of Immunology and Virology* (Oxford University Press, New York, 2000)
24. World Health Organization, Newsroom Fact Sheets HIV/AIDS, in *Newsroom Fact Sheets HIV/AIDS*, https://www.who.int/news-room/fact-sheets/detail/hiv-aids, 14, 2021
25. C. Hadjichrysanthou, E. Cauet E. Lawrence, C. Vegvari, F. de Wolf, R.M. Anderson, Understanding the within-host dynamics of influenza a virus: from theory to clinical implications. J. R. Soc. Interface **13** (2016). Article 20160289
26. World Health Organization, WHO Global influenza strategy 2019–2030, in *WHO Global Influenza Strategy 2019–2030*, https://apps.who.int/iris/handle/10665/311184 (2019)
27. COVID-19 Tracker, Timeline data from Johns Hopkins Center for systems science and engineering, in *COVID-19 Tracker*, https://vac-lshtml.shinyapps.io/ncov_tracker
28. M. Woolhouse, F. Scott, Z. Hudson, R. Howey, M. Chase-Topping, Human viruses: discovery and emergence. Phil. Trans. Royal Soc. B **367**, 2864–2871 (2012)
29. M. Breitbart, F. Rohwer, Here a virus, there a virus, everywhere the same virus? Trends Microbiol. **13**, 278–284 (2005)
30. E. Domingo, *Virus as Populations* (Academic Press, London, 2020)
31. M. Scudellari, How the coronavirus infects our cells. Nature **595**, 640–644 (2021)
32. O. Diekmann, J.A.P. Heesterbeek, *Mathematical Epidemiology of Infectious Diseases* (Wiley, Chichester, 2000)
33. P. van den Driessche, J. Watmough, Reproduction numbers and sub-threshold endemic equilibria for compartment models of disease transmission. Math. Biosci. **180**, 29–48 (2002)
34. M.C. Cross, P.C. Hohenberg, Pattern formation outside of equilibrium. Rev. Mod. Phys. **65**, 851–1112 (1993)
35. J.D. Murray, *Mathematical Biology* (Springer, Berlin, 1993)
36. P.C. Bressloff, J.D. Cowan, M. Golubitsky, P.J. Thomas, M.C. Wiener, Geometric visual hallucinations, Euclidean symmetry and the functional architecture of striate cortex. Phil. Trans. R. Soc. Lond. B **356**, 299–330 (2001)
37. F. Verhulst, *Nonlinear Differential Equations and Dynamical Systems*, 2nd edn. (Springer, Berlin, 1996)
38. T.D. Frank, Simplicity from complexity: on the simple amplitude dynamics underlying COVID-19 outbreaks in China. Adv. Complex Syst. **23** (2020). Article 2050022
39. T.D. Frank, SARS-Coronavirus-2 nonlinear dynamics in patients: three-dimensional state and amplitude state description. J. Phys. Soc. Jpn. **90** (2021). Article 073802

Chapter 2
Nonlinear Physics and Synergetics

This chapter presents in a condensed form several key concepts of nonlinear physics such as fixed points, stability, bifurcations, the linearization matrix, eigenvalues, eigenvectors, and amplitudes. It introduces into the dual description of dynamical systems in state and amplitude spaces. The derivation of amplitude equations is discussed that allow for the description of systems in amplitude spaces. The chapter introduces the order parameter concept of synergetics and discusses why unstable eigenvectors of fixed points are order parameters.

2.1 State and Time

State space

A state of a system is a collection of variables that depend on time t and describe the evolution of the system of interest. Throughout this book, discrete systems [1] will be considered that are described by a finite number of state variables X_i. Let n denote the number of state variables: $X_1, \ldots, X_n$. The vector $\mathbf{X} = (X_1, \ldots, X_n)$ is then referred to as state vector. Let D_i denote the range of definition of X_i such that $X_i \in D_i$. The state space is the collection of all possible states and corresponds to the product space $\prod_{i=1}^{n} D_i$. For example, if $n = 1$ and $X_1 \in \mathbb{R}$ holds, then the state space is the real line. If $n = 2$ and $X_1, X_2 \in \mathbb{R}$ holds, then the state space is the two-dimensional plane $\mathbb{R}^2$. In classical mechanics, a harmonic oscillator that oscillates in a single spatial dimension is described by two variables: the position variable $x \in \mathbb{R}$ and the velocity variable $v \in \mathbb{R}$. Consequently, the state space of the oscillator is given by the plane $\mathbb{R}^2$ with two orthogonal axes that show x and v, respectively.

In the context of epidemics and pandemics, the state $\mathbf{X}$ describes the disease state or health state of a population under consideration. In this context note that in

T. D. Frank, *COVID-19 Epidemiology and Virus Dynamics*,
Understanding Complex Systems, https://doi.org/10.1007/978-3-030-97178-6_2

this book the phrases disease state and health state are considered as synonyms. For example, the special disease state for which there are no infected individuals in a population corresponds to the healthy state or disease-free state of a population. Vice versa, a health state of a population that is characterized by a finite number of infected individuals describes that the health state of the population has moved away from the disease-free state and that the population has been invaded by an infectious disease. On the level of individuals, **X** describes the disease or health state of an individual (e.g., a COVID-19 patient) or affected sites of an individual (e.g., affected areas in the lung of a COVID-19 patient).

Time and dynamical systems

As mentioned above, state variables are time-dependent functions like $X_1(t)$, ... ,$X_n(t)$, where t denotes time. Consequently, the state vector **X** depends on time like $\mathbf{X}(t)$. Throughout this book discrete systems are considered whose dynamics can be described by means of first-order differential equations of the form

$$\frac{\mathrm{d}}{\mathrm{d}t}\mathbf{X} = \mathbf{N}(\mathbf{X}). \tag{2.1}$$

Here the right-hand side (RHS) vector-function **N** describes the change of the state vector (or the change of the state variables or simply the change of the state). The components of **N** are $\mathbf{N} = (N_1, \ldots, N_n)$. Consequently, Eq. (2.1) corresponds to the coupled set of first-order differential equations

$$\begin{aligned}
\frac{\mathrm{d}}{\mathrm{d}t}X_1 &= N_1(X_1, \ldots, X_n), \\
\frac{\mathrm{d}}{\mathrm{d}t}X_2 &= N_2(X_1, \ldots, X_n), \\
&\ldots \\
\frac{\mathrm{d}}{\mathrm{d}t}X_n &= N_n(X_1, \ldots, X_n).
\end{aligned} \tag{2.2}$$

The vector-function **N** may depend explicitly on time: $\mathbf{N}(\mathbf{X}, t)$. Note that Eq. (2.1) describes a dynamical system.

For example, the evolution equation of the harmonic oscillator that oscillates in a single dimension reads

$$\frac{\mathrm{d}}{\mathrm{d}t}x = v \ , \ \frac{\mathrm{d}}{\mathrm{d}t}v = -kx/m, \tag{2.3}$$

where m and k denotes mass and the spring constant, respectively. Using the state vector $\mathbf{X} = (x, v)$, Eq. (2.3) can be written like

$$\frac{\mathrm{d}}{\mathrm{d}t}\mathbf{X} = \mathbf{N} \ , \ \mathbf{N} = \begin{pmatrix} v \\ -kx/m \end{pmatrix} \tag{2.4}$$

and is a special case of Eq. (2.1).

Trajectory and initial conditions

The initial conditions are the values of the state variables at the initial time t_0. That is, they constitute the initial state vector or initial state $\mathbf{X}_0$ with $\mathbf{X}_0 = \mathbf{X}(t_0)$. For example, the harmonic oscillator may be considered for the initial conditions at $t_0 = 0$ s defined by $x(0) = 0$ m, $v(0) = 1$ m/s (here time and space are measured in seconds and meters, respectively). A trajectory of the state of a system or a solution of a dynamical system is defined as the time-dependent vector function $\mathbf{X}(t)$ for a given initial state $\mathbf{X}(t_0) = \mathbf{X}_0$.

Phase portrait and phase curves

The phase portrait of a system is a collection of all possible trajectories or solutions in state space obtained by considering all possible initial conditions. A single trajectory plotted in the state space is also referred to a phase curve. Consequently, a phase portrait is a collection of all possible phase curves. A phase portrait can be drawn using the following steps. First, an initial time is fixed. Second, an initial state is selected and the corresponding trajectory is plotted. Subsequently, another initial state is selected and, again, the corresponding trajectory is plotted. This step, namely, selecting an initial state and plotting the trajectory is repeated again and again. Throughout this book, systems will be considered with continuous state variables (e.g., $X \in \mathbb{R}$). For continuous state variables there are infinitely many initial conditions. Consequently, it is impossible to plot all possible trajectories. Typically, phase portraits involving continuous variables are drawn that show a certain selection of trajectories. In doing so, phase portraits with arbitrarily fine-grained or course-grained resolutions can be created.

The benefit of a phase portrait is that it shows how states evolve over time for all kind of initial states. Consequently, they are illustrations of attractors and repellors (see below, Sect. 2.4). In the context of this book, a repellor means an instability and vice versa. Consequently, instabilities that play a key role for the underlying physics of infectious diseases (see Chap. 1) can be illustrated by phase portraits. In addition, phase portraits can be used to visualize order parameters that again are key elements in the nonlinear physics perspective of infectious diseases (see Chap. 1). The disadvantage of a phase portrait is that it does not present time. A phase portrait does not show at what point in space the system is located at a particular time.

An example of a phase portrait that was obtained in [2] is shown in Fig. 2.1. Figure 2.1 shows a phase portrait of an epidemiological model that involves infectious (I) and exposed (E) individuals (see Sect. 1.2). Trajectories for various initial conditions are shown. The arrows indicate the flow (i.e., dynamics) along the trajectories. States evolve away from the disease-free state $E = I = 0$. In particular, the states approach in time a particular direction indicated by the dotted gray line. In fact, the phase portrait shown in Fig. 2.1 captures the COVID-19 outbreak in the beginning of the year 2020 in Wuhan City, China (see Sect. 1.5). The phase portrait demonstrates that the disease-free fixed point in 2020 was an instability (or repellor) [2–4]. The disease dynamics followed an order parameter (dotted gray straight line). This order parameter is also shown in panel (b) of Fig. 1.5. That is, the dotted black

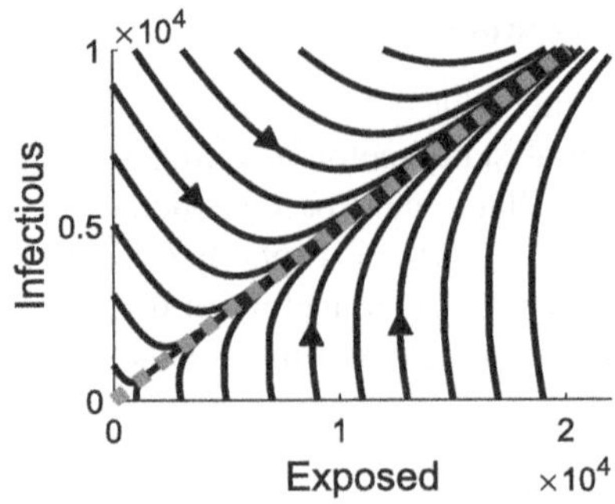

Fig. 2.1 Phase portrait of the dynamical system given in terms of exposed and infectious individuals that supposedly determined the COVID-19 outbreak in Wuhan city, China, in the beginning of the year 2020 (for details see Sects. 1.5 and 6.3)

line in panel (b) of Fig. 1.5 corresponds to the dotted gray line in Fig. 2.1. In other words, as already pointed out in the context of Fig. 1.5, the phase portrait shown in Fig. 2.1 and the underlying model-based analysis [2–4] suggest that there was a certain order parameter that determined the course of the COVID-19 outbreak in Wuhan city, China. This order parameter of the Wuhan city COVID-19 outbreak will be discussed in Sect. 6.3.

2.2 Structure

Structure refers to all quantities that describe the system and do not change over time [1]. Structure is described in terms of parameters. For example, the structure of the harmonic oscillator oscillating in a one-dimensional space is given by the mass m and the spring constant k. That is, the system exhibits two parameters that describe its structure. However, there are situations in which the distinction between structure (constant over time) and state (changing over time) is not so obvious [1]. A quantity Z may be constant over time under certain circumstances A. Under alternative circumstances B, the same quantity Z may change over time. In such situations, under B the quantity Z becomes time-dependent. It becomes part of the state of the system under consideration. For a more detailed discussion of structure variables that become time-dependent see Sect. 1.6.2 and Chap. 7 in [1].

Let us illustrate this issue by means of a damped harmonic oscillator that oscillates in one-dimension. The evolution equations read

$$\frac{\mathrm{d}}{\mathrm{d}t}x = v \, , \, \frac{\mathrm{d}}{\mathrm{d}t}v = -kx/m - \gamma v, \tag{2.5}$$

where γ denotes the friction parameter and the term $-\gamma v$ describes the friction force. Let us assume that due to friction the mass m decays over time. Consequently, the structure parameter m becomes time-dependent and the above equations read

$$\frac{\mathrm{d}}{\mathrm{d}t}x = v \,, \ \frac{\mathrm{d}}{\mathrm{d}t}v = -kx/m(t) - \gamma v. \tag{2.6}$$

A possible interpretation of the model is to introduce a three-dimensional state with the state vector $\mathbf{X} = (x, v, m)$. If so, we would add another evolution equation that describes how the mass changes over time and that formally would read $\mathrm{d}m/\mathrm{d}t = N_3(x, v, m)$, where N_3 is the RHS function of the mass variable.

Structure variables that become time-dependent play a key role in epidemics under the impacts of intervention measures. Intervention measures may change parameters of human-virus systems such that the viral infectious diseases in those systems subside. This issue will be addressed in Chap. 8.

2.3 Fixed Points and Stability

Fixed points and stationary states

In this book, the phrases fixed point and stationary state are considered as synonyms. A fixed point or stationary state of a dynamical system described by Eq. (2.1) is defined by [1, 5–7]

$$\mathbf{N}(\mathbf{X}_{st}) = 0 \Rightarrow \frac{\mathrm{d}}{\mathrm{d}t}\mathbf{X} = 0 \tag{2.7}$$

and is denoted by $\mathbf{X}_{st}$. Accordingly, a fixed point is a state for which the state does not change over time. A system may exhibit no fixed point at all, one fixed point, a number m of fixed points, or infinitely many fixed points (e.g., all points that line up on a particular axis in state space). Fixed points exhibit stability properties. Accordingly, they can be classified into asymptotically stable, stable, and unstable fixed points [6, 8–10].

Stable fixed point

A stable fixed point in the sense of Lyapunov is a fixed point $\mathbf{X}_{st}$ such that for any arbitrarily small neighborhood A of $\mathbf{X}_{st}$ an even smaller neighborhood B can be found such any solution $\mathbf{X}(t)$ starting in B does not leave A over time [6, 8, 10].

Asymptotically stable fixed point

An asymptotically stable fixed point is a stable fixed point $\mathbf{X}_{st}$ such that any state in any arbitrary small neighborhood converges back to $\mathbf{X}_{st}$ [6, 8, 10]. This property may be formulated like [10]

$$\forall \mathbf{X}(t_0) : \{\mathbf{X}(t_0) \neq \mathbf{X}_{st} \wedge |\mathbf{X}(t_0) - \mathbf{X}_{st}| < \epsilon\} \Rightarrow \mathbf{X}(t \to \infty) = \mathbf{X}_{st} \tag{2.8}$$

for any sufficiently small ϵ. That is, ϵ is used to describe initial states $\mathbf{X}(t_0)$ in arbitrarily small neighborhoods of $\mathbf{X}_{st}$ like $|\mathbf{X}(t_0) - \mathbf{X}_{st}| < \epsilon$. Throughout this book, the symbol $|\cdot|$ means the amount of a vector: $|\mathbf{X}| = \sqrt{\sum_{i=1}^{n} X_i^2}$.

Note that if Eq. (2.8) holds for all states with $\epsilon < 0.001$ but not for states with $\epsilon > 0.001$ (where the explicit number, here 0.001, does not matter), then the fixed point is asymptotically stable. That is, it does not matter whether or not states at a larger distance from the fixed point are evolving towards the fixed point. For example, a one-dimensional system with a fixed point at $X_{st} = 0$ may be setup such that all states for $|X(t_0)| < 0.001$ converge to $X_{st} = 0$ but all states for $|X(t_0)| > 0.001$ converge to either plus or minus infinity. In this case, the fixed point $X_{st} = 0$ is asymptotically stable although from the perspective of fixed point $X_{st} = 0$ all initial states with $\epsilon > 0.001$ do not converge to $X_{st} = 0$. For more illustrations see [1, 5].

Perturbations and relative states

In nonlinear physics, a frequently used concept is the concept of a perturbation. A perturbation $\mathbf{u}$ at time t out of a fixed point $\mathbf{X}_{st}$ is defined by

$$\mathbf{u}(t) = \mathbf{X}(t) - \mathbf{X}_{st} . \tag{2.9}$$

Throughout this book the variable $\mathbf{u}$ will be considered as perturbation or relative state. In the latter context, $\mathbf{u}$ is the state of a system under consideration relative to a given fixed point $\mathbf{X}_{st}$. In this context, no other assumptions are made. In particular, $\mathbf{u}$ does not have to be a small quantity. In contrast, when the variable $\mathbf{u}$ is regarded as a perturbation, it is frequently assumed that its magnitude $|\mathbf{u}|$ reflects a small quantity. Using the concept of small perturbations, an asymptotically stable fixed point is a stable fixed point for which all sufficiently small perturbations out of the fixed point decay to zero over time.

Unstable fixed points

In general, an unstable fixed point is a fixed point that is not stable, that is, a fixed point for which the stability property formulated above does not hold [6, 8, 10]. For practical purposes, a sufficient condition for a fixed point to be unstable is the following. If a fixed point $\mathbf{X}_{st}$ exhibits the property that there exists in any arbitrary small neighborhood of the fixed point at least one perturbation that originates at $\mathbf{X}_{st}$ and increases in magnitude over time, then the fixed point is unstable. Mathematically speaking, if there exists at least one solution $\mathbf{X}(t)$ that originates at $\mathbf{X}_{st}$ and satisfies

$$\{\mathbf{X}(t_0) \neq \mathbf{X}_{st} \wedge |\mathbf{X}(t_0) - \mathbf{X}_{st}| < \epsilon \wedge \frac{\mathrm{d}}{\mathrm{d}t}|\mathbf{X}(t) - \mathbf{X}_{st}| > 0 \text{ for } t \in [t_0, t_0 + \delta]\,\} \tag{2.10}$$

for any sufficiently small ϵ, then the fixed point $\mathbf{X}_{st}$ is unstable. The requirement that $\mathbf{X}(t)$ originates at $\mathbf{X}_{st}$ means that if time is reversed than the solution $\mathbf{X}(t)$ converges to $\mathbf{X}_{st}$ like $\lim_{t \to -\infty} \mathbf{X}(t) = \mathbf{X}_{st}$. In Eq. (2.10) the parameter δ is positive and can be arbitrarily small. Graphically speaking there must be at least one way out of the fixed point (i.e., an "escape route") such that at least for a small period δ the state $\mathbf{X}$ of the system can get away from the fixed point $\mathbf{X}_{st}$. The sufficient condition follows from the general definition of an unstable fixed point because if the sufficient condition holds then the stability property formulated above (under the subsection "stable fixed point") cannot be satisfied.

An example of an unstable fixed point is illustrated in Fig. 2.1. The disease-free state $E = I = 0$ or $\mathbf{X}_{st} = (0, 0)$ under the conditions illustrated in Fig. 2.1 is an unstable fixed point. As can be seen from the phase portrait shown in Fig. 2.1, if the state $\mathbf{X}$ is perturbed out of the fixed point $\mathbf{X}_{st} = (0, 0)$, then it will never return to the fixed point. Consequently, the fixed point is not stable, which (in view of the general definition of being unstable) implies that the fixed point is unstable. In particular, if the stationary state is perturbed along the dotted gray line, then $\mathbf{X}(t)$ corresponds to a solution that originates from $\mathbf{X}_{st}$. In this case $\mathbf{X}(t)$ also evolves away from $\mathbf{X}_{st}$. Consequently, the more applied sufficient condition for instability formulated in Eq. (2.10) applies. The dotted line describes a way out of the fixed point.

Neutrally stable fixed point

In this book, fixed points that are stable but not asymptotically stable will be referred to as neutrally stable [8]. Examples of neutrally stable fixed points will be given in the following chapters.

2.4 Attractors and Repellors

Attractors and repellors are generalizations of stable and unstable fixed points. While fixed points are described by single points in state spaces, in general, attractors and repellors are described by certain sets of points of state spaces. For example, a limit cycle attractor is described by the set of points that form a closed line or curve in the state space [1, 6, 10]. Stable and unstable fixed points in turn are special cases of attractors and repellors. An asymptotically stable fixed point is a fixed point attractor. An unstable fixed point is a fixed point repellor.

Figure 2.2 provides an overview over some attractors and repellors. Accordingly, some attractors and repellors come as fixed points. Fixed points, in turn, correspond to nodes, saddles, or foci. While nodes and foci can be asymptotically stable or unstable, saddle points correspond to unstable fixed points in any case. Nodes and saddle points are fixed points for which solutions do not exhibit oscillations close to the fixed points. Unstable nodes and saddle points are both unstable fixed points. The difference between an unstable node and a saddle is given in terms of the number of directions, in which perturbations increase. For unstable nodes perturbations increase in all directions. For saddles there is a mixture of stable and unstable directions in the sense that there is at least one direction in the state space under consideration along which perturbations decrease in magnitude over time. Foci are fixed points for which solutions exhibit oscillations close to the fixed points. For more details the reader may consult [1, 5]. In Sect. 2.7 it will be shown that fixed points can conveniently be classified by means of its eigenvalues.

As indicated in Fig. 2.2, some attractors and repellors come as limit cycles. Limit cycles in turn can be stable or unstable. Consequently, a stable limit cycle and a

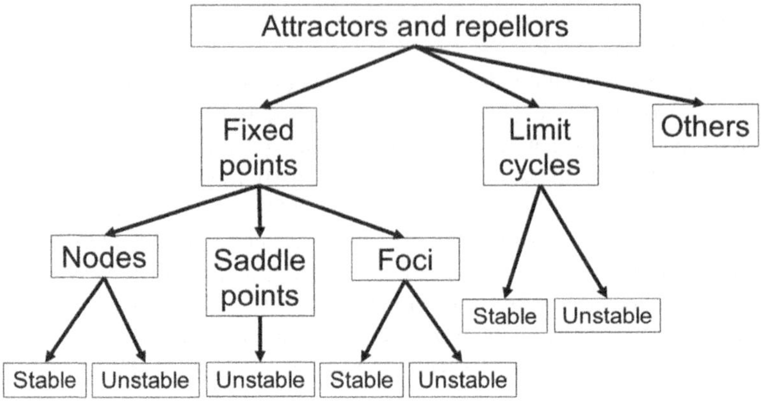

Fig. 2.2 Overview over different types of attractors and repellors

Table 2.1 Fixed points in one- and two-dimensional state spaces and their terminology

Attractor/repellor	Dimension of state space	Synonym	Subtype
Fixed point attractor	1	Asymp. stable fixed point	Stable node
Fixed point repellor	1	Unstable fixed point	Unstable node
Fixed point attractor	≥ 2	Asymp. stable fixed point	
			Stable node
			Stable focus
Fixed point repellor	≥ 2	Unstable fixed point	
			Unstable node
			Saddle
			Unstable focus
Limit cycle attractor	≥ 2	Stable limit cycle	
Limit cycle repellor	≥ 2	Unstable limit cycle	

limit cycle attractor denote the same physical or mathematical object. Likewise, the phrases unstable limit cycle and limit cycle repellor have the same meaning.

Table 2.1 illustrates that certain attractors and repellors only exists in two-dimensional or higher-dimensional state spaces. Systems described by a single variable (i.e., featuring a one-dimensional state space) can exhibit only fixed point attractors and repellors. Let X denote the state variable of a one-dimensional system. Let us assume that Eq. (2.1) for the system reads

$$\frac{\mathrm{d}}{\mathrm{d}t}X = -aX \tag{2.11}$$

with parameter $a > 0$. The fixed point is given by $X_{st} = 0$. Solutions for $t_0 = 0$ are given by $X(t) = X(0)\exp\{-at\}$. Accordingly, all perturbations starting with $X(0) \neq 0$ decay in the amount. Consequently, $X_{st} = 0$ is an asymptotically stable fixed point. Since the solutions do not exhibit oscillations, we may refer to the fixed point as a stable node. Next, let us assume the dynamics satisfies

$$\frac{\mathrm{d}}{\mathrm{d}t}X = aX \tag{2.12}$$

with $a > 0$ again. Note that there is no minus sign in front of the parameter. Consequently, solutions $X(t) = X(0)\exp\{at\}$ for $X(0) \neq 0$ increase in magnitude over time. The fixed point is unstable and corresponds to a fixed point repellor. We may refer to it as an unstable node.

While for single variable systems ($n = 1$) with one-dimensional state spaces, solutions close to fixed points can only be non-oscillatory, for systems described by two or more than two variables ($n \geq 2$) exhibiting two-dimensional or higher-dimensional state spaces, solutions close to fixed points can be non-oscillatory or oscillatory. Moreover, in such spaces fixed points exhibiting mixtures of stable and unstable directions can exist. Consequently, two types of fixed point attractors exist: nodes and foci. Moreover, three types of fixed point repellors exist: nodes, foci, and saddle (see Table 2.1 again).

2.5 Phase Transitions and Bifurcations

About jumps and kinks

When structural properties of a system are changed, the state of the system can change as well. There are quantitative and qualitative state changes. In order to distinguish between quantitative and qualitative changes the concept of jumps and kinks can be used [1]. In short, if a structural change induces a state change and there is a jump or kink of an appropriately defined system variable, then the state change is qualitative. Otherwise, the state change is quantitative. Let us dwell on this issue.

Let α denote a structural property of a system that can be changed in a continuous way. Let $\mathbf{X}$ denote the state of the system and Y a physical quantity that corresponds to one of the components of $\mathbf{X}$ or can be computed from $\mathbf{X}$: $Y = Y(\mathbf{X})$. It what follows it is assume that $\mathbf{X}$ and Y are selected in such as way that they depend on α like

$$Y = f(\alpha). \tag{2.13}$$

The function f describes the dependency of Y on α. Let us consider the case in which α is varied at α_0 by a small amount z. There are two scenarios.

First, the change z induces a smooth change in Y such that Y and all of its derivatives (first, second, third order an so on) with respect to α at α_0 are continuous functions. For example, for $Y = Y_0 + c\ \alpha$ (i.e., a linear dependency), any change in α results in a smooth change of Y. If for all physical properties Y such smooth changes take place, then the change of $\mathbf{X}$ induced by α at α_0 is quantitative.

Second, the change z induces a jump in Y defined by

$$Y(\alpha_0 - z) \neq Y(\alpha_0 + z) \quad \text{for } z \to 0 \tag{2.14}$$

or a kink in Y defined by

$$\frac{\mathrm{d}}{\mathrm{d}\alpha} Y(\alpha_0 - z) \neq \frac{\mathrm{d}}{\mathrm{d}\alpha} Y(\alpha_0 + z) \quad \text{for } z \to 0. \tag{2.15}$$

As indicated above, what matters are small changes in α at α_0. If a jump or kink takes place, then the system changes at α_0 from a state $\mathbf{X}$ to a qualitatively different state $\mathbf{X}$ when α is increased from values smaller than α_0 to values larger than α_0. In this case, α_0 is denoted by α_{crit} and referred to as critical value of the parameter α.

In summary, the first scenario that involves only smooth changes of physical quantities describes changes of $\mathbf{X}$ within the same type of state. The first scenario describes quantitative changes of $\mathbf{X}$. In contrast, the second scenario that involves jumps and kinks describes transition between qualitatively different states $\mathbf{X}$.

Equilibrium phase transitions are such qualitative changes that take place in systems in thermal equilibrium with their environments. For example, transitions between aggregate states such as solid, liquid, and gas and magnetic phase transitions between magnetic and non-magnetic phases of materials exhibit kinks and jumps at the transition points and belong to the class of equilibrium phase transitions [1]. In particular, the bifurcation of the state of the magnetic material shown in panel (a) of Fig. 1.3 is characterized by a kink of the magnetization Y. In this example α corresponds to the temperature T. In general, in thermodynamics, transitions between states that exhibit a discontinuity, meaning a jump, are referred to as first-order phase transitions [11]. The aforementioned solid-liquid transitions are first-order phase transitions. In contrast, transitions between states that are continuous but exhibit a kink are referred to as second-order phase transitions or continuous phase transitions [11–13]. The aforementioned magnetic phase transitions are second-order phase transitions.

By analogy to equilibrium phase transitions, non-equilibrium phase transitions take place in systems that are not in thermodynamic equilibrium with their environments. For example, layers of fluids and gases heated on one side are such non-equilibrium systems, see Sect. 1.5. In such layers various kinds of spatio-temporal patterns can emerge such as cloud patterns, tornados, hurricanes, and Earth mantle and Earth atmospheric rotating cells (Sect. 1.5). They correspond to states that are qualitatively different from the respective resting or stationary states. As far as jumps and kinks as indicators for qualitatively new states are concerned, as exemplified in panel (b) of Fig. 1.3, the roll velocity Y exhibits a kink when rotating roll patterns in fluids and gases emerge [1]. In this context, α corresponds to the temperature difference across the fluid or gas layer. Lasers exhibit a kink in the light intensity at the threshold at which they start to produce laser light [5]. The kink indicates that the laser light is a state qualitatively different from the ordinary light that lasers produce at pumping currents below the threshold. Lasers with two different laser modes exhibit jumps in output power when the lasers switch between the modes [14]. When plotting the oxygen consumption of horses over locomotion speed, then the oxygen consumption functions of horses exhibit kinks at the critical locomotion speeds at which they switch gaits from walk to trot [1].

Bifurcations, bifurcation parameters, and critical values

The notion of a bifurcation was introduced in Sect. 1.5. Let us dwell on this issue. A bifurcation is a qualitative change of the state of a system induced by a change of the structure of the system and related to a change in the set of attractors and repellors characterizing the system. That is, in the case of a bifurcation it is required that the qualitative change of the state of the system is related to a change in the circumstances regarding the attractors and repellors of the system. For example, a stable fixed point may become unstable. An attractor may disappear. A new attractor may appear. In summary, a bifurcation is a jump-like or kink-like change of an appropriately defined variable characterizing the (state of the) system induced by a change of a structural parameter and related to

1. A change of the stability of attractors and repellors
2. And/or the appearance and disappearance of attractors and repellors

(see Sect. 3.2.1 in [1], see also [6, 8, 10]). The structural parameter is called bifurcation parameter. The value of the bifurcation parameter (structural parameter) at which the bifurcation occurs is referred to as critical value or bifurcation point.

2.6 The Linear Domain: Basic Concepts

2.6.1 Linearization

Given the fundamental equation $\mathrm{d}\mathbf{X}/\mathrm{d}t = \mathbf{N}(\mathbf{X})$, see Eq. (2.1), and a fixed point $\mathbf{X}_{st}$ defined by $\mathbf{N}(\mathbf{X}_{st}) = 0$, see Eq. (2.7), any state $\mathbf{X}$ can be seen relative to a fixed point $\mathbf{X}_{st}$ with the help of the difference vector $\mathbf{u} = \mathbf{X} - \mathbf{X}_{st}$. The vector $\mathbf{u}$ was introduced in Eq. (2.9) as a perturbation out of the fixed point $\mathbf{X}_{st}$. As stated there, in general, $\mathbf{u}$ describes a relative state. Let us consider states close to the fixed point $\mathbf{X}_{st}$ such that $\mathbf{u}$ is small in the amount. In this case, the vector-function $\mathbf{N}$ that in general expresses nonlinear dependencies can be approximated by means of linear relationships. In other words, $\mathbf{N}$ can be linearized at $\mathbf{X}_{st}$. In doing so, the fundamental Eq. (2.1) becomes an evolution equation for the relative state or perturbation $\mathbf{u}$ and reads

$$\frac{\mathrm{d}}{\mathrm{d}t}\mathbf{u} = L\mathbf{u} \tag{2.16}$$

where L is the $n \times n$ matrix derived from $\mathbf{N}$. Explicitly, the components L_{ik} are defined by

$$L_{ik} = \left. \frac{\partial N_i}{\partial X_k} \right|_{\mathbf{X}_{st}} . \tag{2.17}$$

For example, let us consider a state vector $\mathbf{X} = (X_1, X_2, X_3)$ for a $n = 3$ variable system and a perturbation vector $\mathbf{u} = (u_1, u_2, u_3)$. In this case, Eq. (2.16) reads explicitly

$$\frac{\mathrm{d}}{\mathrm{d}t} \begin{pmatrix} u_1 \\ u_2 \\ u_3 \end{pmatrix} = \begin{pmatrix} L_{11} & L_{12} & L_{13} \\ L_{21} & L_{22} & L_{23} \\ L_{31} & L_{32} & L_{33} \end{pmatrix} \begin{pmatrix} u_1 \\ u_2 \\ u_3 \end{pmatrix} , \tag{2.18}$$

where L_{ik} correspond to the matrix elements of L. As stated above, the right-hand side of Eq. (2.16) is linear with respect to $\mathbf{u}$, which can be seen explicitly in Eq. (2.18). For example, from Eq. (2.18) it follows that the evolution equation for u_1 reads $\mathrm{d}u_1/\mathrm{d}t = L_{11}u_1 + L_{12}u_2 + L_{13}u_3$.

While Eq. (2.17) gives a short computational equation for linearizing Eq. (2.1), the linearization procedure can also be carried out in a step by step fashion. Let us illustrate the step by step approach for the single-variable dynamical system

$$\frac{\mathrm{d}}{\mathrm{d}t} X = \alpha X - X^3 . \tag{2.19}$$

The model describes a so-called pitchfork bifurcation [1, 6, 10]. For $\alpha < 0$ there exists a single fixed point $X_{st} = 0$. Consequently, the relative state is given by $u = X$. Substituting $u = X$ into Eq. (2.19) and separating linear and nonlinear terms, we obtain

$$\frac{\mathrm{d}}{\mathrm{d}t} u = \alpha u - u^3 = \alpha u + \text{nonlinear terms} . \tag{2.20}$$

The linearized equation is obtained by neglecting nonlinear terms. Accordingly, it reads

$$\frac{\mathrm{d}}{\mathrm{d}t} u = \alpha u \tag{2.21}$$

and describes an asymptotically stable fixed point at $u = 0$ (or $X_{st} = 0$), see Eq. (2.11) and note that $\alpha < 0$. Next, let us consider the case $\alpha > 0$. In this case the model may be written like

$$\frac{\mathrm{d}}{\mathrm{d}t} X = \alpha X - X^3 = X(\alpha - X^2) \tag{2.22}$$

and exhibits $X_{st} = 0$ and $X_{st} = \pm\sqrt{\alpha}$ as fixed points. The linearized equation for the fixed point $X_{st} = 0$ is again given by Eq. (2.21). Since $\alpha > 0$ the fixed point $X_{st} = 0$ is unstable, see Eq. (2.12). Let us consider the fixed point $X_{st} = \sqrt{\alpha}$ for which perturbations read like $u = X - \sqrt{\alpha} \Rightarrow X = u + \sqrt{\alpha}$. Substituting these relations into (2.19), we obtain

$$
\begin{aligned}
\frac{\mathrm{d}}{\mathrm{d}t}u &= \alpha(u+\sqrt{\alpha}) - (u+\sqrt{\alpha})^3 \\
&= \alpha u + \alpha\sqrt{\alpha} - \left[u^3 + 3u^2\sqrt{\alpha} + 3u\alpha + \sqrt{\alpha}^3\right] \\
&= -2\alpha u - u^3 - 3u^2\sqrt{\alpha} = -2\alpha u + \text{nonlinear terms.}
\end{aligned}
\tag{2.23}
$$

Again, neglecting the nonlinear terms, we obtain the linearized equation

$$\frac{\mathrm{d}}{\mathrm{d}t}u = -2\alpha u. \tag{2.24}$$

This equation describes an asymptotically stable fixed point located at $u = 0$ (see Eq. (2.11) and note that $\alpha > 0$), which corresponds to an asymptotically stable fixed point at $X_{st} = \sqrt{\alpha}$. Alternatively, Eqs. (2.21) and (2.24) can be obtained from Eq. (2.17). For the pitchfork bifurcation model, N reads $N = \alpha X - X^3$, which implies that $\mathrm{d}N/\mathrm{d}X = \alpha - 3X^2$. Consequently, for $X_{st} = 0$ the linearization coefficient $L = \mathrm{d}N/\mathrm{d}X_{st}$ is given by $L = \alpha$, which yields Eq. (2.21). In contrast, for $X_{st} = \sqrt{\alpha}$ the coefficient L is given by $L = \alpha - 3\alpha = -2\alpha$, which yields Eq. (2.24).

2.6.2 *Eigenvalues and Eigenvectors*

The matrix L defined by Eq. (2.17) has vectors $\mathbf{v}_i$ that satisfy

$$L\mathbf{v}_i = \lambda_i \mathbf{v}_i \tag{2.25}$$

with $i = 1, 2, \ldots n$. The vector $\mathbf{v}_i$ is called the ith eigenvector and the variable λ_i is called the ith eigenvalue [1, 6, 8–10]. Note that in what follows the assumption is made that in fact the matrix L is such that n eigenvectors exist that can be considered as different from each other. Mathematically speaking, this means that it is assumed that the matrix L exhibits n linearly independent eigenvectors.

The eigenvalues λ_i may be real-valued or complex. Accordingly, let us defined type I and II eigenvectors by

$$\text{Type I eigenvector } \mathbf{v}_i : \lambda_i \in \mathbb{R} \Rightarrow \text{Imag}(\lambda_i) = 0\,, \tag{2.26}$$

$$
\begin{aligned}
&\text{Type II eigenvector } \mathbf{v}_i, \mathbf{v}_k \;: \\
&\text{Re}(\lambda_i) = \text{Re}(\lambda_k) \;\wedge\; \text{Imag}(\lambda_i) = -\text{Imag}(\lambda_k) \neq 0.
\end{aligned}
\tag{2.27}
$$

As indicated, type II eigenvectors only occur in pairs. That is, if λ_i is complex with eigenvector $\mathbf{v}_i$, then there exists also an eigenvalue λ_k with $\lambda_k = \lambda_i^*$ and eigenvector $\mathbf{v}_k = \mathbf{v}_i^*$, where z^* denotes the complex-conjugate of z (and z may be a scalar or vector). Eigenvectors are normalized such that $|\mathbf{v}_i| = 1$ holds. Note that if $\mathbf{v}_i$ corresponds to a complex-valued vector this implies that the scalar product (dot product) of $\mathbf{v}_i$ with its complex-conjugate vector $\mathbf{v}_i^*$ is used like $\mathbf{v}_i\mathbf{v}_i^* = 1$.

Under the aforementioned assumption that there exists a set of n linearly independent eigenvectors $\mathbf{v}_i$, there exists a second set of special vectors $\mathbf{w}_i$ with $i = 1, \ldots, n$ that are orthogonal to $\mathbf{v}_i$ like

$$\mathbf{w}_i\mathbf{v}_k = \delta_{ik}, \tag{2.28}$$

where δ_{ik} is the Kronecker symbol. Typically, the vectors $\mathbf{w}_i$ are not normalized. This is, in general, for the length or magnitude of $\mathbf{w}_i$ we obtain $|\mathbf{w}_i| \neq 1$. However, each vector $\mathbf{w}_i$ is normalized with respect to its corresponding eigenvector $\mathbf{v}_i$ like

$$\mathbf{w}_i\mathbf{v}_i = 1. \tag{2.29}$$

The vectors $\mathbf{w}_i$ are also called the left eigenvectors because they satisfy

$$\mathbf{w}_i L = \lambda_i \mathbf{w}_i. \tag{2.30}$$

In this context, $\mathbf{v}_i$ are also referred to as right eigenvectors. The two sets of vectors $\mathbf{v}_1, \ldots, \mathbf{v}_n$ and $\mathbf{w}_1, \ldots, \mathbf{w}_n$ are said to form a biorthogonal system. For this reason, in what follows the vectors $\mathbf{w}_1, \ldots, \mathbf{w}_n$ will be referred to as biorthogonal vectors of the eigenvectors $\mathbf{v}_1, \ldots, \mathbf{v}_n$.

The biorthogonal vectors $\mathbf{w}_k$ can be used to derive the amplitude descriptions of systems defined by Eq. (2.1), epidemiological models, and virus dynamics models (as will be shown in Sect. 2.9.3). The vectors can be derived from Eq. (2.28). To this end, let us define the $n \times n$ matrix M of (right) eigenvectors $\mathbf{v}_i$ like

$$M = (\mathbf{v}_1\mathbf{v}_2 \ldots \mathbf{v}_n). \tag{2.31}$$

In Eq. (2.31) the vectors $\mathbf{v}_i$ are considered as column vectors. Since $\mathbf{v}_i$ are linearly independent, the inverse matrix M^{-1} exists and satisfies

$$M^{-1}M = E, \tag{2.32}$$

where E is the identity matrix (i.e., the diagonal matrix with diagonal elements given by 1). Let us cast the inverse matrix M^{-1} (which is again a $n \times n$ matrix) in the form

$$M^{-1} = \begin{pmatrix} \mathbf{w}_1 \\ \mathbf{w}_2 \\ \ldots \\ \mathbf{w}_n \end{pmatrix}. \tag{2.33}$$

Here $\mathbf{w}_k$ are row vectors. From Eqs. (2.31) and (2.32) it follows that the row vectors $\mathbf{w}_k$ satisfy Eq. (2.28).

In summary, the biorthogonal vectors $\mathbf{w}_k$ can be derived by forming the matrix M out of the vectors $\mathbf{v}_i$, computing the inverse M^{-1} of M, and identifying the rows of M^{-1} as vectors $\mathbf{w}_k$. Note that in linear algebra the matrix M is called the diagonalization matrix because the matrix product $M^{-1}LM$ yields a diagonal matrix, whose diagonal elements are the eigenvalues λ_i. This property will be used in Sect. 2.9.4 in the context of the matrix method to derive amplitude equations.

2.6.3 Amplitudes, Amplitude Description, and Amplitude Space

Any relative state $\mathbf{u}$ can be expressed in terms of the eigenvectors $\mathbf{v}_i$ by means of the superposition

$$\mathbf{u} = \sum_{i=1}^{n} A_i \mathbf{v}_i , \tag{2.34}$$

where A_i are called amplitudes [1]. Since $\mathbf{u} = \mathbf{X} - \mathbf{X}_{st}$ holds (see Eq. (2.9)) any state $\mathbf{X}$ can be expressed like

$$\mathbf{X} = \mathbf{X}_{st} + \sum_i A_i \mathbf{v}_i . \tag{2.35}$$

The vectors $\mathbf{v}_i$ span a new basis in the state space. The amplitudes A_i measure distances along the directions defined by $\mathbf{v}_i$. That is, they can be considered as coordinates. Importantly, the location $A_1 = \cdots = A_n = 0$ corresponds to the fixed point under consideration: $\mathbf{X} = \mathbf{X}_{st}$. That is, the axes $\mathbf{v}_i$ of the new basis are attached to the fixed point $\mathbf{X}_{st}$. The fixed point corresponds to the point of origin in the new basis. For detailed graphical illustrations see [1].

The eigenvectors are taken as dimensionless quantities. Consequently, the amplitudes $A_1, \ldots, A_n$ have the unit of the state variables $X_1, \ldots, X_n$. For epidemiological models the state variables typically count humans or animals and, consequently, are given in units of human individuals or animals. If so, the amplitudes $A_1, \ldots, A_n$ reflect individuals or animals (e.g., see Chaps. 4 and 5). For virus dynamical models we will return to the issue of the units of amplitudes in Chaps. 9 and 10.

In the context of the new basis, recall that a state vector $\mathbf{X} = (X_1, \ldots, X_n)$ can be expressed by means of the orthogonal basis vectors

$$\mathbf{e}_1 = \begin{pmatrix} 1 \\ 0 \\ \ldots \\ 0 \end{pmatrix} , \mathbf{e}_2 = \begin{pmatrix} 0 \\ 1 \\ 0 \\ \ldots \end{pmatrix} \ldots \mathbf{e}_n = \begin{pmatrix} 0 \\ \ldots \\ 0 \\ 1 \end{pmatrix} \tag{2.36}$$

like

$$\mathbf{X} = X_1\mathbf{e}_1 + X_2\mathbf{e} + \cdots + X_n\mathbf{e}_n. \tag{2.37}$$

Equation (2.35) exhibits a structure similar to Eq. (2.37). However, in general, the eigenvectors $\mathbf{v}_1, \ldots, \mathbf{v}_n$ are not orthogonal to each other. Consequently, Eq. (2.35) describes the decomposition of a state with respect to a non-orthogonal basis.

The decomposition defined by Eq. (2.35) allows to introduce an amplitude description of a system. Just like the state variables $X_1, \ldots, X_n$ describe a system in state space, the amplitude variables $A_1, \ldots, A_n$ describe the system in amplitude space. The two spaces (i.e., state space and amplitude space) are connected to each other by means of two mappings: a mapping from state space to amplitude space and, vice versa, a mapping from amplitude space to state space. Equation (2.35) describes the mapping $(A_1, \ldots, A_n) \to (X_1, \ldots, X_n)$ from amplitude space to state space. Multiplying Eq. (2.35) with a biorthogonal vector $\mathbf{w}_i$, the amplitude A_i can be expressed in terms of the state variables $X_1, \ldots, X_n$ like

$$A_i = \mathbf{w}_i(\mathbf{X} - \mathbf{X}_{st}). \tag{2.38}$$

This equation describes the mapping $(X_1, \ldots, X_n) \to (A_1, \ldots, A_n)$ from state space to amplitude space. Finally, by means of the relative state vector $\mathbf{u}$ and the amplitude vector $\mathbf{A} = (A_1, \ldots, A_n)$ the two mappings defined by Eqs. (2.34) and (2.38) can alternatively be expressed like

$$\mathbf{u} = M\mathbf{A}\ , \ \mathbf{A} = M^{-1}\mathbf{u}, \tag{2.39}$$

where $\mathbf{u}$ and $\mathbf{A}$ are column vectors.

2.7 Linear Domain Dynamics and Characterization of Fixed Points

Close to the fixed point $\mathbf{X}_{st}$ the evolution of the state $\mathbf{X}(t)$ is determined by Eq. (2.16). Substituting Eq. (2.34) into Eq. (2.16), we obtain

$$\sum_i \mathbf{v}_i \frac{\mathrm{d}}{\mathrm{d}t} A_i = \sum_i A_i L \mathbf{v}_i = \sum_i \lambda_i \mathbf{v}_i A_i, \tag{2.40}$$

where also Eq. (2.25) has been used. Multiplying this equation with $\mathbf{w}_k$ and exploiting the orthogonality relation given by Eq. (2.28), we obtain

$$\frac{\mathrm{d}}{\mathrm{d}t} A_i = \lambda_i A_i. \tag{2.41}$$

Equation (2.41) describes the evolution of amplitudes when the state is close to its fixed point, that is, when all amplitudes are small ($A_i \approx 0$). Equation (2.41) is a linear equation, just like the linearized Eq. (2.16) describing the evolution of the state $\mathbf{X}$. In other words, Eqs. (2.16) and (2.41) are counterparts to each other. Both describe the evolution of the system under consideration close to the fixed point of interest. Both are linear. However, Eq. (2.16) provides the state space description, while Eq. (2.41) provides the amplitude space description. Solutions $\mathbf{X}(t)$ and $\mathbf{A}(t)$ can be mapped to each other using the mappings defined by Eqs. (2.34), (2.35), (2.38), and (2.39).

Importantly, the amplitude space description given by Eq. (2.41) is given in terms of a diagonal form. The amplitudes A_i evolve independent form each other like

$$A_i(t) = A_i(t_0)\exp\{\lambda_i(t - t_0)\}, \tag{2.42}$$

with the initial amplitudes $A_i(t_0)$ defined by

$$A_i(t_0) = \mathbf{w}_i(\mathbf{X}(t_0) - \mathbf{X}_{st}), \tag{2.43}$$

see Eq. (2.38).

As mentioned above, the evolution of the amplitudes $A_1, \ldots, A_n$ determines the evolution of the state variables $X_1, \ldots, X_n$ and vice versa. Consequently, the explicit solutions for $A_i(t)$ can be used to construct explicit solutions for the state variables. Substituting Eq. (2.42) into Eqs. (2.34) and (2.35), we obtain

$$\mathbf{u}(t) = \sum_{i=1}^{n} A_i(t_0)\exp\{\lambda_i(t - t_0)\}\mathbf{v}_i \tag{2.44}$$

and

$$\mathbf{X}(t) = \mathbf{X}_{st} + \sum_{i=1}^{n} A_i(t_0)\exp\{\lambda_i(t - t_0)\}\mathbf{v}_i, \tag{2.45}$$

respectively. In particular, the initial state $\mathbf{X}(t_0)$ is related to the initial amplitudes $A_1(t_0), \ldots, A_n(t_0)$ like

$$\mathbf{X}(t_0) = \mathbf{X}_{st} + \sum_{i=1}^{n} A_i(t_0)\mathbf{v}_i. \tag{2.46}$$

Equation (2.46) is the inverse mapping of Eq. (2.43).

The analytical expressions given by Eqs. (2.44) and (2.45) for the dynamics of perturbations $\mathbf{u}(t)$ and states $\mathbf{X}(t)$, respectively, close to fixed points can be used to classify fixed point attractors and repellors by means of the eigenvalues λ_i. Using this approach, Table 2.2 describes nodes, foci, and saddles for single-variable and two-variable systems in terms of eigenvalues λ_i [1, 6, 9, 10]. In particular, when eigenvalues are complex, then solutions exhibit an oscillatory dynamics. Consequently, the fixed point is a stable or unstable focus depending on whether the real part of the pair of eigenvalues under consideration is negative (stable case) or positive (unstable case).

Table 2.2 Eigenvalue characterization of fixed point attractors and repellors in one- and two-dimensional state spaces

Dimension	Attractor/repellor	Subtype	Eigenvalue(s) λ
1	Attractor	Stable fixed point	λ negative
	Repellor	Unstable fixed point	λ positive
2	Attractor	Stable node	λ_1, λ_2 real and negative
		Stable focus	λ_1, λ_2 complex real parts negative
	Repellor	Unstable node	λ_1, λ_2 real and positive
		Saddle	λ_1, λ_2 real, one positive, one negative
		Unstable focus	λ_1,λ_2 complex, real parts same and positive

Table 2.3 Eigenvalue characterization of fixed point attractors and repellors for arbitrary dimensions n

Dimension	Attractor/repellor	Subtype	Eigenvalue(s)
n	Attractor		$\forall i\ :\ \{\lambda_i \in \mathbf{R} \wedge \lambda_i < 0\}$ or $\mathrm{Re}(\lambda_i) < 0$
n	Attractor	Stable node	$\forall i\ :\ \lambda_i \in \mathbf{R} \wedge \lambda_i < 0$
n	Repellor		$\exists i\ :\ \{\lambda_i \in \mathbf{R} \wedge \lambda_i > 0\}$ or $\mathrm{Re}(\lambda_i) > 0$

For systems described by an arbitrary number n of variables, the eigenvalues can be used to distinguish between fixed point attractors and repellors, as shown in Table 2.3. The characterization of a node is shown in Table 2.3 as well. A fixed point that exhibits only negative real-valued eigenvalues or eigenvalues that are complex and show a negative real part is a fixed point attractor. In contrast, a fixed point that exhibit at least one positive eigenvalue or at least a pair of complex eigenvalues with a positive real part is a fixed point repellor.

Special case of a single positive eigenvalue

Let us assume that only one eigenvalue is positive, say, λ_k. All other eigenvalues are negative or complex-valued with negative real parts. In this case, the fixed point corresponds to a saddle with a single unstable direction $\mathbf{v}_k$. If furthermore, the negative eigenvalues (or negative real parts of the eigenvalues) are relative large in the amount such that $|\lambda_i| > \lambda_k$ (or $|\mathbb{R}(\lambda_i)| > \lambda_k$) for $i \neq k$ holds, then during an intermediate period T_i all amplitudes A_i decay in magnitude relatively quickly to zero. If the period during which the linearized evolution equations hold T_L is sufficiently long $T_L > T_i$, then after the intermediate period, the exponential increase of A_k entirely determines the evolution of the system under consideration. In this case, Eq. (2.45) becomes

$$\mathbf{X}(t) \approx \mathbf{X}_{st} + A_k(t_0)\exp\{\lambda_k(t-t_0)\}\mathbf{v}_k. \tag{2.47}$$

The approximative description given by Eq. (2.45) holds during the interval $[T_i, T_L]$ and becomes inaccurate when nonlinear terms become relevant, that is, for $t > T_L$. However, under certain circumstances this dominance of the unstable amplitude (or a set of unstable amplitudes) can be generalized such that it holds for $t > T_L$, which will be discussed in Sect. 2.10.

Diagonal form of linear amplitude equations: matrix calculations

The diagonal form of the linear part of amplitude equations is given by Eq. (2.41) and has been derived using vector calculations. In what follows Eq. (2.41) will be derived using matrix calculations. As will be shown in Sect. 2.9 vector calculations and matrix calculations are two methods that can also be used to derive the full, nonlinear amplitude equations of a system.

To simplify the presentation, the matrix calculation method will be demonstrated for the case of a two-variable system with state vector $\mathbf{X} = (X_1, X_2)$ and fixed point $\mathbf{X}_{st} = (X_{1,st}, X_{2,st})$. The following calculations are taken from [15]. Let $\mathbf{u} = (u_1, u_2)$ denote the relative state vector $\mathbf{u} = \mathbf{X} - \mathbf{X}_{st}$. Then, Eq. (2.16) reads

$$\frac{\mathrm{d}}{\mathrm{d}t}\begin{pmatrix} u_1 \\ u_2 \end{pmatrix} = L \begin{pmatrix} u_1 \\ u_2 \end{pmatrix}. \tag{2.48}$$

The eigenvalues $\lambda_{1,2}$ and eigenvectors $\mathbf{v}_{1,2}$ of L are defined by Eq. (2.25). It is assumed that $\mathbf{v}_1$ and $\mathbf{v}_2$ exist and that they are linearly independent from each other. Using $\mathbf{v}_{1,2}$ the relative state $\mathbf{u}$ can be cast into the form

$$\begin{pmatrix} u_1 \\ u_2 \end{pmatrix} = \sum_{i=1,2} A_i \mathbf{v}_i, \tag{2.49}$$

see Eq. (2.34). Using the aforementioned matrix M composed of eigenvectors, Eq. (2.49) can be equivalently expressed as

$$\begin{pmatrix} u_1 \\ u_2 \end{pmatrix} = M \begin{pmatrix} A_1 \\ A_2 \end{pmatrix}, \quad M = \begin{pmatrix} \mathbf{v}_1 & \mathbf{v}_2 \end{pmatrix}, \tag{2.50}$$

see also Eqs. (2.31) and (2.39). The inverse matrix M^{-1} is composed of row vectors that will be denoted by $\mathbf{w}_1$ and $\mathbf{w}_2$ such that

$$M^{-1} = \begin{pmatrix} \mathbf{w}_1 \\ \mathbf{w}_2 \end{pmatrix}, M^{-1}M = \begin{pmatrix} 1 & 0 \\ 0 & 1 \end{pmatrix} \Rightarrow \mathbf{w}_k \mathbf{v}_i = \delta_{ik}. \tag{2.51}$$

Using M^{-1}, the mapping (2.50) can be inverted and reads

$$\begin{pmatrix} A_1 \\ A_2 \end{pmatrix} = M^{-1} \begin{pmatrix} u_1 \\ u_2 \end{pmatrix}, \tag{2.52}$$

see again Eq. (2.39).

The actual matrix calculations are conducted in the following steps. Applying M^{-1} to the left-hand side of Eq. (2.48), we obtain

$$\frac{\mathrm{d}}{\mathrm{d}t} M^{-1} \begin{pmatrix} u_1 \\ u_2 \end{pmatrix} = \frac{\mathrm{d}}{\mathrm{d}t} \begin{pmatrix} A_1 \\ A_2 \end{pmatrix} . \tag{2.53}$$

Applying M^{-1} to the right-hand side of Eq. (2.48) and using Eq. (2.50), we obtain

$$M^{-1} L \begin{pmatrix} u_1 \\ u_2 \end{pmatrix} = M^{-1} L M \begin{pmatrix} A_1 \\ A_2 \end{pmatrix} . \tag{2.54}$$

As mentioned above, the product $M^{-1}LM$ of the three matrices yields a diagonal matrix D. Explicitly, D reads

$$\begin{aligned} LM &= L \left(\mathbf{v}_1 \ \mathbf{v}_2 \right) = \left(\lambda_1 \mathbf{v}_1 \ \lambda_2 \mathbf{v}_2 \right) \\ \Rightarrow \ D = M^{-1} L M &= \begin{pmatrix} \mathbf{w}_1 \\ \mathbf{w}_2 \end{pmatrix} \left(\lambda_1 \mathbf{v}_1 \ \lambda_2 \mathbf{v}_2 \right) = \begin{pmatrix} \lambda_1 & 0 \\ 0 & \lambda_2 \end{pmatrix} , \end{aligned} \tag{2.55}$$

as expected. That is, as expected, the matrix M is the diagonalization matrix that transforms L into diagonal form. Finally, from Eqs. (2.53), (2.54), and (2.55) it follows that

$$\frac{\mathrm{d}}{\mathrm{d}t} \begin{pmatrix} A_1 \\ A_2 \end{pmatrix} = \begin{pmatrix} \lambda_1 & 0 \\ 0 & \lambda_2 \end{pmatrix} \begin{pmatrix} A_1 \\ A_2 \end{pmatrix} . \tag{2.56}$$

As can be seen from Eq. (2.56), the evolution equation for A_1 does not contain an A_2 term. Likewise, the evolution equation for A_2 does not contain an A_1 term. Consequently, in components, the linear parts of the amplitude equations read $\mathrm{d}A_1/\mathrm{d}t = \lambda_1 A_1$ and $\mathrm{d}A_2/\mathrm{d}t = \lambda_2 A_2$. In summary, the matrix approach yields the same result as the vector calculation approach: in the linear domain, amplitudes evolve independent of each other and amplitude equations assumes diagonal form.

2.8 Stable and Unstable Amplitudes and Eigenvectors, Order Parameters Amplitudes, and Order Parameters

Amplitudes A_i that are related to eigenvalues λ_i that are positive or complex with positive real parts increase in magnitude over time as long as they describe states sufficiently close to the fixed point of interest (i.e., as long as they are sufficiently small in magnitude). Therefore, these amplitudes are referred to as unstable amplitudes. In contrast, amplitudes A_i that are related to eigenvalues λ_i that are negative or complex with negative real parts decrease in magnitude over time sufficiently close to the fixed point of interest. Perturbations described by those amplitudes decay. The

Table 2.4 Terminology used to distinguish between amplitudes and eigenvectors related to positive and negative eigenvalues (or related to eigenvalues with positive and negative real parts)

Physical quantity	Eigenvalue λ_i positive or positive real part	Eigenvalue λ_i negative or negative real part
Amplitude A_i	Unstable amplitude Order parameter amplitude	Stable amplitude
Eigenvector $\mathbf{v}_i$	Unstable eigenvector Order parameter	Stable eigenvector

amplitudes describe a dynamics that is stable against perturbations. Accordingly, the amplitudes are referred to as stable amplitudes. By analogy, the eigenvectors are called in the same way as stable and unstable eigenvectors. Since an eigenvector describes a direction in a state space, an unstable eigenvector describes an unstable direction. That is, the vector describes a direction in which the state escapes away from the fixed point of interest. In contrast, a stable eigenvector describes a stable direction in which a perturbation decays such that the state returns to the fixed point (for more details see [1, 5]).

Table 2.4 summarizes the terminology introduced so far. As pointed out in Sect. 1.5, there are directions that determine the order in which the state of a system evolves away from an instability. These directions are the order parameters using the terminology of synergetics (see Sect. 1.5). In short, order parameters correspond to the unstable eigenvectors. Likewise, unstable amplitudes are referred to as order parameter amplitudes using the terminology of synergetics.

Outstanding properties of positive eigenvalues and unstable eigenvectors and amplitudes

Positive eigenvalues or eigenvalues with positive real parts and their corresponding unstable eigenvectors and amplitudes (or order parameters and order parameter amplitudes) exhibit the following outstanding properties.

1. The evolution of a state away from an unstable fixed point is determined in state space by the directions defined by the unstable eigenvectors (order parameters). The unstable amplitudes (order parameter amplitudes) describe explicitly how the state evolves in time along those directions.
2. The evolution of a state away from an unstable fixed point is determined in amplitude space by the increase of the unstable amplitudes.
3. The positive eigenvalues or the positive eigenvalue real parts determine the relevant time scales on which the state evolves away from the unstable fixed point. That is, they determine the speed of the initial dynamics that takes the system away from the unstable fixed point.

2.9 The Linear and Nonlinear Domain: Amplitude Equations

2.9.1 *Where We are and Where We Go Next*

So far, it has been worked out that in general there is a mapping between amplitude space and state space like $\mathbf{X}(t) = \mathbf{X}_{st} + \sum_{i=1}^{n} A_i(t)\mathbf{v}_i$ (see Eq. (2.35)) such that if the amplitudes $A_i(t)$ as functions of time are given, then the evolution of the state $\mathbf{X}(t)$ can be computed from those functions. Moreover, it has been worked out that the states close to a fixed point evolve according to linearized equations. Exact analytical solutions for the amplitudes can be found in terms of exponential functions $A_i(t) = A_i(t_0)\exp\{\lambda_i(t - t_0)\}$ (see Eq. (2.42)), which implies that the evolution of the state $\mathbf{X}(t)$ can be computed from a superposition of such exponential functions (see Eq. (2.45)). At issue is to answer the question: what happens next? What happens when a state is no longer in the vicinity of a fixed point? What happens when amplitudes can no longer be considered to be small?

Primarily, we are interested into two related phenomena: the dynamics of systems close to instabilities and the dynamics of systems in which bifurcations take place (see Sect. 1.5).

Instability scenario

When studying instability phenomena, systems are considered that exhibit an instability. For our purposes to discuss the time course of epidemics in populations and infectious diseases in individuals, it is sufficient to consider instabilities that correspond to fixed point repellors such as unstable nodes, saddles, and foci. The systems under consideration simply exist with the instabilities at hand. That is, we are confronted with systems exhibiting instabilities that come "out of the blue". In fact, this situation describes COVID-19 outbreaks in populations and SARS-CoV-2 infections in individuals as will be shown in the remaining chapters of this book. For a novel virus that spreads out in a population, it typically can be assumed that the initial disease state of the population is close to the disease-free state. Likewise, it might be plausible to assume that a virus that invades certain regions in the body of an individual shifts the disease or health state of the individual only by relatively small amount such that at the onset of the disease the state of the individual or the affected regions can be considered to be close to the virus-free state. If the initial state is sufficiently close to the unstable fixed point, then during an initial period the linear amplitude dynamics as described in Sect. 2.7 holds. Subsequently, the linear approximation will become inaccurate such that more accurately the dynamics is described by a full amplitude equation model that contains both linear and nonlinear terms. In contrast, if for a given situation the aforementioned assumption of an initial state close to its respective fixed point does not hold, then the linear amplitude dynamics as given in terms of exponential functions does not accurately describe the dynamics of the system. However, the dynamics of the system can accurately be described by amplitude equations that contain both linear and nonlinear terms.

Bifurcation scenario

The bifurcation scenario most relevant for discussing epidemics in populations and viral infections in individuals is the following. The system under consideration under circumstances A exhibits a fixed point attractor (e.g., a stable node or stable focus). Moreover, the system is prepared in the stationary state defined by the fixed point attractor at hand. Subsequently, the bifurcation parameter is changed such that it reaches a critical value at which a bifurcation occurs. The bifurcation turns the asymptotically stable fixed point into an unstable one (i.e., turns it into an unstable node, unstable focus, or saddle). The system is then considered under circumstances B in which the bifurcation parameter is slightly above the critical value. Detailed discussions of this scenario can be found in [1, 5]. There are two key issue of the bifurcation scenario. First, the initial state can be considered to be close to the unstable fixed point. That is, unlike the instability scenario, it follows that all amplitudes are initially sufficiently small such that the linear dynamics $A_i(t) = A_i(t_0)\exp\{\lambda_i(t - t_0)\}$ is a good approximation. Second, if the bifurcation parameter is larger than the critical value but sufficiently close to the critical value, then from theoretical consideration [1, 5] it follows that there exist only a single positive eigenvalue or a relative small number of eigenvalues that are real and positive or complex and exhibit positive real parts. If so, the initial dynamics is determined by the corresponding unstable eigenvector or the small number of unstable eigenvectors. The bifurcation phenomenon under consideration exhibits an order parameter or a small number of order parameters. This issue will be discussed in Sect. 2.10.

In both scenarios, at best, it can only be assumed that for a certain initial period the state of the system under consideration is close to the vicinity of the fixed point under consideration and that the amplitudes are relatively small. In order to describe the dynamics of the system under consideration over a longer period by means of the amplitude description, amplitude equations can be used that exhibit both linear and nonlinear terms.

Bifurcation and instability scenarios

In both scenarios the full description of the dynamics can either be obtained by solving the state space model defined by Eq. (2.1) or an amplitude equation model that goes beyond Eq. (2.41) and contains nonlinear terms. Formally, such an amplitude equation model can be expressed like

$$\frac{\mathrm{d}}{\mathrm{d}t}A_i = N_{A,i}(A_1, \ldots, A_n), \tag{2.57}$$

where $N_{A,i}$ are nonlinear functions of the amplitudes. That is, the amplitude equation model corresponds to a set of n coupled first-order differential equations. The subindex A of $N_{A,i}$ indicates that the functions $N_{A,i}$ are related to the amplitude description of the system of interest. Using the amplitude vector $\mathbf{A}$ and the vector-function $\mathbf{N}_A = (N_{A,1}, \ldots, N_{A,n})$, Eq. (2.57) can be written like

$$\frac{d}{dt}\mathbf{A} = \mathbf{N}_A(A_1, \ldots, A_n). \tag{2.58}$$

Equation (2.58) is the counterpart to Eq. (2.1). Likewise, Eq. (2.57) is the counterpart to Eq. (2.2). Eq. (2.1) (or in components Eq. (2.2)) describes the evolution of the state of a system under consideration in state space. Likewise, Eq. (2.58) (or in components Eq. (2.57)) describes the evolution of the system in amplitude space. In view of the linear amplitude dynamics defined by Eq. (2.41), the functions $N_{A,i}$ can be decomposed into a linear part and purely nonlinear part. Using this approach, Eq. (2.57) becomes [1]

$$\frac{d}{dt}A_i = \lambda_i A_i + G_i(A_1, \ldots, A_n), \tag{2.59}$$

where λ_i are the eigenvalues occurring in Eq. (2.41) and G_i are functions that do not exhibit any terms linear in $A_1, \ldots, A_n$. That is, the functions G_i contain terms like A_1^2 or $A_1 A_2$. There are several methods to derive the explicit expressions of the purely nonlinear parts G_i.

2.9.2 *Method 1: Scalar Calculations*

This method uses all state variables $X_1, \ldots, X_n$ and amplitude variables $A_1, \ldots, A_n$ one at a time. That is, the method works with scalars. The mappings $\mathbf{A} \to \mathbf{X}$ and $\mathbf{X} \to \mathbf{A}$, see Eqs. (2.34), (2.35), (2.38), and (2.39), are written out explicitly for each state and amplitude like

$$X_i = X_{i,st} + \sum_{k=1}^{n} A_k v_{k,i}, \quad A_i = \sum_{k=1}^{n} w_{i,k}(X_k - X_{st,k}), \tag{2.60}$$

where $v_{k,i}$ is the ith component of $\mathbf{v}_k$ and $w_{i,k}$ is the kth component of $\mathbf{w}_i$ like

$$\mathbf{v}_k = \begin{pmatrix} v_{k,1} \\ \ldots \\ v_{k,n} \end{pmatrix}, \quad \mathbf{w}_i = \begin{pmatrix} w_{i,1} \\ \ldots \\ w_{i,n} \end{pmatrix}. \tag{2.61}$$

Subsequently, using the mapping $\mathbf{X} \to \mathbf{A}$ described in components in Eq. (2.60), A_1 is differentiated with respect to time to arrive at

$$\frac{d}{dt}A_1 = \sum_{k=1}^{n} w_{1,k}\frac{d}{dt}X_k = \sum_{k=1}^{n} w_{1,k}N_k(X_1, \ldots, X_n), \tag{2.62}$$

where $dX_k/dt = N_k$ (see Eq. (2.2)) has been used as well. The state variables occurring in the functions N_k are expressed in terms of amplitudes using the mapping

$\mathbf{A} \to \mathbf{X}$ described in components in Eq. (2.60) again. In doing so, formally, we arrive at

$$\frac{\mathrm{d}}{\mathrm{d}t} A_1 = \sum_{k=1}^{n} w_{1,k} N_k(X_1(A_1, \ldots, A_n), \ldots, X_n(A_1, \ldots, A_n)) = N_{A,1} \tag{2.63}$$

for A_1. When carrying out the calculations, the function $N_{A,1}$ will show a structure as indicated in Eq. (2.59): $N_{A,1} = \lambda_1 A_1 + G_1$. The calculations may be simplified by taking this structure into account. That is, the calculations of terms linear in A_1, A_2 and so on, may be skipped since the result is given and reads: $\lambda_1 A_1$. That is, in the evolution equation of A_1 all coefficients of terms involving only A_1 add up to λ_1. In the evolution equation of A_1, all coefficients of terms involving only A_2 will add up to zero. All coefficients of terms involving only A_3 will add up to zero, and so on. The steps presented above to derive the evolution equation of A_1 can then be repeated in a similar vein to obtain the evolution equations of the remaining amplitudes $A_2, \ldots, A_n$.

Illustration for an epidemiological model

Following [3, 15], let us use the scalar method to derive the amplitude equations of a fundamental epidemiological model, the susceptible-exposed-infectious-removed (SEIR) model. The SEIR model will be discussed in detail in Chap. 3. The amplitude equations of the SEIR model will be discussed in detail in Chap. 5.

The following calculations primarily demonstrate methodological aspects, namely, how the scalar method works in an application. Other aspects will be addressed in Chaps. 3 and 5. Let S denote the susceptible individuals, E describe the exposed individuals, I describe the infected individuals, and R describe the number of recovered individuals. The total population is given by $N = S + E + I + R$. The reader should watch out. N does not refer to a component of the RHS vector-function $\mathbf{N}$ of Eq. (2.1). The SEIR model assumes the form of Eq. (2.1) with $\mathbf{X} = (S, E, I, R)$ and reads [15, 16]

$$\frac{\mathrm{d}}{\mathrm{d}t} S = -\frac{\beta}{N} IS \, , \; \frac{\mathrm{d}}{\mathrm{d}t} E = \frac{\beta}{N} IS - \alpha E \, , \; \frac{\mathrm{d}}{\mathrm{d}t} I = \alpha E - \gamma I \, , \; \frac{\mathrm{d}}{\mathrm{d}t} R = \gamma I . \tag{2.64}$$

The model exhibits the parameters $\beta > 0$, $\alpha > 0$, and $\gamma > 0$ that will be explained in Chap. 3. Since N is constant, N is considered to be another parameter. However, since N is constant, the variable R can be computed from $R = N - S - E - I$. It is sufficient to consider the three-variable model with state vector $\mathbf{X} = (S, E, I)$ that was also briefly introduced in Sect. 1.2. The fixed point under consideration is the disease-free state with $\mathbf{X}_{st} = (N, 0, 0)$.

In line with the general index notation of state variables as $X_1, \ldots X_n$, the variables S, E, and I may be referred to as X_1, X_2, and X_3, respectively. However, in this demonstration, an alternative notation will be used and S, E, and I will be referred to as X_S, X_E, and X_I such that $\mathbf{X} = (X_S, X_E, X_I)$.

Linearization of Eq. (2.64) at $\mathbf{X}_{st}$ taking only the first three variables into account, yields the linear model defined by Eq. (2.16) with the relative state $\mathbf{u} = (u_S, u_E, u_I)$, $u_S = S - N$, $u_E = E$, $u_I = I$, and the 3×3 matrix

$$L = \begin{pmatrix} 0 & 0 & -\beta \\ 0 & -\alpha & \beta \\ 0 & \alpha & -\gamma \end{pmatrix}. \tag{2.65}$$

The matrix is singular and exhibits the eigenvalues $\lambda_1 = 0$ and [17]

$$\lambda_{2,3} = \frac{\hat{T}}{2} \pm \sqrt{\frac{\hat{T}^2}{4} + \hat{D}_{neg}}. \tag{2.66}$$

In Eq. (2.66) the trace $\hat{T} = -(\alpha + \gamma) < 0$ is computed from the non-singular submatrix L_{sub} defined by

$$L_{\text{sub}} = \begin{pmatrix} -\alpha & \beta \\ \alpha & -\gamma \end{pmatrix}. \tag{2.67}$$

Moreover, $\hat{D}_{neg} = \alpha(\beta - \gamma)$ is the negative value of the determinant of L_{sub}. The eigenvalue $\lambda_1 = 0$ is associated with the eigenvector

$$\mathbf{v}_1 = \begin{pmatrix} 1 \\ 0 \\ 0 \end{pmatrix}. \tag{2.68}$$

In this application, the amplitude space description will only be applied to the two-dimensional subspace spanned by the variables E and I. The motivation for this step will be explained in Chap. 6. This subspace is the plane orthogonal to $\mathbf{v}_1$. The eigenvectors in the E-I subspace can be obtained from the submatrix L_{sub}. In doing so, the E and I components of $\mathbf{v}_2$ and $\mathbf{v}_3$ can be obtained. The S components of $\mathbf{v}_2$ and $\mathbf{v}_3$ are zero (because $\mathbf{v}_2$ and $\mathbf{v}_3$ are in the plane orthogonal to $\mathbf{v}_1$ as mentioned above). A detailed calculation [3, 15] (see also Sect. 6.2.2) yields

$$\mathbf{v}_2 = \frac{1}{Z_2} \begin{pmatrix} 0 \\ \beta \\ \lambda_2 + \alpha \end{pmatrix}, \quad \mathbf{v}_3 = \frac{1}{Z_3} \begin{pmatrix} 0 \\ \beta \\ \lambda_3 + \alpha \end{pmatrix} \tag{2.69}$$

with $Z_j = \sqrt{\beta^2 + (\lambda_j + \alpha)^2}$ for $j = 2, 3$. The vectors $\mathbf{v}_1$, $\mathbf{v}_2$, and $\mathbf{v}_3$ are linearly independent of each other provided that $\lambda_2 \neq \lambda_3$ holds, which is always the case because $\hat{T}^2/4 + \hat{D}_{\text{neg}} > 0$ holds for $\alpha, \beta, \gamma > 0$, see [3]. Since the assumption of linear independency is satisfied, in analogy to Eq. (2.35), the state $\mathbf{X}$ can be decomposed like

$$\mathbf{X} = \mathbf{X}_{st} + u_S\mathbf{v}_1 + A_2\mathbf{v}_2 + A_3\mathbf{v}_3 = \mathbf{X}_{st} + \sum_{i=1}^{3} A_i\mathbf{v}_i. \tag{2.70}$$

As indicated in Eq. (2.70), the relative state u_S will be denoted as A_1 in order to simplify the presentation. In order to conduct the scalar method, the mappings $\mathbf{A} \rightarrow \mathbf{X}$ and $\mathbf{X} \rightarrow \mathbf{A}$ should be given in components. The mapping $(A_1, A_2, A_3) \rightarrow (S, E, I)$ defined by Eq. (2.70) reads in components

$$S = N + A_1 \ , \ \ E = A_2 v_{2,E} + A_3 v_{3,E} \ , \ \ I = A_2 v_{2,I} + A_3 v_{3,I}. \tag{2.71}$$

with $\mathbf{v}_j = (0, v_{j,E}, v_{j,I})$ for $j = 2, 3$ defined by Eq. (2.69). The inverse mapping $(S, E, I) \rightarrow (A_1, A_2, A_3)$ reads [3]

$$A_1 = S - N \ , \ \ A_2 = (v_{3,I}E - v_{3,E}I)/B \ , \ \ A_3 = (v_{2,E}I - v_{2,I}E)/B. \tag{2.72}$$

In Eq. (2.72) the parameter B is computed from the determinant of the diagonalization matrix

$$M = \begin{pmatrix} v_{2,E} & v_{3,E} \\ v_{2,I} & v_{3,I} \end{pmatrix} \tag{2.73}$$

and reads $B = |M| = \beta(\lambda_3 - \lambda_2)/(Z_2 Z_3) < 0$ for $\alpha > 0, \beta > 0$ [3].

Before conducting the key steps of the scalar calculation method (centered around Eqs. (2.62) and (2.63)), it is convenient to discuss the nonlinear term $\beta SI/N$ of the SEIR model defined by Eq. (2.64). As indicated in Eq. (2.63), eventually, the state variables X_i that occur in the nonlinear terms $N_{A,k}$ are expressed in terms of amplitude variables. Accordingly, substituting the relations of Eq. (2.71) into the $\beta SI/N$ term, we obtain

$$\frac{\beta}{N} SI = \frac{N + A_1}{N} \beta \left(v_{2,I} A_2 + v_{3,I} A_3 \right) = (N + A_1) k_0(A_2, A_3) \tag{2.74}$$

with

$$k_0 = \frac{\beta}{N} (v_{2,I} A_2 + v_{3,I} A_3). \tag{2.75}$$

Now, the steps related to Eqs. (2.62) and (2.63) can be carried out. From $A_1 = S - N$ (see Eq. (2.72)) and $\mathrm{d}S/\mathrm{d}t = -\beta SI/N$ (see Eq. (2.64)), it follows that

$$\frac{\mathrm{d}}{\mathrm{d}t} A_1 = \frac{\mathrm{d}}{\mathrm{d}t} S = -\frac{\beta}{N} SI \tag{2.76}$$

Using Eq. (2.74), the evolution equation

$$\frac{\mathrm{d}}{\mathrm{d}t} A_1 = -(N + A_1) k_0(A_2, A_3) \tag{2.77}$$

for the amplitude A_1 can be obtained. As far as A_2 is concerned, differentiating the variable transformation $A_2 = (v_{3,I}E - v_{3,E}I)/B$ (see Eq. (2.72)) with respect to time, leads to

$$\frac{\mathrm{d}}{\mathrm{d}t}A_2 = \frac{v_{3,I}}{B}\frac{\mathrm{d}}{\mathrm{d}t}E - \frac{v_{3,E}}{B}\frac{\mathrm{d}}{\mathrm{d}t}I. \tag{2.78}$$

The evolution equation of I is linear (see Eq. (2.64)). Let us first focus on the evolution equation of E that involves the nonlinear term $\beta SI/N$. Replacing $\mathrm{d}E/\mathrm{d}t$ by N_E defined by $N_E = \beta SI/N - \alpha E$, Eq. (2.78) becomes

$$\frac{\mathrm{d}}{\mathrm{d}t}A_2 = \frac{v_{3,I}}{B}\frac{\beta}{N}SI + \text{linear terms}. \tag{2.79}$$

Using Eq. (2.74) again, Eq. (2.79) can be written as

$$\frac{\mathrm{d}}{\mathrm{d}t}A_2 = \frac{v_{3,I}}{B}A_1 k_0(A_2, A_3) + \frac{v_{3,I}N}{B}k_0(A_2, A_3) + \text{linear terms}. \tag{2.80}$$

The second term on the right-hand side of the equals sign is a linear term. From Eq. (2.41) it follows that all linear terms taken together yield the term $\lambda_2 A_2$. Consequently, Eq. (2.80) can be cast into the form

$$\frac{\mathrm{d}}{\mathrm{d}t}A_2 = \lambda_2 A_2 + \frac{v_{3,I}}{B}A_1 k_0(A_2, A_3). \tag{2.81}$$

Conducting the analogous steps for A_3, the following result can be obtained [3]:

$$\frac{\mathrm{d}}{\mathrm{d}t}A_3 = \lambda_3 A_3 - \frac{v_{2,I}}{B}A_1 k_0(A_2, A_3). \tag{2.82}$$

Note that there is a minus sign in front of the nonlinear term, which is due to the fact that A_3 depends on E like $A_3 = (v_{2,E}I - v_{2,I}E)/B$, see Eq. (2.72). The evolution equations for A_2 and A_3 assume diagonal form. That is, the linear parts are given by $\mathrm{d}A_i/\mathrm{d}t = \lambda_i A_i$. In contrast, the evolution equation of A_1 exhibits in addition to the term $\lambda_1 A_1$ that equals zero (because of $\lambda_1 = 0$) two linear terms in A_2 and A_3 like

$$\begin{aligned}\frac{\mathrm{d}}{\mathrm{d}t}A_1 &= Nk_0(A_2, A_3) + \text{nonlinear terms}\\ &= \beta(v_{2,I}A_2 + v_{3,I}A_3) + \text{nonlinear terms}.\end{aligned} \tag{2.83}$$

The reason for this is that A_1 actually corresponds to the relative state u_S. That is, A_1 is not an amplitude derived from a diagonalization procedure. As stated above, the amplitude space description was only applied to the E-I subspace. Therefore, only A_2 and A_3 are amplitudes. Using u_S rather than A_1, the SEIR model defined by Eq. (2.64) in the partial amplitude space description worked out in this example reads

$$\frac{d}{dt}u_S = -(N + u_S)k_O(A_2, A_3) ,$$
$$\frac{d}{dt}A_2 = \lambda_2 A_2 + \frac{v_{3,I}}{B}u_S k_O(A_2, A_3) ,$$
$$\frac{d}{dt}A_3 = \lambda_3 A_3 - \frac{v_{2,I}}{B}u_S k_O(A_2, A_3). \quad (2.84)$$

In summary, the SEIR model is defined in state space and amplitude space by Eqs. (2.64) and (2.84), respectively. The two descriptions are connected to each other by the mappings defined by Eqs. (2.71) and (2.72), in which A_1 is replaced by u_S. The model Eq. (2.84) will be discussed in a more general context in Sect. 6.2 and applied to the COVID-19 epidemic of the year 2020 in Wuhan city, China, in Sect. 6.3.

2.9.3 Method 2: Vector Calculations

The vector calculations method makes explicit use of the biorthogonal vectors $\mathbf{w}_i$. First of all, the right-hand side vector-function **N** of Eq. (2.1) is decomposed into a linear and nonlinear part. To illustrate this decomposition, let us consider a function $f(x)$ depending on the coordinate x that vanishes at a particular value x_{st} like $f(x_{st}) = 0$. The Taylor expansion of f using the relative state $u = x - x_{st}$ reads

$$f(x) = f(x_{st}) + Lu + \frac{1}{2}f''u^2 + \frac{1}{6}f'''u^3 + \dots , \quad (2.85)$$

where $L = df/dx$ is the first derivative of f at x_{st} and f'' and f''' are the second and third derivatives of f at x_{st}, respectively. If the Taylor expansion is truncated after a particular term, then the truncated Taylor expansion of f differs from f by a remainder term R. Adding the remainder term to the truncated Taylor expansion, the original function f is re-obtained. In particular, truncating after the linear term gives $f(x) = f(x_{st}) + Lu$ such that the remainder term is $R = f(x) - [f(x_{st}) + Lu]$. In short, $f(x)$ can be written like

$$f(x) = f(x_{st}) + Lu + R = f(x_{st}) + Lu + R(u, x_{st}). \quad (2.86)$$

While R originally is a function of x, using $x = x_{st} + u$, the function R can also be expressed as a function of u, as indicated above. Moreover, the term $f(x_{st})$ equals zero by definition of x_{st}. Applying this type of decomposition to **N**, gives us

$$\mathbf{N}(\mathbf{X}) = \mathbf{N}(\mathbf{X}_{st} + \mathbf{u}) = \mathbf{N}(\mathbf{X}_{st}) + L\mathbf{u} + \mathbf{R}(\mathbf{u}, \mathbf{X}_{st}) = L\mathbf{u} + \mathbf{R}(\mathbf{u}, \mathbf{X}_{st}), \quad (2.87)$$

where **R** does not exhibit any linear terms. Equation (2.87) may be considered as the definition of the remainder term like

$$\mathbf{R}(\mathbf{u}, \mathbf{X}_{st}) = \mathbf{N}(\mathbf{X}_{st} + \mathbf{u}) - L\mathbf{u}. \tag{2.88}$$

Substituting into the left-hand side of Eq. (2.1) the relative state $\mathbf{u} = \mathbf{X} - \mathbf{X}_{st}$, from Eqs. (2.1) and (2.87), it follows that

$$\frac{\mathrm{d}}{\mathrm{d}t}\mathbf{u} = L\mathbf{u} + \mathbf{R}(\mathbf{u}, \mathbf{X}_{st}). \tag{2.89}$$

This equation is the departure point for the vector calculation method and the matrix calculation method to derive amplitude equations. In the context of the vector calculation method, the left-hand and right-hand sides of Eq. (2.89) are multiplied with the vector $\mathbf{w}_i$ such that

$$\mathrm{LHS} \; : \; \mathbf{w}_i \frac{\mathrm{d}}{\mathrm{d}t}\mathbf{u} = \frac{\mathrm{d}}{\mathrm{d}t}\mathbf{w}_i \sum_{k=1}^{n} \mathbf{v}_k \frac{\mathrm{d}}{\mathrm{d}t} A_k = \frac{\mathrm{d}}{\mathrm{d}t} A_i, \tag{2.90}$$

where Eqs. (2.28) and (2.34) have been used, and

$$\mathrm{RHS} \; : \; \mathbf{w}_i \left(L\mathbf{u} + \mathbf{R}\right) = \mathbf{w}_i \left(\sum_{k=1}^{n} \lambda_k A_k \mathbf{v}_k\right) + \mathbf{w}_i \mathbf{R} = \lambda_i A_i + \mathbf{w}_i \mathbf{R}, \tag{2.91}$$

where Eqs. (2.28) and (2.34) have been used again. By putting LHS = RHS and taking again advantage of Eq. (2.34), the amplitude equations of the form

$$\frac{\mathrm{d}}{\mathrm{d}t} A_i = \lambda_i A_i + \mathbf{w}_i \mathbf{R} \left(\sum_{k=1}^{n} A_k \mathbf{v}_k, \mathbf{X}_{st}\right) \tag{2.92}$$

are obtained. A comparison of Eq. (2.92) with Eq. (2.59) shows that within the framework of the vector-calculation method the purely nonlinear terms G_i are computed from

$$G_i = \mathbf{w}_i \mathbf{R} \left(\sum_{k=1}^{n} A_k \mathbf{v}_k, \mathbf{X}_{st}\right). \tag{2.93}$$

In Sect. 2.9.2 the amplitude equations of the SEIR model defined by Eq. (2.64) have been derived for a partial two-dimensional amplitude space description using the scalar calculation method. In Sect. 6.1.2 it will be shown how to apply the vector calculation method (and the matrix calculation method) to derive Eq. (2.64). Moreover, in Sect. 5.7.1 the amplitude equations of the SEIR model in the full three-dimensional amplitude space will be derived using the vector calculation method.

2.9.4 Method 3: Matrix Calculations

According to the vector calculation method, Eq. (2.89) is multiplied by $\mathbf{w}_1$ and the amplitude equation for A_1 is obtained. Subsequently, Eq. (2.89) is multiplied by $\mathbf{w}_2$ and the amplitude equation for A_2 is obtained. This procedure is repeated for all vectors $\mathbf{w}_i$ and corresponding amplitudes A_i. Consequently, the method involves n steps. The matrix calculation method formally takes these n steps together. To this end, vector multiplications are replaced by matrix multiplications [18]. Just as for the vector calculation method, the departure point is Eq. (2.89). Then, differentiating the mapping $\mathbf{A} = M^{-1}\mathbf{u}$ (see Eq. (2.39)) with respect to time and replacing $\mathbf{du}/\mathrm{d}t$ by the right-hand side of Eq. (2.89), the intermediate result

$$\frac{\mathrm{d}}{\mathrm{d}t}\mathbf{A} = M^{-1}\frac{\mathrm{d}}{\mathrm{d}t}\mathbf{u} = M^{-1}[L\mathbf{u} + \mathbf{R}(\mathbf{u}, \mathbf{X}_{st})] \tag{2.94}$$

is obtained. Substituting $\mathbf{u} = M\mathbf{A}$ (see Eq. (2.39) again) into Eq. (2.94), leads to

$$\frac{\mathrm{d}}{\mathrm{d}t}\mathbf{A} = M^{-1}LM\mathbf{A} + M^{-1}\mathbf{R}(M\mathbf{A}, \mathbf{X}_{st}) . \tag{2.95}$$

Using the fact that M is the diagonalization matrix of L such that $M^{-1}LM = D$, Eq. (2.95) can equivalently be expressed as

$$\frac{\mathrm{d}}{\mathrm{d}t}\mathbf{A} = D\mathbf{A} + M^{-1}\mathbf{R}(M\mathbf{A}, \mathbf{X}_{st}) \ , \ D = \begin{pmatrix} \lambda_1 & \dots & 0 \\ \dots & \dots & \dots \\ 0 & \dots & \lambda_n \end{pmatrix} . \tag{2.96}$$

Consequently, Eq. (2.96) assumes the general from of Eq. (2.58). Within the framework of the matrix calculation method the nonlinear term $\mathbf{N}_A$ of Eq. (2.58) can be decomposed into the linear part involving the diagonal matrix D and the vector-function $\mathbf{G}$ that contains only nonlinear terms:

$$\frac{\mathrm{d}}{\mathrm{d}t}\mathbf{A} = \mathbf{N}_A = D\mathbf{A} + \mathbf{G}(\mathbf{A}) \ , \ \mathbf{G} = M^{-1}\mathbf{R}(M\mathbf{A}, \mathbf{X}_{st}) . \tag{2.97}$$

In components, Eq. (2.97) reads

$$\frac{\mathrm{d}}{\mathrm{d}t}A_i = \lambda_i A_i + \sum_{k=1}^{n} M_{ik}^{-1} R_k , \tag{2.98}$$

where M_{ik}^{-1} are the matrix elements of M^{-1} and R_k are the components of $\mathbf{R}$ like $\mathbf{R} = (R_1, \dots, R_n)$. Therefore, it follows that the components of the purely nonlinear terms G_i can be computed from

$$G_i = \sum_{k=1}^{n} M_{ik}^{-1} R_k \left(\sum_{j=1}^{n} A_j \mathbf{v}_j, \mathbf{X}_{st} \right) . \tag{2.99}$$

Since (as it was stated in the context of Eq. (2.33)) the ith row of M^{-1} is given by the vector $\mathbf{w}_i$ (when considering $\mathbf{w}_i$ as a row vector), Eq. (2.99) is equivalent to Eq. (2.93).

2.10 Reduced Amplitude Spaces

The two scenarios addressed in Sect. 2.9.1, the instability and bifurcation scenarios, both are such that we are confronted with systems that exhibit a number of positive real-valued eigenvalues or eigenvalues that are complex with positive real parts. For instabilities arising due to bifurcations, it can be argued that the number of such eigenvalues is relatively small [1]. In particular, under some relatively weak assumptions that are described in detail in [1], systems close to their bifurcation points exhibit a single real-valued positive eigenvalue. In what follows, we assume that among the n eigenvalues there are m eigenvalues that are either real-valued and positive or complex with positive real parts. For sake of simplicity, let denote $\lambda_1, \ldots, \lambda_m$ positive eigenvalues of this kind. The corresponding eigenvectors and amplitudes are the unstable ones (see Sect. 2.8). Moreover, it is assumed that $\lambda_{m+1}, \ldots, \lambda_n$ are real-valued and negative or complex with negative real parts such that the corresponding eigenvectors and amplitudes are the stable ones (see Sect. 2.8). In summary, we have

$$\begin{aligned} i = 1, \ldots, m \; &: \; (\lambda_i \in \mathbb{R} \wedge \lambda_i > 0) \text{ or } \mathrm{Re}(\lambda_i) > 0 \;\Rightarrow\; \mathbf{v}_i, A_i \text{ unstable} \\ i = m+1, \ldots, n\,, \; &: \; (\lambda_i \in \mathbb{R} \wedge \lambda_i < 0) \text{ or } \mathrm{Re}(\lambda_i) < 0 \;\Rightarrow\; \mathbf{v}_i, A_i \text{ stable.} \end{aligned} \tag{2.100}$$

In the context of the bifurcation scenario, it can also be argued that there is a time-scale separation. The eigenvalues that are positive or exhibit positive real parts are close to zero and in magnitude smaller than the negative eigenvalues or the eigenvalues with negative real parts such that

$$|\mathrm{Re}(\lambda_i)| \gg \mathrm{Re}(\lambda_k) \text{ for } i = m+1, \ldots, n\,, \; k = 1, \ldots, m. \tag{2.101}$$

This implies that in the linear domain, in which Eq. (2.41) holds, the amplitudes A_i with $i = 1, \ldots, m$ are slowly evolving, whereas the amplitude A_k with $k = m + 1, \ldots, n$ are fast evolving. There is a time-scale separation [1, 5] at least during the initial dynamics away from the fixed point under consideration for which all amplitudes are relatively small:

$$i = 1, \ldots, m \; : \; A_i \text{ slowly evolving} \, ,$$
$$i = m+1, \ldots, n \; : \; , A_i \text{ fast evolving.} \tag{2.102}$$

That is, the stable amplitudes are fast evolving, whereas the unstable amplitudes are slowly evolving.

In what follows, it is further assumed that the amplitude equations defined by Eq. (2.59) for the stable amplitudes exhibit nullcline points. That is, they exhibit fixed points if the values of the slowly evolving amplitudes are assumed to be constants. If so, the key idea that has been frequently used in the literature (adiabatic elimination [1, 5, 19]) is to assume that the fast variables quickly converge to these fixed points (nullcline points) even when the slow variables are actually slowly changing in time. That is, the amplitude A_i for $i = m + 1, \ldots, n$ can be replaced by their fixed point (or nullcline) values. More explictly, putting $\mathrm{d}A_i/\mathrm{d}t = 0$ in Eq. (2.59) for $i = m + 1, \ldots, n$, we obtain

$$A_i = \frac{G_i(A_1, \ldots, A_n)}{|\lambda_i|}, \tag{2.103}$$

when $\lambda_i < 0$ is real and

$$A_i = -\lambda_i^* \frac{G_i(A_1, \ldots, A_n)}{|\lambda_i|^2}, \tag{2.104}$$

when λ_i is complex. Eqs. (2.103) and (2.104) describes a set of $n - m$ equations for $n - m$ stable amplitudes A_i. Of particular interest is the case in which this set of implicit algebraic equations can be solved for the $n - m$ amplitudes $A_{m+1}, \ldots, A_n$. If so, the result reads

$$A_i = f_i(A_1, \ldots, A_m), \tag{2.105}$$

where f_i are certain functions. Eq. (2.105) describes a mapping $(A_1, \ldots, A_m) \rightarrow (A_{m+1}, \ldots, A_n)$ from the slowly evolving amplitudes to the fast evolving amplitudes in terms of $n - m$ functions f_i. Accordingly, the fast evolving amplitudes are considered as dependent variables. They depend on the slowly evolving variables. Importantly, substituting these relations into the amplitude equations defined by Eq. (2.59) again, but this time for the slowly evolving amplitude, then we obtain

$$\frac{\mathrm{d}}{\mathrm{d}t} A_i = \lambda_i A_i + G_{i,eff}(A_1, \ldots, A_m) \; , \; i = 1, \ldots, m \tag{2.106}$$

with the effective purely nonlinear functions

$$G_{i,eff}(A_1, \ldots, A_m) = G_i(A_1, \ldots, A_m, f_{m+1}, \ldots, f_n), \tag{2.107}$$

where f_i only depend on $A_1, \ldots, A_m$, see Eq. (2.105).

Equation (2.106) is a closed set of m coupled first-order differential equations for the unstable amplitudes $A_1, \ldots, A_m$. The unstable amplitudes $A_1, \ldots, A_m$ span the reduced amplitude space. The dynamics in this reduced amplitude space plays the key role for the dynamics of the system in the entire original n-dimensional state space. The dynamics of the state $\mathbf{X}$ can be reconstructed from the dynamics in this m-dimensional reduced amplitude space using the mappings f_i from $(A_1, \ldots, A_m) \rightarrow (A_{m+1}, \ldots, A_n)$ and the general mapping $\mathbf{A} \rightarrow \mathbf{X}$. In summary, the reduced amplitude space allows for

1. Simplification of the analysis of the system dynamics under consideration (by reducing the dimensionality of the system).
2. Identification of the relevant dynamics and mechanisms (by pointing out the subspace in which the relevant dynamics takes place).

As mentioned in the beginning of this section, a system under consideration featuring an instability might exhibit only a single positive, real-valued eigenvalue. If so, there is a single order parameter $\mathbf{v}_1$ and a single order parameter amplitude A_1. The order parameter amplitude satisfies the equation

$$\frac{\mathrm{d}}{\mathrm{d}t} A_1 = \lambda_1 A_1 + G_{1,eff}(A_1). \tag{2.108}$$

For such a system, the key aspects of the phenomena of interest can be captured by a nonlinear evolution equation for a single variable. In fact, for the SEIR model defined by Eq. (2.64) it can be shown that only one of the two eigenvalues $\lambda_{2,3}$ defined by Eq. (2.66) can be positive [3]. That is, the SEIR model is an example of models that describe populations with instabilities characterized by single positive eigenvalues.

In the context of the reduced amplitude space, the dependency of the stable amplitudes on the unstable amplitudes and vice versa may be studied in more detail. For a given system, we may ask the question: Do the stable amplitudes depend on the unstable ones? We may distinguish between the de-coupled case in which they do not and the coupled case in which they do so [1]. Likewise, for a given system, we may ask the question: Do the unstable amplitudes depend on the stable ones? We may distinguish again between the de-coupled case in which they do not and the coupled case in which they do so [1]. For a detailed discussion of this matter the reader is referred to [1].

Haken-Zwanzig model

Originally the Haken-Zwanzig model was used in thermodynamics to discuss how heat bath variables can be eliminated [20–23]. Nowadays, it is frequently used in nonlinear physics to demonstrate the elimination of the stable amplitudes and the construction of reduced amplitude spaces [1, 5, 24–27]. The model involves two amplitudes with one stable and the other unstable such that $n = 2$, $m = 1$, $\lambda_1 > 0$, $\lambda_2 < 0$. In the case of the Haken-Zwanzig model, Eq. (2.59) reads [1, 5]

$$\frac{\mathrm{d}}{\mathrm{d}t} A_1 = \lambda_1 A_1 - A_1 A_2 \,, \quad \frac{\mathrm{d}}{\mathrm{d}t} A_2 = \lambda_2 A_2 + A_1^2. \tag{2.109}$$

Consequently, the purely nonlinear terms are given by

$$G_1 = -A_1 A_2 \, , \; G_2 = A_1^2 . \tag{2.110}$$

Eliminating the stable amplitude A_2 by putting $\mathrm{d}A_2/\mathrm{d}t = 0$, yields the mapping f_2 in form of

$$G_2 = A_1^2 \Rightarrow A_2 = f_2(A_1) = G_2/|\lambda_2| = A_1^2/|\lambda_2| . \tag{2.111}$$

Subsequently, the effective function $G_{1,eff}$ can be obtained:

$$G_1 = -A_1 A_2 \Rightarrow G_{1,eff} = -A_1 f_2(A_1) = -A_1^3/|\lambda_2| . \tag{2.112}$$

The reduced amplitude space description is given by Eq. (2.108), which reads explicitly

$$\frac{\mathrm{d}}{\mathrm{d}t} A_1 = \lambda_1 A_1 - \frac{1}{|\lambda_2|} A_1^3 . \tag{2.113}$$

The amplitude equation for A_1 describes a pitchfork bifurcation, see Sect. 2.6.1 and Eq. (2.19).

References

1. T. Frank, *Determinism and Self-Organization of Human Perception and Performance* (Springer, Berlin, 2019)
2. T.D. Frank, COVID-19 outbreaks follow narrow paths: a computational phase portrait approach based on nonlinear physics and synergetics. Int. J. Modern Phy. C **32**, 2150110 (2021)
3. T.D. Frank, COVID-19 order parameters and order parameter time constants of Italy and China: a modeling approach based on synergetics. J. Biol. Syst. **28**, 589–608 (2020)
4. L. Pang, S. Liu, X. Zhang, T. Tian, Z. Zhao, Transmission dynamics and control strategies of COVID-19 in Wuhan. China. J. Biol. Syst. **28**, 543–561 (2020)
5. H. Haken, *Synergetics: Introduction and Advanced Topics* (Springer, Berlin, 2004)
6. G. Nicolis, *Introduction to Nonlinear Sciences* (Cambridge University Press, Cambridge, 1995)
7. Y. Oono, *The Nonlinear World: Conceptual Analysis and Phenomenology* (Springer, Berlin, 2013)
8. J. Guckenheimer, P. Holmes, *Nonlinear Oscillations, Dynamical Systems, and Bifurcations of Vector Fields* (Springer, Berlin, 1983)
9. H. Haken, *Synergetics* (An Introduction. Springer, Berlin, 1977)
10. F. Verhulst, *Nonlinear Differential Equations and Dynamical Systems*, 2nd edn. (Springer, Berlin, 1996)
11. S.J. Blundell, K.M. Blundell, *Concepts in Thermal Physics* (Oxford University Press, Oxford, 2008)
12. D.G. Dotov, S. Kim, T.D. Frank, Non-equilibrium thermodynamical description of rhythmic motion patterns of active systems: a canonical-dissipative approach. BioSystems **128**, 26–36 (2015)
13. G. Strobl, *Condensed Matter Physics I Crystals, Liquids, Liquid Crystals, and Polymers* (Springer, Berlin, 2004)
14. S. Chiangga, T.D. Frank, Stochastic properties in bistable region of single-transverse-mode vertical-surface emitting lasers. Nonlin. Phenom. Complex Syst. **13**, 32–37 (2010)

15. T.D. Frank, Simplicity from complexity: on the simple amplitude dynamics underlying COVID-19 outbreaks in China. Adv. Compl. Syst. **23**(2050022) (2020)
16. K. Rock, S. Brand, J. Moir, M.J. Keeling. Dynamics of infectious diseases. Rep. Prog. Phys. **77**(026602) (2014)
17. T.D. Frank, Unstable modes and order parameters of bistable signaling pathways at saddle-node bifurcations: a theoretical study based on synergetics. Adv Mathem Phy. **2016**(8938970) (2016)
18. T.D. Frank. SARS-Coronavirus-2 nonlinear dynamics in patients: three-dimensional state and amplitude state description. J. Phy. Soc. Japan 90(073802) (2021)
19. T.D. Frank. Formal derivation of Lotka-Volterra-Haken amplitude equations of task-related brain activity in multiple, consecutively preformed tasks. Int. J. Bifurc. Chaos. 10(1650164) (2016)
20. H. Fujisaka, H. Mori, Reduction of variables and processes in macrodynamics. Prog. Theor. Phys. **56**, 754–772 (1976)
21. H. Haken, Cooperative phenomena in systems far from thermal equilibrium and in nonphysical systems. Rev. Mod. Phys. **47**, 67–121 (1975)
22. H. Mori, T. Morita, K.T. Mashiyama, Contration of state variables in non-equilibrium open systems. I. Prog. Theoret. Phy. **63**, 1865–1884 (1980)
23. R. Zwanzig, Nonlinear generalized Langevin equations. J. Stat. Phy. **9**, 215–220 (1973)
24. S. Chiangga, S. Pitakwongsaporn, T.D. Frank. Photon entanglement on a chip, optical instability, and Haken-Zwanzig model. Physica D. **415**(132760) (2021)
25. R. Friedrich, H. Haken, Exact stationary probability distribution for a simple model of a nonequilibrium phase transition. Z. Physik. **71**, 515–517 (1988)
26. H.K. Leung, B.C. Lai, Stochastic transient of a noise-perturbed Haken-Zwanzig model. Phy. Rev. E. **47**, 3043–3047 (1993)
27. F. Marchesoni, P. Grigolini, On the contraction of fast variables in stochastic processes: the influence of pumping on relaxation. Zeitschrift fuer Physik B: Cond. Matter. **55**, 257–262 (1984)

Chapter 3
Epidemiological Models and COVID-19 Epidemics

This chapter focuses on epidemiological models and their applications to the COVID-19 pandemic. It begins by introducing four basic compartments of epidemiological models. Subsequently, two fundamental compartmental models, the SIR and SEIR models, are discussed in detail. Stability properties of these models are analyzed in an ad-hoc fashion. The chapter also presents applications to COVID-19 outbreaks in Italy and China during the year 2020.

3.1 Type of Models and Some Definitions

Compartments and compartmental models

As discussed in Sect. 1.2, populations such as the population of a country can be decomposed into different subpopulations, groups, or classes. In this context, the phrase compartments is frequently used [1, 2]. Compartments describe individuals with certain characteristic properties. A compartment may be used to describe the under 60 years old people of a population. A second compartment may be used to describe the remaining people who are at the age of 60 years or older. In particular, compartments may describe individuals with certain disease-related characteristics. For example, a fundamental compartment are healthy, disease-free individuals, where the phrase disease-free frequently just means that they have not been infected by the virus under consideration. They may have other health issues. On the one hand, compartments are non-overlapping in the sense that an individual can only belong to one compartment. On the other hand, compartments should be complete in the sense that any individual of the population can be assigned to a compartment. Importantly, individuals typically make transitions between compartments. The epidemiological models that will be studied throughout this book describe populations in terms of compartments and the number of individuals belonging to those compartments. In

T. D. Frank, *COVID-19 Epidemiology and Virus Dynamics*,
Understanding Complex Systems, https://doi.org/10.1007/978-3-030-97178-6_3

particular, the models describe how these numbers change over time, for example, during a COVID-19 outbreak. These models are referred to as compartmental models [1–3]. Let X_i denote the number of individuals in compartment i, then the vector $\mathbf{X} = (X_1, \ldots, X_n)$ describes the state of the population in terms of n compartments.

The number of individuals of the whole population under consideration (i.e., the population size) is denoted by N. If death and birth processes are neglected, then N is constant. If so, N corresponds to the sum of the individuals in all compartments like $N = \sum_{i=1}^{n} X_i$.

Basic compartments

Some basic compartments are those addressed in Sect. 1.2. They will be denoted by certain symbols rather than by X_1, X_2, and so on. Let us consider the following four compartments [2–4]:

- Susceptible individuals are healthy individuals who can be infected. They have not been infected before or they have previously been infected and have recovered. If after recovery they achieved some kind of immunity, then only those recovered individuals who lost their immunity against the virus infection belong to the compartment of susceptibles. The symbol S is used for susceptibles and has a qualitative and quantitative aspect. On the one hand, S is used as compartment label for the susceptibles. On the other hand, S describes the number of susceptible individuals.
- Exposed individuals are individuals that are infected but not yet infectious. That is, they cannot infect others. The symbol E is used for exposed individuals.
- Infectious individuals are infected individuals who can infect others. They are denoted by I.
- Recovered individuals are individuals recovered from an infection and are no longer infectious. They may have achieved some level of immunity against the virus. For example, they cannot get infected at all or the chances that they get infected are relatively low. If so, this level of immunity may not hold for a lifetime [5]. It may lasts only for some finite period. Recovered individuals are denoted by the symbol R.

Epidemic versus endemic models

From a purely modeling point of view, epidemic and endemic models may be distinguished as follows [1]. Epidemic models describe rapid outbreaks of an infectious disease in a population during relatively short periods (for example, a few months or less than one year). On such time scales the impacts of births and deaths by natural causes can be neglected and are not taken into account in the models. In this context, deaths by natural causes are understood as deaths that are not related to the infectious disease. Deaths associated with the disease may or may not be considered depending on the aims of the study at hand. In this context, birth rates and mortality rates in the absence of the infectious disease are called demographic terms and describe the vital dynamics of populations. In summary, when constructing an epidemic model the vital dynamics of a population in terms of demographic terms is ignored because the outbreak or epidemic under consideration usually is more rapid than the vital dynamics.

As far as the population variable N of epidemic models is concerned, if the number of deaths associated with the infectious disease is small relative to N, then N can be regarded as a constant.

Endemic models describe the spread of infectious diseases in populations over longer periods (e.g., several years or generations). Therefore, it is taken into account that new susceptibles show up by birth. Likewise, the death of individuals due to causes unrelated to the infectious disease is taken into account as well. In short, an endemic model features birth and death terms that describe the vital dynamics of a population in the absence of the infectious disease.

Taking a phenomenological point of view, we may distinguish between infectious diseases that are short-lived and disappear in populations after relatively short periods and infectious diseases that persist in populations for longer periods such as years and decades. For example, seasonal influenza infections are short lived, whereas HIV infections persist in the world population since 1981 (see Sect. 1.2). From this perspective, endemic models are models that exhibit stable fixed points (or, in more general terms, attractors) and, in doing so, describe that under appropriate initial conditions infectious diseases persist in populations forever [2, 3, 5].

These two perspectives (i.e., the time-scale-focused birth and death term perspective and the phenomenologically-motivated endemic fixed point perspective) are to some extent consistent with each other. Compartmental models without birth and death terms frequently do not exhibit stable, endemic fixed points, that is, fixed points that describe that portions of populations suffer permanently from infectious diseases. In contrast, taking birth and death terms into account when constructing epidemiological models, is frequently a necessary (but not sufficient) step to arrive at models that exhibit stable, endemic fixed points.

In the COVID-19 literature, various epidemiological models describing COVID-19 outbreaks have been proposed and used to fit COVID-19 data. Some of those models take the vital dynamics of populations into account by considering explicitly birth processes and death processes due to natural causes. It is questionable if this makes sense, in particular, when only relatively short periods of one or two months are considered. For studies that describe the evolution of COVID-19 data over relatively short periods of a few months, epidemic models that do not feature vital dynamics terms may be more appropriate than endemic models. Having said that, since from a mathematical point of view the effects of vital dynamics terms on the solutions of the models are negligibly small, taking these terms into account does not do any harm.

Time-discrete versus time-continuous models

Time-discrete models study the dynamics of epidemics on a discrete time grid. Frequently, the grid size is one day. For example, it is modeled how the number I of infectious individuals varies from day to day. In general, time-discrete models correspond to iterative maps (also called difference equations). For example, the dynamics of $I(t)$ within the framework of a time-discrete model may be given by

$$I(t+1) = I(t) + N_I, \tag{3.1}$$

where N_I is a function that depends on all compartment variables at time t. In contrast, time-continuous models involve a time variable t that is a real number. The models are frequently given in terms of differential equations. For example, in the context of a time-continuous model, the evolution of $I(t)$ may be described like

$$\frac{\mathrm{d}}{\mathrm{d}t} I(t) = N_I, \tag{3.2}$$

where N_I is again a function that depends on all compartment variables taken at time t. Throughout this book only time-continuous models will be considered. More precisely, time-continuous models will be considered that assume the general form of dynamical systems defined by Eq. (2.1).

3.2 Effective Contact Rate, Rate Constant k_0, and "Force of Infection"

A key concept in epidemiology is the effective contact rate [4]. The effective contact rate β describes the rate (or frequency per time unit) of contacts between individuals that lead to infections of individuals. As such the rate of contacts is the number of contacts of an individual with other individuals per time unit (e.g., per day). The effective contact rate can be computed from two quantities. Let ν denote the contact rate between people of a population. That is, ν describes how many contacts a person has with other people per time unit. Let p denote the probability of a susceptible to get infected when having a contact with an infectious person. Then, the effective contact rate β is given by [1, 2]

$$\beta = p\nu. \tag{3.3}$$

For example, a susceptible individual has 5 contacts per day with other people ($\nu =$ 5/d) and all contact people are infectious. Moreover, let us assume that on average 1 out of 10 contacts with an infectious individual leads to an infection ($p = 10\%$). This implies $\beta = p\nu = 0.1 \cdot 5 = 0.5$/d. That is, within two days the individual gets infected. In the literature, β is also called the number of adequate contacts of a person per time unit, where an adequate contact is a contact that leads to a transmission of the virus [1, 6]. Frequently, the adjectives effective or adequate are dropped and β is simply referred to as contact rate. Note however that (as explained above) the parameter β accounts for the number of contacts per time unit as well as the chance that a transmission of the virus takes place during such a contact. That is, the parameter β describes more than just a contact rate. Finally, the parameter β has also been called the transmission rate [3, 7].

The parameter β describes the rate with which a susceptible individual becomes infected due to contacts with others assuming that the contacts are exclusively contacts with infectious individuals. Let us assume that there is a complete mixing of the individuals of a population. Any susceptible individual can make contacts with other individuals of all sub-populations or compartments. Then, the probability of a susceptible individual to make contact with an infectious individual is I/N. Consequently, the rate with which a susceptible individual becomes infected due to contacts that involve both infectious individuals and other individuals is smaller than β and is given by

$$k_0 = \beta \frac{I}{N}. \tag{3.4}$$

As such k_0 is the rate constant of the transformation reaction $S \to Y$ that transforms susceptible individuals S into infected individuals Y. The infected individuals Y may correspond to exposed individuals E or infectious individuals I or other kind of infected individuals. The quantity k_0 has also been called the "force of infection" [1, 2] although k_0 does not correspond to any force in the sense of classical mechanics. Rather, in analogy to the rate constants of chemical reactions that describe how quickly reaction products are built, k_0 describes the rate with which a susceptible individual becomes infected. In a population with S susceptibles the rate with which those individuals become infected is given by multiplying k_0 by S such that

$$k_0 S = \beta \frac{IS}{N} \tag{3.5}$$

describes the rate with which the number of infected individuals increases. Let Y denote this number. Within the time-continuous modeling framework, it then follows that [1, 2]

$$\frac{\mathrm{d}}{\mathrm{d}t} Y = k_0 S = \beta \frac{IS}{N} \tag{3.6}$$

when neglecting all other impacts that cause a change of the number of infected individuals. Note that in some applications the population size N is included in the effective contact rate. That is, Eq. (3.3) may be replaced by

$$\beta = \frac{p\nu}{N}. \tag{3.7}$$

In this case, the rate constant k_0 (i.e., the "force of infection") reads

$$k_0 = \beta I. \tag{3.8}$$

3.3 Continuity Equations

Compartmental models typically describe transitions between compartments. If individuals leave a compartment X_1 at a rate U and enter a compartment X_2 at the same rate U and all other impacts on the two compartments X_1 and X_2 can be ignored, then the number of individuals is constant across X_1 and X_2. In such a situation

$$\frac{\mathrm{d}}{\mathrm{d}t}X_1 = -U, \ \frac{\mathrm{d}}{\mathrm{d}t}X_2 = U \Rightarrow X_1 + X_2 = \text{const} \tag{3.9}$$

holds. As a proof, the two differential equations listed in Eq. (3.9) can be added together such that

$$\frac{\mathrm{d}}{\mathrm{d}t}(X_1 + X_2) = -U + U = 0. \tag{3.10}$$

Continuity equations like $X_1 + X_2 = \text{const}$ may be used to reduce the number of variables. In the example, X_2 may be expressed in terms of X_1 like $X_2 = C - X_1$, where C is a constant.

3.4 Linear Terms and Their Coefficients

Epidemiological models frequently involve terms that are linear in the state variables $X_1, \ldots, X_n$ such as aX_1, bX_2, and so on. These terms involve coefficients (e.g., a and b). In particular, if the evolution of a state variable X_i involves a term cX_i, then the coefficient c can be interpreted in terms of an exponential law. For example, let us consider

$$\frac{\mathrm{d}}{\mathrm{d}t}I = -\gamma I \ \Rightarrow I(t) = I_0 \exp\{-\gamma t\}, \tag{3.11}$$

where I denotes the number of infectious individual and $I_0 = I(t = 0)$ is the initial state (or number). The parameter γ is positive. Only the impact of the linear term on the evolution is considered. The model describes an exponential decay due to recovery of the infectious individuals. The parameter γ corresponds to the recovery rate, that is, the rate at which individuals recover and leave the compartment I. The reciprocal of γ is a time constant $\tau = 1/\gamma$. The parameter τ can be interpreted in two ways. First of all, τ is the characteristic time at which the number of infectious individuals decays by factor e (with $e \approx 2.7$). That is, $I(\tau) = I_0/e \approx I_0/2.7$.

Alternatively, in order to get insights into the nature of the parameter γ, an analogous problem from atomic physics may be exploited. In order to compute the mean life time of a radioactive atom, it is typically assumed that the probability density $P(t_l)$ of life times t_l is an exponential function and reads

$$P(t_l) = \gamma \exp\{-\gamma t_l\}. \tag{3.12}$$

Note that the pre-factor γ has been chosen such that the function is normalized like $\int_0^\infty P(t_l)\,\mathrm{d}t_l = 1$. The mean life time $t_{l,\text{mean}}$ can be computed from

$$t_{l,\text{mean}} = \int_0^\infty t_l P(t_l)\,\mathrm{d}t_l = \frac{1}{\gamma} \tag{3.13}$$

and equals $\tau = 1/\gamma$. Next, let us relate the life time distribution $P(t_l)$ to the law that describes the decay of radioactive atoms. To this end, let $p_{\text{surv}}(t_l > t)$ denote the probability to have atoms with life times t_l larger than t. These are the radioactive atoms that still exist at time t. They are the survivor atoms. Accordingly, p_{surv} is a survival probability, as indicated. We obtain

$$p_{\text{surv}}(t_l > t) = \int_t^{\infty} P(t_l)\,\mathrm{d}t_l = \exp\{-\gamma t\}. \tag{3.14}$$

This holds for a single atom. If there are $N_{0,\text{atom}}$ radioactive atoms initially at time $t = 0$ and the decays of the atoms can be considered as independent events, then the number of radioactive atoms $N_{\text{atom}}(t)$ at time t is $N_{\text{atom}}(t) = N_{0,\text{atom}}\, p_{\text{surv}}(t_l > t)$. Consequently, the law of radioactive decay reads

$$N_{\text{atom}}(t) = N_{0,\text{atom}} \exp\{-\gamma t\}. \tag{3.15}$$

Note again that $N(t)$ describes the number of atoms that have not yet decayed at time t (i.e., that are still radioactive).

If we interprete Eq. (3.11) in the sense of the radioactive decay law defined by Eq. (3.15) and assume that the recovery times of the infectious individuals satisfy an exponential waiting time distribution [1, 3] as defined by Eq. (3.12), then the time constant $\tau = 1/\gamma$ of the linear coefficient γ may be interpreted as the mean duration of individuals being infectious. In other words, τ is the duration that individuals spend on average in the compartment I. It is the mean recovery time of infectious individuals. In general, linear coefficients (like γ) may be interpreted as reciprocals of time constants τ that describe mean durations of individuals in compartments [1, 3]. Importantly, if data is available about such mean durations τ, then the linear coefficients can be estimated on the basis of those time constants. For example, if the mean recovery time is $\tau = 14$ d then $\gamma = 1/\tau \approx 0.07$ d^{-1}.

3.5 SIR Model

In this section the susceptible-infectious-recovered (SIR) model is introduced. It involves only three compartments. The susceptible individuals S, the infected and infectious individuals I, and the recovered individuals R, see Sect. 3.1. That is, there are no exposed individuals. If an individual gets infected, then the individual immediately can infect others. The SIR model was introduced in a benchmark study by Kermack and McKendrick in 1927 [8].

3.5.1 Without Demographic Terms

The model

Let us consider an epidemic (Sect. 3.1) and, consequently, neglect effects of demographic terms (i.e., vital dynamics). In this case, the SIR model reads [1–3, 7]

$$\frac{\mathrm{d}}{\mathrm{d}t}S = -\beta\frac{IS}{N}, \ \frac{\mathrm{d}}{\mathrm{d}t}I = \beta\frac{IS}{N} - \gamma I, \ \frac{\mathrm{d}}{\mathrm{d}t}R = \gamma I, \tag{3.16}$$

where S, I, R are susceptible, infectious, and recovered individuals, respectively. The parameters $\beta > 0$ and $\gamma > 0$ denote the effective contact rate and the mean recovery rate of individuals, respectively. The total population N is given by $N = S + I + R$ and is constant:

$$\frac{\mathrm{d}}{\mathrm{d}t}N = \frac{\mathrm{d}}{\mathrm{d}t}(S + I + R) = -\beta\frac{IS}{N} + \beta\frac{IS}{N} - \gamma I + \gamma I = 0. \tag{3.17}$$

Fig. 3.1 illustrates the three compartments S, I, and R and the transitions performed by individuals between the compartments. Taking the three variables together, the state vector $\mathbf{X} = (S, I, R)$ describes the disease (or health) state of a population of interest.

Fixed points

Putting the left-hand sides in Eq. (3.16) equal to zero, we obtain $\mathrm{d}R/\mathrm{d}t = 0 \Rightarrow I = 0$ and $I = 0 \Rightarrow \mathrm{d}S/\mathrm{d}t = \mathrm{d}I/\mathrm{d}t = 0$. Consequently, the SIR model exhibits fixed points described by

$$I_{st} = 0, \ S_{st} \in [0, N], \ R_{st} = N - S. \tag{3.18}$$

These fixed points describe disease-free states (i.e., states with zero infectious individuals). When applied to COVID-19 epidemics that took place at the beginning of the pandemic in the years 2019/2020, the fixed point of interest is given by

$$S_{st} = N, I_{st} = 0, \ R_{st} = 0. \tag{3.19}$$

Accordingly, COVID-19 was a novel disease at the beginning of the pandemic. Nobody had been infected before by that disease and, consequently, when considering a particular population of N individuals all individuals were susceptibles to COVID-19: $S_{st} = N$. A COVID-19 outbreak that took place at an initial time t_0 is considered as a perturbation that exhibited a small number of infectious individuals $I(t_0) > 0$

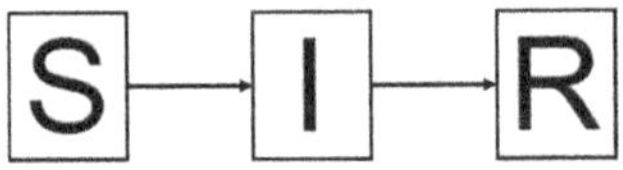

Fig. 3.1 Compartments and flow-chart of the SIR model

and, in doing so, moved the disease (or health) states $\mathbf{X}$ of the population out of the disease-free fixed point (3.19). The initial state $S_0 = S(t_0)$, $I_0 = I(t_0)$, $R_0 = R(t_0)$ of a COVID-19 epidemic that took place at time t_0 was given in terms of a perturbed fixed point of the form

$$I_0 > 0, \text{ but small}, \ S_0 = N - I_0, \ R_0 = 0. \tag{3.20}$$

In applications, typically, N is relatively large as compared to I_0 (e.g., $N = 1,000,000$ and $I_0 = 1$) such that

$$\frac{S_0}{N} \approx 1. \tag{3.21}$$

Consequently, the assumption that local COVID-19 epidemics started as small perturbations, whose initial dynamics can be studied by linearized equations (see Chap. 2), is justified. Although the SIR model (3.16) features fixed points that are located anywhere on the S-axis with $S_{st} \leq N$ (see Eq. (3.18)), in general, for novel infectious diseases (such as COVID-19) Eqs. (3.19) and (3.20) are the disease-free fixed points and the initial disease states of interest.

The SIR model can be reduced to a two variable model by exploiting that $S + I + R = N$ and N is constant. Accordingly, the two-variable model reads

$$\frac{\mathrm{d}}{\mathrm{d}t} S = -\beta \frac{IS}{N}, \ \frac{\mathrm{d}}{\mathrm{d}t} I = \beta \frac{IS}{N} - \gamma I \tag{3.22}$$

with $R(t) = N - S(t) - I(t)$. Comparing Eqs. (3.16) and (3.22), we see that we just dropped the evolution equation for R in Eq. (3.16) and replaced it by the mapping $I, S, N \rightarrow R$ defined by $R(t) = N - S(t) - I(t)$.

Trajectories

Solutions of the SIR model can be visualized as trajectories $S(t)$, $I(t)$ and $R(t)$ (i.e., functions over time). From Eq. (3.16) it follows that $S(t)$ is a monotonically decaying function over time (i.e., $\mathrm{d}S/\mathrm{d}t < 0$ for $S, I > 0$), while R is a monotonically increasing function over time (i.e., $\mathrm{d}R/\mathrm{d}t > 0$ for $I > 0$). The solutions $I(t)$ are of particular interest. Let us write the evolution equation of $I(t)$ of the SIR model (3.16) like

$$\frac{\mathrm{d}}{\mathrm{d}t} I = I \left(\beta \frac{S}{N} - \gamma \right). \tag{3.23}$$

If the expression $Z(t) = \beta S(t)/N - \gamma$ is zero at the initial time t_0 or any time point $t_p > t_0$, then I does not change at that particular moment in time. However, S decays over time which implies that Z becomes negative and remains negative for $t > t_0$ or $t > t_p$. In particular, if $Z = \beta S/N - \gamma < 0$ holds at t_0 then $Z < 0$ for $t > t_0$, which implies that I decays monotonically. In contrast, if initially $Z > 0$ holds, then from Eq. (3.23) it follows that I increases. However, S decreases, which implies that at a certain time point $t_p > t_0$, the expression Z becomes zero. At that time point I reaches its maximum. For $t > t_p$ we have $Z < 0$ and I decays over time.

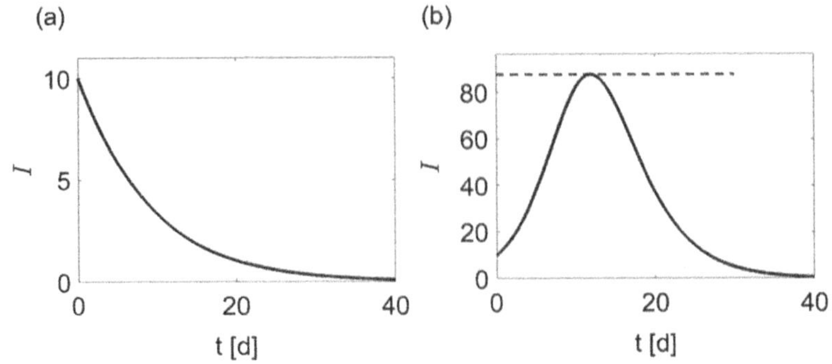

Fig. 3.2 Monotonically decaying (panel (**a**)) and wave-like (panel (**b**)) solutions $I(t)$ of Eq. (3.22) for $\xi < 1$ with $\beta = 0.4$/d (panel (**a**)) and $\xi > 1$ with $\beta = 0.8$/d (panel (**b**)). Parameters and initial conditions for both simulations: $N = 1,000$, $\gamma = 0.5$/d, $I(0) = 10$, and $S(0) = N - I(0)$

Consequently, there are two types of $I(t)$ trajectories [1, 3]. They either describe a monotonic decay or a wave. In the later case, the function $I(t)$ initially increases, reaches a peak at $t_p > t_0$ at which $Z = 0$ holds, and subsequently decays towards zero. In order to distinguish between these two cases, the expression $Z(t)$ for $t = t_0$ may be written like $Z = \gamma(\xi - 1)$, where ξ is defined by [1, 3]

$$\xi = \frac{\beta}{\gamma} \frac{S_0}{N} \tag{3.24}$$

and will be referred to as stability parameter. From the discussion above about the expression Z at t_0 it follows that for $\xi \leq 1$ the epidemic under consideration decays monotonically. In contrast, for $\xi > 1$ there is an epidemic wave. The stability parameter ξ that plays a crucial role for solutions of the SIR model (3.16) will be derive in Chap. 4 in a more systematic way using the nonlinear physics perspective discussed in Chap. 2.

Figure 3.2 presents solutions $I(t)$ for the two qualitatively different cases $\xi \leq 1$ and $\xi > 1$. The reader should pay attention to the following: a monotonic decay of I does not simply mean that the initial infectious individuals recover over time. That is, the graph I does not show how the initial infectious people recover. For example, in panel (a) of Fig. 3.2 there are initially 10 infectious individuals. They recover over time. However, they also infect new individuals such that the cumulative number of infectious individuals during the whole time course is larger than 10. The function $I(t)$ in panel (a) (and also in panel (b)) shows the number of infectious people at a give time point t. Some of them may correspond to the initial infectious individuals. In general, the compartment I of the SIR model does not distinguish between the initial infectious individuals, secondary cases of infectious individuals infected by initial cases, third generation cases of infectious individuals infected by secondary cases, and so on.

Epidemic waves ($\xi > 1$) exhibit a peak value $I_{\max}$ given by [1, 3]

$$I_{max} = I_0 + S_0 \left(1 - \frac{1}{\xi}[1 + \ln(\xi)]\right). \quad (3.25)$$

For $\xi = 1$, we obtain $I_{max} = I_0$. The function $f(\xi) = [1 + \ln(\xi)]/\xi$ occurring Eq. (3.25) is a monotonically decaying function of ξ and decays from $f = 1$ for $\xi = 1$ to $f = 0$ for $\xi \to \infty$. Consequently, the peak value I_{max} of the wave increases from $I_{max} = I_0$ at $\xi = 1$ to $I_{max} = I_0 + S_0$ in the limiting case $\xi \to \infty$, which implies $I_{max} = N$ for $R_0 = 0$. Note that for $\xi = 1$ the function $I(t)$ still corresponds to a monotonically decaying function. However, if a value slightly larger than $\xi = 1$ is chosen, then the SIR dynamics produces a wave with a peak slightly larger than I_0.

In the special case $\gamma = 0$, there is no recovery of infectious individuals (see Eq. (3.16)). Consequently, all susceptible individuals become infectious over time and remain infectious forever. The wave degenerates to a monotonic increase of I towards I_{max}, which is given by $I_{max} = I_0 + S_0$ (and $I_{max} = N$ for $R_0 = 0$). As mentioned above, we obtain this maximum value also from Eq. (3.25) in the limiting case $\xi \to \infty$, which is consistent with the fact that $\gamma \to 0$ for a constant $\beta > 0$ implies $\xi \to \infty$ (see Eq. (3.24)).

Finally, note that in the context of novel infectious diseases, in general, and, in particular, COVID-19 outbreaks during the years 2019/2020, we have $S(0)/N \approx 1$ (see Eq. (3.21)), such that the stability parameter (3.24) reads

$$\xi = \frac{\beta}{\gamma}. \quad (3.26)$$

According to the SIR model, whether an initial group of infectious people $I(t_0) = I_0$ in a population leads to a monotonically decaying infection dynamics or to a wave with $I_{max} > I_0$ depends on the ratio β/γ between the effective contact rate β and the mean recovery rate γ. If infectious individuals recover faster than they induce new infections ($\gamma > \beta$), then, subsequent to that initial perturbation event, the number of infectious individuals decays over time to zero. If not, that is, if infectious individuals infect other individuals faster than they recover ($\beta > \gamma$), then the initial group of infectious people triggers a wave of infections.

Phase portrait and stability of fixed points

In what follows a small population of $N = 1000$ individuals will be considered. First the case $\beta > \gamma$ with $\beta = 0.8$/d and $\gamma = 0.5$/d will be considered. Figure 3.3 illustrates the construction of a phase portrait for the SIR model (3.16) as it was discussed in Sect. 2.1. In line with the reduced SIR model (3.22) only the two-dimensional S-I subspace is considered. Panel (a) shows the solutions S and I of Eq. (3.22) as functions of time for the initial conditions $R_0 = 0$ individuals, $I_0 = 10$ individuals, and $S_0 = N - I_0$ individuals. Panel (b) shows the corresponding phase curve in the two-dimensional S-I subspace.

Figure 3.4 shows two phase portraits that were drawn by repeating the procedure used for panel (b) of Fig. 3.3. For both phase portraits shown in Fig. 3.4 the parameters $\beta = 0.8$/d, $\gamma = 0.5$/d, and $N = 1000$ were used.

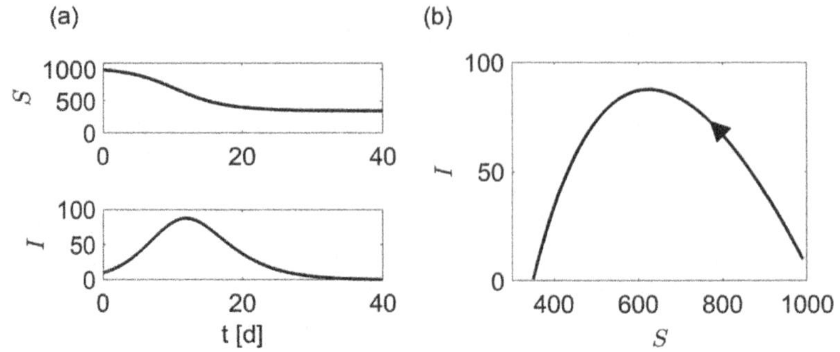

Fig. 3.3 Wave-like solution of the SIR model (3.22) presented in terms of the trajectories $S(t)$ and $I(t)$ (panel (**a**)) and the phase curve $I(S)$ (panel (**b**)). Parameters and initial conditions as for the simulation shown in panel (**b**) of Fig. 3.2

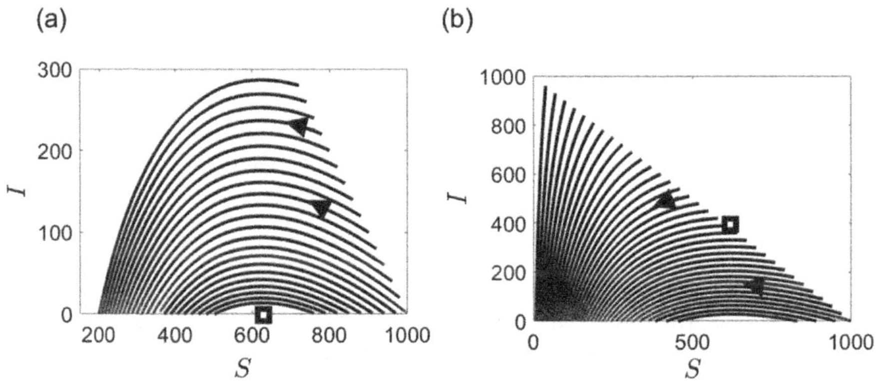

Fig. 3.4 Phase portraits of the SIR model (3.22) for $\beta > \gamma$. Panels (**a**) and (**b**) show a detail of the S-I plane and the full plane, respectively. Parameters as for the simulation shown in panel (**b**) of Fig. 3.2

Let us first consider the phase portrait in panel (a). To this end, let us have a look at the horizontal axis (i.e., the S-axis) and the phase curves that originate at states slightly above the horizontal axis. From Eq. (3.18) it follows that states on the axis correspond to fixed points. Since the states can be shifted along the axis (e.g., the fixed point $S_{st} = 1000$, $R_{st} = 0$ can be shifted to $S_{st} = 999$, $R_{st} = 1$), the fixed points do not satisfy the conditions of asymptotically stable fixed points as defined in Eq. (2.8). They either correspond to unstable fixed points or neutrally stable fixed points. Following the initial states with $I_0 = 1$, $R_0 = 999 - S_0$, and S_0 ranging from $S_0 = 999$ down towards the value indicated by the square, we see that all trajectories depart from the S-axis. More precisely, for those trajectories $I(t)$ increases initially. Consequently, the fixed points do not satisfy the property of stable fixed points as described in Sect. 2.3, which implies that they are unstable. In particular, Eq. (2.10) is satisfied (since solutions $I(t)$ with $I_0 = 1$ originate on the S-axis at $I = 0$, i.e., converge to $I = 0$ when time is reversed), which shows explicitly that the fixed points with $I_{st} = 0$ and $S_{st} > S_{0,crit}$ are unstable. The critical value $S_{0,crit}$ indicated by the square can be obtained from Eq. (3.24). Putting $\xi = 1$ and solving for S, we obtain

$$S_{0,crit} = \frac{\gamma}{\beta} N. \tag{3.27}$$

For $\beta = 0.8$/d, $\gamma = 0.5$/d, and $N = 1000$ Eq. (3.27) yields $S_{0,crit} = 625$. Looking to the fixed points along the S-axis with $S_{st} < S_{0,crit}$ (i.e., to the left of the square), we see that trajectories approach those fixed points on the S-axis. Consequently, if fixed points with $S_{st} < S_{0,crit}$ and $I_{st} = 0$ are perturbed by increasing I to a finite value, then the perturbation of I disappears over time. I decays to zero. States that correspond to fixed points and exhibit perturbations that decay in this manner are consistent with the notion of stable fixed points. However, as mentioned above, there is a direction along which the fixed points can be perturbed (i.e., shifted) such that perturbations do not decay, namely, the direction given by the S-axis. Therefore, fixed points with $S_{st} \leq S_{0,crit}$, $I_{st} = 0$, and $R_{st} = N - S_{st}$ are neutrally stable (rather than asymptotically stable).

Next, let us follow phase curves with initial states defined by $R_0 = 0$ along the line $I_0 = N - S_0$. More precisely, panel (a) shows trajectories $\mathbf{X}(t) = (S(t), I(t))$ in terms of phase curves $I(S)$ that were computed for initial conditions I_0 and $S_0 = 1000 - I_0$, where I_0 was varied in steps of 20 in the range from $I_0 = 1$ up to $I_0 = 281$. Consequently, the initial states $\mathbf{X}_0 = (S_0, I_0)$ when plotted in the S-I subspace form the straight line $I_0(S_0) = 1000 - S_0$. As can be seen when following the phase curves, for all trajectories $I(t)$ initially increases. From Eq. (3.24) it follows that the critical value of β at which $\xi = 1$ holds is given by

$$\beta_{crit} = \gamma \frac{N}{S_0}. \tag{3.28}$$

Alternatively, Eq. (3.28) can be obtained by solving Eq. (3.27) for β. For the most extreme initial condition $I_0 = 281$ we have $S_0 = 719$ and $\beta_{crit} = 0.70$/d. Since $\beta = 0.8$/d holds in the simulations under consideration, for all initial states along the line with $S_0 = 1000 - I_0$ and $I_0 \leq 281$ the effective contact rate $\beta = 0.8$ is larger than the critical values β_{crit} defined by Eq. (3.27) such that the solutions correspond to waves.

Panel (b) of Fig. 3.4 shows an extension of the phase portrait presented in panel (a). In panel (b) the initial state I_0 was varied in steps of 30 individuals from $I_0 = 1$ up to $I_0 = 961$. Given $N = 1000$ this implies that S_0 decayed from $S_0 = 999$ down to $S_0 = 39$. The value $S_{0,crit}$ determined in the previous discussion and defined by Eq. (3.27) also applies in the current context. When S_0 drops below $S_{0,crit} = 625$, that is, I_0 exceeds $I_{crit} = 375$, then the stability parameter ξ becomes smaller than 1. The dynamics of the SIR model is no longer a wave dynamics. Rather, solutions $I(t)$ correspond to monotonically decaying functions. The square in panel (b) of Fig. 3.4 indicates the first initial condition for which $I_0 > I_{crit} = 375$ holds. For this

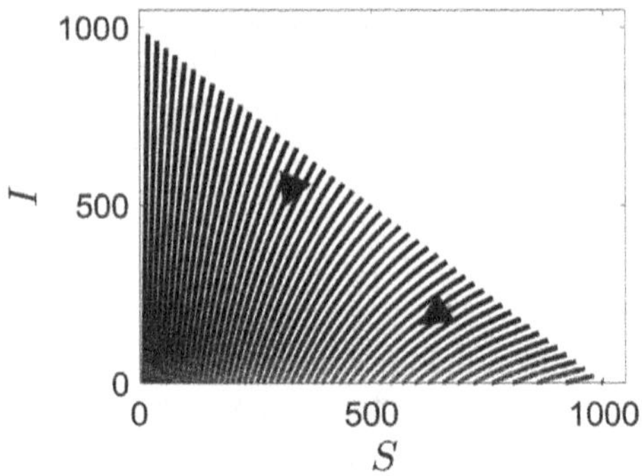

Fig. 3.5 Phase portraits of the SIR model (3.22) for the case $\xi < 1$. Parameters as for the simulation shown in panel (**a**) of Fig. 3.2

initial condition and all remaining initial conditions with even large initial values I_0 visual inspection of the phase curves $I(S)$ shows that the solutions $I(t)$ are given by monotonically decaying functions.

Panel (b) of Fig. 3.4 features a phase portrait of the state dynamics of a SIR model for a scenario in which the SIR model exhibits unstable fixed points on the S-axis in the interval $[S_{0,crit}, N]$ and exhibits under appropriate initial conditions wave-solutions. Similar phase portraits can be obtained for all SIR models with $\beta > \gamma$. In contrast, for $\beta \leq \gamma$ from Eq. (3.24) it follows that $\xi \leq 1$ and, likewise, from Eq. (3.23) it follows that $\mathrm{d}I/\mathrm{d}t < 0$ holds irrespective of the initial conditions for any $S, I > 0$. Consequently, SIR models with $\beta \leq \gamma$ exhibit only monotonically decaying solutions $I(t)$ and neutrally stable fixed points. They do not exhibit unstable fixed points that give rise to wave-solutions. Figure 3.5 present a phase portrait of a SIR model for $\beta < \gamma$ with $\gamma = 0.5$/d again, but $\beta = 0.4$/d. Clearly, $I(t)$ decays over time monotonically for all initial states. All fixed points on the S-axis are neutrally stable.

3.5.2 With Demographic Terms

The model

Let us include vital dynamics in terms of birth processes, on the one hand, and deaths not related to the infectious disease under consideration, on the other hand. In this case, the SIR model (3.16) becomes [1–3, 7]

$$\frac{\mathrm{d}}{\mathrm{d}t}S = -\beta\frac{IS}{N} + B - \mu S, \ \frac{\mathrm{d}}{\mathrm{d}t}I = \beta\frac{IS}{N} - (\gamma + \mu)I, \ \frac{\mathrm{d}}{\mathrm{d}t}R = \gamma I - \mu R. \quad (3.29)$$

The parameter B is referred to as birth rate. The B-term describes a linear increase of the population like $S(t) = S(t_0) + B(t - t_0)$. In contrast, the parameter μ describes a death rate related to an exponential decay of the number of individuals in all compartments like $X_i(t) = X_i(t_0)\exp\{-\mu(t - t_0)\}$, where X_i can be S, I, or R. The total population is given by $N(t) = S(t) + I(t) + R(t)$ and satisfies the dynamics

$$\frac{\mathrm{d}}{\mathrm{d}t}N = \frac{\mathrm{d}}{\mathrm{d}t}(S + I + R) = B - \mu S - \mu I - \mu R = B - \mu N. \tag{3.30}$$

Frequently the situation is considered for which N is constant over time [1, 7] such that

$$B = \mu N. \tag{3.31}$$

In this case, the SIR model reads

$$\frac{\mathrm{d}}{\mathrm{d}t}S = -\beta\frac{IS}{N} + \mu(N - S),\ \frac{\mathrm{d}}{\mathrm{d}t}I = \beta\frac{IS}{N} - (\gamma + \mu)I,\ \frac{\mathrm{d}}{\mathrm{d}t}R = \gamma I - \mu R \tag{3.32}$$

and can be reduced to a two-variable model like

$$\frac{\mathrm{d}}{\mathrm{d}t}S = -\beta\frac{IS}{N} + \mu(N - S),\ \frac{\mathrm{d}}{\mathrm{d}t}I = \beta\frac{IS}{N} - (\gamma + \mu)I \tag{3.33}$$

with $R(t) = N - S(t) - I(t)$. In what follows, the SIR model for a constant population will be discussed either on the level of the three-variable model (3.32) or the two-variable model (3.33). Note that although the SIR model (3.32) with vital dynamics and constant N and the SIR model (3.16) without vital dynamics exhibit both a constant population N, they have different characteristic properties, as will be shown below.

Fixed points

It is plausible to assume that disease-free states with $I = 0$ are fixed points of the model. In order to check if this is the case, let us substitute $I = 0$ into Eq. (3.32), which leads to $\mathrm{d}I/\mathrm{d}t = 0$, $\mathrm{d}S/\mathrm{d}t = \mu(N - S)$, and $\mathrm{d}R/\mathrm{d}t = -\mu R$. The last two relations imply that the fixed point must satisfy $S_{st} = N$ and $R_{st} = 0$ such that

$$S_{st} = N,\ I_{st} = R_{st} = 0. \tag{3.34}$$

The fixed points $S_{st} < N$ and $I_{st} = 0$ of the SIR model (3.16) without demographic terms do not exist. The fixed point (3.34) relates to the state of a population before an epidemic outbreak or long after the epidemic has disappeared. In the latter case, due to the epidemic some individuals have been infectious for some period and have subsequently recovered. On long demographic time scales, however, all of those recovered individuals will die at some point in time. Consequently, after a sufficiently long waiting period, the number of recovered individuals decays to zero. Other fixed points can be obtained by putting $\mathrm{d}I/\mathrm{d}t = 0$ in the reduced two-variable model (3.33). We either have $I = 0$, which leads us to the disease-free fixed point (3.34), or $I > 0$ and

$$\beta\frac{S}{N} - (\gamma + \mu) = 0 \Rightarrow S_{st} = \frac{\gamma + \mu}{\beta}N \tag{3.35}$$

Substituting this result into the evolution equation of S shown in Eq. (3.33), we obtain

$$I_{st} = \frac{\mu}{\beta(\gamma + \mu)} N(\beta - (\gamma + \mu)). \tag{3.36}$$

As such the term $\beta - (\gamma + \mu)$ can be positive or negative. However, in the context of a fixed point with I_{st} infectious individuals only semi-positive values are considered. Consequently, Eq. (3.36) suggests that for the SIR model with demographic terms a stability parameter exists similar to the parameter defined by Eq. (3.26) that describes the emergence of fixed points with $I_{st} > 0$. In fact, it can be shown that the stability parameter of the SIR model (3.32) reads [1]

$$\xi = \frac{\beta}{\gamma + \mu}, \tag{3.37}$$

which reduces to Eq. (3.26) for $\mu = 0$. Note that the term $S(0)/N$ as in Eq. (3.24) does not show up in Eq. (3.37) because the stability parameter defined by Eq. (3.37) does not refer to initial states. It refers to the disease-free fixed point (3.34) for which $S_{st}/N = 1$. The stability parameter (3.37) will be derived in Chap. 4 using the nonlinear physics perspective discussed in Chap. 2.

If $\beta > \gamma + \mu \Rightarrow \xi > 1$ holds, then the fixed point (3.36) with a finite number of infectious individuals exists. The fixed point describes an endemic in the population of interest. The infectious disease under consideration persists in the population forever (see Sect. 3.1). Taking Eqs. (3.35) and (3.36) together, the fixed point $\mathbf{X}_{st} = (S_{st}, I_{st}, R_{st})$ reads (see also Ref. [1])

$$S_{st} = \frac{\gamma + \mu}{\beta} N = \frac{N}{\xi}, \quad I_{st} = \frac{\mu}{\beta(\gamma + \mu)} N(\beta - (\gamma + \mu)) = N\frac{\mu}{\beta}(\xi - 1) \tag{3.38}$$

with $R_{st} = N - S_{st} - I_{st}$.

If $\beta \leq \gamma + \mu \Rightarrow \xi \leq 1$ holds, then only the disease-free fixed point (3.34) exists. Consequently, the SIR model (3.29) features a bifurcation at $\beta = \gamma + \mu$ or $\xi = 1$. At the bifurcation point the fixed point with a finite number of infectious individuals emerges. Let us illustrate this bifurcation with the help of phase portraits and a bifurcation diagram.

Phase portraits

Fig. 3.6 shows solutions $S(t)$ and $I(t)$ for the case $\beta < \gamma + \mu$ (panel (a)) and $\beta > \gamma + \mu$ (panel (b)) for a small population with $N = 1000$ individuals. The SIR model parameters $\gamma = 0.5$/d and $\mu = 0.013$/d will be used, whereas β will be varied from $\beta = 0.4$/d to $\beta = 0.8$/d. As expected, for $\beta < \gamma + \mu$ (i.e., $\beta = 0.4$/d) the solutions converge to zero (panel (a)). For $\beta > \gamma + \mu$ (i.e., $\beta = 0.8$/d) the solutions converge to the non-vanishing fixed point (3.38) describing an endemic in the population of interest (panel (b)). A period of 700 days (about 2 years) is simulated. At the end of the period the number of infectious individuals as obtained from the numerical simulation is $I = 9.28$ (i.e., about 9 people). Substituting the model parameters into

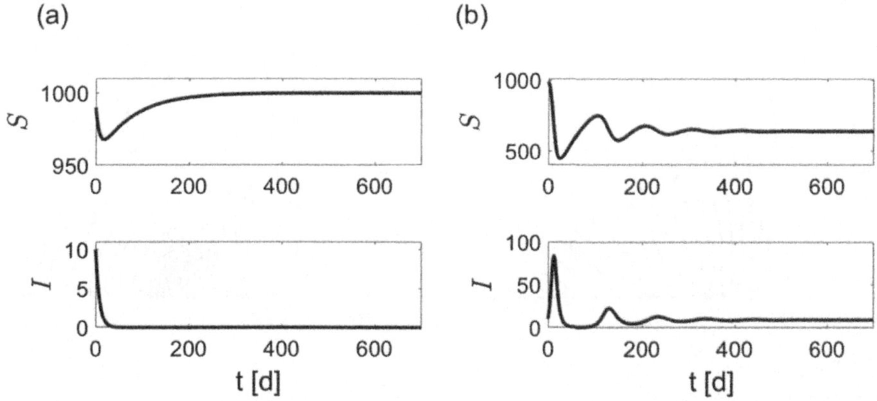

Fig. 3.6 Solutions $S(t)$ and $I(t)$ of the SIR model (3.27) in the absence (panel (**a**)) and presence (panel (**b**)) of an endemic fixed point. Parameters and initial conditions for both simulations: $N = 1,000$, $\gamma = 0.5$/d, $\mu = 0.013$/d, $I(0) = 10$, $S(0) = N - I(0)$. The effective contact rates are $\beta = 0.4$/d for panel (**a**) and $\beta = 0.8$/d for panel (**b**)

Eq. (3.36), we obtain a stationary value $I_{st} = 9.30$. That is, the simulated value obtained at $t = 700$ days is close to the stationary value.

Figure 3.7 presents the phase portraits of the SIR model (3.33) for the two cases $\beta < \gamma + \mu$ and $\beta > \gamma + \mu$ in the two-dimensional S-I subspace. For $\beta < \gamma + \mu$ all trajectories converge to the disease-free fixed point (3.34) indicated by the full circle at $S_{st} = N = 1000$ and $I_{st} = 0$ (see panel a). This suggests that perturbations out of the fixed point satisfy Eq. (2.8) and the fixed point is asymptotically stable [1]. A proof for the asymptotic stability of the fixed point will be given in Chap. 4. In contrast, as shown in panel (b), for $\beta > \gamma + \mu$ trajectories evolve away from the disease-free fixed point $S_{st} = N = 1000$ and $I_{st} = 0$, which is indicated by an open circle in panel (b). That is, perturbations of the disease-free fixed point increase initially in magnitude such that the fixed point does not satisfy the property of stable fixed points as described in Sect. 2.3. In particular, the trajectory drawn above the fixed point (i.e., drawn right above the open circle) indicates (at least in good approximation) an "escape route" of a solution that originates at the fixed point such that Eq. (2.10) is satisfied. In summary, the disease-free fixed point is unstable. For the model parameters used in the simulation, solutions $I(t)$ and $S(t)$ converge in an oscillatory manner to the non-vanishing fixed point (3.38) that describes an endemic. The fixed point is asymptotically stable [1, 7] and corresponds to a stable focus. Substituting the model parameters into Eq. (3.38), we obtain $I_{st} = 9.3$ (as mentioned above) and $S_{st} = 641.7$, which, by visual inspection is consistent with the center of the focus shown in panel (b) of Fig. 3.7.

In summary, for $\beta \leq \gamma + \mu$ (i.e., $\xi \leq 1$), the SIR model (3.32) describes a monostable system. It exhibits a single fixed point, the disease-free state (3.34), which corresponds to an asymptotically stable fixed point. For $\beta > \gamma + \mu$ (i.e., $\xi > 1$), the SIR model (3.32) is monostable as well. However, it exhibits two fixed points. The

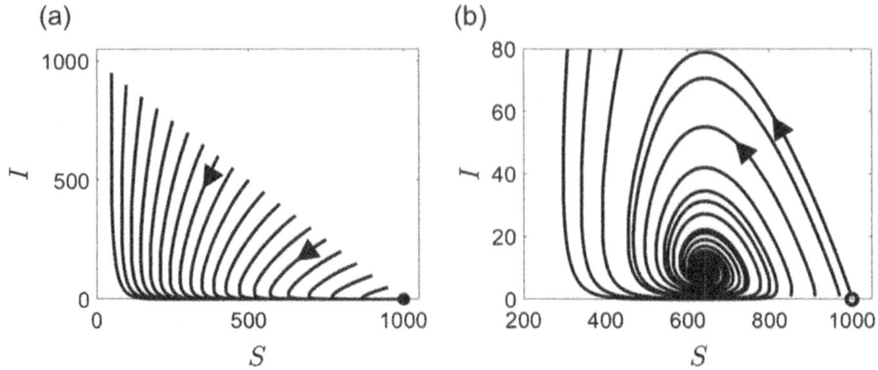

Fig. 3.7 Phase portraits of the SIR model (3.27) in the absence (panel (**a**)) and presence (panel (**b**)) of an endemic fixed point. Parameters as reported in Fig. 3.6

disease-free fixed point (3.34), which is unstable, and the endemic fixed point (3.38), which is asymptotically stable.

Bifurcation diagram

We may vary any of the parameters β, μ, γ or the stability parameter ξ to induce a bifurcation. Let us fix $\gamma = 0.5$/d and $\mu = 0.013$/d (as in the previous examples) and consider β as bifurcation parameter (see Sect. 2.5). Plotting all possible stationary solutions I_{st} versus β yields the bifurcation diagram shown in Fig. 3.8. Here, the critical value for β can be obtained from Eq. (3.37) by putting $\xi = 1$ and reads

$$\beta_{crit} = \gamma + \mu. \tag{3.39}$$

For $\gamma = 0.5$/d and $\mu = 0.013$/d the parameter reads $\beta_{crit} = 0.5133$/d. In line with the previous discussion, Fig. 3.8 shows that for $\beta \geq \beta_{crit}$ there exists only one stationary state I_{st} with $I_{st} = 0$. The state $I_{st} = 0$ for $\beta < \beta_{crit}$ is depicted in Fig. 3.8 by a solid line in order to indicate that it corresponds to an asymptotically stable fixed point. For $\beta > \beta_{crit}$ the fixed point $I_{st} = 0$ still exist as indicated by the dashed line in Fig. 3.8. However, the fixed point is unstable. The switch from a solid to a dashed line should indicate the change of the stability of the disease-free fixed point. At the bifurcation point, a new stationary solution with $I_{st} > 0$ emerges that is given by Eq. (3.36). The states $I_{st} > 0$ corresponds to asymptotically stable fixed points and, consequently, are described by a solid line. They describe endemic states of populations. I_{st} increases monotonically as a function of β, which follows from Eq. (3.36). Importantly, at the bifurcation point $\beta = \beta_{crit}$ the function $I_{st}(\beta)$ that describes the asymptotically stable fixed points (i.e., the solid line) exhibits a kink, which is a key characteristics of phase transitions and bifurcations (see Sect. 2.5). This result was anticipated in Sect. 1.5. Panel (c) of Figure 1.3 corresponds to Fig. 3.8 when presenting I_{st} as a percentage value. As argued in Sects. 1.5 and 2.5, bifurcation diagrams such as shown in Fig. 3.8 demonstrate that bifurcations that describe the emergence of endemic states

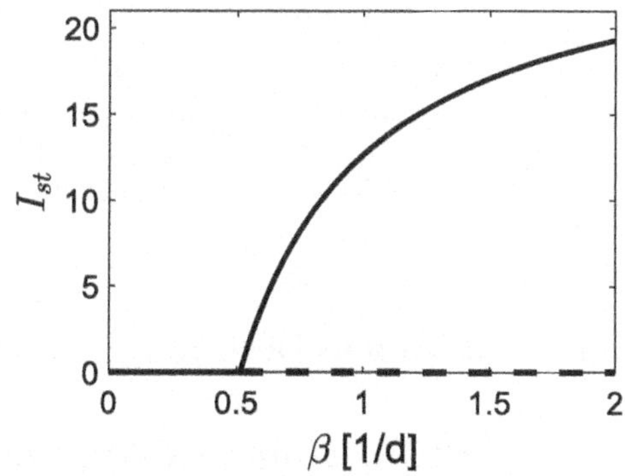

Fig. 3.8 Bifurcation diagram of the SIR model (3.27) for the parameters as reported in Fig. 3.6 except for β. The effective contact rate β is varied in the range $[0, 2]$. Solid and dashed lines indicate stable and unstable fixed points, respectively

in populations belong to the class of instability-, bifurcation-, and phase transition-phenomena that range from the buckling of an iron bar under pressure to magnetic phase transitions and the production of laser light.

Focus or node

The endemic fixed point presented in panel (b) of Fig. 3.6 and, likewise, panel (b) of Fig. 3.7 corresponds to a stable focus. Solutions spiral in an oscillatory fashion towards the fixed point. In general, close to the endemic fixed point (3.38) solutions $I(t)$ and $S(t)$ satisfy linearized equations that assume the form of damped harmonic oscillator equations (e.g., see page 45 in Ref. [7]). The damping constant corresponds to ξ. The spring constant K of the damped oscillator is given by [7]

$$K = \left(1 + \frac{\beta}{\mu}\right)(\xi - 1). \tag{3.40}$$

In general, damped harmonic oscillators feature a so-called damping ratio ζ that is proportional to the ratio between the damping constant and the spring constant. The damping ratio ζ determines whether solutions decay in an oscillatory fashion or decay monotonically. Explicitly, for the endemic fixed point (3.38) of the SIR model (3.32), the damping ratio ζ reads

$$\zeta = \frac{\xi}{2\sqrt{K}} = \frac{\xi}{2\sqrt{(1 + \beta/\mu)\,(\xi - 1)}}. \tag{3.41}$$

For $\zeta < 1$ (when the spring constant is relatively strong) solutions are oscillatory, which implies that the SIR model exhibits an endemic fixed point in terms of a stable focus. In contrast, for $\zeta \geq 1$ (when the damping is relatively strong) solutions are non-oscillatory, which implies that the endemic fixed point of the SIR model corresponds to a stable node. When the stability parameter is sufficiently large such that $\xi - 1 \approx \xi$, then $\xi/\sqrt{(\xi - 1)} \approx 1/\sqrt{\xi} < 1$ which implies $\zeta < 1$. That is, the endemic fixed point is a stable focus. In contrast, sufficiently close to the bifurcation point, when $\xi > 1$ but $\xi \approx 1$ holds, then $\xi/\sqrt{\xi - 1}$ is relatively large and becomes larger than the factor $2\sqrt{(1+\beta/\mu)}$. In this case, the systems exhibits a damping ratio $\zeta > 1$. Consequently, sufficiently close to the bifurcation point the endemic fixed point is a stable node. For the parameters used to produce the solution shown in panel (b) of Fig. 3.6 we have $\zeta = 0.17$, which is consistent with the observed oscillatory character of the solution $I(t)$.

3.6 SIR Models Describing COVID-19 Epidemics

3.6.1 SIR Model-Based COVID-19 Studies

The SIR model has been applied in a number of studies to address epidemiological data of COVID-19 epidemics [9–24]. Such SIR model-based studies frequently aimed to clarify the impact of intervention measures on the spread of COVID-19. In this context, several studies have used the SIR model to shown that intervention measures were implemented successfully and changed infection dynamics parameters such that respective COVID-19 epidemics subsided [10, 20, 22, 23]. Other studies have used the SIR model to predict the time course of COVID-19 epidemics under various types of interventions [13, 14, 17, 19]. Table 3.1 presents a list of SIR model-based studies that have been used for forecasting COVID-19 epidemics and studying the impact of intervention measures in those countries (see also Ref. [25]). The list starts which China and moves from there geographically in a westward direction towards the USA and Brazil. The table reports the purpose of each study. Note that Table 3.1 does not provide a comprehensive review of SIR model-based studies. Rather, Table 3.1 illustrates that SIR models have been applied to describe COVID-19 epidemics in various countries around the globe. While the SIR model (3.16) as such is a three-variable model, some of the studies listed in Table 3.1 involved further variables such as COVID-19 diagnosed, quarantined individuals or individuals deceased due to COVID-19. Some examples in this regard will be presented in Sect. 3.6.2 below and in Sect. 4.4. However, such models may be regarded as SIR-type models as defined in Sect. 4.1 and are included in Table 3.1.

3.6.2 COVID-19 Outbreaks in China and Italy 2020

In early 2020, Fanelli and Piazza [14] conducted a model-based analysis of the emergence of COVID-19 in China and Italy. To this end, the SIR model (3.16) without demographic terms was used. The model parameter were estimated from data. More precisely, the parameter β/N was estimated rather than β. Moreover, the three-compartment SIR model (3.16) was supplemented by a forth compartment that described COVID-19 associated deaths. In what follows, this compartment will be ignored. Nevertheless, it will be taken into account that in the model by Fanelli and

Table 3.1 Studies that have used SIR models and SIR-type models to study COVID-19 epidemics. Studies are grouped into three categories according to their study objectives. The countries addressed by the studies are listed as well

Forecasting	Discussion of interventions	Forecasting and discussion of interventions	Country
Refs. [18, 21, 26]	Ref. [20]	Refs. [14, 15, 22]	China
Ref. [18]		Ref. [23]	S. Korea
Ref. [9]			Japan
		Ref. [17]	Thailand
Ref. [15]			Bangladesh, India, Pakistan
Refs. [24, 26]		Ref. [19]	Iran
Ref. [18]			Russia
Refs. [18, 21, 26]	Ref. [27]	Refs. [10, 14, 22]	European countries (Austria, Belgium, Denmark, France, Germany, Greece, Italy, Netherlands, Spain, Sweden, Switzerland, Turkey, United Kingdom)
Ref. [16]			Morocco
Refs. [18, 21, 26]	Refs. [11, 20]	Ref. [10]	USA
		Refs. [12, 13]	Brazil

Piazza the parameter γ occurring in Eq. (3.16) is given by a mean removal rate that is composed of two terms: the mean recovery rate of individuals that recover and the death rate of individuals who decease due to COVID-19. As mentioned above, Fanelli and Piazza considered the ratio β/N as fixed parameter. Therefore, N will be considered as a constant irrespective of the fact that the model actually considered COVID-19 associated deaths. This can be motivated by noting that for the data under consideration from China and Italy the death numbers D were much smaller than the respective population sizes N: $D/N \ll 1$. Importantly, in Ref. [14] the initial numbers of susceptibles S_0 and infectious individuals I_0 were estimated as well. R_0 was put to zero. Consequently, the relevant population sizes $N = S_0 + I_0$ were estimated. We will return to this issue below.

Let us first review the analysis of the COVID-19 outbreak in China by Fanelli and Piazza [14]. In this analysis the period from January 22 to February 19, 2020, was considered. Data were taken from the data repository of the Johns Hopkins University [28]. It was assumed that the infectious individuals I of the SIR model correspond to the so-called active COVID-19 cases. The active cases are computed

Table 3.2 Parameters of the SIR model (3.16) describing the COVID-19 outbreak in China during January 22 to February 19, 2020

β	γ	$\xi = S_0\beta/(\gamma N)$	$\xi = \beta/\gamma$	I_{max}
0.2671/d	0.0210/d	12.6	12.7	57,935

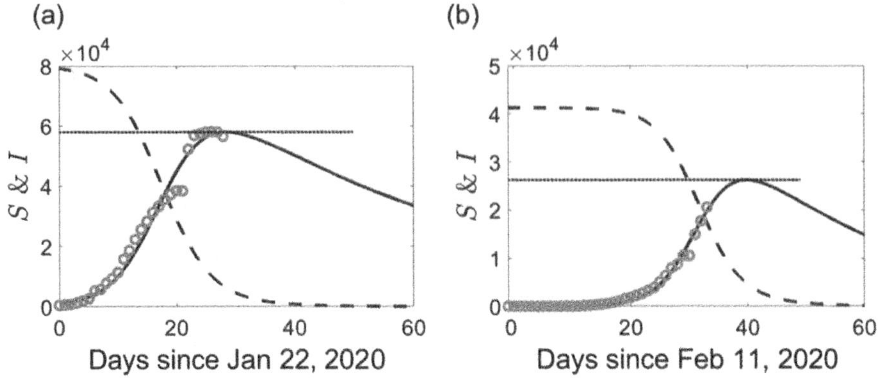

Fig. 3.9 SIR model (3.16) solutions $S(t)$ (dashed lines) and $I(t)$ (solid lines) describing the COVID-19 epidemics in China (panel (**a**)) and Italy (panel (**b**)) in the beginning of the year 2020. Gray circles show reported active cases [29]

by subtracting the recovered cases and reported cumulative deaths from the cumulative confirmed COVID-19 cases. We will return briefly to this issue below. The estimated model parameters were $\beta/N = 3.33 \cdot 10^{-6}$/d and $\gamma = a + d = (1.80 + 0.3) \cdot 10^{-2} = 0.021$/d (where a is the mean recovery rate and d the death rate). The estimated initial states were $S_0 = 79,200$ and $I_0 = 999$. Using $N = S_0 + I_0 = 80,199$ and $\beta = (\beta/N) \cdot N$, leads to the parameters β and γ shown in Table 3.2. Since $S_0/N = 0.9875 \approx 1$, the SIR model stability parameter $\xi = S_0\beta/(\gamma N)$ (see Eq. (3.24)) is approximately given by $\xi = \beta/\gamma$ (see Eq. (3.26)). Both parameters are reported in Table 3.2 as well together with I_{max} computed from Eq. (3.25).

Panel (a) of Fig. 3.9 shows the COVID-19 data and the simulation results. The COVID-19 data taken from [29] are shown as gray circles. The jump of cases between February 12 and February 13 was due to a change of the reporting procedure of the Chinese authorities [14, 30]. The solid and dashed lines show $S(t)$ and $I(t)$ as computed numerically from the SIR model (3.16) for the aforementioned model parameters and initial conditions. The horizontal dotted line corresponds to I_{max} as computed from Eq. (3.25)). By visual inspection, the solution $I(t)$ provides a good fit to the data. As expected the model solution $I(t)$ exhibits a peak at I_{max}. The model-based analysis reveals that during the 2020 COVID-19 outbreak in China the effective contact rate β was by a factor 10 larger than the removal rate γ, which implies that the stability parameter ξ was much larger than the critical value of 1, see Table 3.2. This indicates that the disease-free fixed point was unstable during the

Table 3.3 Parameters of the SIR model (3.16) describing the COVID-19 outbreak in Italy during February 11 to March 15, 2020

β	γ	$\xi = S_0\beta/(\gamma N)$	$\xi = \beta/\gamma$	I_{max}
0.3263/d	0.0376/d	8.7	8.7	26,260

COVID-19 outbreak. Moreover, in this case, for fixed model parameters, the SIR model predicts a wave-solution as it is shown in panel (a) of Fig. 3.9.

In the study by Fanelli and Piazza [14] the COVID-19 outbreak in Italy was analyzed as well. The period from February 11 to March 15, 2020 was considered. Again, COVID-19 active cases from the Johns Hopkins data repository were used [28]. From the active case data the parameters β/N, γ and $S(0)$ were estimated as $\beta/N = 7.90 \cdot 10^{-6}$/d, $\gamma = 0.0376$/d and $S_0 = 41,300$. The initial value $I_0 = 3$ was taken from the active case data set at February 11. Using $N_0 = S_0 + I_0 = 41,303$, the model parameter β can be determined (see above) as well as ξ and I_{max}. Table 3.3 shows the model parameters and characteristic measures thus obtained (with I_{max} computed from Eq. (3.25)).

Panel (b) of Fig. 3.9 shows the COVID-19 data [29] (gray circles) and the simulation results (dashed line for $S(t)$ and solid line for $I(t)$ as computed from Eq. (3.16)) for the COVID-19 outbreak in Italy. The horizontal dotted line corresponds to I_{max} again, as listed in Table 3.3. By visual inspection, the solution $I(t)$ provides a good fit to the data. As expected, $I(t)$ exhibits a peak at I_{max}. The analysis shows that the effective contact rate β was much larger (by a factor 10) than the mean removal rate γ (which implies $\xi > 1$), see Table 3.3. Accordingly, the disease-free fixed point was unstable during the COVID-19 outbreak in Italy. Moreover, in this case, when assuming fixed model parameters, the SIR model predicts a wave-solution as it is shown in panel (b) of Fig. 3.9.

The study by Fanelli and Piazza [14] was published early in the year 2020 and, consequently, considered only the first few months of the COVID-19 pandemic. It might be of interest to compare the SIR model predictions of the study with the actual dynamics of the epidemics in China and Italy in the later months of 2020. Figure 3.10 shows active cases as gray circles reported from Italy during the period from February 11 to October 1, 2020 [29]. The solid line in Fig. 3.10 shows the SIR model solution $I(t)$ that was already presented in panel (b) of Fig. 3.9. The dotted vertical line indicates March 15, which is the end of the period for which the model fit was optimized. While the model predicted a peak of about 26,000 infectious individuals (see Table 3.3), the epidemic in Italy took an even more dramatic turn and reached a four times higher peak of about 110,000 active cases.

Active cases and infectious individuals

Let us return to the issue of how to relate COVID-19 data to model variables. Fanelli and Piazza [14] interpreted the infectious individuals I of the SIR model in terms of active cases. Active cases as reported in the data repository [29] are computed like

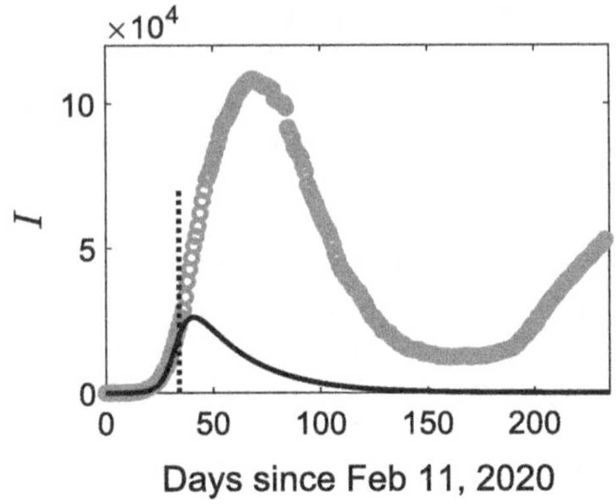

Fig. 3.10 Active COVID-19 cases in Italy (gray circles) [29] reported for a longer period until October 1, 2020, versus the model solution $I(t)$ (solid black line) of the SIR model (3.16) as computer for the parameters of the simulation shown in panel (**b**) of Fig. 3.9

$$\begin{aligned} &\text{active cases at a given day} \\ &= \text{cumulative confirmed COVID cases up to that day} \\ &\quad -\text{recovered cases up to that day} \\ &\quad -\text{cumulative deaths up to that day.} \end{aligned} \tag{3.42}$$

Consequently, the active cases reflected diagnosed COVID-19 cases. In contrast, the infectious individuals I described by the SIR model do not necessarily correspond to diagnosed individuals. Moreover, diagnosed cases are frequently quarantined or hospitalized. It is plausible to assume that both the contact rate ν and transmission probability p (see Eq. (3.3)) for diagnosed cases is lower as compared to non-diagnosed infectious individuals. This implies that it is plausible to assume that the effective contact rate $\beta = p\nu$ is lower for diagnosed infectious individuals as compared to non-diagnosed infectious individuals. The active cases in the data repository reflect diagnosed cases with presumably lower effective contact rates. In view of these difficulties, in the following chapters a variety of studies on COVID-19 epidemics will be presented that used alternative approaches to relate COVID-19 data to variables of compartmental models.

When N is not taken from demographic records

The population size N of a population in a given region is typically taken from demographic records as the number of people living in that region at the time of interest. For example, in the COVID-19 study on China by Tao [20] that is listed in Table 3.1 the population size $N = 1.4 \cdot 10^9$ was used, which is the population of China during 2019/2020. In the absence of vaccination or other intervention measures, that is, for a novel infectious disease such as COVID-19 in the years 2019/2020, the entire population is typically considered to be susceptible. In line with this argument and in the context of the SIR model for a novel infectious disease with $R_0 = 0$, the number of susceptibles S_0 is given by $S_0 = N$ in the absence of infectious individuals and $S_0 = N - I_0$ if there is a number I_0 of infectious individuals. In contrast, Fanelli and Piazza [14] estimated the initial sizes S_0 of susceptible individuals for China and

Italy. In particular, the total number of individuals $N = S_0 + I_0 + R_0$ considered in their SIR models for China and Italy were much smaller than the respective population sizes N of China and Italy (e.g., $N = 80,199$ for the SIR model addressing the COVID-19 epidemic in China versus $N = 1.4 \cdot 10^9$ reflecting the population of China during 2019/2020). Just like Fanelli and Piazza [14], a few other COVID-19 studies [11, 31, 32] estimated also S_0. Therefore, the question arises to what extent the results obtained in the study by Fanelli and Piazza [14] and in similar studies involving S_0 estimates change when N is taken from demographic records and $S_0 = N - I_0$ is used [33].

3.7 SEIR Model

The model

The susceptible-exposed-infectious-recovered (SEIR) model involves all four compartments introduced in Sect. 3.1. The transitions of individuals between compartments are illustrated in the Fig. 3.11.

In what follows the SEIR model without demographic terms will be discussed. The model reads [2]

$$\begin{aligned}
\frac{\mathrm{d}}{\mathrm{d}t}S &= -\beta\frac{IS}{N}, \\
\frac{\mathrm{d}}{\mathrm{d}t}E &= \beta\frac{IS}{N} - \alpha E, \\
\frac{\mathrm{d}}{\mathrm{d}t}I &= \alpha E - \gamma I, \\
\frac{\mathrm{d}}{\mathrm{d}t}R &= \gamma I.
\end{aligned} \tag{3.43}$$

Just as the SIR model (3.16), the SEIR model (3.43) involves as parameters the effective contact rate β and the recovery rate γ. In addition, it involves the rate of progression α from being exposed to being infectious (i.e., the rate of transitions from E to I). In line with the discussion of coefficients of linear terms (see Sect. 3.4) it is frequently assumed that $\tau = 1/\alpha$ corresponds to the latent period of an infectious disease (i.e., the average time it takes to become infectious after being infected). The total population is given by $N = S + E + I + R$. The variable N is constant, which follows from

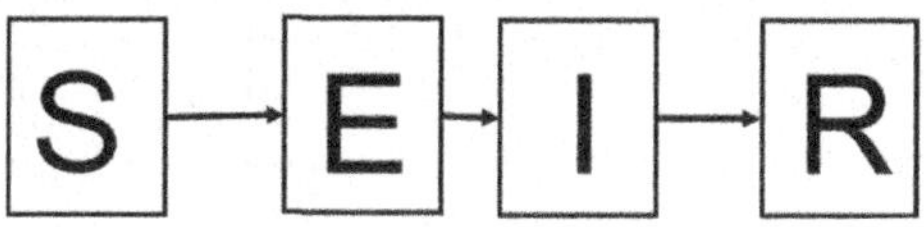

Fig. 3.11 Compartments and flow-chart of the SEIR model

$$\frac{\mathrm{d}}{\mathrm{d}t}N = \frac{\mathrm{d}}{\mathrm{d}t}(S + E + I + R) = -\beta\frac{IS}{N} + \beta\frac{IS}{N} - \alpha E + \alpha E - \gamma I + \gamma I = 0. \tag{3.44}$$

Consequently, the SEIR model can alternatively be expressed in terms of a reduced three-variable model like

$$\begin{aligned}\frac{\mathrm{d}}{\mathrm{d}t}S &= -\beta\frac{IS}{N},\\ \frac{\mathrm{d}}{\mathrm{d}t}E &= \beta\frac{IS}{N} - \alpha E,\\ \frac{\mathrm{d}}{\mathrm{d}t}I &= \alpha E - \gamma I\end{aligned} \tag{3.45}$$

with $R(t) = N - S(t) - I(t) - E(t)$.

Fixed points

Putting the left-hand sides in Eq. (3.43) equal to zero, yields in a first step $\mathrm{d}R/\mathrm{d}t = 0 \Rightarrow I = 0$. From $I = 0$ it follows that $\mathrm{d}I/\mathrm{d}t = 0 \Rightarrow E = 0$. Finally, for any $S \geq 0$ the conclusion $I = E = 0 \Rightarrow \mathrm{d}S/\mathrm{d}t = \mathrm{d}E/\mathrm{d}t = 0$ can be made. Consequently, fixed points are given by

$$I_{st} = E_{st} = 0,\ S_{st} \in [0, N],\ R_{st} = N - S_{st}. \tag{3.46}$$

They describe disease-free states (i.e., states with zero exposed and infectious individuals).

Trajectories

Just as in the case of the SIR model, the number of susceptibles S decays monotonically over time, while the number of recovered individuals R increases monotonically. Moreover, just as for the SIR model, the SEIR model exhibits two types of qualitatively different solutions: monotonically decaying solutions and wave-solutions. Unlike the SIR model, these types of solutions are defined on the basis of the total infected individuals given by

$$C = I + E. \tag{3.47}$$

Differentiating C with respect to time and using the evolution equations for I and E occurring in Eq. (3.43), the evolution equation for C can be obtained. Subsequently, eliminating E in the evolution equation for E like $E = C - I$, the reduced, three-variable SEIR model (3.45) becomes the three-variable model

$$\begin{aligned}\frac{\mathrm{d}}{\mathrm{d}t}S &= -\beta\frac{IS}{N},\\ \frac{\mathrm{d}}{\mathrm{d}t}C &= \left(\beta\frac{S}{N} - \gamma\right)I,\\ \frac{\mathrm{d}}{\mathrm{d}t}I &= \alpha C - (\alpha + \gamma)I\end{aligned} \tag{3.48}$$

with $R(t) = N - S(t) - C(t)$. The SCIR model (3.48) does not describe a compartmental model because infectious individuals show up both in the variables I and C. Irrespectively, at issue is that the evolution equation for C of the SCIR model exhibits a similar structure as Eq. (3.23). Therefore, the discussion about wave-solutions versus monotonically decaying solutions can be carried out in analogy to what was written there.

If $Z = \beta S/N - \gamma < 0$ holds at the initial time point t_0, then $Z < 0$ for $t > t_0$, which implies that the total number of infected individuals C decays monotonically. If the expression $Z(t) = \beta S(t)/N - \gamma$ is zero either at the initial time point t_0 or any later time point $t_p > t_0$, then C does not change at that particular moment in time. However, S is a monotonically decaying function of time, which implies that Z becomes negative and remains negative for $t > t_0$ or $t > t_p$. This implies that if $Z = 0$ at t_0, then C is a monotonically decaying function just as for $Z < 0$ at t_0. In contrast, if $Z > 0$ holds at the initial time t_0, then from the evolution equation of C in Eq. (3.48) it follows that C increases. While C increases, S decreases, which implies that at a certain time point $t_p > t_0$ the expression Z becomes zero. At that time point C reaches its maximum. For $t > t_p$ we have $Z < 0$ and C decays over time. Consequently, for the SEIR model the stability parameter ξ can be defined by Eq. (3.24) again [2]. Moreover, if $S(0)/N \approx 1$ holds, the stability parameter reduces to Eq. (3.26). With respect to the parameter ξ it follows that if $\xi \leq 1$ holds there is a monotonic decay of the variable $C = E + I$ of the SEIR model. During a transient period $I(t)$ may increase at the cost of $E(t)$ (i.e., when exposed individuals become infectious). That is, $I(t)$ does not necessarily decrease monotonically for $\xi \leq 1$. For $\xi > 1$ the SEIR model exhibits wave-solutions that describe a temporary increase in the total number of infected individuals C that is followed by a decrease of the total number of infected individuals C towards zero (i.e., towards one of the stationary fixed points defined by Eq. (3.46) with $C_{st} = 0$). The case $\xi > 1$ describes epidemic waves. In Chap. 5 the stability parameter ξ of the SEIR model will be derive in a more systematic way using the nonlinear physics perspective presented in Chap. 2.

Figure 3.12 exemplified the two type of solutions of the SEIR model (3.43). Panel (a) shows the variables C, E, and I as functions of time as computed from Eq. (3.43) for the case $\xi < 1$. For the selected parameters and initial conditions $I(t)$ is increasing initially. However, $C(t)$ decays over time. Panel (b) presents a wave-solution as obtained by solving Eq. (3.43) numerically for $\xi > 1$. For the selected parameters and initial conditions there is a wave of exposed individuals that is followed by a wave of infectious individuals. That is, the wave in $E(t)$ precedes the wave in $I(t)$.

Applications of the SEIR model and similar models to COVID-19 epidemics will be discussed in Chap. 5.

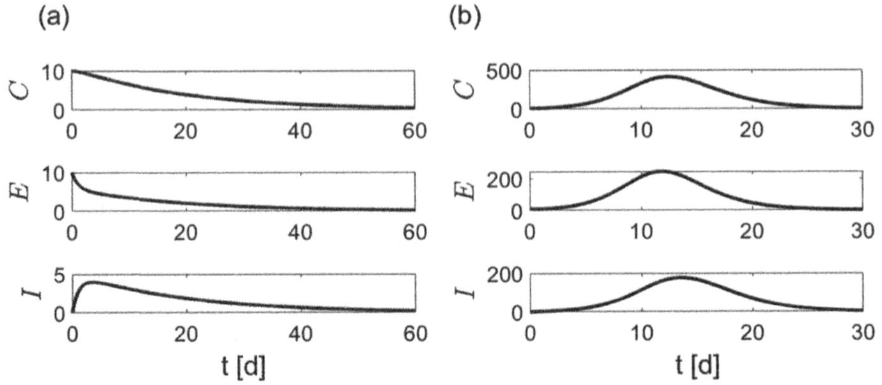

Fig. 3.12 Solutions $C(t)$, $E(t)$, and $I(t)$ of the SEIR model (3.43) for $\xi < 1$ (panel (**a**)) and $\xi > 1$ (panel (**b**)). Parameters and initial conditions for both simulations: $N = 1,000$, $\gamma = 0.5$/d, $\alpha = 0.4$/d, $E(0) = 10$, $I(0) = 0$, $S(0) = N - E(0)$. The effective contact rates are $\beta = 0.4$/d for panel (**a**) and $\beta = 2.0$/d for panel (**b**)

References

1. H.W. Hethcote, The mathematics of infectious diseases. SIAM Rev. **42**, 599–653 (2000)
2. K. Rock, S. Brand, J. Moir, M.J. Keeling. Dynamics of infectious diseases. Rep. Prog. Phys. **77**, article 026602 (2014)
3. M. Martcheva, *An Introduction to Mathematical Epidemiology* (Springer, New York, 2015)
4. E.C. Tolles, T. Luong, Modeling epidemics with compartmental models. JAMA **323**, 2515–2516 (2020)
5. R.M. Anderson, R.M. May, Population biology of infectious diseases: part I. Nature **280**, 361–367 (1979)
6. H.W. Hethcote, Three basic epidemiological models, in *Applied Mathematical Ecology*, ed. by S. Levin, T.G. Hallam, L.J. Gross. (Springer, Berlin, 1989), pp. 119–144
7. O. Diekmann, J.A.P. Heesterbeek, *Mathematical Epidemiology of Infectious Diseases* (Wiley, Chichester, 2000)
8. W.O. Kermack, A.G. McKendrick, A contribution to the mathematical theory of epidemics. Proc. R. Soc. A **115**, 700–721 (1927)
9. N.S. Barlow, S.J. Weinstein, Accurate closed form solution of the SIR epidemic model. Physica D **408**, 132540 (2020)
10. G.D. Barmparis, G.P. Tsironis, Estimating the infection horizon of COVID-19 in eight countries with a data-driven approach. Chaos, Solitons Fractals **135**, 109842 (2020)
11. F. Croccolo, H.E. Roman. Spreading of infections on random graphs: a percolation-type model for COVID-19. Chaos, Solitons Fractals **139**, article 110077 (2020)
12. N. Crokidakis, Modeling the early evolution of the covid-19 in brazil: results from a susceptible-infectious-quarantined-recovered (siqr) model. Int. J. Mod. Phys. C, **31**, article 2050135 (2020)
13. N. Crokidakis, Covid-19 spread in Rio de Janeiro, Brazil: do the policies of social isolation really well. Chaos, Solitons Fractals **136**, article 109930 (2020)
14. D. Fanelli, F. Piazza, Analysis and forecast of COVID-19 spreading in China, Italy, and France. Chaos, Solitons Fractals **134**, article 109761 (2020)
15. M.E. Hoque, An early estimation of the number of affected people in south Asia due to COVID-19 pandemic using susceptible, infected and recovered model. Int. J. Mod. Phys. C **31**, 2050140 (2020)

16. O. Ifguis, M.E. Ghozlani, F. Ammou, A. Moutcine, Z. Abdellah. Simulation. J. Environ. Public Health **2020**, article 9769267 (2020)
17. P. Pornphol, S. Chittayasothorn, System dynamics model of COVID-19 pandemic situation: the case of Phuket Thailand, in *ICCMS 20: Proceedings of the 12th International Conference on Computer Modeling and Simulation* (Association for Computer Machinery, 2020), pp. 77–81
18. E.B. Postnikov. Estimation. Chaos, Solitons Fractals **135**, article 109841 (2029)
19. S. Rafieenasab, A.P. Zahiri, E. Roohi. Prediction. Int. J. Mod. Phys. C **31**, article 2050152 (2020)
20. Y. Tao. Maximum. Phys. Rev. E **102**, article 032136 (2020)
21. G. Vattay. Forecasting. Phys. Biol. **17**, article 065002 (2020)
22. J. Wangping, H. Ke, S. Yang, C. Wenzhe, W. Shengshu, Y. Shanshan, W. Jianwei, K. Fuyin, T. Penggang, L. Jing, L. Miao, H. Yao, Extended SIR prediction of the epidemics trend of COVID-19 in Italy and compared with Hunan, China. Front. Med. **7**, article 169 (2020)
23. M.J. Willis, V.H.G. Diaz, O.A. Prado-Rubio, M. von Stosch, Insights. Chaos, Solitons Fractals **138**, article 109937 (2020)
24. B. Zareie, A. Roshani, M.A. Mansournia, M.A. Rasouli, G. Moradi, A model. Arch. Iran. Med. **23**, 244–248 (2020)
25. T.D. Frank, COVID-19 outbreaks follow narrow paths: a computational phase portrait approach based on nonlinear physics and synergetics. Int. J. Mod. Phys. C **32**, 2150110 (2021)
26. T. Carletti, D. Fanelli, F. Piazza. COVID-19: the unreasonable effectiveness of simple models. Chaos, Solitons Fractals: X, **5**, article 100034 (2020)
27. M.G. Pedersen, M. Meneghini, Data-driven estimation of change points reveals correlation between face mask use and accelerated curtailing of the first wave of COVID-19 epidemic in Italy. Infect. Dis. **53**, 243–251 (2021)
28. E. Dong, H. Du, L. Gardner, An interactive web-based dashboard to track COVID-19 in real time. Lancet: Infect. Dis. **20**, 533–534 (2020)
29. COVID-19 tracker. Timeline data from Johns Hopkins Center for Systems Science and Engineering. *COVID-19 tracker,* https://vac-lshtml.shinyapps.io/ncov_tracker
30. B. Ivorra, M.R. Ferrandez, M. Vela-Perez, A.M. Ramos. Mathematical modeling of the spread of coronavirus disease 2019 (COVID-19) taking into account the undetected infections. the case of China. Commun Nonlinear Sci Numer Simulat **88**, article 105303 (2020)
31. T. Sun, Y. Wang. Modeling COVID-19 epidemic in Heilongjiang province, China. Chaos, Solitons, Fractals **138**, article 109949 (2020)
32. H.B. Taboe, K.V. Salako, J.M. Tison, C.N. Ngonghala, R.G. Kakai, Predicting COVID-19 spread in the face of control measures in West Africa. Math. Biosci. **328**, article 108431 (2020)
33. T.D. Frank. Simplicity from complexity: on the simple amplitude dynamics underlying COVID-19 outbreaks in China. Adv. Complex Syst. **23**, article 2050022 (2020)

Chapter 4
Nonlinear Physics of Epidemics: Part A

This chapter is the first of three chapters that present the nonlinear physics of epidemics and pandemics that can be described in terms of deterministic dynamical systems. The chapter focuses on the SIR model and SIR-type models. The amplitude space description is worked out in detail and the SI order parameter is determined. It is demonstrated that within the SIR modeling framework, the SI order parameter determines the initial stage of epidemics. An application to the COVID-19 outbreak in Italy during the year 2020 is also presented.

4.1 SIR-Type Models and 2D Autonomous Amplitude Descriptions

4.1.1 n-Dimensional Approach

Let $\mathbf{X} = (X_1, \ldots, X_n)$ denote the state vector describing an epidemic whose dynamics satisfies the general evolution equation $\mathrm{d}\mathbf{X}/\mathrm{d}t = \mathbf{N}(\mathbf{X})$ (see Chap. 2). In what follows the notion of susceptible and infectious individuals as used in the SIR model (see Chap. 3) will be used. More precisely, the population under consideration is decomposed in healthy individuals S, who have not been infected but can be infected, and individuals I, who have been infected and are infectious. In addition, other individuals who have been infected but cannot infected others are considered. This latter group of individuals can be decomposed in an arbitrary number of subgroups. In what follows, the individuals in the compartment I will be referred to as infectious individuals. Let $X_1 = S$ denote the susceptibles and $X_2 = I$ denote the infectious individuals. Then, $X_3, \ldots, X_n$ denote $n - 2$ compartments of individuals who have been infected at some point in time but cannot infect others any more. For example, hospitalized individuals (assuming a perfect isolation such that they cannot infect others) and recovered individuals belong to these compartments $k = 3, \ldots, n$. Since

T. D. Frank, *COVID-19 Epidemiology and Virus Dynamics*,
Understanding Complex Systems, https://doi.org/10.1007/978-3-030-97178-6_4

there is only one group of individuals who can infect others, it is assumed that the rate constant k_0 is given by Eq. (3.4). Models that satisfy all of the above assumptions read

$$\frac{d}{dt}S = -k_0 S + B - \mu S,$$
$$\frac{d}{dt}I = k_0 S - (\gamma + \mu)I,$$
$$\frac{d}{dt}X_k = N_k(X_1, \ldots, X_n) \text{ for } k = 3, \ldots, n \tag{4.1}$$

with $k_0 = \beta I/N$ (see Eq. (3.4)), where N_k are the standard right-hand side functions of dynamical systems discussed in Chap. 2. They include the death terms $-\mu X_k$ provided demographic effects are considered and X_k does not correspond to a compartment of deceased individuals. Models that can be cast into the form of Eq. (4.1) will be referred to as SIR-type models.

If the functions N_k only reflect linear transition mechanisms, Eq. (4.1) becomes

$$\frac{d}{dt}S = -k_0 S + B - \mu S,$$
$$\frac{d}{dt}I = k_0 S - (\gamma + \mu)I,$$
$$\frac{d}{dt}X_3 = \sum_{k=1}^{n} a_{3,k} X_k,$$
$$\ldots$$
$$\frac{d}{dt}X_n = \sum_{k=1}^{n} a_{n,k} X_k, \tag{4.2}$$

In Eq. (4.1) the coefficients $a_{i,k}$ describe transition rates of the aforementioned linear transition mechanisms. They also account for the death terms $-\mu X_k$ provided demographic effects are considered and X_k does not correspond to a compartment of deceased individuals. The models (4.1) and (4.2) involve birth and death terms, where the death terms reflects only death to causes other than the infectious disease under consideration. For epidemics over short periods we put $B = \mu = 0$.

Let us discuss two examples. The SIR model without demographic terms given by $dS/dt = -\beta IS/N$, $dI/dt = \beta IS/N - \gamma I$, $dR/dt = \gamma I$ (see Eq. (3.16)) corresponds to Eq. (4.2) with $\mathbf{X} = (S, I, R)$ and $n = 3$, $X_3 = R$, $B = \mu = 0$, $k_0 = \beta I/N$, and $a_{3,2} = \gamma$. All other coefficients $a_{i,k}$ are zero. The study by Fanelli and Piazza [1] discussed in Sect. 3.6.2 used the SIR-type model

$$\frac{d}{dt}S = -rIS,\ \frac{d}{dt}I = rIS - (a+b)I,\ \frac{d}{dt}R = aI,\ \frac{d}{dt}D = bI. \tag{4.3}$$

In this model D denotes the individuals deceased due to COVID-19. The state vector of the model is given by $\mathbf{X} = (S, I, R, D)$. The model can be cast into the form of a SIR-type model (4.2) with $n = 4$, $B = \mu = 0$, $r = \beta/N$, $\gamma = a + b$, $a_{3,2} = a$, $a_{4,2} = b$. All other coefficients $a_{i,k}$ are zero.

Let us return to the general case of SIR-type models defined by Eq. (4.1). The disease-free fixed point of the model (4.1) for novel infectious diseases (i.e., all individuals are susceptible) is given by $\mathbf{X}_{st} = (N, 0, 0, \ldots, 0)$ assuming $N_k(\mathbf{X}_{st}) = 0$ holds for $k = 3, \ldots, n$ and $B = \mu N$ holds (see Sect. 3.5.2) if demographic terms are considered. Linearizing the model (4.1) with the help of the perturbation $\mathbf{u} = \mathbf{X} - \mathbf{X}_{st}$ at the fixed point $\mathbf{X}_{st}$ yields the linearized evolution equation $\mathrm{d}\mathbf{u}/\mathrm{d}t = L\mathbf{u}$ (see Eq. (2.16)). From the matrix L the eigenvalues λ_k and eigenvectors $\mathbf{v}_k$ can be obtained (see Sect. 2.6). It is assumed that $\mathbf{v}_k$ constitute a set of n linearly independent vectors. The state $\mathbf{X}$ can then be expressed with the help of the eigenvectors like $\mathbf{u}(t) = \sum_{k=1}^{n} A_k(t)\mathbf{v}_k$ and $\mathbf{X}(t) = \mathbf{X}_{st} + \mathbf{u} = \mathbf{X}_{st} + \sum_{k=1}^{n} A_k(t)\mathbf{v}_k$ (see Eqs. (2.34) and (2.35)), where $A_1, \ldots, A_n$ denote the amplitudes of the SIR-type model under consideration. The amplitudes evolve according to the amplitude equations $\mathrm{d}A_k/\mathrm{d}t = \lambda_k A_k + G_k(A_1, \ldots, A_n)$ for $k = 1, \ldots, n$ introduced in Sect. 2.9, see Eq. (2.59). Accordingly, SIR-type models of the form (4.1) are described in amplitude space by the amplitudes $A_1, \ldots, A_n$ satisfying amplitude equations of the general form $\mathrm{d}A_k/\mathrm{d}t = \lambda_k A_k + G_k(A_1, \ldots, A_n)$. Importantly, the amplitude equations holds for the whole course of the epidemic under consideration (i.e., also far away from the disease-free fixed point). The state space description via $\mathbf{X} = (S, I, X_3, \ldots, X_n)$ and Eq. (4.1) and the amplitude space description via $\mathbf{A} = (A_1, \ldots, A_n)$ and $\mathrm{d}A_k/\mathrm{d}t = \lambda_k A_k + G_k(A_1, \ldots, A_n)$ are equivalent, see Chap. 2. In contrast, the linearized dynamics $\mathrm{d}\mathbf{u}/\mathrm{d}t = L\mathbf{u}$ only holds close to the fixed point. In this case, the amplitude equations decouple from each other and reduce to $\mathrm{d}A_k/\mathrm{d}t = \lambda_k A_k$ as discussed in Chap. 2.

4.1.2 Two-Dimensional Approach

In the previous section, the description of SIR-type models in n-dimensional state spaces and n-dimensional amplitude spaces has been considered. SIR-models of the form (4.1) can be decomposed into two subsystems: a two-dimensional closed or autonomous subsystem that describes the S-I dynamics and a $n-2$ dimensional subsystems that is not closed or non-autonomous and depends on the dynamics of the S-I subsystem. Importantly, typically the stability properties of the S-I subsystem determine the stability of the entire system. For example, if the S-I subsystem exhibits an unstable fixed point ($S_{st} = N$, $I_{st} = 0$) related to the disease-free fixed point $\mathbf{X}_{st} = (N, 0, \ldots, 0)$, then the disease-free fixed point $\mathbf{X}_{st} = (N, 0, \ldots, 0)$ corresponds to an unstable fixed point of the entire system. Likewise, if ($S_{st} = N$, $I_{st} = 0$) corresponds to a (neutrally or asymptotically) stable fixed point of the S-I subsystem, then, frequently, it can be shown that the disease-free fixed point $\mathbf{X}_{st} = (N, 0, \ldots, 0)$ corresponds to a stable fixed point of the entire system.

In view of the pivot role of the S-I subsystem, the amplitude equation description may be only derived for that specific subsystem. In doing so, a two-dimensional amplitude space is constructed spanned by the amplitudes A_1 and A_2 that is mapped to the S-I space like $(A_1, A_2) \leftrightarrow (S, I)$. The remaining state variables $X_3, \ldots, X_n$ are not transformed. For SIR-models of the form (4.1) amplitude space descriptions using such two-dimensional amplitude spaces read

$$\frac{\mathrm{d}}{\mathrm{d}t}A_1 = \lambda_1 A_1 + G_1(A_1, A_2),$$
$$\frac{\mathrm{d}}{\mathrm{d}t}A_2 = \lambda_2 A_2 + G_2(A_1, A_2),$$
$$\frac{\mathrm{d}}{\mathrm{d}t}X_k = N_k(S(A_1, A_2), I(A_1, A_2), X_3, \ldots, X_n) \text{ for } k = 3, \ldots, n, \tag{4.4}$$

where $S(A_1, A_2)$ and $I(A_1, A_2)$ describe the mapping $(A_1, A_2) \rightarrow (S, I)$ from amplitude space to state space. Note that the coupled amplitude equations for A_1 and A_2 describe a closed or autonomous dynamical system and can be solved independently of the state variables X_k with $k = 3, \ldots, n$. The amplitudes A_1 and A_2 span a two-dimensional amplitude space and provide a two-dimensional autonomous amplitude description. The explicit expression for λ_1, λ_2, G_1, G_2 will be derived in the subsequent sections and summarized in Sect. 4.4. With the help of A_1 and A_2, the state vector $\mathbf{X}$ can be expressed like

$$\mathbf{X} = \begin{pmatrix} S_{st} \\ I_{st} \\ 0 \\ \ldots \\ 0 \end{pmatrix} + \sum_{k=1}^{2} A_k \tilde{\mathbf{v}}_k + \sum_{k=3}^{n} X_k \mathbf{e}_k, \tag{4.5}$$

where $\mathbf{e}_k = (\delta_{1,k}, \delta_{2,k}, \ldots, \delta_{n,k})$ are the orthogonal basis vectors listed in Eq. (2.36). In Eq. (4.5) the vectors $\tilde{\mathbf{v}}_k$ are the two-dimensional eigenvectors $\mathbf{v}_k = (v_{k,S}, v_{k,I})$ of the S-I subspace extended to the n-dimensional space like

$$\tilde{\mathbf{v}}_k = \begin{pmatrix} v_{k,S} \\ v_{k,I} \\ 0 \\ \ldots \\ 0 \end{pmatrix}. \tag{4.6}$$

4.2 SIR Model Without Demographic Terms

The two-variable version of the SIR model (3.16) reads, see Eq. (3.22),

$$\frac{\mathrm{d}}{\mathrm{d}t}S = -\frac{\beta}{N}IS, \quad \frac{\mathrm{d}}{\mathrm{d}t}I = \frac{\beta}{N}IS - \gamma I \tag{4.7}$$

with $R(t) = N - S - I(t)$. Let us introduce the two-dimensional state vector $\mathbf{X} = (S, I)$. As discussed in Sect. 3.5, the model exhibits the fixed points $\mathbf{X}_{st} = (S_{st}, 0)$, $S_{st} \in [0, N]$, $R_{st} = N - S_{st}$ that describe disease-free states. For novel virus infections such as COVID-19 outbreaks in the year 2020, the fixed point $\mathbf{X}_{st} = (N, 0)$ corresponds to the disease-free state with all individuals susceptible.

4.2.1 Eigenvalues and Eigenvectors

Let us linearize Eq. (4.7) at $\mathbf{X}_{st} = (S_{st}, 0)$. To this end, I is assumed to be small and we put $S = S_{st} + \delta$, where δ is a small quantity. In doing so, we obtain $IS = I(S_{st} + \delta) = S_{st}I + I\delta = S_{st}I +$ nonlinear terms. Moreover, it follows that $\mathrm{d}S/\mathrm{d}t = \mathrm{d}\delta/\mathrm{d}t$. Substituting these results into Eq. (4.7), we obtain

$$\frac{\mathrm{d}}{\mathrm{d}t}\delta = -\beta\frac{S_{st}}{N}I, \quad \frac{\mathrm{d}}{\mathrm{d}t}I = \left(\beta\frac{S_{st}}{N} - \gamma\right)I. \tag{4.8}$$

Using vector and matrix notations, Eq. (4.8) can equivalently be expressed like

$$\frac{\mathrm{d}}{\mathrm{d}t}\begin{pmatrix}\delta\\ I\end{pmatrix} = L\begin{pmatrix}\delta\\ I\end{pmatrix}, \quad L = \begin{pmatrix}0 & -\beta S_{st}/N\\ 0 & \beta S_{st}/N - \gamma\end{pmatrix} \tag{4.9}$$

and exhibits the form $\mathrm{d}\mathbf{u}/\mathrm{d}t = L\mathbf{u}$ (see Eq. (2.16)). The eigenvalues λ and eigenvectors $\mathbf{v} = (v_S, v_I)$ of L correspond to the solutions of $L\mathbf{v} = \lambda\mathbf{v}$, which implies $L\mathbf{v} - \lambda\mathbf{v} = 0$ and

$$\begin{pmatrix}-\lambda & -\beta S_{st}/N\\ 0 & \beta S_{st}/N - \gamma - \lambda\end{pmatrix}\begin{pmatrix}v_S\\ v_I\end{pmatrix} = \begin{pmatrix}0\\ 0\end{pmatrix}. \tag{4.10}$$

From Eq. (4.10) it follows that the determinant of the matrix must equal zero like

$$\begin{vmatrix}-\lambda & -\beta S_{st}/N\\ 0 & \beta S_{st}/N - \gamma - \lambda\end{vmatrix} = 0. \tag{4.11}$$

Explicitly, this requirement reads

$$\begin{vmatrix}-\lambda & -\beta S_{st}/N\\ 0 & \beta S_{st}/N - \gamma - \lambda\end{vmatrix} = -\lambda(\beta S_{st}/N - \gamma - \lambda) = 0 \tag{4.12}$$

such that the two eigenvalues are given by

$$\lambda_1 = 0, \ \lambda_2 = \beta\frac{S_{st}}{N} - \gamma. \tag{4.13}$$

The vanishing eigenvalue reflects the property that the fixed point $(S_{st}, 0)$ can be shifted along the S-axis in the SIR model without demographic terms (see Sect. 3.5.1). The eigenvector of the eigenvalue $\lambda_1 = 0$ can be obtained by substituting $\lambda = 0$ into Eq. (4.10) such that

$$\begin{pmatrix} 0 & -\beta S_{st}/N \\ 0 & \beta S_{st}/N - \gamma \end{pmatrix} \begin{pmatrix} v_S \\ v_I \end{pmatrix} = \begin{pmatrix} 0 \\ 0 \end{pmatrix} \Rightarrow \begin{pmatrix} v_S \\ v_I \end{pmatrix} = \begin{pmatrix} 1 \\ 0 \end{pmatrix} \tag{4.14}$$

Likewise, the eigenvector associated with the eigenvalue $\lambda_2 = \beta S_{st}/N - \gamma$ can be obtained by substituting $\lambda = \lambda_2$ into Eq. (4.10). Thus, we obtain

$$\begin{pmatrix} -\lambda_2 & -\beta S_{st}/N \\ 0 & 0 \end{pmatrix} \begin{pmatrix} v_S \\ v_I \end{pmatrix} = \begin{pmatrix} 0 \\ 0 \end{pmatrix} \Rightarrow \lambda_2 v_S + \frac{\beta S_{st}}{N} v_I = 0. \tag{4.15}$$

Solutions can be written as $v_S = -Z \cdot \beta S_{st}/N \Rightarrow v_I = Z\lambda_2$, where Z is an arbitrary parameter. The normalization requirement (i.e., $v_S^2 + v_I^2 = 1$) determines the value of Z such that

$$\mathbf{v}_2 = \frac{1}{\sqrt{(\beta S_{st}/N)^2 + (\lambda_2)^2}} \begin{pmatrix} -\beta S_{st}/N \\ \lambda_2 \end{pmatrix}. \tag{4.16}$$

In order to simplify the notation, let us introduce the parameter

$$g = \frac{\beta S_{st}/N - \gamma}{\beta S_{st}/N} = \frac{\lambda_2}{\beta S_{st}/N} = \frac{\lambda_2}{\beta} \frac{N}{S_{st}}. \tag{4.17}$$

Then

$$\begin{pmatrix} -\beta S_{st}/N \\ \lambda_2 \end{pmatrix} = \beta \frac{S_{st}}{N} \begin{pmatrix} -1 \\ g \end{pmatrix} \Rightarrow \mathbf{v}_2 = \frac{1}{\sqrt{1+g^2}} \begin{pmatrix} -1 \\ g \end{pmatrix}. \tag{4.18}$$

COVID-19 Outbreaks at the Beginning of the Pandemic and Outbreaks of Other Novel Infectious Diseases

For the special case $S_{st} = N \Rightarrow S_{st}/N = 1$ that describes COVID-19 outbreaks at the beginning of the pandemic in the years 2019/2020 (and, in general, outbreaks of novel infectious diseases), we obtain

$$\lambda_1 = 0, \ \lambda_2 = \beta - \gamma, \ g = (\beta - \gamma)/\beta \tag{4.19}$$

and

$$\mathbf{v}_2 = \frac{1}{\sqrt{\beta^2 + (\lambda_2)^2}} \begin{pmatrix} -\beta \\ \lambda_2 \end{pmatrix} = \frac{1}{\sqrt{1+g^2}} \begin{pmatrix} -1 \\ g \end{pmatrix}. \tag{4.20}$$

4.2.2 State Space and Amplitude Space

The mapping from amplitude space to state space $(A_1, A_2) \rightarrow (S, I)$ is given by Eq. (2.35) for $n = 2$ and reads

$$\mathbf{X} = \mathbf{X}_{st} + A_1\mathbf{v}_1 + A_2\mathbf{v}_2. \tag{4.21}$$

Using Eq. (4.18), the mapping reads explicitly

$$\begin{pmatrix} S \\ I \end{pmatrix} = \begin{pmatrix} S_{st} \\ 0 \end{pmatrix} + A_1 \begin{pmatrix} 1 \\ 0 \end{pmatrix} + A_2 \frac{1}{\sqrt{1+g^2}} \begin{pmatrix} -1 \\ g \end{pmatrix}. \tag{4.22}$$

In components, the mapping reads

$$S = S_{st} + A_1 - \frac{A_2}{\sqrt{1+g^2}}, \quad I = \frac{g}{\sqrt{1+g^2}} A_2. \tag{4.23}$$

The sign of the amplitude A_2 is determined by the sign of λ_2 irrespective of the initial conditions S_0 and I_0. For $\beta S_{st}/N > \gamma \Rightarrow \lambda_2 > 0$ the inequality $g > 0$ holds, which implies that $A_2 \geq 0$. In contrast, for $\beta S_{st}/N < \gamma \Rightarrow \lambda_2 < 0$ we have $g < 0$, which implies that $A_2 \leq 0$.

In order to obtain the inverse mapping $S, I \rightarrow A_1, A_2$, first, the mapping $A_2 \rightarrow I$ can be inverted like $A_2 = \sqrt{1+g^2} I/g$. Second, $A_2/\sqrt{1+g^2} = I/g$ can be substituted into the equation for S in Eq. (4.23) to eliminate A_2, which leads to $S = S_{st} + A_1 - I/g$. Solving this result for A_1, the mapping

$$A_1 = S + \frac{I}{g} - S_{st}, \quad A_2 = \frac{\sqrt{1+g^2}}{g} I \tag{4.24}$$

can be obtained. The mappings (4.23) and (4.24) hold for any states S, I and amplitudes A_1, A_2. They are not limited to states close to the fixed point $\mathbf{X}_{st}$ or small amplitudes.

4.2.3 Stability Analysis

As argued in Sect. 2.7, the eigenvalues λ_1 and λ_2 determine the stability of the fixed point $\mathbf{X}_{st} = (S_{st}, 0)$. In particular, for $n = 2$ substituting Eq. (4.21) expressed like $\mathbf{u} = \mathbf{X} - \mathbf{X}_{st} = A_1\mathbf{v}_1 + A_2\mathbf{v}_2$ into the linearized equation $d\mathbf{u}/dt = L\mathbf{u}$, we obtain

$$LHS : \frac{\mathrm{d}}{\mathrm{d}t}\mathbf{u} = \mathbf{v}_1 \frac{\mathrm{d}}{\mathrm{d}t} A_1 + \mathbf{v}_2 \frac{\mathrm{d}}{\mathrm{d}t} A_2$$

$$RHS : L\mathbf{u} = A_1 L\mathbf{v}_1 + A_2 L\mathbf{v}_2 = A_1\lambda_1\mathbf{v}_1 + A_2\lambda_2\mathbf{v}_2$$

$$LHS = RHS : \mathbf{v}_1 \frac{\mathrm{d}}{\mathrm{d}t} A_1 + \mathbf{v}_2 \frac{\mathrm{d}}{\mathrm{d}t} A_2 = \lambda_1 A_1 \mathbf{v}_1 + \lambda_2 A_2 \mathbf{v}_2. \tag{4.25}$$

Let us consider the case $\lambda_2 \neq 0$ (i.e., $\beta S_{st}/N \neq \gamma$) in which the eigenvectors $\mathbf{v}_1, \mathbf{v}_2$ are linearly independent. From the last relation in Eq. (4.25) and the linear independency of $\mathbf{v}_1$ and $\mathbf{v}_2$ (or by constructing biorthogonal eigenvectors $\mathbf{w}_1$ and $\mathbf{w}_2$ and multiplying the last relation in Eq. (4.25) by $\mathbf{w}_k$ as in Sect. 2.9.3) it follows

$$\frac{\mathrm{d}}{\mathrm{d}t} A_1 = \lambda_1 A_1 = 0 \;\Rightarrow\; A_1 = \text{const},$$

$$\frac{\mathrm{d}}{\mathrm{d}t} A_2 = \lambda_2 A_2 \;\Rightarrow\; A_2(t) = A_2(0)\exp\{\lambda_2 t\} \tag{4.26}$$

for $t_0 = 0$. Using $\lambda_2 = \beta S_{st}/N - \gamma$, the bifurcation parameter

$$\alpha = \frac{\beta}{\gamma}\frac{S_{st}}{N} \tag{4.27}$$

can be defined that exhibits a critical value of 1. The bifurcation parameter α allows to distinguish between the cases

$$\begin{aligned} &\alpha > 1 \Rightarrow \lambda_2 > 0 \\ &\Rightarrow A_2(t) \text{ initially increases } \Rightarrow\; I(t) \text{ initially increases} \\ &\Rightarrow \mathbf{X}_{st} \text{ unstable} \end{aligned} \tag{4.28}$$

and

$$\begin{aligned} &\alpha < 1 \Rightarrow\; \lambda_2 < 0 \\ &\Rightarrow |A_2(t)| \text{ decreases for arbitrarily small } |A_2(0)| \\ &\Rightarrow\; I(t) \text{ decreases for arbitrarily small } I_0 \\ &\Rightarrow\; \mathbf{X}_{st} \text{ neutrally stable.} \end{aligned} \tag{4.29}$$

The two conclusions presented in Eqs. (4.28) and (4.29) that (i) an increase in A_2 implies an increase in I and (ii) a decrease in $|A_2|$ for $A_2 < 0$ implies a decrease in I follow from the mapping $A_2 \to I$ shown in Eq. (4.23). The conclusion presented in Eq. (4.29) that $\lambda_2 < 0$ implies that $\mathbf{X}_{st} = (S_{st}, 0)$ is neutrally stable cannot be drawn from Eq. (4.26) alone. Since $\lambda_1 = 0$ holds, it must be shown that for $\lambda_2 < 0$ the nonlinear terms determining the evolution of A_1 result in a decay of a perturbation of A_1 out of $A_{1,st} = 0$. For example, if $\lambda_2 < 0$ holds but the nonlinear terms in the evolution equation A_1 are such that A_1 evolves away from $A_{1,st} = 0$, then the fixed point would be a saddle, that is, an unstable fixed point. As it will be shown in Sect. 4.2.7, for $\lambda_2 < 0$ the nonlinear terms are indeed such that perturbations of A_1 out of $A_{1,st} = 0$ decay. Therefore, as stated in Eq. (4.29), a negative eigenvalue λ_2 indeed implies that the fixed point characterized by $A_{1,st} = A_{2,st} = 0$, that is, the fixed point $\mathbf{X}_{st} = (S_{st}, 0)$, is neutrally stable.

From Eq. (4.28) it follows that for $\alpha > 1$ there is an increase of the number of infectious individuals I that initiates an epidemic wave. In contrast, from Eq. (4.29) it follows that for $\alpha < 1$ the epidemic given in terms of the number of infectious individuals $I(t)$ subsides initially. Since the disease dynamics described by the SIR model (4.7) does not exhibit any fixed points with $I_{st} > 0$, the epidemic not only initially subsides in the population under consideration but subsides entirely, that is, $I(t)$ decays monotonically to zero over time. The bifurcation parameter defined by Eq. (4.27) corresponds to the stability parameter $\xi = \beta S_0/N$ discussed in Sect. 3.5.1 (see Eq. (3.24)) provided the case $S_{st} = S_0$ is considered and I_0 can be regarded as a small quantity. If I_0 is sufficiently small, then an initial state S_0 and I_0 describes a small perturbation out of the fixed point defined by $S_{st} = S_0$ and $I_{st} = 0$ such that the linear stability analysis applies. Mathematically speaking, the small perturbation shifts the disease state out of $\mathbf{X}_{st} = (S_{st}, 0)$ by increasing I_0 (and reducing R_0 such that $N = S_{st} + I_0 + R_0$ is still satisfied), while S is kept constant, which implies that the initial disease state after the perturbation is given by $S_0 = S_{st}$ and $I_0 > 0$. From the linear stability analysis (conducted above) it follows that the dynamics of the SIR model (4.7) under the initial condition S_0 and I_0 exhibits a wave-solution for $\alpha > 1$ with $S_{st} = S_0$, which means for $\alpha = \beta S_{st}/N = \beta S_0/N = \xi > 1$. Likewise, the dynamics under that initial condition S_0 and I_0 exhibits a monotonically decaying solution $I(t)$ for $\alpha < 1$ with $S_{st} = S_0$, which means for $\alpha = \beta S_{st}/N = \beta S_0/N = \xi < 1$. That is, the systematic stability analysis of the disease-free fixed point $\mathbf{X}_{st} = (S_{st}, 0)$ conducted within the nonlinear physics perspective of the SIR model reproduces the stability parameter ξ that was derived in an ad hoc manner in Sect. 3.5.1.

COVID-19 Outbreaks at the Beginning of the Pandemic

For the epidemics of novel infectious diseases for which all individuals of a population can be considered to be susceptibles and, in particular, for COVID-19 outbreaks at the beginning of the pandemic (i.e., during the years 2019/2020) the disease-free fixed point is given by $\mathbf{X}_{st} = (N, 0)$, that is, $S_{st} = N$. In this special case, we have $\lambda_1 = 0$ and $\lambda_2 = \beta - \gamma$ (see Eq. (4.19)) and

$$\alpha = \frac{\beta}{\gamma}. \tag{4.30}$$

The case $\alpha > 1$ or $\beta > \gamma$ implies $\lambda_2 > 0$ and corresponds to the case $\xi > 1$ of the stability parameter. The disease-free fixed point is unstable. An epidemic wave emerges. The alternative case $\beta < \gamma$ or $\alpha < 1$ implies $\lambda_2 < 0$ and corresponds to the case $\xi < 1$. In this case, the disease-free fixed point is neutrally stable and $I(t)$ decays monotonically as a function of time.

4.2.4 Special Case $\lambda_2 = 0$

For $\beta S_{st}/N = \gamma$ the second eigenvalue vanishes $\lambda_2 = 0$. The corresponding eigenvector $\mathbf{v}_2$ becomes $\mathbf{v}_2 = (-1, 0)$ (see Eq. (4.18)) and is linearly dependent with $\mathbf{v}_1$. Consequently, the two vectors do no longer span a two-dimensional plane and Eq. (4.21) does not hold. In this special case, the three-variable SIR model (3.16) reads $\mathrm{d}S/\mathrm{d}t = -\beta IS/N$, $\mathrm{d}I/\mathrm{d}t = \beta IS/N - \gamma I$, and $\mathrm{d}R/\mathrm{d}t = \gamma I$ and may be analyzed in a direct manner. Substituting $S = S_{st} + \delta$ into the evolution equation for I, where δ, I are assumed to be small quantities, we obtain

$$\frac{\mathrm{d}}{\mathrm{d}t} I = \left(\frac{\beta}{N} S_{st} - \gamma \right) I + \text{nonlin. terms.} \tag{4.31}$$

Consequently, the linearized equation reads $\mathrm{d}I/\mathrm{d}t = 0$, as expected from the fact that $\lambda_2 = 0$. This implies that I is constant over time (as long as the linear approximation holds). Accordingly, let us substitute $I = I_{const}$, where I_{const} is a constant, and $S = S_{st} + \delta$ into the evolution equation for δ as derived from $\mathrm{d}S/\mathrm{d}t = -\beta IS/N$ and let us neglect nonlinear terms, we obtain

$$\frac{\mathrm{d}}{\mathrm{d}t}\delta = -\frac{\beta}{N}(S_{st} I_{const} + \delta I_{const}) = -\gamma I_{const}, \quad \frac{\mathrm{d}}{\mathrm{d}t} R = \gamma I_{const}. \tag{4.32}$$

Above, the evolution equation of $R(t)$ is also presented. Summarizing Eq. (4.32), the dynamics of the linearized system is such that the number of infectious I remains constant. Susceptibles S decay linearly. Recovered R increase linearly. That is, there is a flow from susceptibles to infectious and from infectious to recovered individuals like

$$I = I_{const}, \quad S(t) = S_{st} - \gamma I_{const} t, \quad R(t) = R_{st} + \gamma I_{const} t. \tag{4.33}$$

This dynamics featuring linear changes over time holds as long as δ is small, that is, $\gamma I_{const} t \ll S_{st}$ holds. When there is a substantial decay of S away from S_{st}, then the linearized model fails and nonlinear terms become important. In particular, let us assume that at the time point t^* there is a substantial decay of S with $S(t^*) < S_{st}$. This implies that $\beta S(t^*)/N < \gamma$. Consequently, if we consider the fixed point $\mathbf{X}_{st} = (S(t^*), 0)$ with $S_{st} = S(t^*)$, then the fixed point exhibits an eigenvalue $\lambda_2 < 0$. Consequently, as soon as the linear flow approximation with $I = I_{const}$ does no longer hold, the epidemic subsides in the sense that $I(t)$ decays over time.

4.2.5 Nonlinear Parts G_k: Scalar Calculation Method

As discussed in Chap. 2 and reiterated in Sect. 4.1.1, the SIR model (4.7) can be equivalently expressed in terms of amplitude equations of the form $\mathrm{d}A_k/\mathrm{d}t = \lambda_k A_k + G_k$. In Eq. (4.26) the linear parts of those amplitude equations are shown. Let us derive the nonlinear parts G_k for the special case of infectious disease outbreaks in completely susceptible populations: $S_{st} = N$. To this end, we will use the scalar calculation method presented in Sect. 2.9.2.

To begin with, let us differentiate the mapping $A_1 = S + I/g - N$ (see Eq. 4.24)) with respect to time to determine the evolution equation of A_1, which leads to $\mathrm{d}A_1/\mathrm{d}t = \mathrm{d}S/\mathrm{d}t + \mathrm{d}I/\mathrm{d}t/g$. Substituting the evolution equations of the SIR model (4.7) into the terms on the right-hand side, it follows that

$$\frac{\mathrm{d}}{\mathrm{d}t}A_1 = -\frac{\beta}{N}IS + \frac{1}{g}\frac{\beta}{N}IS - \frac{1}{g}\gamma I = (-1 + \frac{1}{g})\frac{\beta}{N}IS - \frac{\gamma}{g}I. \tag{4.34}$$

Using the mappings $S = N + A_1 - A_2/\sqrt{1+g^2}$ and $I = gA_2/\sqrt{1+g^2}$ (see Eq. (4.23) for $S_{st} = N$), the product term IS can be expressed as

$$IS = \frac{g}{\sqrt{1+g^2}}A_2\left(N + A_1 - \frac{A_2}{\sqrt{1+g^2}}\right). \tag{4.35}$$

Substituting Eq. (4.35) into Eq. (4.34), we obtain

$$\begin{aligned}\frac{\mathrm{d}}{\mathrm{d}t}A_1 &= (\frac{1-g}{g})\frac{\beta}{N}\frac{g}{\sqrt{1+g^2}}A_2\left(N + A_1 - \frac{A_2}{\sqrt{1+g^2}}\right) - \frac{\gamma}{\sqrt{1+g^2}}A_2 \\ &= \left(\frac{(1-g)\beta}{\sqrt{1+g^2}} - \frac{\gamma}{\sqrt{1+g^2}}\right)A_2 + \frac{1-g}{\sqrt{1+g^2}}\frac{\beta}{N}A_2\left(A_1 - \frac{A_2}{\sqrt{1+g^2}}\right).\end{aligned} \tag{4.36}$$

Note that from the previous consideration it follows that the linearized evolution equation reads $\mathrm{d}A_1/\mathrm{d}t = \lambda_1 A_1 = 0$. That is, the equation for A_1 does not exhibit a linear term in A_2. Consequently, the coefficient of the linear A_2 term occurring in Eq. (4.36) must vanish. Let us check that this is indeed the case by computing

$$(1-g)\beta = (1 - \frac{\beta-\gamma}{\beta})\beta = \gamma. \tag{4.37}$$

Consequently, from Eq. (4.36) it follows

$$\frac{\mathrm{d}}{\mathrm{d}t}A_1 = \frac{1-g}{\sqrt{1+g^2}}A_2 p(A_1, A_2) = C_1 A_2 p(A_1, A_2), \tag{4.38}$$

where p is a function linear in A_1 and A_2 defined by

$$p = \frac{\beta}{N}\left(A_1 - \frac{A_2}{\sqrt{1+g^2}}\right) \leq 0. \tag{4.39}$$

As indicated $p \leq 0$ holds, which follows from $S = N + A_1 - A_2/\sqrt{1+g^2} \Rightarrow p = \beta(S-N)/N$ (see Eq. (4.23) for $S_{st} = N$) and $S - N \leq 0$. Moreover, the factor $C_1 = (1-g)/\sqrt{1+g^2}$ occurring in Eq. (4.38) reads $C_1 = \beta/[\gamma\sqrt{1+g^2}] > 0$ and is positive (as indicated) for $\beta, \gamma > 0$. Alternatively, Eq. (4.38) can be written like

$$\frac{\mathrm{d}}{\mathrm{d}t}A_1 = G_1(A_1, A_2),\ G_1 = \frac{1-g}{\sqrt{1+g^2}}A_2\frac{\beta}{N}\left(A_1 - \frac{A_2}{\sqrt{1+g^2}}\right). \tag{4.40}$$

The evolution equation for A_2 can be derived by analogy. To begin with, let us differentiate the mapping $A_2 = \sqrt{1+g^2}I/g$, which gives $\mathrm{d}A_2/\mathrm{d}t = g^{-1}\sqrt{1+g^2}\mathrm{d}I/\mathrm{d}t$. Substituting the evolution equation for $I(t)$ of the SIR model (4.7) into this result, we obtain

$$\begin{aligned}\frac{\mathrm{d}}{\mathrm{d}t}A_2 &= \frac{\sqrt{1+g^2}}{g}\frac{\mathrm{d}}{\mathrm{d}t}I = \frac{\sqrt{1+g^2}}{g}\left(\frac{\beta}{N}IS - \gamma I\right)\\ &= \frac{\sqrt{1+g^2}}{g}\frac{\beta}{N}IS - \gamma A_2.\end{aligned} \tag{4.41}$$

Substituting Eq. (4.35) into Eq. (4.41), we obtain

$$\begin{aligned}\frac{\mathrm{d}}{\mathrm{d}t}A_2 &= \frac{\beta}{N}A_2\left(N + A_1 - \frac{A_2}{\sqrt{1+g^2}}\right) - \gamma A_2\\ &= (\beta-\gamma)A_2 + \frac{\beta}{N}A_2\left(A_1 - \frac{A_2}{\sqrt{1+g^2}}\right)\\ &= \lambda_2 A_2 + A_2 p(A_1, A_2),\end{aligned} \tag{4.42}$$

where p is again given by Eq. (4.39). Alternatively, Eq. (4.42) can be written like

$$\frac{\mathrm{d}}{\mathrm{d}t}A_2 = \lambda_2 A_2 + G_2(A_1, A_2),\ G_2 = \frac{\beta}{N}A_2\left(A_1 - \frac{A_2}{\sqrt{1+g^2}}\right). \tag{4.43}$$

The amplitude equations may be expressed with the help of the the rate constant k_0 and the relative state δ that read

$$k_0(A_2) = \frac{\beta}{N}I = \frac{\beta}{N}\frac{g}{\sqrt{1+g^2}}A_2 \tag{4.44}$$

and

$$\delta(A_1, A_2) = S(A_1, A_2) - N = A_1 - \frac{A_2}{\sqrt{1+g^2}}, \tag{4.45}$$

respectively. From Eqs. (4.40) and (4.25) it then follows that

$$\begin{aligned} \frac{\mathrm{d}}{\mathrm{d}t}A_1 &= U_1 k_0(A_2)\delta(A_1, A_2), \\ \frac{\mathrm{d}}{\mathrm{d}t}A_2 &= \lambda_2 + U_2 k_0(A_2)\delta(A_1, A_2) \end{aligned} \tag{4.46}$$

with $U_1 = (1-g)/g$ and $U_2 = \sqrt{1+g^2}/g$.

4.2.6 SIR Model State Space and Amplitude Equations: Equivalence, SI Order Parameter, and Case $\lambda_2 > 0$

Equivalence

Let us summarize the state space equations and amplitude equations of the SIR model. The state space equations for S and I of the SIR model read (4.7)

$$\frac{\mathrm{d}}{\mathrm{d}t}S = -\frac{\beta}{N}IS, \quad \frac{\mathrm{d}}{\mathrm{d}t}I = \frac{\beta}{N}IS - \gamma I. \tag{4.47}$$

The third model variable R can be computed from $R(t) = N - S(t) - I(t)$. Within the state space description, the initial conditions at time t_0 are given by $S(t_0) = S_0$ and $I(t_0) = I_0$. The amplitude equations of the SIR model as presented above and derived in Refs. [2, 3] read (see Eqs. (4.38), (4.39), and (4.42))

$$\begin{aligned} \frac{\mathrm{d}}{\mathrm{d}t}A_1 &= C_1 A_2 p(A_1, A_2), \\ \frac{\mathrm{d}}{\mathrm{d}t}A_2 &= \lambda_2 A_2 + A_2 p(A_1, A_2), \\ p &= \frac{\beta}{N}\left(A_1 - \frac{A_2}{\sqrt{1+g^2}}\right) \end{aligned} \tag{4.48}$$

with $p \leq 0$, $C_1 = (1-g)/\sqrt{1+g^2} > 0$, $g = \lambda_2/\beta$, and

$$\lambda_2 = \beta - \gamma = \gamma(\alpha - 1) \tag{4.49}$$

(see Eqs. (4.19) and (4.30)), where α is the bifurcation parameter. As far as the amplitude space description is concerned, the initial conditions can be computed from

$$A_1(t_0) = S_0 + \frac{I_0}{g} - N, \ A_2(t_0) = \frac{\sqrt{1+g^2}}{g} I_0 \tag{4.50}$$

(see Eq. (4.24)). Solutions $S(t)$ and $I(t)$ computed from Eq. (4.47) for S_0 and I_0 are equivalent to solutions $A_1(t)$ and $A_2(t)$ computed from Eq. (4.48) for $A_1(t_0)$ and $A_2(t_0)$ obtained from Eq. (4.50) in the sense that they describe the same epidemic. They just view the epidemic from two different perspectives: the state space perspective and the amplitude space perspective. In other words, they describe the same spread of the infectious disease by means of two different coordinate systems: a coordinate system with orthogonal basis $\mathbf{e}_1 = (1, 0)$ and $\mathbf{e}_2 = (0, 1)$ and another coordinate system with a non-orthogonal basis given by $\mathbf{e}_1 = (1, 0)$ and $\mathbf{v}_2$ (see Eq. (4.20)) and an origin shifted to $\mathbf{X}_{st} = (N, 0)$. This equivalence of the two model equations (4.47) and (4.48) or the two different perspectives or descriptions, can be illustrated by computing $S(t)$, $I(t)$ by means of $A_1(t)$, $A_2(t)$ via the mapping Eq.(4.23) like

$$S(t) = N + A_1(t) - \frac{A_2(t)}{\sqrt{1+g^2}}, \ I(t) = \frac{g}{\sqrt{1+g^2}} A_2(t). \tag{4.51}$$

Since the nonlinear amplitude equation model (4.48) is not an approximation and in this sense is an exact model, solutions $S(t)$, $I(t)$ of the state space equations (4.47) and solutions $S(t)$, $I(t)$ obtained by means of Eq. (4.51) and the amplitude equations (4.48) are identical. However, close to the fixed point $\mathbf{X}_{st} = (N, 0)$, that is, for $\mathbf{X} = (S, I) \approx \mathbf{X}_{st} = (N, 0)$, the amplitude equations listed in Eq. (4.48) may be replaced by their linear approximations given in terms of $\mathrm{d}A_1/\mathrm{d}t = 0$ and $\mathrm{d}A_2/\mathrm{d}t = \lambda_2 A_2$. Solutions $S(t)$ and $I(t)$ obtained via the mapping (4.51) are then approximations to the solutions computed directed from the state space equations (4.47).

Figure 4.1 present simulation results of the state space equations (4.47) and amplitude equations (4.48) of the SIR model. Panel (a) of Fig. 4.1 shows simulation results for an SIR model characterized by a stable disease-fixed point $\mathbf{X}_{st} = (N, 0)$ with $\alpha < 1$. The solutions $S(t)$ and $I(t)$ computed from the state space equations (4.47) are shown as solid lines. The model parameters and initial conditions are used as in panel (a) of Fig. 3.2. Therefore, the functions $I(t)$ in panel (a) of Fig. 4.1 and panel (a) of Fig. 3.2 are identical. Importantly, panel (a) of Fig. 4.1 presents the functions $S(t)$ and $I(t)$ as obtained by solving the amplitude equations (4.48) under the appropriate initial conditions and mapping the solutions $A_1(t)$ and $A_2(t)$ to $S(t)$ and $I(t)$ via Eq. (4.51). The solutions thus obtained are shown as full circles. As expected, they are identical with the solutions $S(t)$ and $I(t)$ computed directly from the state space equations.

Panel (b) shows the simulation results for a SIR model with an unstable disease-free fixed point $\mathbf{X}_{st} = (N, 0)$ characterized by $\alpha > 1$. The solid lines show again the solutions $S(t)$ and $I(t)$ computed from Eq. (4.47), whereas the full circles show the solutions $S(t)$ and $I(t)$ computed from Eqs. (4.48) and (4.51). As expected, the two approaches produced identical results. Note that for the simulation results presented in panel (b) the same model parameters were used as in panel (a) of Fig. 3.3. Therefore, curves $S(t)$ and $I(t)$ for the epidemic wave shown in panel (b) Fig. 4.1 are identical to the curves $S(t)$ and $I(t)$ presented in panel (a) of Fig. 3.3.

The SIR model in the state space description (4.47) exhibits in the two-dimensional S-I state space the fixed points $\mathbf{X}_{st} = (S_{st}, 0)$ that correspond to disease-free states. The SIR model in the amplitude space description (4.48) exhibits the fixed points $\mathbf{A}_{st} = (A_{1,st}, 0)$. That is, the stationary value of A_2 is $A_{2,st} = 0$, whereas A_1 can assume any value $A_{1,st}$ (see Eq. (4.48)). However, since the amplitude equations (4.48) are interpreted in the context of the state space equations (4.47) the stationary value $A_{1,st}$ is subjected to some limitations. From the mapping $A_1, A_2 \to S$ shown in Eq. (4.51) with $A_1 = A_{1,st}$, $A_2 = A_{2,st} = 0$, and $S = S_{st}$ it follows that $A_{1,st} = S_{st} - N$, which implies $A_{1,st} \in [-N, 0]$ (since $S_{st} \in [0, N]$). In summary, the fixed points of the SIR model in state space and amplitude space read and are related to each other like

$$\mathbf{X}_{st} = (S_{st}, 0),\ S_{st} \in [0, N],\ \mathbf{A}_{st} = (A_{1,st}, 0),\ A_{1,st} \in [-N, 0],$$
$$A_{1,st} = S_{st} - N. \tag{4.52}$$

Note that the amplitude equations (4.48) have been derived by taking the fixed point $\mathbf{X}_{st} = (N, 0)$ as reference point. This does not mean that they only hold if the system is initially close to $\mathbf{X}_{st} = (N, 0)$. As pointed out above, the amplitude equations (4.48) are mathematically equivalent to the state space equations (4.47) regardless of the initial values S_0 and I_0 under consideration. That is, if the initial

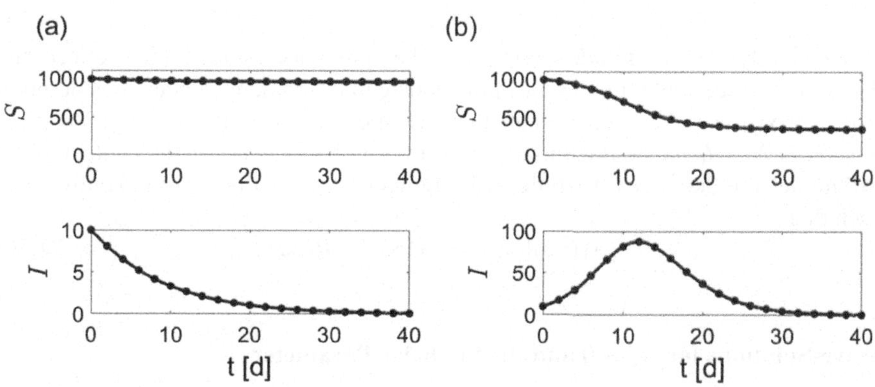

Fig. 4.1 Monotonically decaying solution (panel (**a**)) and wave-solution (panel (**b**) of the SIR model as obtained from state space (4.48) and amplitude space (4.51) descriptions. Solutions $S(t)$ and $I(t)$ computed from Eq. (4.48) (solid lines) and indirectly by means of Eq. (4.51) (full circles) are shown for $\alpha < 1 \Rightarrow \lambda_2 < 0$ (panel (**a**)) and $\alpha > 1 \Rightarrow \lambda_2 > 0$ (panel (**b**)). Parameters and initial conditions as in Figures 3.3 and 3.4: $N = 1,000$, $\gamma = 0.5$/d, $I(0) = 10$, $S(0) = N - I(0)$, $\beta = 0.4$/d (for $\alpha < 1$), and $\beta = 0.8$/d (for $\alpha > 1$)

state does not correspond to a state close to $\mathbf{X}_{st} = (N, 0)$, then the two descriptions still yield identical solutions $S(t)$ and $I(t)$ just like in Fig. 4.1. However, if the initial state is located in the vicinity of $\mathbf{X}_{st} = (N, 0)$, then this implies that during an initial period the linearized equations $\mathrm{d}A_1/\mathrm{d}t = 0$ and $\mathrm{d}A_2/\mathrm{d}t = \lambda_2 A_2$ describe a good approximation of the dynamics of the system (as mentioned above). Moreover, the SIR model order parameter $\mathbf{v}_2$ determines in the case of an instability with $\lambda_2 > 0$ the system dynamics, as will be shown below. In contrast, if the initial state is not located in the vicinity of $\mathbf{X}_{st} = (N, 0)$, then the full nonlinear amplitude equations must be considered in order to arrive within the amplitude space perspective at an accurate description of the dynamics of the system. For sake of completeness, let us point out that the reference point of the amplitude equation description (4.48) is defined in state space and amplitude space like

$$\mathbf{X}_{st} = (N, 0) \Leftrightarrow \mathbf{A} = (A_1, A_2) = (0, 0). \tag{4.53}$$

In general, both for monotonically decaying solutions $I(t)$ and wave-solutions $I(t)$ (see Sect. 3.5.1) as exemplified in Fig. 4.1, the dynamics starts at an initial location that is described in state space and amplitude space like

$$\mathbf{X}_0 = (S_0, I_0) \Leftrightarrow \mathbf{A}_0 = (A_1(t_0), A_2(t_0)), \tag{4.54}$$

where Eq. (4.50) holds. From this location, the dynamics evolves towards one of the fixed points of the SIR model as described in Eq. (4.52). That is, for $t \to \infty$ the dynamics converges to a location that is described in state space and amplitude space like

$$\mathbf{X}(t \to \infty) = (S(\infty), 0), \ \mathbf{A}(t \to \infty) = (A_1(\infty), 0), \ A_1(\infty) = S(\infty) - N. \tag{4.55}$$

Consequently, the magnitude $|A_1(\infty)|$ of the stationary value $A_1(\infty)$ reflects the decrease of susceptibles during the entire course of the epidemic under consideration like $|A_1(\infty)| = N - S(\infty)$ (which includes the drop in S from $S = S_{st} = N$ to $S = S_0 = N - I_0$ due to the initially infectious individuals I_0). In the context of the SIR model, this decreases also reflects the number of eventually recovered individuals such that

$$|A_1(\infty)| = N - S(\infty) = R(\infty). \tag{4.56}$$

Wave-Solutions for $\lambda_2 > 0$ and the SI Order Parameter

For bifurcation parameters $\alpha > 1$, we have $\lambda_2 > 0$ and an unstable fixed point $\mathbf{X}_{st} = (N, 0)$. In what follows we assume that the initial state $\mathbf{X}_0 = (S_0, I_0)$ is located in the vicinity of the unstable fixed point. It is plausible to assume that this situation was given during COVID-19 outbreaks at the beginning of the pandemic in the years 2019/2020 (some examples will be presented in Sect. 4.5 and Chap. 5). Let us follow the considerations in Ref. [2].

For $\alpha > 1$ it follows $\beta > \gamma$ and $g > 0$. This implies that $A_2 \geq 0$ holds (see the mapping $I \to A_2$ in Eq. (4.24) that also holds for $S_{st} = N$ and take into account that $I \geq 0$). However, the amplitude A_1 can be positive or negative (see the mapping $S, I \to A_1$ in Eq. (4.24) and put $S_{st} = N$). For example, for $I = 1$ and $S_{st} = N$ Eq. (4.24) reads $A_1 = S - N + 1/g$ such that in the limiting case $\beta \to \gamma \Rightarrow g \to 0$ we have $A_1 > 0$ (irrespective of S and N). In contrast, for $\beta \to \infty \Rightarrow g \to 1$ we have $A_1 < 0$ if $S < N - 1$.

For initial states $\mathbf{X}_0 = (S_0, I_0)$ close to $\mathbf{X}_{st}$ during an initial period the trajectory $\mathbf{X}(t) = (S(t), I(t))$ remains close to $\mathbf{X}_{st}$. We introduce a small parameter ϵ with $\epsilon \ll N$ and assume that during the initial period $I \propto O(\epsilon)$, $N - S \propto O(\epsilon)$. From Eq. (4.24) with $S_{st} = N$ it then follows that $A_1 \propto O(\epsilon)$ and $A_2 \propto O(\epsilon)$. From the definition of p (see Eq. (4.48)) it follows that $p \propto O(\epsilon)\beta/N$, where β/N is typically a relatively small quantity. In particular, in applications typically $\beta/N \ll \lambda_2$ holds (e.g., see Sect. 4.5). Consequently, as far as the initial dynamics of A_1 is concerned, from Eq. (4.48) it follows that $\mathrm{d}A_1/\mathrm{d}t \propto O(\epsilon^2)\beta/N$. In contrast, with respect to the initial dynamics of A_2 we have $\mathrm{d}A_2/\mathrm{d}t \propto \lambda_2 O(\epsilon)$ and we assume that $\lambda_2 \gg \beta/N$. This implies that the amplitude A_1 varies initially to a small amount as compared to the amplitude A_2. Consequently, the dynamics along $\mathbf{v}_2$ as described by A_2 dominates initially the overall dynamics and, in doing so, determines the course of the disease outbreak under consideration. As discussed in Sect. 2.9, in general, the unstable eigenvector and the corresponding unstable amplitude determine the dynamics of a system close to an instability. In this context, the unstable eigenvector is called the order parameter and the unstable amplitude is called the order parameter amplitude. Accordingly, $\mathbf{v}_2$ is the order parameter [2, 3] in the susceptible-infectious two-dimensional state space. For sake of brevity, $\mathbf{v}_2$ will be referred to as SI order parameter. The amplitude A_2 is the order parameter amplitude [2, 3].

As argued above, the order parameter amplitude A_2 increases initially in an exponential manner satisfying the linear dynamics $\mathrm{d}A_2/\mathrm{d}t = \lambda_2 A_2$, while variations in A_1 can be neglected such that A_1 can be regarded as constant. From Eq. (4.21) it then follows that [3]

$$\frac{\mathrm{d}}{\mathrm{d}t}\mathbf{X} \approx \mathbf{v}_2 \frac{\mathrm{d}}{\mathrm{d}t} A_2 \Rightarrow \Delta\mathbf{X} \approx \mathbf{v}_2 \Delta A_2 \tag{4.57}$$

with $\Delta\mathbf{X} = \mathbf{X}(t) - \mathbf{X}_0$ and $\Delta A_2 = A_2(t) - A_2(t_0)$, where $\mathbf{X}_0$ and t_0 are the initial state and time point, respectively. Equation (4.57) may be expressed like

$$\mathbf{X}(t) \approx \mathbf{v}_2 \Delta A_2 + \mathbf{X}_0. \tag{4.58}$$

It is worth noting that Eq. (4.58) can be derived from Eq. (4.21) in an alternative way. Assuming that A_1 is approximately constant such that $A_1(t) \approx A_1(0)$ with $t_0 = 0$, then from Eq. (4.21) it follows that $\mathbf{X}(t) \approx \mathbf{X}_{st} + \mathbf{v}_2 A_2(t) + \mathbf{v}_1 A_1(0)$. Putting $A_2(t) = A_2(t) - A_2(0) + A_2(0) = \Delta A_2 + A_2(0)$, we obtain $\mathbf{X}(t) \approx \mathbf{v}_2 \Delta A_2 + \mathbf{X}_{st} + \mathbf{v}_1 A_1(0) + \mathbf{v}_2 A_2(0)$. This result is equivalent to Eq. (4.58) because

$$\mathbf{X}_0 = \mathbf{X}_{st} + \mathbf{v}_1 A_1(0) + \mathbf{v}_2 A_2(0) \tag{4.59}$$

holds (see Eq. (4.21) again and put $t = t_0 = 0$).

If the initial state $\mathbf{X}_0$ corresponds to a small perturbation out of the fixed point $\mathbf{X}_{st}$, then in Eq. (4.58) the initial state approximately corresponds to the fixed point ($\mathbf{X}_0 \approx \mathbf{X}_{st}$) and $A_1(t_0) = A_2(t_0) \approx 0 \Rightarrow \Delta A_k \approx A_k$ such that Eq. (4.58) becomes

$$\mathbf{X}(t) \approx \mathbf{v}_2 A_2(t) + \mathbf{X}_{st}. \tag{4.60}$$

Let us return to the somewhat more general expression (4.57). Equation (4.57) implies that the direction given by $\mathbf{v}_2$ determines the relative changes in the sizes of compartments S and I like

$$\frac{\Delta S}{\Delta I} \approx \frac{v_{2,S}}{v_{2,I}} = -\frac{\beta}{\beta - \gamma}. \tag{4.61}$$

Figure 4.2 illustrates the considerations above and shows a wave-solution of the SIR model in state space (panel (a)) and amplitude space (panel (b)) and the corresponding trajectory together with the order parameter in the S-I plane (panel (c)). Panel (a) of Fig. 4.2 show $S(t)$ and $I(t)$ computed from Eq. (4.47). The same parameters as used in Fig. 3.3 and panel (b) of Fig. 4.1 were used (i.e., the curves shown in panel (a) of Fig. 3.3, panel (b) of Fig. 4.1, and panel (a) of Fig. 4.2 are identical). As can be seen in panel (a) of Fig. 4.2, $S(t)$ decays monotonically over time, while $I(t)$ increases initially, reaches a maximum $I_{\max}$, and subsequently decays monotonically to zero (see also Sect. 3.5.1). Panel (b) shows A_1 and A_2 computed from Eq. (4.48).

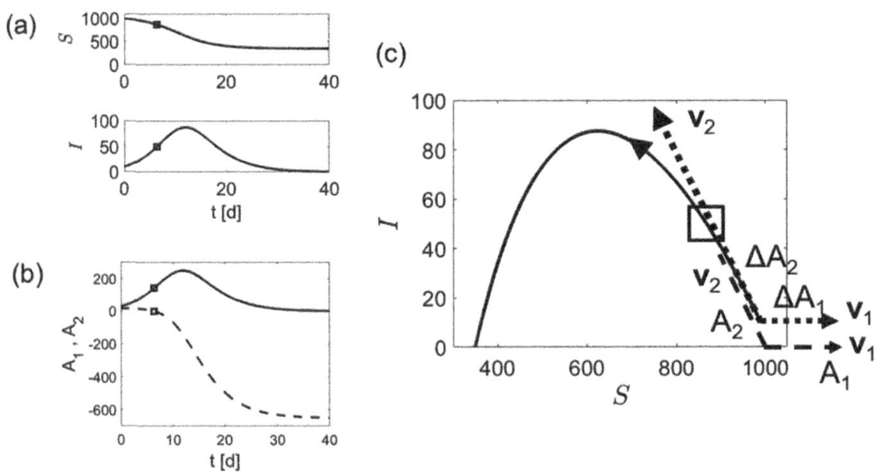

Fig. 4.2 Characteristic features of wave-solutions of the SIR model for $\lambda_2 > 0$ (i.e., $\alpha > 1$). Panel (**a**): wave-solution in terms of functions $S(t)$ and $I(t)$. Panel (**b**): wave-solution in terms of amplitudes $A_1(t)$ (dashed line) and $A_2(t)$ (solid line). Panel (**c**): wave-solution as phase curve $I(S)$ determined initially by $\mathbf{v}_2$ and $A_2(t)$. Parameters and initial conditions as for the simulation shown in panel (**b**) of Fig. 4.1

As expected, $A_2(t) > 0$ holds any time. Importantly, during an initial period, A_2 increases monotonically (and exponentially), while A_1 does not vary substantially over time. Only at a later time point t^*, A_1 starts to decay. A_1 decays monotonically, which follows from Eq. (4.48). More precisely, from Eq. (4.48) it follows that

$$\frac{\mathrm{d}}{\mathrm{d}t}A_1 = C_1 A_2\, p \leq 0 \tag{4.62}$$

because $p \leq 0$ (see Eq. 4.39), $C_1 > 0$ (see the comment below Eq. (4.39)), and $A_2 > 0$. Returning to A_2, the function A_2 increases to a maximum $A_{2,\max}$ at a time point t_p at which

$$t_p \ : \ p(A_1(t_p), A_2(t_p)) = -\lambda_2,\ A_2(t_p) = A_{2,\max} \tag{4.63}$$

holds. When A_1 decays (becomes "more negative") for $t > t_p$, then p decays as well (note that $p = \beta(S - N)/N$, which implies that p monotonically decreases). The decay in p implies $|p| > \lambda_2$ with $p < 0$ such $\lambda_2 + p < 0$ and A_2 decays like

$$\frac{\mathrm{d}}{\mathrm{d}t}A_2 = (\lambda_2 + p)A_2 < 0. \tag{4.64}$$

As a result, for any $t > t_p$ we have $\lambda_2 + p < 0$ and $\mathrm{d}A_2/\mathrm{d}t < 0$ (see Eq (4.64)) as long as $A_2 > 0$, which implies that $A_2(t)$ decays monotonically towards the fixed point $A_{2,st} = 0$.

Panel (c) presents the plot of $I(t)$ versus $S(t)$ in the S-I plane. The plot was shown previously in panel (b) of Fig. 3.3. In panel (c) of Fig. 4.2 the order parameter $\mathbf{v}_2$ as computed from Eq. (4.20) is shown two times located at two different positions: the reference fixed point $\mathbf{X}_{st} = (N, 0)$ and the initial state $\mathbf{X}_0 = (S_0, I_0)$. The eigenvector $\mathbf{v}_1 = (1, 0)$ is shown as well for both cases and forms the corresponding horizontal axes of the non-orthogonal basis systems spanned by $\mathbf{v}_1$ and $\mathbf{v}_2$. In the initial period, the trajectory $\mathbf{X}(t)$ follows the direction of the order parameter as stated by Eq. (4.57). When shifting $\mathbf{v}_2$ to the location $\mathbf{X}_0$ then distances along $\mathbf{v}_2$ correspond to ΔA_2, (see Eq. (4.58)) and the same holds for $\mathbf{v}_1$ and ΔA_1. As can be seen in panel (c), the changes ΔA_2 completely determine the time course of the epidemic in the initial state. At a certain time point t^* the trajectory $\mathbf{X}(t)$ branches off from the order parameter $\mathbf{v}_2$. A precise definition of the time point t^* will not be given here. Rather, the square shown in panel (c) indicates the region in the S-I plane obtained by visual inspection in which the branching off event takes place. The square shown in panel (c) corresponds to the squares shown in panels (a) and (b) and, in doing so, identifies approximately the time point t^*. As can be seen in panel (b), at the time point t^* the amplitude A_1 begins to increase in magnitude (i.e., starts to decay). This increase in A_1 results in the branching off dynamics of the phase curve $I(S)$ from $\mathbf{v}_2$ illustrated in panel (c). Subsequent to this branching off event, $I(t)$ reaches its maximum. From that maximum value $I(t)$ decays monotonically. During the whole time course $S(t)$ decays monotonically. The combination of an "up and down movement" of $I(t)$ and a "leftwards movement" of $S(t)$ creates an inverted parabola in the S-I plane.

Panel (c) illustrates that $\mathbf{v}_2$ determines the relationship between size changes ΔS and ΔI of S and I, respectively, during the initial outbreak as described by Eq. (4.61). Moreover, the time course of S and I is quantitatively determined by the change ΔA_2 of the order parameter amplitude that involves an exponential increase of A_2. Equation (4.58) reads explicitly

$$\begin{aligned}
S(t) &\approx S_0 + v_{2,S}[A(t) - A(t_0)] = S_0 + v_{2,S}A_2(t_0)\,(\exp\{\lambda_2(t-t_0)\} - 1)\,,\\
I(t) &\approx I_0 + v_{2,I}[A(t) - A(t_0)] = I_0 + v_{2,I}A_2(t_0)\,(\exp\{\lambda_2(t-t_0)\} - 1)\,,\\
\Rightarrow\ I(t) &\approx v_{2,I}A_2(t_0)\exp\{\lambda_2(t-t_0)\} = I_0\exp\{\lambda_2(t-t_0)\}. \qquad (4.65)
\end{aligned}$$

That is, the number of infectious individuals increases exponentially just like the order parameter amplitude A_2. Note that the expression for $S(t)$ cannot be simplified in the same way. The reason for this is that $I_0 = v_{2,I}A_2(t_0)$ holds (see Eq. 4.23) or Eq. (4.59)). In contrast, $S_0 = S_{st} + A_1(t_0) + v_{2,S}A_2(t_0)$ holds (see Eq. 4.23) or Eq. (4.59)). That is, while the term $I_0 - v_{2,I}A_2(t_0)$ equals zero, the corresponding term $S_0 - v_{2,I}A_2(t_0) = S_{st} + A_1(t_0)$ does not necessary vanish.

As such Eq. (4.61) can be obtained in various alternative ways. For example, using the linearized versions of Eq. (4.7) at $\mathbf{X}_{st}$, Eq. (4.61) can be derived like

$$\mathrm{d}S = -\beta I\mathrm{d}t,\ \mathrm{d}I = (\beta-\gamma)I\mathrm{d}t\ \Rightarrow \frac{\mathrm{d}S}{\mathrm{d}I} = -\frac{\beta}{\beta-\gamma}. \qquad (4.66)$$

However, the systematic nonlinear physics approach via Eqs. (4.21) and (4.57) allows to interpret Eqs. (4.57) and (4.61) in the broader context of bifurcation-phenomena and the dynamics of systems close to instabilities (see Sects. 1.4 and 1.5).

4.2.7 Case $\lambda_2 < 0$ and the Impact of Nonlinear Terms

In the pervious Sect. 4.2.6, the case $\lambda_2 > 0$ leading to wave-solutions was discussed in the context of the amplitude space description of the SIR model. Let us turn to the case $\lambda_2 < 0$ related to bifurcation parameters $\alpha < 1$ and parameters $g < 0$. The linear stability analysis conducted in Sect. 4.2.3 suggest that the fixed point $\mathbf{X}_{st} = (N, 0)$ is a neutrally stable fixed point such that small perturbations out of the fixed point do not increase in magnitude. They decay to some degree over time and converge to one of the other neutrally stable fixed points. However, this conclusion was drawn under the assumption that the nonlinear terms in the evolution equation of A_1 for the case $\lambda_2 < 0$ do not result in a destabilization of the fixed point $\mathbf{X}_{st} = (N, 0)$. Therefore, let us discuss this case $\lambda_2 < 0$ in more detail.

First of all, from the mapping $I \to A_2$ in Eq. (4.24) (which also holds for $S_{st} = N$) and $g < 0$ it follows that $A_2 \leq 0$. Likewise, from the mapping $S, I \to A_1$ in Eq. (4.24), $S_{st} = N$, and $g < 0$, it follows that $A_1 = S - N - I/|g|$, which implies

$A_1 \leq 0$. Consequently, for $\lambda_2 < 0$ both amplitudes A_1 and A_2 are negative or zero. From the evolution equation of A_2 in Eq. (4.48) it then follows that

$$\frac{\mathrm{d}}{\mathrm{d}t} A_2 = \underbrace{\lambda_2 A_2}_{\geq 0} + \underbrace{A_2\, p}_{\geq 0} \tag{4.67}$$

(where we have used $\lambda_2 < 0$ and $p < 0$) such that $\mathrm{d}A_2/\mathrm{d}t > 0$ for $A_2 < 0$ and $\mathrm{d}A_2/\mathrm{d}t = 0$ for $A_2 = 0$. Consequently, A_2 increase monotonically and exhibits the limiting case $A_2(t \to \infty) = A_{2,st} = 0$. Likewise, from the evolution equation of A_1 in Eq. (4.48) it follows that

$$\frac{\mathrm{d}}{\mathrm{d}t} A_1 = C_1 A_2\, p \geq 0 \tag{4.68}$$

because of $C_1 > 0$, $p < 0$, and $A_2 \leq 0$. In particular, we see that $\mathrm{d}A_1/\mathrm{d}t \geq 0$ holds for $A_2 < 0$ and $\mathrm{d}A_1/\mathrm{d}t = 0$ holds for $A_2 = 0$. The nonlinear term $C_1 A_2 p$ in the case $\lambda_2 < 0$ indeed does not result in a increase of the magnitude $|A_1|$ of the amplitude A_1. Rather, the term results in a decrease of $|A_1|$ over time. Eventually, the limiting case $A_1(t \to \infty) = A_{1,st} \leq 0$ holds. In summary, the amplitudes A_1, A_2 are both negative and monotonically increasing functions of time:

$$\begin{aligned} \lambda_2 < 0 \Rightarrow\ & A_1, A_2 \leq 0,\ \frac{\mathrm{d}}{\mathrm{d}t} A_1 \geq 0,\ \frac{\mathrm{d}}{\mathrm{d}t} A_2 \geq 0, \\ & A_1(t \to \infty) = A_{1,st} \leq 0,\ A_2(t \to \infty) = A_{2,st} = 0. \end{aligned} \tag{4.69}$$

In other words, while the linear term in the evolution equation for A_1 vanishes such that in linear approximation we have $\mathrm{d}A_1/\mathrm{d}t = 0$, the nonlinear term G_1 in the evolution equation of A_1 for $\lambda_2 < 0$ makes that any perturbation of A_1 out of $A_{1,st} = 0$ (related to $S = S_{st} = N$) to an initial value $A_1(0) < 0$ decays in magnitude or remains constant: $\mathrm{d}|A_1(t)|/\mathrm{d}t \leq 0$. This was anticipated in Sect. 4.2.3 when conducting the stability analysis of the SIR model (4.7).

Figure 4.3 illustrates the amplitude space description of the SIR model for the case $\lambda_2 < 0$. Panel (a) of Fig. 4.3 shows the solutions $S(t)$ and $I(t)$ computed from Eq. (4.47) as presented earlier in panel (a) of Fig. 4.1. Unlike Fig. 4.1, panel (b) of Fig. 4.3 presents the corresponding solutions $A_1(t)$ (dashed line) and $A_2(t)$ (solid line) as computed from Eq. (4.48). As summarized in Eq. (4.69), both amplitudes A_1 and A_2 are negative and increases monotonically as functions of time. A_2 converges to $A_{2,st} = 0$. A_1 converges to a finite stationary value $A_{1,st} < 0$. The magnitude $|A_{1,st}|$ corresponds to the decrease of susceptibles $|A_{1,st}| = \Delta S = N - S(\infty)$ (see Eq. (4.55)) during the course of the simulated epidemic and the final cumulative number of recovered individuals $|A_{1,st}| = R(\infty)$.

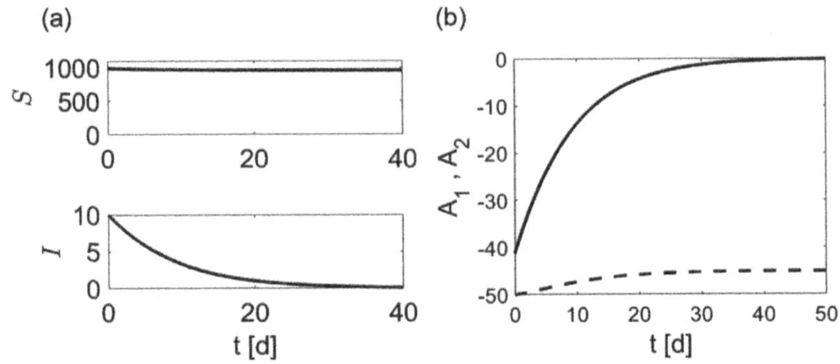

Fig. 4.3 Solutions of the SIR model in state space (panel (**a**)) and amplitude space (panel (**b**)) as obtained from Eqs. (4.47) and (4.48), respectively, for $\lambda_2 < 0$ (i.e., $\alpha < 1$). $S(t)$ decreases from $S_0 = 990$ to a stationary value $S(\infty) = 955$. The total decay $\Delta S = 45$ from $S = S_{st} = N = 1000$ to $S = S(\infty) = 955$ corresponds to the final number of recovered individuals $\Delta S = R(\infty)$. The stationary value $A_{1,st} = -45$ is the negative of that number. The solid (dashed) line in panel (**b**) stands for $A_2(t)$ ($A_1(t)$)

4.2.8 Fixed Points with $S_{st} < N$ and Nonlinear Parts G_k

Let us consider fixed points $\mathbf{X}_{st} = (S_{st}, 0)$ with $S_{st} < N$. In this case, the nonlinear terms G_k can be obtained following the approach presented in Sect. 4.2.5 for $\mathbf{X}_{st} = (N, 0)$. In doing so, it can be shown that for $\mathbf{X}_{st} = (S_{st}, 0)$ with $S_{st} < N$ the amplitude equations for A_1 and A_2 are defined by Eq. (4.48) again. However, for λ_2 and g the general definitions presented in Sect. 4.2.1 must be applied. That is, $\lambda_2 = \beta S_{st}/N - \gamma$ (see Eq. (4.13)) and $g = (\beta S_{st}/N - \gamma)/(\beta S_{st}/N)$ (see Eq. (4.17)). Note that p is defined as in Eq. (4.48). That is, the factor β/N occurring in p is not changed. The SIR model amplitude equations (4.48) for fixed points $\mathbf{X}_{st} = (S_{st}, 0)$ are solved under initial conditions computed from

$$A_1(0) = S_0 + \frac{I_0}{g} - S_{st}, \quad A_2(0) = \frac{\sqrt{1+g^2}}{g} I_0 \tag{4.70}$$

(see Eq. (4.24)). The solutions $S(t)$ and $I(t)$ of the state space equations (4.47) can be re-obtained by solving the amplitude equations (4.48) under the initial conditions (4.70) to arrive at $A_1(t)$ and $A_2(t)$ and computing $S(t)$ and $I(t)$ from $A_1(t)$ and $A_2(t)$ like

$$S(t) = S_{st} + A_1(t) - \frac{A_2(t)}{\sqrt{1+g^2}}, \quad I(t) = \frac{g}{\sqrt{1+g^2}} A_2(t) \tag{4.71}$$

(see Eq. (4.23)). In short, the state space equations (4.47) and amplitude equations (4.48) of the SIR model are equivalent descriptions irrespective of the reference fixed point $\mathbf{X}_{st} = (S_{st}, 0)$ that is considered.

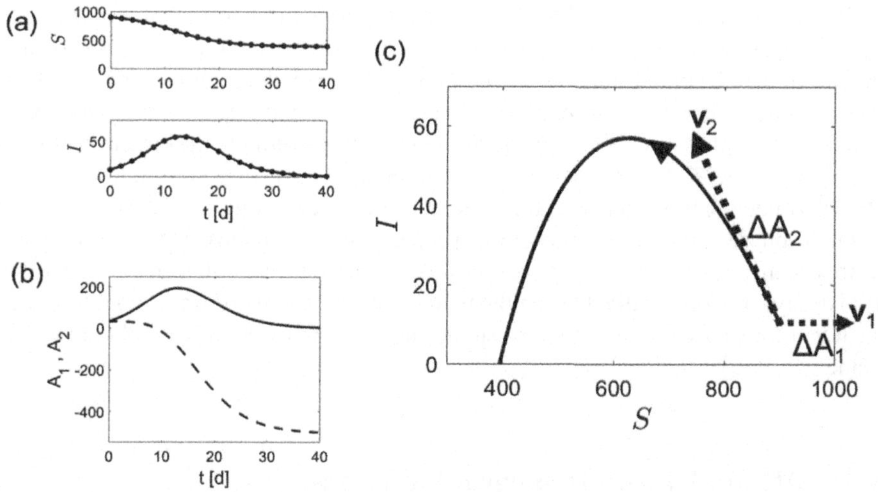

Fig. 4.4 Simulation results as in Fig. 4.2 for $\lambda_2 > 0$ (i.e., $\alpha > 1$) but for $S_{st} < N$. Panels (**a**), (**b**), and (**c**) show the same quantities as in Fig. 4.2. All parameters as for the simulation shown in Fig. 4.2: $N = 1,000$, $\beta = 0.8$/d, and $\gamma = 0.5$/d. Fixed point and initial conditions: $R_{st} = 100$, $S_{st} = N - R_{st}$, $I_{st} = 0$, $I(0) = 10$, $S(0) = S_{st}$, $R(0) = R_{st} - I(0)$

Equation (4.54) holds also for amplitude equations models with $\mathbf{X}_{st} = (S_{st}, 0)$. Equations (4.53) and (4.55) become

$$\mathbf{X}_{st} = (S_{st}, 0) \Leftrightarrow \mathbf{A} = (A_1, A_2) = (0, 0) \tag{4.72}$$

and

$$\mathbf{X}(t \to \infty) = (S(\infty), 0), \ \mathbf{A}(t \to \infty) = (A(\infty), 0), \ A_1(\infty) = S(\infty) - S_{st}, \tag{4.73}$$

respectively. The arguments made in Sects. 4.2.6 and 4.2.7 about wave-solutions for $\lambda_2 > 0$ and monotonically-decaying solutions for $\lambda_2 < 0$ hold. In particular, Eqs. (4.58) to (4.69) hold for the SIR amplitude equation model (4.48) based on reference fixed points $\mathbf{X}_{st} = (S_{st}, 0)$ with $S_{st} < N$.

Figure 4.4 illustrates the considerations for a reference fixed point $\mathbf{X}_{st} = (S_{st}, 0)$ with $S_{st} < N$ and a population with an unstable disease-free fixed point with $\lambda_2 > 0$ (i.e., $\alpha > 1$). Panels (a), (b), and (c) show the same quantities as panels (a), (b), and (c) of Fig. 4.2. Panel (a) of Fig. 4.4 shows the functions $S(t)$ and $I(t)$ computed from Eq. (4.47) as solid lines. Panel (b) shows the amplitudes A_1 and A_2 computed from Eq. (4.48). Using these amplitudes $A_1(t)$ and $A_2(t)$, the original state variables $S(t)$ and $I(t)$ were reconstructed by means of Eq. (4.71) and were plotted as circles in panel (a). As expected, the two approaches produced identical results.

Panel (c) shows $I(t)$ versus $S(t)$ as phase curve $I(S)$ in the S-I plane. The unstable eigenvector (i.e., order parameter) $\mathbf{v}_2$ computed from Eq. (4.18) is shown at the location of the initial state $\mathbf{X}_0 = (S_0, I_0)$ together with the eigenvector $\mathbf{v}_1 = (1, 0)$ (pointing in the horizontal direction). As can be seen in panel (c), the trajectory $\mathbf{X}(t)$ in terms of the phase curve $I(S)$ in the S-I plane follows initially the direction of $\mathbf{v}_2$ as stated by Eq. (4.57). During that initial period, changes ΔA_2 of the unstable amplitude A_2 determine the time course of the epidemic simulated in Fig. 4.4. At a certain time point the phase curve $I(S)$ branches off from the unstable eigenvector $\mathbf{v}_2$. Subsequent to this branching off event, $I(t)$ reaches its maximum. From that maximum value $I(t)$ decays monotonically. During the whole time course $S(t)$ decays monotonically such that the phase curve $I(S)$ corresponds again to an inverted parabola in the S-I plane.

4.3 SIR Model with Demographic Terms

The two-variable version of the SIR model with demographic terms (3.29) reads (see Eq. (3.33))

$$\frac{\mathrm{d}}{\mathrm{d}t}S = -\frac{\beta}{N}IS + \mu(N - S), \ \frac{\mathrm{d}}{\mathrm{d}t}I = \frac{\beta}{N}IS - (\gamma + \mu)I \tag{4.74}$$

for a population whose size is constant over time. Consequently, the third variable of the SIR model (i.e., the recovered individuals) can be obtained from $R(t) = N - S(t) - I(t)$. The two-dimensional state vector of Eq. (4.74) reads $\mathbf{X} = (S, I)$. The model exhibits the fixed point $\mathbf{X}_{st} = (N, 0)$ reflecting the disease-free state (see Sect. 3.5.2). In addition, there might be an endemic fixed point with $I_{st} > 0$ (see Sect. 3.5.2 again). Using the approach discussed in Sect. 4.2, for the SIR model (4.74) the amplitude equations

$$\begin{aligned} \frac{\mathrm{d}}{\mathrm{d}t}A_1 &= \lambda_1 A_1 + \frac{1-g}{\sqrt{1+g^2}} A_2 p(A_1, A_2), \\ \frac{\mathrm{d}}{\mathrm{d}t}A_2 &= \lambda_2 A_2 + A_2 p(A_1, A_2) \end{aligned} \tag{4.75}$$

with

$$\begin{aligned} &\lambda_1 = -\mu, \ \lambda_2 = \beta - \gamma - \mu, \\ &g = \frac{\beta - \gamma}{\beta}, \ p = \frac{\beta}{N}\left(A_1 - \frac{A_2}{\sqrt{1+g^2}}\right) \end{aligned} \tag{4.76}$$

can be derived. That is, comparing the amplitude equations (4.75) with the amplitude equations (4.48) for the SIR model without demographic terms, we see that formally in the evolution equation for A_1 the linear term $\lambda_1 A_1$ must be added. The eigenvalue λ_1 is negative and corresponds to the death rate $-\mu$. Furthermore, the eigenvalue λ_2 is modified by the death rate μ. The parameter g is identical to the g parameter of the SIR model without demographic terms based on the reference fixed point $\mathbf{X}_{st} = (N, 0)$. The linear function p is identical to the function p of the SIR model without demographic terms (see Eq. (4.48)). In summary, the eigenvalues λ_1 and λ_2 depend on the death rate μ and, consequently, are different for the cases $\mu = 0$ and $\mu > 0$. In contrast, g and p do not depend on μ and, consequently, correspond to the respective quantities derived for the SIR model without demographic terms based on $\mathbf{X}_{st} = (N, 0)$.

The amplitude equation description in terms of Eqs. (4.75) and (4.76) of the SIR model with demographic terms includes as special case the amplitude equation description (4.48) of the SIR model without demographic terms based on the stationary state $\mathbf{X}_{st} = (N, 0)$ when putting $\mu = 0$.

The initial conditions for the amplitude space description are given by Eq. (4.50). The amplitude equations (4.75) are equivalent to the state space equations (4.74) in the sense that the solution $S(t)$, $I(t)$ of Eq. (4.74) can be obtained by solving the amplitude equations (4.75) for the initial conditions (4.50) to obtain the solutions $A_1(t)$, $A_2(t)$ and computing $S(t)$, $I(t)$ from $A_1(t)$, $A_2(t)$ with the help of Eq. (4.51).

In view of the eigenvalues listed in Eq. (4.76) and the general discussion of the stability of fixed points presented in Sect. 2.7, the bifurcation parameter

$$\alpha = \frac{\beta}{\gamma + \mu} \tag{4.77}$$

with a critical value of 1 can be introduced and the following two cases can be distinguished. First, for $\alpha < 1$, which implies $\beta < \gamma + \mu$ and $\lambda_2 < 0$ the fixed point $\mathbf{X}_{st} = (N, 0)$ is asymptotically stable because both eigenvalues λ_1, λ_2 are negative. The fixed point corresponds to a stable node. Second, for $\alpha > 1$, which implies $\beta > \gamma + \mu$ and $\lambda_2 > 0$, the fixed point $\mathbf{X}_{st} = (N, 0)$ corresponds to an unstable saddle point because it is characterized by a negative and a positive eigenvalue. The bifurcation parameter α corresponds to the stability parameter $\xi = \beta/(\gamma + \mu)$ that was introduced in Eq. (3.37) and discussed in Sect. 3.5.2. In particular, as discussed in Sect. 3.5.2, in the case of $\alpha > 1$ (i.e., $\xi > 1$) the SIR model (4.74) exhibits an endemic fixed point $\mathbf{X}_{st}$ with $I_{st} > 0$. Fixing γ and μ and using β as bifurcation parameter (instead of α), the two cases ($\alpha < 1 \Rightarrow \beta < \gamma + \mu$ and $\alpha > 1 \Rightarrow \beta > \gamma + \mu$) can be connected with each other in the bifurcation diagram shown in Fig. 3.8. The bifurcation diagram shows the bifurcation of the SIR model at $\alpha = 1 \Rightarrow \beta = \gamma + \mu$ in terms of a change of the stability of the disease-free state and fixed point $\mathbf{X}_{st} = (N, 0)$ from a stable node to an unstable saddle. In addition, at the bifurcation point an attractor in terms of an endemic stable fixed point emerges.

4.4 SIR-Type Models Revisited: 2D Autonomous Amplitude Descriptions

In Sect. 4.1.2 the amplitude equation description of SIR-type models was formally defined by Eq. (4.4) using the two-dimensional amplitude space approach. We are now in the position to present the explicit form of Eq. (4.4). To this end, we consider the model (4.1) in the absence of demographic terms $\mu = B = 0$ but for an arbitrary reference fixed point $\mathbf{X}_{st} = (S_{st}, 0)$. In this case, from Eq. (4.48) and the discussion in Sect. 4.2.8 it follows that the amplitude equation description (4.4) reads explicitly

$$\begin{aligned} \frac{\mathrm{d}}{\mathrm{d}t} A_1 &= C_1 A_2 p(A_1, A_2), \\ \frac{\mathrm{d}}{\mathrm{d}t} A_2 &= \lambda_2 A_2 + A_2 p(A_1, A_2), \\ \frac{\mathrm{d}}{\mathrm{d}t} X_k &= N_k(S(A_1, A_2), I(A_2), X_3, \ldots, X_n) \end{aligned} \tag{4.78}$$

with

$$\lambda_2 = \beta \frac{S_{st}}{N} - \gamma, \; g = \frac{\lambda_2 N}{\beta S_{st}}, \; C_1 = \frac{1-g}{\sqrt{1+g^2}}, \; p = \frac{\beta}{N}\left(A_1 - \frac{A_2}{\sqrt{1+g^2}}\right) \tag{4.79}$$

and

$$\begin{aligned} S(A_1, A_2) &= S_{st} + v_{1,S} A_1 + v_{2,S} A_2 = S_{st} + A_1 - \frac{A_2}{\sqrt{1+g^2}}, \\ I(A_2) &= v_{2,I} A_2 = \frac{g}{\sqrt{1+g^2}} A_2, \end{aligned} \tag{4.80}$$

where the relations in Eq. (4.80) are copied from Eq. (4.71) and $v_{k,S}$, $v_{k,I}$ for $k = 1, 2$ are the eigenvector components ($v_{1,S} = 1$, $v_{1,I} = 0$, $v_{2,S} = -1/\sqrt{1+g^2}$, and $v_{2,I} = g/\sqrt{1+g^2}$) that have been derived in Sect. 4.2.1 (see Eq. (4.18)).

For example, the SIRD model (4.3) for $r = \beta/N$ involving the compartment D of individuals deceased from COVID-19 reads in the amplitude space description

$$\begin{aligned} \frac{\mathrm{d}}{\mathrm{d}t} A_1 &= C_1 A_2 p(A_1, A_2), \\ \frac{\mathrm{d}}{\mathrm{d}t} A_2 &= \lambda_2 A_2 + A_2 p(A_1, A_2), \\ \frac{\mathrm{d}}{\mathrm{d}t} R &= a\, v_{2,I} A_2, \\ \frac{\mathrm{d}}{\mathrm{d}t} D &= b\, v_{2,I} A_2 \end{aligned} \tag{4.81}$$

with $\gamma = a + b$. In Eq. (4.81) $N = S + I + R + D$ is constant. As such the variable N occurring in the factor β/N should reflect the size of a population and should not include disease-related deaths. Therefore, it is assumed that D makes only a negligibly small contribution to N.

As another application let us consider the SIQR model that was used by Pedersen and Meneghini to describe the COVID-19 first-wave epidemic in Italy [4]. They used the model [4]

$$\frac{\mathrm{d}}{\mathrm{d}t}S = -\frac{\beta}{N}IS, \ \frac{\mathrm{d}}{\mathrm{d}t}I = \frac{\beta}{N}IS - (a+b)I, \ \frac{\mathrm{d}}{\mathrm{d}t}Q = bI - k_Q Q, \ \frac{\mathrm{d}}{\mathrm{d}t}R = k_Q Q. \tag{4.82}$$

The compartment I describes non-diagnosed infectious individuals. In contrast, the compartment Q describes individuals diagnosed with COVID-19 that have been quarantined and cannot infected others. The compartment R describes recovered individuals or individuals deceased from COVID-19 out of the group of quarantined individuals. That is, according to the model, it is assumed that any individual who is diagnosed with COVID-19 is immediately quarantined and put in isolation such that the individual cannot infect others. Pedersen and Meneghini argued that this was at least in good approximation the case during the first wave of COVID-19 that hit Italy in February/March 2020.

The model (4.82) does not describe explicitly the group of non-diagnosed individuals recovered or deceased from COVID-19. This group, say R_2, satisfies the equation $\mathrm{d}R_2/\mathrm{d}t = aI$. That is, the coefficient a describes the transition rate from I to R_2, whereas the coefficient b describes the diagnosis rate and, consequently, the transition rate from I to Q. Finally, the coefficient k_Q describes both the combined recovery rate and death rate of quarantined individuals. The model has the advantage that it can be conveniently linked to data of confirmed COVID-19 cases [4]. Let $C = Q + R$ denote the diagnosed cases (quarantined, recovered, or deceased from COVID-19). Then, from Eq. (4.82) it follows that

$$\frac{\mathrm{d}}{\mathrm{d}t}C = bI. \tag{4.83}$$

The model involving S, I, and C formally has a similar structure as a SIR model. However, $S + I + Q + R = S + I + C$ is not constant. If the evolution equation $\mathrm{d}R_2/\mathrm{d}t = aI$ is added to the SIC model to obtain a SIC-R_2 model, then $N = S + I + C + R_2$ is constant and all individuals are mapped to one of the four compartments. In this context is should again be assumed that the number of disease-related deaths makes only a negligible contribution to N. The SIQR model (4.82) and the corresponding extended SIQR-R_2 model are SIR-type models of the form (4.2). From Eqs. (4.78)-(4.80) it follows that the amplitude equation descriptions of the SIQR and SIQR-R_2 models read

$$
\begin{aligned}
\frac{\mathrm{d}}{\mathrm{d}t}A_1 &= C_1 A_2 p(A_1, A_2),\\
\frac{\mathrm{d}}{\mathrm{d}t}A_2 &= \lambda_2 A_2 + A_2 p(A_1, A_2),\\
\frac{\mathrm{d}}{\mathrm{d}t}Q &= b\, v_{2,I} A_2 - k_Q Q,\\
\frac{\mathrm{d}}{\mathrm{d}t}R &= k_Q Q,\\
\frac{\mathrm{d}}{\mathrm{d}t}R_2 &= a\, v_{2,I} A_2
\end{aligned}
\tag{4.84}
$$

with $\gamma = a + b$, where the R_2 dynamics is ignored in the SIQR model. The SIQR model (4.82) has also been used in studies on the COVID-19 epidemic in Brazil during the year 2020 [5, 6]. A slightly revised version was used [5, 6]:

$$
\begin{aligned}
&\frac{\mathrm{d}}{\mathrm{d}t}S = -\frac{\beta}{N} IS - wS,\ \frac{\mathrm{d}}{\mathrm{d}t}I = \frac{\beta}{N} IS - (a+b)I,\\
&\frac{\mathrm{d}}{\mathrm{d}t}Q = bI - k_Q Q,\ \frac{\mathrm{d}}{\mathrm{d}t}R = k_Q Q + aI.
\end{aligned}
\tag{4.85}
$$

Here, the parameter $w \geq 0$ describes the impact of intervention measures that are assumed to reduce the number of susceptibles. However, it is plausible to assume that during the initial stage of the COVID-19 epidemic in Brazil the parameter w was zero or close to zero [5]. Putting $w = 0$, the key difference between the models (4.82) and (4.85) is that in Eq. (4.85) in the evolution equation for R the term aI occurs. Consequently, the compartment R and the lump variable $C = Q + R$ are somewhat difficult to relate to reported COVID-19 data. The compartment R contains both diagnosed and non-diagnosed cases. Likewise, the variable $C = Q + R$ that satisfies $\mathrm{d}C/\mathrm{d}t = (a + b)I$ reflects both diagnosed and non-diagnosed cases. The issue at hand is that non-diagnosed cases typically do not show up in COVID-19 records.

4.5 COVID-19 Outbreak in Italy 2020 and Its SI Order Parameter

The COVID-19 pandemic reached Italy in February 2020 (see Sect. 3.6.2). Active COVID-19 cases (defined as the number of individuals who were diagnosed with COVID-19 and had not yet recovered or died) increased dramatically, see Fig. 3.10. The active case count reached a peak of 108,000 individuals in April 2020, see Fig. 3.10. In Sect. 3.6.2 the study by Fanelli and Piazza [1] was presented that analyzed the very early period from February 11 to March 15 of the epidemic in Italy by means of a SIR model. As discussed in Sect. 3.6.2 the study had three disadvantages. First, only data from February 11 to March 15 were considered such that the SIR model describes a peak 4 times less than the actual peak of 108,000 cases. Second, the size of the susceptible population was not taken from demographic records and was considerably smaller than the population of Italy in the year 2020. Third, the variable I of the SIR model was identified with the active COVID-19 cases such that the presumably large group of non-diagnosed individuals with COVID-19 were not reflected by I. In what follows these three disadvantages will be overcome within the SIR and SIR-type modeling framework.

4.5.1 Active Cases Within the SIR Model Interpretation by Fanelli and Piazza (2020)

Let us address the first two issues listed above but keep the interpretation by Fanelli and Piazza of the variable I as active cases. Accordingly, let us apply the SIR model (4.7) to reported active cases [7] from Italy. The modeling results are summarized in Fig. 4.5. Panel (a) of Fig. 4.5 shows the active cases since February 7, 2020, over a period of 100 days. In general, when modeling COVID-19 epidemics in countries around the world, it has been acknowledged that structure parameters (like β, γ, λ_2) of populations, in which the epidemics under consideration take place, change over time, in particular, due to the implementations of intervention measures. We will return to this issue in Chap. 8. Therefore, in what follows it is assumed that the SIR model parameters β, γ and, consequently, λ_2 were approximately constant only during an initial stage of the COVID-19 epidemic in Italy. Under this assumption and the assumption that the linear approximation of the SIR model holds during the initial epidemic stage, the solution $I(t) = I_0 \exp\{\lambda_2(t - t_0)\}$ (see Eq. (4.65)) was fitted to the data. March 15 was used as end date of the initial stage (which happens to be the last day of the modeling period in the study by Fanelli and Piazza) and is indicated as vertical line in panel (a) of Fig. 4.5. The March 15 date can be put into the perspective of the whole first 100 days of the COVID-19 epidemic in Italy and can be identified at least by visual inspection as the date at which the exponential increase ended and switched over to a more linear increase. The linear increase is consistent with a bifurcation point characterized by $\lambda_2 = 0$ (see Sect. 4.2.4). The issue of a switch from $\lambda_2 > 0$ to a critical stage with $\lambda_2 = 0$ will be discussed in detail in Chap. 8.

The initial date t_0 was varied in the range from January 31 to February 20 to find the best fit of the data to the exponential function $I(t) = I_0 \exp\{\lambda_2(t - t_0)$. The best fit was obtained for t_0 corresponding to February 7. As a result, the period from February 7 to March 15 was fitted, which is a 38 days period. Panel (b) of Fig. 4.5 shows the reported active cases again during the 38 days period from February 7 to March 15 (gray circles). The best fit solution $I(t) = I_0 \exp\{\lambda_2(t - t_0)\}$ is shown as well (solid line; we will return to the full black circles below). Note that I_0 was not fitted but taken from the data [7] as the active cases on February 7 (i.e., $I_0 = 3$ as reported in Ref. [7]). As can be seen in panel (b) the model solution fits the data with moderate accuracy.

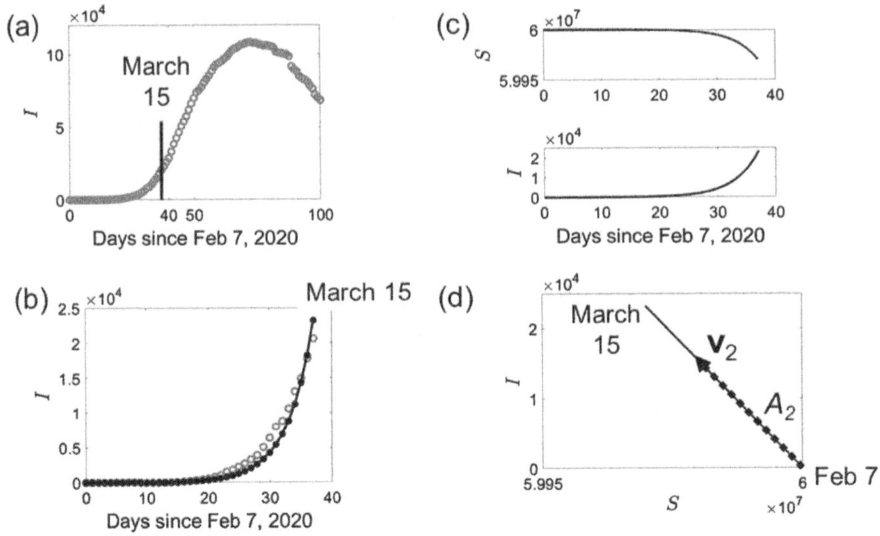

Fig. 4.5 Intermediate modeling step to describe active COVID-19 case data from Italy during the first COVID-19 wave in 2020 via the SIR model (4.7). Panel (**a**): active case data [7]. Panel (**b**): initial stage data (gray circles) and exponential fit (solid line). Panels (**c**) and (**d**): solutions S and I as function of time and as phase curve $I(S)$. In panel (**d**) the unstable eigenvector $\mathbf{v}_2$ is shown as well

The aforementioned data fitting procedure produced an eigenvalue $\lambda_2 = 0.24$/d. It was further assumed that the recovery rate γ was related to the mean recovery duration $T = 14$ days like $\gamma = 1/T = 0.07$/d (see Sect. 3.4). Consequently, $\lambda_2 = \beta - \gamma \Rightarrow \beta = \lambda_2 + \gamma = 0.31$/d. Importantly, the susceptible population was given by the population of Italy during the year 2020, which is about 60 million. That is, $N = S_{st} = 60,000,000$ was used. Using those model parameters and $I_0 = 3$, $S_0 = N - I_0$, the SIR model (4.7) was solved numerically. The solutions thus obtained are considered as the exact solutions of the SIR model that go beyond the linear approximation. The solutions $S(t)$ and $I(t)$ are shown in panel (c). As expected, $S(t)$ decayed monotonically. $I(t)$ increased in an exponential manner as expected from the linear approximation to the SIR model. In fact, the exact solution $I(t)$ obtained from the numerical simulation of the SIR model is shown in panel (b) as well (black full circles). By visual inspection, the difference between the exact solution (black full circles) and the solution $I(t) = I_0 \exp\{\lambda_2 t\}$ (black solid line) of the linear approximation of the SIR model is negligibly small. In this context, note that the initial state $\mathbf{X}_0 = (N - 3, 3)$ can indeed be considered as a small perturbation to the stationary point $\mathbf{X}_{st} = (N, 0)$. Furthermore, note that during the initial 38 days period $S(t)$ dropped by about 30,000 (see panel (c) of Fig. 4.5). Consequently, the change in size $\Delta S(t) = S(t) - N$ was relatively small during the initial stage such that $S(t)/N \approx 1$. Therefore, the linear approximation of the SIR model is justified for the whole initial 38 days period under consideration.

As discussed in Sect. 4.2.6, COVID-19 outbreaks in the year 2019/2020 and, in general, outbreaks of other novel infectious diseases, are characterized by exponentially increasing amplitudes A_2 and amplitudes A_1 that are approximately constant provided the modeling framework of SIR-type models can be applied. In order to arrive at that conclusion, it was assumed that $\beta/N \ll \lambda_2$ holds. For the SIR model describing active cases data of the COVID-19 outbreak in Italy this assumption is satisfied because we found $\beta = 0.31$/d, $N = 60,000,000$ and $\lambda_2 = 0.24$/d.

Importantly, the analysis presented above reveals that λ_2 was positive during the COVID-19 outbreak in Italy. In doing so, the model-based analysis indicates that the disease-free fixed point $\mathbf{X}_{st} = (N, 0)$ in Italy during February and March 2020 was an unstable fixed point. According to the nonlinear physics description discussed in Sect. 4.2.6, the course of the epidemic followed the unstable eigenvector (or order parameter) $\mathbf{v}_2$. The vector $\mathbf{v}_2$ was computed from Eq. (4.20) and the coefficients $v_{2,S} = -0.79$ and $v_{2,I} = 0.61$ were found. Panel (d) presents $\mathbf{v}_2$ (dotted line) in the S-I state space. The vector was magnified for the sake of visibility. Panel (d) also presents $I(t)$ versus $S(t)$ (solid line), that is, the phase curve $I(S)$. It can be seen that the phase curve followed the order parameter during the whole initial 38 days period. That is, the nonlinear physics analysis of the COVID-19 data from Italy suggests that the COVID-19 outbreak in Italy in February/March 2020 was determined by an unstable disease-free fixed point $\mathbf{X}_{st} = (N, 0)$ and a susceptible-infectious order parameter (SI order parameter) with $v_{2,S} = -0.79$ and $v_{2,I} = 0.61$. The dynamics was determined by the corresponding order parameter amplitude A_2 with eigenvalue $\lambda_2 = 0.24$/d and a time constant $\tau = 1/\lambda_2 = 4.12$ d. Accordingly, $A_2(t)$ increased exponentially along $\mathbf{v}_2$ with a time constant (or e-folding time) of 4.12 days, which resulted in an exponential increase of active cases $I(t)$ with the same time constant.

4.5.2 *Confirmed Cases and SIQR Modeling*

As argued in Sect. 3.6.2, it is questionable whether the infectious individuals I described by the SIR model should be identified with the number of diagnosed active COVID-19 cases. A better method to relate the SIR model to reported data is the SIQR approach described in Sect. 4.4. Therefore, in what follows, let us apply the SIQR model (4.82) to COVID-19 data from Italy. Figure 4.6 summarizes some of the basic modeling results in this regard. Panel (a) shows the confirmed cumulative COVID-19 cases (i.e., the diagnosed cumulative cases) in Italy over a 250 days period starting on January 31, 2020 [7]. As indicated, these cases are identified with the variable $C = Q + R$ of the SIQR model (4.82) that describes quarantined and removed (i.e., recovered or deceased) individuals, who have been diagnosed with COVID-19 (see Sect. 4.4). From the graph shown in panel (a) it follows that there was an initial stage with an exponential-like increase in confirmed COVID-19 cases that was followed by a more or less linear increase. Subsequently, the COVID-19 cases reached a plateau (with a slight positive slope). At the end of this plateau (i.e., about 200 days after January 31, which is in the days around August 18, 2020)

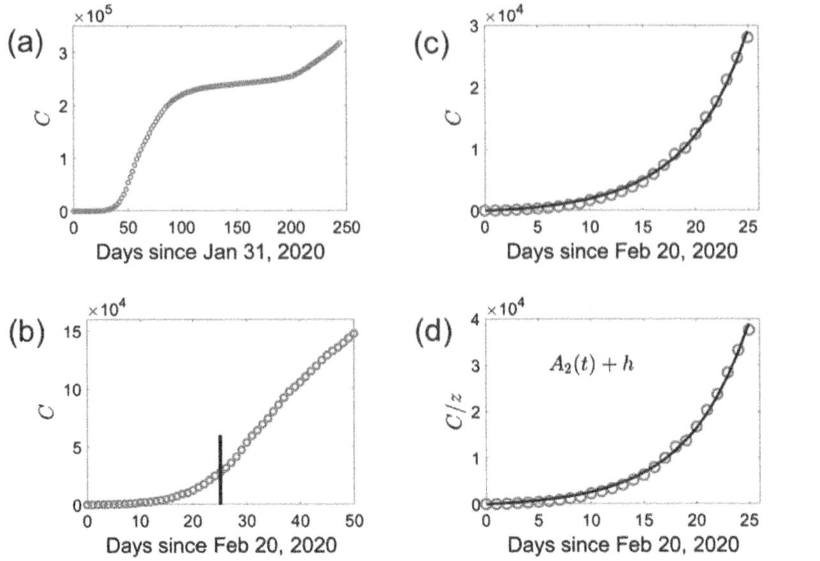

Fig. 4.6 Final step to describe COVID-19 case data from Italy during the first COVID-19 wave in 2020 within a SIQR modeling framework (see Eq. (4.82)) that allows to use the class of diagnosed cases $C = Q + R$. Panel (**a**): cumulative confirmed COVID-19 cases of Italy over the first 250 days of the epidemic in Italy [7]. Panel (**b**): initial stage definition. Panel (**c**): fit (solid black line) of Eq. (4.86) to the initial stage data (gray circles). Panel (**d**): explanation of the cumulative cases C when expressed in units z by means the order parameter amplitude A_2 (solid line describes $A_2(t) + h$, see Eq. (4.88) and text)

the confirmed COVID-19 cases started to increase again. Italy was hit by a second COVID-19 wave. The objective is to model the initial COVID-19 outbreak during February/March 2020 using the SIQR model (4.82). During such an initial stage from the linearized SIQR model it follows that the number of infectious individuals increases exponentially like $I(t) = I_0 \exp\{\lambda_2(t - t_0)\}$ (see Eq. (4.65)). Substituting this function into Eq. (4.83), we obtain

$$C(t) = C_0 + V\left(\exp\{\lambda_2(t - t_0)\} - 1\right), \tag{4.86}$$

with $V = bI_0/\lambda_2$. The coefficient V may be interpreted as slope parameter determining the initial linear increase of $C(t)$ for $t \approx t_0$ like $C(t) \approx C_0 + V(t - t_0)/\tau$ with $\tau = 1/\lambda_2$. In order to fit the analytical solution (4.86) to the data shown in panel (a), the initial time point t_0 was defined as the first day for which diagnosed cumulative cases increased every day for at least one week. This day was February 20, 2020, for the data reported in Ref. [7]. Panel (b) shows the diagnosed cumulative cases in Italy during a 50 days period starting February 20 (i.e., the graph in panel (b) is a zoom-in detail of the graph shown in panel (a)). The data shown in panel (b) was then fitted to Eq. (4.86) for a 26 days period. By visual inspection at the end of that period the graph started to increase more or less in a linear manner. The end of the 26 days period is indicated by a vertical line. More sophisticated data fitting procedures will be presented in Chap. 8.

The best-fit estimates of λ_2 and V were $\lambda_2 = 0.17/\mathrm{d}$ and $V = 441.46$, respectively. Panel (c) shows the diagnosed cases again (as gray circles) during the 26 days fitting period from February 20 to March 16. The analytical solution (4.86) is shown as well (solid black line) for the best-fit parameters. The solution (4.86) produced a fit to the data with moderate accuracy.

Importantly, within the SIQR framework the nonlinear physics of COVID-19 outbreaks is described in terms of the amplitude equations listed in Eq. (4.84). From Eq. (4.84) and $C = Q + R$ it follows that

$$\frac{\mathrm{d}}{\mathrm{d}t}C = bv_{2,I}A_2, \tag{4.87}$$

which is consistent with $I(t) = v_{2,I}A_2(t)$ and Eq. (4.83). During the initial stage, the linear approximation holds with $\mathrm{d}A_2/\mathrm{d}t = \lambda_2 A_2$ and

$$A_2(t) = A(t_0)\exp\{\lambda_2(t - t_0)\}. \tag{4.88}$$

Substituting Eq. (4.88) into Eq. (4.87) and solving for $C(t)$ (i.e., integrating over time), we arrive again at Eq. (4.86) with $I_0 = v_{2,I}A_0$. Consequently, Eq. (4.86) reads in terms of A_2 like

$$C(t) = C_0 + zA_2(t_0)\left(\exp\{\lambda_2(t - t_0)\} - 1\right) \Rightarrow \frac{C(t)}{z} = A_2(t) + h \tag{4.89}$$

with $z = bv_{2,I}/\lambda_2$ and $h = C_0/z - A_2(t_0)$. In words, the order parameter amplitude (or unstable amplitude) A_2 determines up to an additive constant h the diagnosed cumulative COVID-19 cases C during the initial stage of an epidemic provided that C is rescaled by the factor z. The parameters b, $v_{2,I}$ and the initial amplitude $A_2(t_0)$ can be determined as described below. In doing so, z, h, and the analytical solution $A_2(t) = A(t_0)\exp\{\lambda_2(t - t_0)\}$ can be obtained. Panel (d) of Fig. 4.6 presents the expression C/z as function of time. The shifted order parameter amplitude, that is, the expression $A_2(t) + h$ with $A_2(t) = A(t_0)\exp\{\lambda_2(t - t_0)\}$, is plotted as well (solid black line). Panel (d) demonstrates explicitly for the COVID-19 outbreak in Italy during February/March 2020 the role of the unstable amplitude A_2. Accordingly, A_2 determined the confirmed COVID-19 cases. Note that from a mathematical point of view, the curves shown in panel (d) are the same as the curves shown in panel (c) except for the scales of the vertical axes.

With the help of the estimates λ_2 and V the remaining parameters $a, b, \gamma = a + b, \beta$ of the SIQR model (4.82) and the initial values I_0 and $A_2(t_0)$ were determined as follows. Since $V = bI_0/\lambda_2$ the approach conducted above only yields the product $bI_0 = V\lambda_2$ and does not provide separate estimates for b and I_0. The following analysis was conducted for an assumed initial group of $I_0 = 100$ infec-

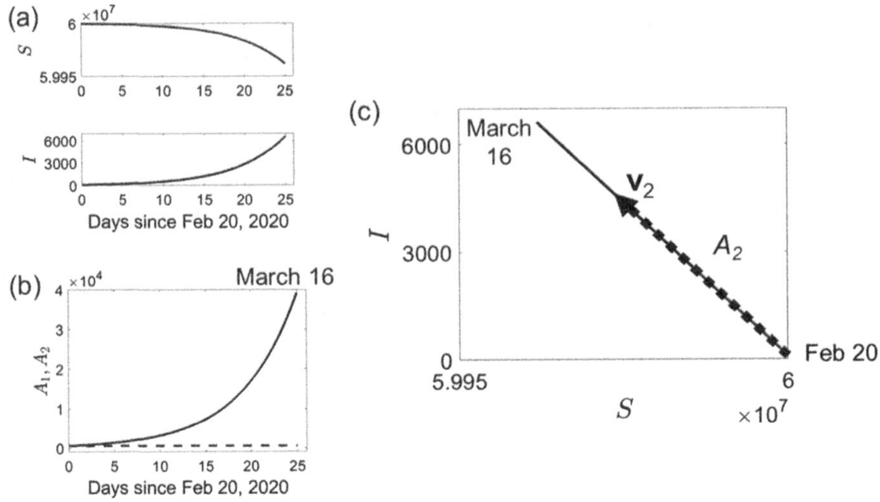

Fig. 4.7 Characteristic features of the COVID-19 outbreak in Italy during February/March 2020: Panel (**a**): state space functions $S(t)$ and $I(t)$. Panel (**b**): amplitudes $A_1(t)$ (dashed line) and $A_2(t)$ (solid line). Panel (**c**): COVID-19 outbreak as phase curve $I(S)$ determined by $\mathbf{v}_2$ and $A_2(t)$

tious individuals. This selected value affects a, β, g, and the order parameter coefficients $v_{2,S}$ and $v_{2,I}$. Therefore, the following analysis results only reflect the inferred dynamics of the epidemic for that particular selected value. From $I_0 = 100$ and $b = V\lambda_2/I_0$ we obtain a diagnoses rate b of $b = 0.74$/d for a single infectious individual. Assuming a recovery rate $a = 1/T = 0.07$/d with $T = 14$d, we obtain $\gamma = a + b = 0.81$/d, $\beta = \lambda_2 + \gamma = 0.98$/d, and the order parameter coefficients $v_{2,S} = -0.99$ and $v_{2,I} = 0.17$ (more precisely: $v_{2,S} = -0.9857$ and $v_{2,I} = 0.1688$). The SIQR model (4.82) for S, I and C (see Eq. (4.83)) rather than Q and R was solved numerically using $N = S_{st} = 60,000,000$, $S_0 = N - I_0$ and the parameters mentioned above. Figure 4.7 shows the simulation results. Panel (a) shows $S(t)$ and $I(t)$ as functions of time. As expected, $S(t)$ decayed monotonically. $I(t)$ increased in an exponential manner.

Furthermore, the amplitude equations for A_1 and A_2 occurring in Eq. (4.84) were solved numerically using the initial conditions computed from Eq. (4.50). In particular, Eq. (4.50) for $A_2(t_0)$ reads $A_2(t_0) = I_0/v_{2,I}$. Panel (b) of Fig. 4.7 shows the numerical solutions. As expected, A_2 increased in an exponential manner up. During the period under consideration A_2 increased to a value of about 40,000 individuals. In contrast, A_1 varied only to a small degree. A_1 decayed from 581 individuals to 522 individuals, that is, it varied by about 60 individuals (see panel (b) of Fig. 4.7 again).

Let us identify again the COVID-19 outbreak in Italy as an instability-phenomenon characterized by a SI order parameter. Panel (c) shows $\mathbf{v}_2$ (dotted line; $\mathbf{v}_2$ was magnified) in the S-I state space. In addition, the functions $S(t)$ and $I(t)$ shown in panel (a) are plotted in panel (c) as $I(S)$ phase curve (solid line). As can be seen, the trajectory of the disease state when plotted as phase curve in the S-I state space followed the order parameter $\mathbf{v}_2$ during the entire initial 26 days period. In summary, the SIQR modeling analysis based on confirmed COVID-19 cases suggests that the COVID-19 outbreak in Italy in February/March 2020 was determined by an unstable disease-free fixed point $\mathbf{X}_{st} = (N, 0)$ and a SI order parameter $\mathbf{v}_2$. The dynamics of the outbreak was determined by the order parameter amplitude A_2 with an eigenvalue $\lambda_2 = 0.17$/d and a time constant $\tau = 1/\lambda_2 = 5.88$d. Accordingly, the amplitude $A_2(t)$ increased exponentially along $\mathbf{v}_2$ with a time constant of 5.88 days, which resulted in an exponential increase of infectious individuals $I(t)$ and a decrease of susceptibles $S(t)$. When making the additional assumption that the epidemic was induced by an initial group of $I_0 = 100$ infectious individuals, then the SI order parameter was given by $v_{2,S} = -0.99$ and $v_{2,I} = 0.17$ and the evolution of the epidemic along the order parameter can be illustrated as shown in panel (c).

Let us compare the analytical solutions defined by Eqs. (4.86) and (4.88) for $C(t)$ and $A_2(t)$, respectively, with the corresponding solutions of the full nonlinear SIQR model. The solutions obtained from the simulation correspond to exact solutions that go beyond the analytical solutions obtained from the linear approximation of the model. In order to compare the exact model solutions with the analytical solutions obtained for the linearized model, we only considered the variable $C(t)$. $C(t)$ as computed from Eq. (4.82) and (4.83) is shown in panel (a) of Fig. 4.8 as full black circles. The analytical solution defined by Eq. (4.86) is presented in panel (c) of Fig. 4.6 and is plotted in panel (a) of Fig. 4.8 as solid line. By visual inspection, there was no difference between the exact solution $C(t)$ (full black circles) and the approximative solution $C(t)$ (solid line) as obtained from Eq. (4.86). This observation does not come as a surprise. The initial state $\mathbf{X}_0 = (N - I_0, I_0)$ in the S-I subspace can be considered as a small perturbation to the disease-free fixed point $\mathbf{X}_{st} = (N, 0)$ even for an initial value of $I_0 = 100$. The reason for this is that the scale is characterized by the enormous population size of $N = 60,000,000$. Moreover, according to the model-based analysis, $S(t)$ decreased from $S_0 \approx S_{st} = N$ during the 26 days period

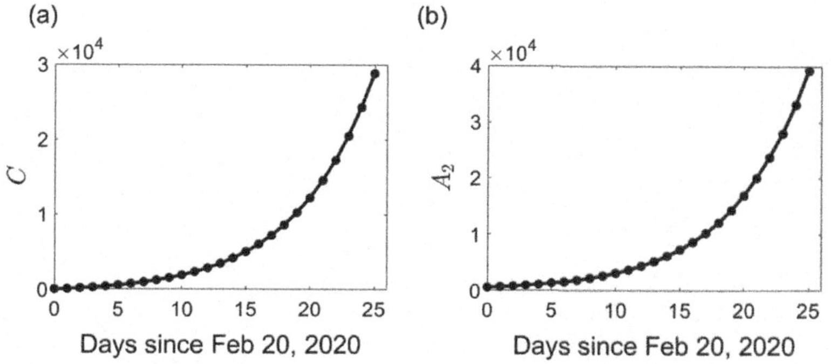

Fig. 4.8 Equivalence of the state and amplitude space descriptions of the COVID-19 outbreak in Italy during February/March 2020. Panels (**a**) and (**b**) shows $C(t)$ and $A_2(t)$, respectively, when computed in a direct way (solid lines) and indirectly (full circles)

by about 40,000 (see panel (a) of Fig. 4.7). The change $\Delta S = S(t) - N$ of about 40,000 at the end of this period was still relatively small such that $S(t)/N \approx 1$ was satisfied during the whole initial 26 days period. The exact solution A_2 obtained by the numerical simulation is shown in panel (b) of Fig. 4.7. It is also shown in panel (b) of Fig. 4.8 as full black circles. The analytical solution defined by Eq. (4.88) is plotted as solid black line in panel (b) of Fig. 4.8. Comparing the exact solution (full black circles) with the approximative exponential solution (solid black line), we see that differences between these solutions are negligibly small on the scale of interest. The reason for this is again that the COVID-19 epidemic in Italy during the COVID-19 outbreak in February/March 2020 evolved relatively close to the disease-free fixed point. In this context, also note that the assumption $\beta/N \ll \lambda_2$ discussed in Sect. 4.2.6 holds because we have $\beta = 0.98$/d, $N = 60,000,000$ and $\lambda_2 = 0.17$/d.

References

1. D. Fanelli, F. Piazza. Analysis and forecast of COVID-19 spreading in China, Italy, and France. Chaos, Solitons, Fractals **134**, article 109761 (2020)
2. T. D. Frank. Simplicity from complexity: on the simple amplitude dynamics underlying COVID-19 outbreaks in China. Adv. Complex Syst. **23**, article 2050022 (2020)
3. T.D. Frank, COVID-19 outbreaks follow narrow paths: a computational phase portrait approach based on nonlinear physics and synergetics. Int. J. Mod. Phys. C **32**, 2150110 (2021)
4. M.G. Pedersen, M. Meneghini, A simple method to quantify country-specific effects of COVID-19 containment measures. MedRxiv (2020). Article 2020.04.07.20057075
5. N. Crokidakis. Covid-19 spread in Rio de Janeiro, Brazil: do the policies of social isolation really well. Chaos, Solitons Fractals **136**, article 109930 (2020)
6. N. Crokidakis. Modeling the early evolution of the covid-19 in brazil: results from a susceptible-infectious-quarantined-recovered (siqr) model. Int. J. Mod. Phys. C **31**, article 2050135 (2020)
7. COVID-19 tracker. Timeline data from Johns Hopkins Center for Systems Science and Engineering. COVID-19 tracker, https://vac-lshtml.shinyapps.io/ncov_tracker

Chapter 5
Nonlinear Physics of Epidemics: Part B

This chapter focuses on the SEIR model, SEIR-type models, and more comprehensive models. It begins with the classification of variables into those variables that are minimally necessary to obtain an autonomous description in amplitude space and all remaining variables. The nonlinear physics approach is then presented for this minimal closed space. Amplitude equations for the SEIR model and SEIR-type models are explicitly derived and their SEI order parameters are determined. It is demonstrated that within the SEIR modeling framework, the SEI order parameter determines the initial stage of epidemics. An application to the COVID-19 outbreak in Wuhan, China, during the year 2020 is also presented.

5.1 Grouping Compartment Variables into Two Classes

Our departure point is again an epidemic model involving n variables described by the state vector $\mathbf{X} = (X_1, \ldots, X_n)$. The dynamics of the state satisfies the general evolution equation $\mathrm{d}\mathbf{X}/\mathrm{d}t = \mathbf{N}(\mathbf{X})$, see Chap. 2. There are several ways to group the state variables X_k into certain classes. In what follows two methods will be presented.

Grouping into compartments of infected and non-infected individuals while ignoring whether or not they are infectious

Variables may be grouped into variables describing infected people, on the one hand, and variables describing non-infected people, on the other hand. This distinction can be used to figure out how many individuals are infected by a single infected person during the course of the infectious disease of that person. This measure can be used as a substitute for the eigenvalues discussed in Chap. 2. The basic idea is that if a person during the course of his or her infection infects more than one other person, then the disease-free fixed point is unstable and an epidemic wave is triggered. In contrast, if a single infected person can infect less than one other person (i.e., if, say, 100

T. D. Frank, *COVID-19 Epidemiology and Virus Dynamics*,
Understanding Complex Systems, https://doi.org/10.1007/978-3-030-97178-6_5

infected individuals, on average, can only infect 80 individuals), then the disease-free fixed point is stable and the infectious disease dies out in the population under consideration. A model-based approach to determine this number will be discussed in Chap. 7. Importantly, this approach does not distinguish between infectious and non-infectious individuals. That is, for this classification approach, it is irrelevant whether or not an infected individual can infect another individual.

Grouping into compartments of infectious and non-infectious individuals and adding individuals necessary to obtain closed, minimal descriptions

The second method is to follow the ideas presented in Sect. 4.1.2 and to decompose the dynamical system $d\mathbf{X}/dt = \mathbf{N}(\mathbf{X})$ into the following two subsystems. The first subsystem contains the compartment variables of all infectious individuals that actually are in the position to infect others. Let us refer to this kind of infectious individuals as actually infectious individuals. For example, infectious individuals who are in perfect isolation do not belong to the class of actually infectious individuals. In addition, variables are added to the first subsystem in order to obtain a closed description in terms of an autonomous dynamical system. The key idea is to add only those variables that are minimally necessary to obtain such a description. In doing so, a closed, minimal description of the epidemic model can be obtained that involves all kind of actually infectious individuals addressed by the model. For example, susceptibles are added to the first subsystem. Moreover, certain infected individuals who cannot infect others (e.g., exposed individuals as defined in Sect. 3.1) are added to the first subsystem if they are necessary to obtain a closed dynamical system. The second subsystem contains all remaining variables. This second subsystem depends on the variables of the first subsystem (if this is not the case, we simply deal with two completely separated systems that can be studied independently from each other). Consequently, the second subsystem is a driven (i.e., non-autonomous) system.

For example, the SIR-type models defined by Eq. (4.1) are systems that can be decomposed into a two-dimensional S-I subsystem, which allows for a closed description of the variables S and I, and subsystems that describe all remaining variables X_k with $k = 3, \ldots, n$.

This second approach to decompose populations into two subsystems provides a basis for the derivation of closed amplitude space descriptions (or autonomous amplitude descriptions) that exhibit dimensions r that are smaller than the dimensions n of the state spaces at hand. Such a reduction of the dimensionality of the systems under consideration helps to carry out analytical approaches. In particular, it can frequently be shown that the stability of the whole dynamical system is determined by the stability of the first subsystem. For example, it frequently can be shown that if the autonomous subsystem dynamics exhibits a neutrally or asymptotically stable disease-free fixed point, then the whole system dynamics exhibits such a fixed point (for an explicit example see Sect. 4.1.2). In view of the second classification method, it is useful to describe populations in terms of models that immediately reveal the two aforementioned subsystems. SIR-type models defined by Eq. (4.1) and SEIR-type epidemic models that will be defined below are examples in this regard.

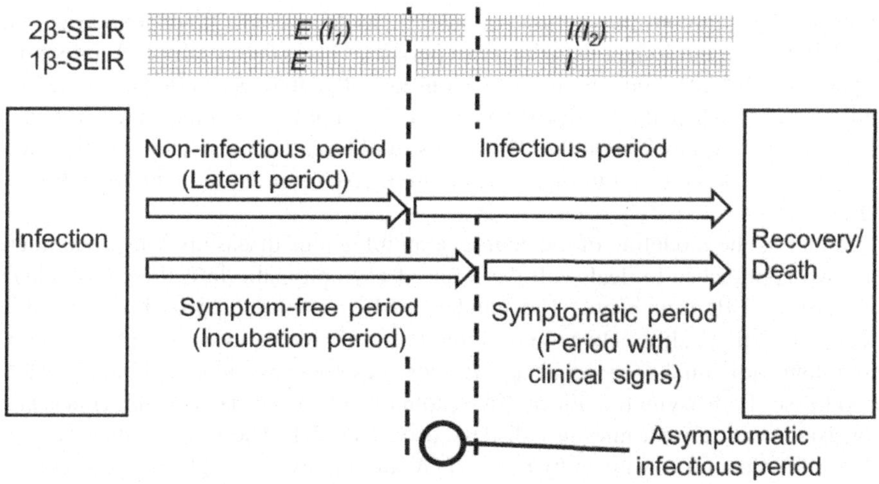

Fig. 5.1 Schematic figure demonstrating different time periods during the course of an infection

5.2 SEIR-Type Models

5.2.1 Latent Versus Incubation Period and SEIR-Type Models

In the context of SEIR-type models the distinction between the latent and incubation period becomes in particular relevant. Figure 5.1 provides a scheme of the typical course of an infection. The scheme involves three levels. The modeling level is shown in the top part. The level describing the infectiousness of individuals is shown in the middle part and the level describing the degree to which individuals show symptoms of an infectious disease is shown in the bottom part.

According to the scheme and as stated in Sect. 3.1 infectious diseases may involve a period during which infected individuals do not infect others. This non-infectious period is called the latent period. For example, during the latent period the virus may move without multiplying within bodies of individuals to specific locations or organs with target cells. Subsequent to the latent period, the individuals typically become infectious. Accordingly, they enter the infectious period of their disease. As far as the symptoms related to the infectious disease under consideration is concerned, after being infected, individuals typically enter a period in which they do not show any symptoms. This symptom-free period is called the incubation period. Subsequent to the incubation period, individuals typically show symptoms (i.e., clinical signs) of the disease (for exceptions see below). As illustrated in Fig. 5.1, being infectious or not is not necessarily related in a one-to-one mapping with experiencing symptoms (i.e., clinical signs) of the infectious disease. That is, the latent period and the incubation period may or may not correspond to each other. Figure 5.1 illustrates the case in which the symptom-free period is longer than the non-infectious period (i.e., the incubation period is longer than the latent period). This is the case for HIV infections [1] and SARS-CoV-2 infections [2–6]. Consequently, there is a period during which individuals (to which the scheme shown in Fig. 5.1 applies) can infect others although they do not show any symptoms. This period will be called the asymptomatic infectious period, as indicated in Fig. 5.1. An alternative terminology will be addressed below.

As far as the modeling of the course of an infectious disease is concerned, there are two approaches to deal with the issue of asymptomatic infectious individuals within the SEIR framework. The standard SEIR modeling approach discussed in Sect. 3.7 distinguishes between non-infectious individuals in the latent period and infectious individuals. Consequently, E addresses non-infectious individuals, while I addresses both asymptomatic and symptomatic infectious individuals. This is also consistent with the terminology discussed in Sect. 3.1. The rate constant k_0 (i.e., "force of infection") is given by $k_0 = \beta I/N$ and involves a single effective contact rate because there is only one compartment of infectious individuals (see also Sect. 3.7). Consequently, we deal with a model that involves a single β parameter. In Fig. 5.1 this kind of SEIR modeling approach is illustrated and called the 1β SEIR approach.

In the context of the second approach, the compartment E is used to denote individuals that are infected and do not show symptoms. In the case of a latent period shorter than the incubation period, as illustrated in Fig. 5.1, this implies that the individuals of compartment E correspond to infected and possibly asymptomatic infectious individuals [7, 8]. In contrast, the compartment I denotes symptomatic infectious individuals. Since this approach involves two compartments with infectious individuals, the rate constant k_0 of the reaction transforming susceptibles into infected individuals (i.e., the "force of infection") reads [7–10]

$$k_0 = \frac{\beta_I I + \beta_E E}{N} , \tag{5.1}$$

and involves two effective contact rates β_I and β_E with

$$\beta_I = p_I \nu_I , \; \beta_E = p_E \nu_E , \tag{5.2}$$

where μ_I, μ_E denote contact rates between susceptibles and individuals of the compartments I and E, respectively, and p_I, p_E denote the corresponding probabilities that during such contacts susceptibles get infected (see also Sect. 3.2). The rate constant k_0 involves two parameters β. Consequently, the SEIR-type models may be referred to as 2β SEIR-type models. Figure 5.1 illustrates the assignment of the compartments E and I to the stages of infectious diseases for these 2β SEIR-type models (see top part).

In summary, there are two key differences between the two approaches. First, the compartment E of 1β models is related to the latent period, whereas the compartment E of 2β models is related to the incubation period. The 1β models involve a single parameter β and a single term occurring in k_0, whereas the 2β models feature two β parameters and two terms in k_0. Models that involves a compartment E of infected and possibly asymptomatic infectious individuals have been used in several epidemiological studies on COVID-19 [9–17].

In the context of the modeling of COVID-19 epidemics, it is frequently assumed that the probability of infection p_A due to contact with asymptomatic cases is lower as compared to the probability p_I of symptomatic cases: $p_A < p_I$. Since E is a mixture of non-infectious and asymptomatic cases, we have $p_E \leq p_A \Rightarrow p_E < p_I$. Consequently, if the contact frequencies of susceptibles with individuals of compartments E and I are the same ($\nu_E = \nu_I$), then $\beta_E < \beta_I$ holds. In this context, in Ref. [12] it has been suggested to put $\beta_E = \theta\beta_I$ and to relate the factor $\theta \leq 1$ to the durations T_L and T_I of the latent and incubation periods. If $T_L = T_I$ then $\theta = 0$ and $\beta_E = 0$. We deal with a 1β model because the latent period corresponds to the incubation period. In contrast, if $T_L = 0$ we have $\theta = 1$ and $\beta_E = \beta_I$. In this special case the compartment E actually does not exist because $T_L = 0$. The SEIR model reduces to a SIR model and the compartment I describes individuals who become infectious immediately after being infected. For $0 < \theta < 1$ the inequality $\beta_E < \beta_I$ holds. In this case, the latent period is finite but shorter than the incubation period.

Finally, as indicated in Fig. 5.1, the 2β SEIR-type models may be written as compartment models involving two compartments I_1 and I_2 of infectious individuals with a different degree of infectiousness. Infectious individuals of the first infectious compartment I_1 progress to the second infectious compartment I_2. In doing so, the degree of infectiousness changes. The interpretation will be addressed in Sect. 5.2.3.

Asymptomatic individuals who do not develop symptoms at all

The scheme illustrated in Fig. 5.1 suggest that every individual eventually enters the symptomatic period. However, this is not necessarily the case for all infectious diseases. Virus infections may take place in the bodies of individuals without the individuals noticing any symptoms. That is, individuals can become infectious and, subsequently, non-infectious again, while experiencing no clinical signs of their disease. SEIR-type models may address asymptomatic infectious individuals of this kind: individuals who can infect others but do not show symptoms during the course of their infections. COVID-19 is a disease that can be transmitted by this kind of asymptomatic individuals [2, 6, 18].

Epidemic models as described in this book and based on deterministic differential equations as discussed in Chap. 2 describe how individuals on average make transitions from one stage (i.e., compartment) to another stage (i.e., compartment) during their course of disease. The model parameters reflect averages (see also Sect. 3.4). For example, as far as COVID-19 patients are concerned, incubation and infection periods vary across individuals [19].

In the context of 1β SEIR-type models (see top part of Fig. 5.1 for the assignment of E and I to the stages of an infectious disease), we first note again that on the level of individuals asymptomatic and symptomatic periods vary. Model parameters refer to averaged transition rates and, consequently, reflect asymptomatic and symptomatic periods when averaged across all kind of individuals. The asymptomatic cases defined above who stay symptom-free correspond to individuals of the compartment I for which the symptomatic period equals zero. That is, the compartment I includes asymptomatic individuals who remain symptom-free as special cases. For those individuals the symptom-free period equals the sum of the non-infectious and infectious periods. The model parameters of 1β SEIR-type models reflect averages across individuals and take into account that in the compartment I there are individuals with vanishing symptomatic periods.

In the context of 2β SEIR-type models, asymptomatic cases who remain symptom-free are modeled by individuals of the compartment E that switch from being non-infectious to being infectious while they belong to the compartment E and show no symptoms. Eventually, these individuals make transitions from E to I but, subsequently, immediately leave the compartment I. For those individuals, the duration in the compartment I (i.e., the symptomatic period) is zero. Consequently, the decay term $-\gamma I$ of the SEIR model (3.43) that occurs also in the 2β SEIR-type models (see Sect. 5.2.2 below) is interpreted as a term that describes how fast individuals on average are removed out of the compartment I and takes as a special case the asymptomatic individuals who remain symptom-free during their whole disease history into account.

In short, for 1β models the asymptomatic and permanently symptom-free cases can be found in the compartment I, whereas for 2β models they can be found in the compartment E.

Presymptomatic versus asymptomatic period

In the context of the role of infections due to contact with infected but symptom-free individuals, the distinction between presymptomatic and asymptomatic periods can be made [6]. Accordingly, for individuals who become in a first stage infectious without symptoms but later develop systems, the symptom-free infectious period is referred to as presymptomatic infectious period. In contrast, for individuals who become infectious during their course of disease but never develop any symptoms, the infectious period is referred to as asymptomatic infectious period. The rational is that the degree to which individuals can infect others during their symptom-free periods depends on whether or not they develop in a later stage symptoms. Some COVID-19 epidemic models have been proposed that specifically take this distinction into account [20, 21]. In Sect. 5.3.2 the model by Gatto et al. [21] that involves presymptomatic cases will be briefly addressed. As such, in this book a distinction between presymptomatic and asymptomatic periods or individuals will not be made except when reviewing the literature in Sect. 5.3.2.

Asymptomatic cases in models beyond SEIR-type models

As discussed above, SEIR-type models allow to address asymptomatic infectious individuals in two ways. They belong either to the compartment I (1β models) or E (2β models). Alternative modeling approaches use a separate compartment that is typically denoted by A or I_a to describe asymptomatic infectious cases. In doing so,

SEIAR models are obtained [22] that have been used in COVID-19 research as well [11, 21, 23–27]. Some examples of such alternative approaches will be presented below in Sect. 5.3.2.

5.2.2 SEIR-Type Models and 3D Autonomous Amplitude Descriptions

Let us consider epidemic models that involve n compartment variables $X_1, \ldots, X_n$, where $X_1 = S$ denotes the susceptibles, $X_2 = E$ denotes exposed individuals, and $X_3 = I$ denotes infectious individuals. The compartments E and I describe individuals as discussed in Sect. 5.2.1 and illustrated in Fig. 5.1. In the context of 1β models, E and I describe infected, non-infectious individuals (E) and asymptomatic or symptomatic infectious individuals (I). In the context of 2β models, E and I describe infected and possibly asymptomatic infectious individuals (E) and symptomatic infectious individuals (I). The remaining variables X_k denote individuals that cannot infect others. For example, such a variable could describe infected individuals that are as such infectious but are quarantined and in perfect isolation such that they cannot infect others. In analogy to the standard SEIR model (see Sect. 3.7 and Eq. (3.43)), let us define 1β and 2β SEIR-type models by

$$\begin{aligned}
\frac{\mathrm{d}}{\mathrm{d}t}S &= -k_0 S + B - \mu S \, , \\
\frac{\mathrm{d}}{\mathrm{d}t}E &= k_0 S - (\alpha + \mu) E \, , \\
\frac{\mathrm{d}}{\mathrm{d}t}I &= \alpha E - (\gamma + \mu) I \, , \\
\frac{\mathrm{d}}{\mathrm{d}t}X_k &= N_k(S, E, I, X_4, \ldots, X_n) \text{ for } k = 4, \ldots, n \, ,
\end{aligned} \tag{5.3}$$

where B and μ denote birth and death rate parameters, respectively. Furthermore we have $k_0 = \beta I/N$ for 1β models and k_0 defined by Eq. (5.1) for 2β models. The remaining model parameters are: the rate of progression α from compartment E to I and the recovery rate γ. As discussed in Sect. 3.4, α may be related to the mean duration T that individuals spend time in the compartment E like $\alpha = 1/T$. Consequently, for 1β SEIR-type models, we may put $\alpha = 1/T$, where T is the latent period. In contrast, for 2β SEIR-type models, we may put again $\alpha = 1/T$ but consider T as incubation period [10]. Frequently, the right-hand side functions N_k in Eq. (5.1) involve only linear terms and read

$$N_k = \sum_{i=1}^{n} a_{ki} X_i \tag{5.4}$$

(see also Eq. (4.2)). When considering epidemics over short periods demographic terms can be neglected such that $B = \mu = 0$.

SEIR-type models exhibit a closed or autonomous three-dimensional subsystem characterized by the variables S, E, and I. The remaining variables $X_4, \ldots, X_n$ form a driven (i.e., non-autonomous) system. The three-dimensional SEI subspace can be equivalently described by a three-dimensional amplitude space using the techniques described in Chap. 2. The corresponding amplitude equations describing the evolution of the amplitudes A_1, A_2, and A_2 in this space form a closed and autonomous system. In this sense SEIR-type models exhibit 3D autonomous amplitude descriptions and can be described by

$$\begin{aligned}
\frac{\mathrm{d}}{\mathrm{d}t} A_1 &= \lambda_1 A_1 + G_1(A_1, A_2, A_3) \ , \\
\frac{\mathrm{d}}{\mathrm{d}t} A_2 &= \lambda_2 A_2 + G_2(A_1, A_2, A_3) \ , \\
\frac{\mathrm{d}}{\mathrm{d}t} A_3 &= \lambda_3 A_3 + G_3(A_1, A_2, A_3) \ , \\
\frac{\mathrm{d}}{\mathrm{d}t} X_k &= N_k(S(\mathbf{A}), E(\mathbf{A}), I(\mathbf{A}), X_4, \ldots, X_n) \ , \ \text{for } k = 4, \ldots, n
\end{aligned} \tag{5.5}$$

with $\mathbf{A} = (A_1, A_2, A_3)$ denoting the three-dimensional amplitude vector. The explicit form of Eq. (5.5) will be derived in Sect. 5.7 and presented in Sect. 5.7.3.

Standard SEIR model revisited

The standard SEIR model for $B = \mu = 0$ was discussed in Sect. 3.7 and is defined by Eq. (3.43). The model is a special case of Eq. (5.3) with $n = 4$, $k_0 = \beta I/N$, and $\mathbf{X} = (S, E, I, R)$. Using vector notation, the SEIR model (3.43) can be written like

$$\frac{\mathrm{d}}{\mathrm{d}t}\begin{pmatrix} S \\ E \\ I \\ R \end{pmatrix} = \begin{pmatrix} 0 & 0 & 0 & 0 \\ 0 & -\alpha & 0 & 0 \\ 0 & \alpha & -\gamma & 0 \\ 0 & 0 & \gamma & 0 \end{pmatrix}\begin{pmatrix} S \\ E \\ I \\ R \end{pmatrix} + k_0 S \begin{pmatrix} -1 \\ 1 \\ 0 \\ 0 \end{pmatrix} \ , \tag{5.6}$$

where the first term on the right-hand side describes the linear part, while the second term on the right-hand side corresponds to the nonlinear part of the model. Note that this distinction between linear and nonlinear part refers to the original state space variables $\mathbf{X}$. In the context of the relative state $\mathbf{u}$ and the linearization matrix L elements of the aforementioned nonlinear part $k_0 S(-1, 1, 0, 0)$ show up in L. The linear part involves a matrix composed of matrix elements a_{ik}: $a_{11} = a_{12} = a_{13} = a_{14} = 0$, $a_{21} = 0$, $a_{22} = -\alpha$, $a_{23} = a_{24} = 0$, $a_{31} = 0$, $a_{32} = \alpha$, $a_{33} = -\gamma$, $a_{34} = 0$, $a_{41} = a_{42} = 0$, $a_{43} = \gamma$, $a_{44} = 0$. The matrix also includes the coefficients that show up in N_4 as defined by Eq. (5.4) like $\mathrm{d}R/\mathrm{d}t = N_4 = a_{43} I = \gamma I$.

5.2.3 SEIR-Type Models as Staged-Progression or Age-Structured Models

When replacing formally the labels E and I by I_1 and I_2, models of SEIR-type read

$$\frac{\mathrm{d}}{\mathrm{d}t} S = -k_0 S + B - \mu S ,$$

$$\frac{\mathrm{d}}{\mathrm{d}t} I_1 = k_0 S - (\alpha_1 + \mu) I_1 ,$$

$$\frac{\mathrm{d}}{\mathrm{d}t} I_2 = \alpha_1 I_1 - (\alpha_2 + \mu) I_2 ,$$

$$\frac{\mathrm{d}}{\mathrm{d}t} X_k = N_k(S, I_1, I_2, X_4, \ldots, X_n) \text{ for } k = 4, \ldots, n ,$$

$$k_0 = \frac{\beta_1 I_1 + \beta_2 I_2}{N} , \tag{5.7}$$

where we have also replaced β_I and β_E by β_1 and β_2 and the parameters α and γ by α_1 and α_2. The models describe individuals that pass through two different phases 1 and 2 (or stages) during the course of the infectious disease under consideration. During the phases individuals exhibit different degrees of infectiousness, which is accounted for in the model with the help of the two effective contact rate parameters β_1 and β_2. Such models have been referred to as staged-progression models [28, 29]. In general, staged-progression models may involve more than just two phases 1 and 2. The models defined by Eq. (5.7) also correspond formally to aged-structured models [30] that involve two infectious age groups 1 and 2 such as children and adults. In this example, the parameter α_1 describe the transition from childhood to adulthood. In general, infectious individuals of different age groups may differ in their degrees of infectiousness, which, again, can be accounted for by the two effective contact rate parameters β_1 and β_2. Equation (5.7) reflects a relatively simple age-structured model that exhibits only a single compartment of susceptibles. Frequently, when discussing age-structured models, separate compartments of susceptibles are defined for all age groups under consideration [30]. Again, in general, age-structured epidemic model may feature more than just two age groups. Finally, note that in the context of aged-structures models the overall death rate parameter μ should be replaced by a set of age-specific parameters [30].

5.3 Beyond SEIR-Type Models

5.3.1 *r < n-Dimensional Approaches: Epidemic Models with r-Dimensional Autonomous Amplitude Descriptions*

SIR-type models exhibit 2D autonomous amplitude descriptions (see Sect. 4.4). SEIR-type models exhibit 3D autonomous amplitude descriptions (see Sect. 5.2.2 above). It is clear that more comprehensive models may exhibit autonomous amplitude descriptions beyond the dimension 3. To demonstrate this point, let us consider disease states $\mathbf{X}$ of infectious diseases in populations that are described by n state variables $X_1, \ldots, X_n$ and exhibit $r < n$ variables that from the smallest closed system that involves all variables referring to actually infectious individuals (in the sense of Sect. 5.1, namely, infectious individuals that are actually in the position to infect others). Without loss of generality, these variables X_k can be labeled with indices $k = 1, \ldots, r$. Furthermore, the state variable $X_1 = S$ is assumed to denote the susceptibles. Since the first r compartment variables involve all types of actually infectious individuals, the rate constant k_0 of $S \to Y$ reactions (where Y are infected individuals) reads

$$k_0 = \frac{1}{N} \sum_{k=2}^{r} \beta_k X_k \tag{5.8}$$

and maximally involves $r - 1$ effective contact rates and $r - 1$ terms $\beta_k X_k$ that refer to the variables $X_2, \ldots, X_r$ of the first subsystem. If compartments of non-infectious individuals have been added to the first subsystem in addition to $X_1 = S$ in order to arrive at a closed description, then the corresponding effective contact rates β_k vanish. Having defined k_0 (i.e., the "force of infection"), the evolution equations of the compartments X_k are defined by

$$\begin{aligned}
\frac{\mathrm{d}}{\mathrm{d}t} S &= -k_0 S + B - \mu S \,, \\
\frac{\mathrm{d}}{\mathrm{d}t} X_k &= N_k(S, X_2, \ldots, X_r) \text{ for } k = 2, \ldots, r \,, \\
\frac{\mathrm{d}}{\mathrm{d}t} X_k &= N_k(S, X_2, \ldots, X_n) \text{ for } k = r+1, \ldots, n \,.
\end{aligned} \tag{5.9}$$

As can be seen from Eq. (5.9), the first r equations provide a closed set of r coupled first-order differential equations. They describe the first subsystem composed of susceptibles, the various kinds of actually infectious individuals, and possibly some additional non-infectious individuals. The first r equations describe the disease state in state space (more precisely, in a r-dimensional subspace of the state space). The state space description can be transformed into an amplitude space description (see Chap. 2). In doing so, a set of r coupled amplitude equations can be obtained.

That is, the first subsystem of the epidemic model (5.9) spanned by $S, X_2, \ldots, X_r$ can alternatively be expressed in terms of a r-dimensional autonomous amplitude description involving the amplitudes $A_1, \ldots, A_r$. The evolution equations for the remaining $n-r$ variables X_k of the second subsystem can then be described in terms of the $n-r$ variables $X_{r+1}, \ldots, X_n$ of the second subsystem and the r amplitudes $A_1, \ldots, A_r$. In summary, the model (5.9) can be expressed by means of a r-dimensional autonomous amplitude description and a $r-n$-dimensional driven (i.e., non-autonomous) system like

$$\frac{\mathrm{d}}{\mathrm{d}t} A_k = \lambda_k A_k + G_k(A_1, \ldots, A_r)\ , \quad \text{for } k = 1, \ldots, r\ ,$$
$$\frac{\mathrm{d}}{\mathrm{d}t} X_k = N_k(X_1(\mathbf{A}), \ldots, X_r(\mathbf{A}), X_{r+1}, \ldots, X_n)\ , \quad \text{for } k = r+1, \ldots, n \tag{5.10}$$

with $\mathbf{A} = (A_1, \ldots, A_r)$ denoting the r-dimensional amplitude vector.

5.3.2 *Examples*

Epidemic model by Arcede et al. (2020)

Arcede et al. [11] studied COVID-19 epidemics during the year 2020 in several countries and regions. They focused on France, Italy, Philippines, Spain and the United Kingdom, on the one hand, and the Hubei province (China) and the state of New York (USA), on the other hand. They proposed a model involving the following compartments: susceptibles (S), exposed and possibly infectious but symptom-free individuals (E), asymptomatic infectious individuals (I_a), symptomatic infectious individuals (I_s), individuals under treatment (U), recovered individuals (R), and individuals deceased due to the disease (D). In particular, the model by Arcede et al. exhibits explicitly a compartment I_a denoting asymptomatic cases in the sense of individuals who can infect others during their course of disease but never develop symptoms. Furthermore, as stated above, the model assumes that at least some portion of the individuals in the compartment E can infect others. The infections due to contact with those subset of individuals are characterized by a certain effective contact rate β_E^*. The effective contact rate for all individuals of the compartment E is given by β_E with $\beta_E < \beta_E^*$. Since in the model individuals from the compartment E can make a transition to I_a or I_s, that is, they either remain asymptomatic or become symptomatic, this portion of infectious individuals in the compartment E reflects asymptomatic or presymptomatic cases. The distinction between individuals in E and I_a is that they differ with respect to their degree of infectiousness. Accordingly, the model features contact rate parameters β_E^*, β_E, and β_a. Effectively, the parameter β_E^* does not appear in the model. Only β_E and β_a are used.

While the number of reported COVID-19 deaths from the countries and regions studied by Arcede et al. are tragic numbers, these numbers are negligibly small compared to the population sizes N (ranging between 20 and 100 millions [11]) of the respective countries or regions. In order to simplify the presentation, in what follows, we will neglect the COVID-19 associated deaths and N will be taken as a constant.

The infectious compartments of the model are: E, I_a, I_s, U. Arcede et al. assumed that individuals under treatment are in perfect isolation such that they cannot infect others. Consequently, the compartments describing individuals who are infectious and in the position to infect others (i.e., the actually infectious individuals) are: E,I_a,I_s. This implies that the "force of infection" of the model reads

$$k_0 = \frac{\beta_E E + \beta_a I_a + \beta_s I_s}{N} \tag{5.11}$$

and exhibits three terms and three effective contact rate parameters β_E, β_a, and β_s. As can be seen below, the variables E, I_a, I_s together with the variable S form a closed description and, consequently, describe the first subsystem. The second subsystem is given by the remaining variables U and R. In order to simplify the presentation again, let us introduce the compartment of removed individuals R_2 that contains both the recovered individuals and those under treatment like $R_2 = U + R$. This implies, that the second subsystem is composed of a single variable: R_2. Taking the two simplifications (i.e., ignoring deaths and using R_2 rather than R and U) into account, the state vector of the model reads $\mathbf{X} = (S, E, I_a, I_s, R_2)$. The model equations read

$$\begin{aligned} &\frac{\mathrm{d}}{\mathrm{d}t}S = -k_0 S\,, \quad \frac{\mathrm{d}}{\mathrm{d}t}E = k_0 S - (\alpha + \nu_E)E\,, \quad \frac{\mathrm{d}}{\mathrm{d}t}I_a = f\alpha E - \gamma_a I_a\,, \\ &\frac{\mathrm{d}}{\mathrm{d}t}I_s = (1-f)\alpha E - (\gamma_s + \nu_s)I_s\,, \quad \frac{\mathrm{d}}{\mathrm{d}t}R_2 = \nu_E E + \gamma_a I_a + (\gamma_s + \nu_s)I_s\,. \end{aligned} \tag{5.12}$$

The parameters γ_a and γ_s denote recovery rates for asymptomatic and symptomatic individuals, respectively. The parameters ν_E and ν_S reflect impacts of interventions that put exposed and symptomatic individuals under treatment (note that a similar parameter for the asymptomatic individuals is missing in the model). That is, ν_E and ν_S describe transition rates from E and I_s to U (and, consequently, to R_2) due to interventions (e.g., contact tracing, testing, and quarantining). Importantly, while α describes the leaving rate with which individuals make transitions out of the compartment E, the parameter $f \in [0, 1]$ is used to describe to which compartment those transitions take place. Accordingly, the product $f\alpha$ describes the rate of transitions from E to I_a (exposed and possibly infectious symptom-free individuals become asymptomatic), while $(1 - f)\alpha$ describes the rate of transitions from E to I_s (exposed and possibly infectious symptom-free individuals become symptomatic). The model (5.12) can be considered as a SEIAR model, for which the variable notations I_s, I_a, R_2 are used rather than I, A, and R. Using vector and matrix notations, the model (5.12) can be written like

$$\frac{\mathrm{d}}{\mathrm{d}t}\begin{pmatrix} S \\ E \\ I_a \\ I_s \\ R_2 \end{pmatrix} = \begin{pmatrix} 0 & 0 & 0 & 0 & 0 \\ 0 & -(\alpha+\nu_E) & 0 & 0 & 0 \\ 0 & f\alpha & -\gamma_a & 0 & 0 \\ 0 & (1-f)\alpha & 0 & -(\gamma_s+\nu_s) & 0 \\ 0 & \nu_E & \gamma_a & (\gamma_s+\nu_s) & 0 \end{pmatrix} \begin{pmatrix} S \\ E \\ I_a \\ I_s \\ R_2 \end{pmatrix} + k_0 S \begin{pmatrix} -1 \\ 1 \\ 0 \\ 0 \\ 0 \end{pmatrix} . \tag{5.13}$$

From Eqs. (5.12) and (5.13) it follows that the first four variables form a closed set of differential equations (as anticipated above). Consequently, the model by Arcede et al. (as presented in our simplified way) exhibits a four-dimensional closed (i.e., autonomous) and minimal subsystem that captures how the numbers of infectious individuals of the three types E, I_a, and I_s vary over time. The evolution equations for S, E, I_a, I_s provide the state space description for the autonomous subsystem. The evolution of the second subsystem is given in state space by the last equation in Eq. (5.12) that can also be written like $\mathrm{d}R_2/\mathrm{d}t = N_5 = a_{52}E + a_{53}I_a + a_{54}I_s$ with $a_{52} = \nu_E$, $a_{53} = \gamma_a$, and $a_{54} = \gamma_s + \nu_s$, when interpreting the linear part of Eq. (5.13) as a matrix with coefficients a_{ik}. According to Eq. (5.10), the model (5.12) can equivalently be expressed with the help of a four-dimensional autonomous amplitude description of the form

$$\frac{\mathrm{d}}{\mathrm{d}t}A_k = \lambda_k A_k + G_k(A_1, \ldots, A_4) \tag{5.14}$$

for $k = 1, \ldots, 4$ and the driven system

$$\frac{\mathrm{d}}{\mathrm{d}t}R_2 = a_{52}E(\mathbf{A}) + a_{53}I_a(\mathbf{A}) + a_{54}I_s(\mathbf{A}) , \tag{5.15}$$

where $\mathbf{A} = (A_1, \ldots, A_4)$ denotes the four-dimensional amplitude vector.

Epidemic model by Gatto et al. (2020)

Gatto et al. (2020) studied the first wave of COVID-19 in Italy during February and March 2020 [21]. To this end, an epidemic model with 9 compartments was used. The model can be decomposed into two subsystems according to the second method described in Sect. 5.1. The first subsystem forms a closed description and involves the following compartments: susceptibles (S), exposed (E), presymptomatic (P), asymptomatic infectious (A), and symptomatic infectious (I). The second subsystem describes a driven (i.e., non-autonomous) system and involves only compartments of individuals that cannot infect other individuals (although they might be infectious). The compartments of this second subsystem are: hospitalized individuals (H), quarantined individuals (Q), recovered individuals (R), and individuals deceased from COVID-19 (D). Again, in what follows a simplified version of the model will be presented. The initial susceptible population $S(0)$ was taken from resident population records of Italy [21]. When applying the model to the overall COVID-19 epidemic in Italy, the initial value $S(0)$ is close to 60 millions. In contrast, the confirmed COVID-19 cases increased up to a number of 100,000 (or 0.1 millions) during the study period considered by Gatto et al. (and the number eventually increased to a plateau value of 200,000 cases, i.e., 0.2 millions, during the summer 2020, as shown in Fig. 4.6, see also Sect. 4.5). The number of 0.1 millions is relatively small as compared to the initial number of 60 millions susceptibles. Likewise, while the number of COVID-19 associated deaths is a tragic number that motivated in Italy the implementation of intervention measures, the number is again relatively small as compared to the number of susceptibles. In summary, in what follows the model is simplified by acknowledging that during the COVID-19 outbreak in Italy in February and March 2020 the size of the population N was approximately 60 millions and did not show considerable variations due to COVID-19 associated deaths. In this context, note that the model was in fact applied to several local regions of Italy. It would be open for debate if the argument made above when considering the population of Italy as a whole also holds for this local regions.

The epidemic model proposed by Gatto et al. comes with the actually infectious compartments P, I, and A. Note that in the study by Gatto et al. exposed individuals were regarded as in 1β SEIR models as individuals in the latent period and, consequently, as individuals who cannot infect others. When taking the simplification about a constant population size into account, the rate constant k_0 reads

$$k_0 = \frac{\beta_P P + \beta_A A + \beta_I I}{N} \tag{5.16}$$

and exhibits three different effective contact rate parameters β_P, β_A, and β_I. According to the study by Gatto et al. [21], The dynamics of the compartment P satisfies the evolution equation $\mathrm{d}P/\mathrm{d}t = aE - bP$ with $a, b > 0$. If E is fixed, then P converges to $P = aE/b$ (which is a stationary point for fixed E and, in general, a nullcline point). Let us consider the following simplification. Let us consider the case in which the time constant $T = 1/b$ that describes how fast P converges to a fixed value of E is small relative to the time constant that describes changes of E. In other words, let us consider the case in which P exhibits a fast dynamics relative to E. In this case, we can put $P(t) = aE(t)/b$ as an approximation to eliminate the compartment P. In this case, Eq. (5.16) becomes

$$k_0 = \frac{\beta_E E + \beta_A A + \beta_I I}{N} \tag{5.17}$$

with $\beta_E = a\beta_P/b$. When eliminating P in this way, the first subsystem of the model consists of the variables S, E, A, I. The second subsystem remains unchanged. However, it can be shown that the variables H, Q, R, and D can be taking together to a single variable R_2 like $R_2 = R + Q + H + D$ describing the removed individuals that cannot infect others. If so, the second subsystem is given by a single variable: R_2. Using the simplifications discussed above, the original nine-variable model can be described in terms of the five variables S, E, A, I, and R_2. The model becomes a SEIAR model. The state vector reads $\mathbf{X} = (S, E, A, I, R_2)$. From the original study [21] it follows that the dynamics of $\mathbf{X}$ is given by

$$\frac{\mathrm{d}}{\mathrm{d}t}S = -k_0 S\,,\ \frac{\mathrm{d}}{\mathrm{d}t}E = k_0 S - \alpha E\,,\ \frac{\mathrm{d}}{\mathrm{d}t}A = (1-f)\alpha E - \gamma_A A\,,$$
$$\frac{\mathrm{d}}{\mathrm{d}t}I = f\alpha E - \gamma_I I\,,\ \frac{\mathrm{d}}{\mathrm{d}t}R_2 = \gamma_A A + \gamma_I I \qquad (5.18)$$

with k_0 defined by Eq. (5.17). The parameter α describes the rate of transitions from E via the eliminated compartment P to A and I. These transitions occur with rates defined by the products $(1-f)\alpha$ and $f\alpha$, respectively, where $f \in [0, 1]$. Comparing the models (5.12) and (5.18), we see that the terms $(1-f)\alpha$ and $f\alpha$ are switched. That is, a value of $f = 0.2$ in the model (5.12) corresponds to a value of $f = 0.8$ in the model (5.18). The coefficients γ_A and γ_I occurring in Eq. (5.18) describe transition rates of the asymptomatic and symptomatic individuals, respectively, into the various compartments H, Q, R, and D that are all captured by the single variable R_2. Equation (5.18) can be written using vector and matrix notations like

$$\frac{\mathrm{d}}{\mathrm{d}t}\begin{pmatrix} S \\ E \\ A \\ I \\ R_2 \end{pmatrix} = \begin{pmatrix} 0 & 0 & 0 & 0 & 0 \\ 0 & -\alpha & 0 & 0 & 0 \\ 0 & (1-f)\alpha & -\gamma_A & 0 & 0 \\ 0 & f\alpha & 0 & -\gamma_I & 0 \\ 0 & 0 & \gamma_A & \gamma_I & 0 \end{pmatrix} \begin{pmatrix} S \\ E \\ A \\ I \\ R \end{pmatrix} + k_0 S \begin{pmatrix} -1 \\ 1 \\ 0 \\ 0 \\ 0 \end{pmatrix}. \qquad (5.19)$$

Comparing the models by Arcede et al. and Gatto et al. taking all the aforementioned simplifications of both studies into account, we see from Eqs. (5.13) and (5.19) that they involve identical autonomous four-variable subsystems if we put $\gamma_a = \gamma_A$, put $\nu_E = \nu_s = 0$, put $\gamma_s = \gamma_I$, and replace f by $1-f$. Irrespective of this analogy between the two models, the four-variable subsystem S, E, A, I described by Eq. (5.18) can be addressed from the amplitude space perspective presented in Chap. 2. In doing so, the model (5.18) can be expressed in terms of Eq. (5.9) with the help of a four-dimensional autonomous amplitude description given again in terms of Eq. (5.14) for $k = 1, \ldots, 4$ and a driven system defined by

$$\frac{\mathrm{d}}{\mathrm{d}t}R_2 = \gamma_A A(\mathbf{A}) + \gamma_I I(\mathbf{A}) \qquad (5.20)$$

with $\mathbf{A} = (A_1, \ldots, A_4)$.

Stability analyses of the Arcede et al. (2020) and Gatto et al. (2020) models

From the amplitude space descriptions in terms of Eqs. (5.14), (5.15), and (5.20) involving four-dimensional autonomous amplitude systems it follows that the stability of the disease-free fixed point of the SEIAR models (5.12) and (5.18) is determined by the stability of the four-dimensional amplitude systems. The stability of those amplitude systems, that is, the stability of the disease-free fixed point with $A_1 = A_2 = A_3 = A_4 = 0$ in amplitude space is in turn determined by the eigenvalues $\lambda_1, \ldots, \lambda_4$. From the evolution equation $\mathrm{d}S/\mathrm{d}t = -k_0 S$ occurring in Eqs. (5.12) and (5.18) it follows that in the four-dimensional subspace spanned by $X_1, \ldots, X_4$ at the fixed point $X_1 = N$, $X_2 = X_3 = X_4 = 0$ the systems exhibit the eigenvalue $\lambda_1 = 0$

related to the eigenvector $\mathbf{v}_1 = (1, 0, 0, 0)$. The remaining eigenvalues $\lambda_2, \ldots, \lambda_4$ need to be determined analytically by more detailed calculations or with the help of numerical methods.

5.4 Eigenvalues and Eigenvectors Revisited: Explicit Approaches

5.4.1 Road Map: Asking and Solving Nonlinear Physics Questions

In the context of the standard SEIR model (5.6) and SEIR-like models (5.3), in the subsequent sections, the following questions will be asked and the corresponding solutions will be derived.

About the stability analysis of the disease-free fixed point of SEIR models

We ask

- Under what conditions is there an infectious disease outbreak? That is, under which conditions is the disease-free fixed point unstable.
- Under what conditions does an initial epidemic subside? That is, under which conditions is the disease-free fixed point neutrally stable or asymptotically stable?
- What is the initial direction of an infectious disease outbreak in the SEIR state space? That is, what is the order parameter of the system?
- How do the compartment sizes evolve relative to each other during an outbreak?

A solution for the first two questions has been presented in Sect. 3.7 using an ad-hoc analysis method. In the following sections, a more systematic approach will be presented based on the nonlinear physics principles presented in Chap. 2 and similar to the one presented in Chap. 4 for the SIR model. The advantage of the systematic approach is that it can be applied to all kind of epidemic models (e.g., the SEIAR models as reviewed in Sect. 5.3.2).

About the mathematical tools to conduct the stability analysis and identify the order parameter

In order to conduct the stability analysis, at issue is what tools are needed to determine explicitly eigenvalues and eigenvectors for a given dynamical system such as the SEIR model (5.6)? In particular, how can the unstable eigenvector of the SEIR model be derived that corresponds to the order parameter?

About the SEIR dynamics in the amplitude space

We may ask

- How do the amplitudes evolve subsequent to the initial stage that is addressed by the linear stability analysis? In other words, if one takes a step beyond the stability analysis, which addressed only the linearized SEIR model, how do the amplitudes evolve when nonlinearities play a role? How do amplitudes evolve in the general case?

- How do the functions G_k of SEIR-like models occurring in Eq. (5.5) look like?

About the mathematical tools to derive the SEIR model amplitude equations

In order to determine explicitly amplitude equations, the three approaches discussed in Sect. 2.9 can be used. In the context of the SIR model, for that purpose, the scalar calculation method was used (see Sect. 4.2.5). While the scalar calculation methods as such works for any kind of problem, it becomes inconvenient when state spaces of more than two dimensions are considered. The vector calculation method provides a more convenient approach to derive analytical results in such cases. In order to conduct the vector calculation method, at issue is to determine the biorthogonal vectors related to the eigenvectors that are determined in the first place in the context of the stability analysis.

5.4.2 *Case n*

Let us begin with the derivation of eigenvalues and eigenvectors. As in Chap. 2, a dynamical system of dimension n is considered described by a state vector $\mathbf{X}$. The state evolves like $\mathrm{d}\mathbf{X}/\mathrm{d}t = \mathbf{N}(\mathbf{X})$. The evolution equation describes an epidemic model such as the SEIR model (5.6). The fixed point of interest is given by $\mathbf{X}_{st}$ and relative states with respect to the fixed point are written like $\mathbf{u} = \mathbf{X} - \mathbf{X}_{st}$. Linearization of the dynamical system at the fixed point yields the linear dynamical system $\mathrm{d}\mathbf{u}/\mathrm{d}t = L\mathbf{u}$, (see Eq. (2.16)) with the linearization matrix L defined by Eq. (2.17). Any eigenvalue λ and its corresponding eigenvector $\mathbf{v}$ of the linearized system satisfy Eq. (2.25), which can be written like

$$L\mathbf{v} = \lambda\mathbf{v} \Rightarrow (L - \lambda E)\mathbf{v} = 0 \tag{5.21}$$

with the $n \times n$ identity matrix E given by

$$E = \begin{pmatrix} 1 & 0 & \dots & 0 \\ 0 & 1 & \dots & 0 \\ \dots & \dots & \dots & \dots \\ 0 & 0 & \dots & 1 \end{pmatrix} . \tag{5.22}$$

The eigenvalues λ are obtained by putting the determinant of $L - \lambda E$ to zero like

$$|L - \lambda E| = 0 \Rightarrow \lambda^n + c_{n-1}\lambda^{n-1} + \dots c_1\lambda + c_0 = 0 . \tag{5.23}$$

As indicated, this leads to a polynomial of order n. The roots (solutions) of the polynomial are the eigenvalues of interest: $\lambda_1, \dots, \lambda_n$. Substituting a given eigenvalue

λ_i into the second relation in Eq. (5.21), and solving for $\mathbf{v}$, yields the eigenvector $\mathbf{v}_i$ associated with the eigenvalue λ_i. In general, the analytical treatment for $n = 3$ and in particular for $n \geq 4$ is mathematically involved.

Numerical approach

If the explicit numerical values of all parameters of an epidemic model are given, then the coefficients of the matrix L can typically be determined in terms of numerical values as well. Once the numerical values of matrix elements of L are given, numerical approaches can be used to determine eigenvalues and eigenvectors. Such numerical methods work well for dimensions $n = 3, n = 4$ or higher dimensions for which analytical methods become mathematically involved.

5.4.3 Case $n = 2$

In order to illustrate the analytical approach to determine eigenvalues and eigenvectors let us consider the simplest, non-trivial case, which is the case $n = 2$. In this case, the state vector reads $\mathbf{X} = (X_1, X_2)$. A perturbation $\mathbf{u}$ out of a fixed point $\mathbf{X}_{st}$ can be written like $\mathbf{u} = (\delta, \epsilon) = \mathbf{X} - \mathbf{X}_{st}$. The linear dynamical system $\mathrm{d}\mathbf{u}/\mathrm{d}t = L\mathbf{u}$ becomes

$$\frac{\mathrm{d}}{\mathrm{d}t}\begin{pmatrix}\delta \\ \epsilon\end{pmatrix} = L\begin{pmatrix}\delta \\ \epsilon\end{pmatrix} = \begin{pmatrix}L_{11} & L_{12} \\ L_{21} & L_{22}\end{pmatrix}\begin{pmatrix}\delta \\ \epsilon\end{pmatrix}, \tag{5.24}$$

where L_{ik} denote the matrix elements of L. As stated above, the eigenvector $\mathbf{v} = (v_1, v_2)$ can be obtained from $(L - \lambda D)\mathbf{v} = 0$, which reads explicitly

$$\begin{pmatrix}L_{11} - \lambda & L_{12} \\ L_{21} & L_{22} - \lambda\end{pmatrix}\begin{pmatrix}v_1 \\ v_2\end{pmatrix} = \begin{pmatrix}0 \\ 0\end{pmatrix}. \tag{5.25}$$

Putting the determinant to zero like

$$\begin{vmatrix}L_{11} - \lambda & L_{12} \\ L_{21} & L_{22} - \lambda\end{vmatrix} = 0, \tag{5.26}$$

a quadratic equation in λ can be obtained. The quadratic equation can be solved for λ and produces two eigenvalues λ_1 and λ_2. The result reads

$$\lambda_{1,2} = \frac{\hat{T}}{2} \pm \sqrt{\frac{\hat{T}^2}{4} - \hat{D}} \tag{5.27}$$

with the trace $\hat{T}$ of the matrix L defined by $\hat{T} = L_{11} + L_{22}$ and the determinant of L given by $\hat{D} = L_{11}L_{22} - L_{12}L_{21}$. The upper sign (i.e., plus sign) holds for λ_1. The lower sign (i.e., minus sign) holds for λ_2. In order to compute the eigenvector $\mathbf{v}$, the first row of the matrix equation (5.25) can be evaluated like

$$(L_{11} - \lambda)v_1 + L_{12}v_2 = 0 \Rightarrow v_1 = \frac{L_{12}}{\lambda - L_{11}} v_2 . \tag{5.28}$$

Let us put $v_2 = Z(\lambda - L_{11})$, where $Z \neq 0$ is an arbitrary factor, such that

$$\mathbf{v} = Z \begin{pmatrix} L_{12} \\ \lambda - L_{11} \end{pmatrix} . \tag{5.29}$$

The vector $\mathbf{v}$ must be normalized to 1 like $v_1^2 + v_2^2 = 1$ (or alternatively $\sqrt{v_1^2 + v_2^2} = 1$). Taking the normalization condition into account and acknowledging that there are two eigenvectors $\mathbf{v}_1$ and $\mathbf{v}_2$ for the two eigenvalues $\lambda = \lambda_1$ and $\lambda = \lambda_2$, we obtain

$$\mathbf{v}_i = \frac{1}{\sqrt{(\lambda_i - L_{11})^2 + L_{12}^2}} \begin{pmatrix} L_{12} \\ \lambda_i - L_{11} \end{pmatrix} \tag{5.30}$$

with $i = 1, 2$. Using the eigenvectors, any time-dependent relative state $\mathbf{u}(t)$ or state $\mathbf{X}(t)$ can be expressed with the help of the time-dependent amplitudes $A_1(t)$ and $A_2(t)$ like

$$\mathbf{u}(t) = A_1(t)\mathbf{v}_1 + A_2(t)\mathbf{v}_2 , \ \mathbf{X}(t) = \mathbf{X}_{st} + A_1(t)\mathbf{v}_1 + A_2(t)\mathbf{v}_2 . \tag{5.31}$$

The relations in Eqs. (5.31) describe mappings from the amplitude space (A_1, A_2) to the state space (X_1, X_2) and correspond to special cases of the general mappings defined by Eqs. (2.34) and (2.35) discussed in Sect. 2.6.3. For initial states $\mathbf{X}(t_0)$ close to the fixed point $\mathbf{X}_{st}$ (i.e., for initial amplitudes $A_1(t_0), A_2(t_0) \approx 0$) the linearized model (5.24) holds, which implies that $\mathrm{d}A_1/\mathrm{d}t = \lambda_1 A_1$ and $\mathrm{d}A_2/\mathrm{d}t = \lambda_2 A_2$ (see Eq. (2.41)). Consequently (using the initial time $t_0 = 0$), we obtain

$$\mathbf{u}(t) = \mathbf{v}_1 A_1(0) \exp\{\lambda_1 t\} + \mathbf{v}_2 A_2(0) \exp\{\lambda_2 t\} . \tag{5.32}$$

From Eq. (5.32) and the definitions of stable and unstable fixed points (see Sect. 2.3 and Eqs. (2.8) and (2.10) or Sect. 2.7 and Table 2.2) it follows that

- The fixed point $\mathbf{X}_{st}$ is asymptotically stable if
  (a) Both eigenvalues λ_1 and λ_2 are real and negative or
  (b) the two eigenvalues are complex and exhibit negative real parts.
- The fixed point $\mathbf{X}_{st}$ is unstable if
  (a) both eigenvalues λ_1 and λ_2 are real and there is at least one positive eigenvalue or
  (b) the two eigenvalues are complex and exhibit positive real parts.

Discussion of the eigenvalue equation (5.27)

Equation (5.27) can be expressed like

$$\lambda_{1,2} = \frac{\hat{T}}{2} \pm \sqrt{\frac{\hat{T}^2}{4} - \hat{D}} = \frac{\hat{T}}{2} \pm \sqrt{U} \ , \ U = \frac{\hat{T}^2}{4} - \hat{D} \ , \tag{5.33}$$

where the plus sign holds for λ_1 and the minus sign holds for λ_2, as stated in the context of Eq. (5.27). At this stage, we would like to reiterate that $\hat{T} = L_{11} + L_{22}$, while the determinant $\hat{D} = L_{11}L_{22} - L_{12}L_{21}$ depends on all four matrix coefficients. Therefore, in general, it may not be possible to vary the determinant $\hat{D}$ independently from the trace $\hat{T}$. Nevertheless, in what follows, let us consider the situation in which $\hat{T}$ can be fixed, while $\hat{D}$ can be decreased from a large positive value to a large negative value. Table 5.1 provides an overview of the six possible cases that can occur in this kind of scenario.

Let us assume that $\hat{T} < 0$ holds. For $\hat{D} > \hat{T}^2/4$ (see case A1) we have $U < 0$ and, consequently, both eigenvalues are complex. The real parts of the eigenvalues are given by $\hat{T}/2 < 0$, which implies that the fixed point is a stable focus. Decreasing $\hat{D}$ such that $\hat{T}^2/4 > \hat{D} > 0$ (see case A2) the eigenvalues become real-valued numbers. They differ from each other. However, both are negative such that in case A2 the fixed point corresponds to a stable node. Decreasing $\hat{D}$ even further such that $\hat{D} < 0$ (see case A3) the eigenvalues remain real-valued. However, λ_1 becomes positive, while λ_2 is still negative. For case A3 the fixed point is unstable. It corresponds to a saddle.

Table 5.1 Overview of different cases of stability that occur when $\hat{D}$ is decreased from a large positive value to a large negative value, while $\hat{T}$ is fixed

Case	Trace $\hat{T}$	Determinant $\hat{D}$	U	λ_i real/compl.	λ_1, λ_2	Fixed point type
A1	<0	$\hat{D} > \hat{T}^2/4 \Rightarrow$	$U < 0$	Compl.	$\mathbf{R}(\lambda_1) = \mathbf{R}(\lambda_2) = \hat{T}/2 < 0$	Stable focus
A2	<0	$0 < \hat{D} < \hat{T}^2/4 \Rightarrow$	$0 < U < \hat{T}^2/4$	Real	$\lambda_1 < 0, \lambda_2 < 0$	Stable node
A3	<0	$\hat{D} < 0 \Rightarrow$	$U > \hat{T}^2/4$	Real	$\lambda_1 > 0, \lambda_2 < 0$	Saddle
B1	>0	$\hat{D} > \hat{T}^2/4 \Rightarrow$	$U < 0$	Compl.	$\mathbf{R}(\lambda_1) = \mathbf{R}(\lambda_2) = \hat{T}/2 > 0$	Unstable focus
B2	>0	$0 < \hat{D} < \hat{T}^2/4 \Rightarrow$	$0 < U < \hat{T}^2/4$	Real	$\lambda_1 > 0, \lambda_2 > 0$	Unstable node
B3	>0	$\hat{D} < 0 \Rightarrow$	$U > \hat{T}^2/4$	Real	$\lambda_1 > 0, \lambda_2 < 0$	Saddle

Next, let us assume that $\hat{T} > 0$ holds. For $\hat{D} > \hat{T}^2/4$ (see case B1) we have $U < 0$ again (just as for case A1) and, consequently, both eigenvalues are complex. The real parts of the eigenvalues are given again by $\hat{T}/2$. However, for case B1 this implies that the real parts are positive such that the fixed point is an unstable focus. Decreasing $\hat{D}$ such that $\hat{T}^2/4 > \hat{D} > 0$ (see case B2) the eigenvalues become real-valued, positive numbers. Consequently, in case B2 the fixed point corresponds to an unstable node. Decreasing $\hat{D}$ even further such that $\hat{D} < 0$ holds (see case B3) the eigenvalues remain real-valued. However, only λ_1 is positive, while λ_2 is negative. Consequently, in case B3 the fixed point corresponds to a saddle.

On a bifurcation scenario relevant for the SEIR model

The eigenvalue equation (5.33) in combination with Table 5.1 allows to discuss bifurcations that occur in two-dimensional dynamical systems with evolution equations $\mathrm{d}\mathbf{X}/\mathrm{d}t = \mathbf{N}(\mathbf{X})$. Such bifurcations describe how asymptotically stable fixed points $\mathbf{X}_{st}$ become unstable fixed points and vice versa. Let us consider a particular bifurcation. Let us assume that the system parameters that constitute the diagonal elements L_{11} and L_{22} can be fixed such that $\hat{T} < 0$ holds and remains constant, while other parameters related to the off-diagonal elements L_{12} and L_{21} can be freely varied. Let us assume they are varied such that $\hat{D}$ decreases from a positive to a negative value. If $\hat{D} > 0$ holds for sufficiently small values of $\hat{D}$ (i.e., $\hat{D} < \hat{T}^2/4$) the system exhibits an asymptotically stable fixed point in terms of a stable node. Varying the system parameters such that $\hat{D}$ becomes zero, we reach the bifurcation point at which the stability of the fixed point changes. For $\hat{D} < 0$ the fixed point becomes unstable and is given by a saddle point. As will be shown in Sect. 5.5 (see below) and Chap. 8 this scenario is relevant for the SEIR-model and SEIR-type models.

5.5 Application: Stability Analysis of SEIR Models

5.5.1 Eigenvalues and Stability of Disease-Free States

In this section SEIR-type models are considered with the rate constant k_0 (i.e., "force of infection") defined by

$$k_0 = \frac{\beta_I I + \beta_E E}{N} \tag{5.34}$$

(see Eq. (5.1)) and the evolution equations for S, E, I that read

$$\frac{\mathrm{d}}{\mathrm{d}t} S = -k_0 S \, , \; \frac{\mathrm{d}}{\mathrm{d}t} E = k_0 S - \alpha E \, , \; \frac{\mathrm{d}}{\mathrm{d}t} I = \alpha E - \gamma I \, . \tag{5.35}$$

If, in addition, $R(t) = N - S(t) - E(t) - I(t)$ holds, then 1β and 2β SEIR models are considered, and, in this context, for $\beta_E = 0$ the standard SEIR model (3.43) is considered. In general, Eq. (5.35) describes the autonomous subsystem of 1β and

2β SEIR-type models (5.3) when demographic terms are neglected. In particular, 1β models correspond to the special case when in Eq. (5.34) the parameter β_E is put to zero.

The disease-free fixed point $\mathbf{X}_{st} = (N, 0, 0)$ is considered. Let δ, E, and I denote a perturbation out of $\mathbf{X}_{st}$ with $S = N + \delta$ and $\delta \leq 0$, then substituting $S = N + \delta$ into the evolution equation of E shown in Eq. (5.35), the intermediate result

$$\frac{\mathrm{d}}{\mathrm{d}t}E = \beta_I I + \beta_E E - \alpha E + \frac{\beta_I I + \beta_E E}{N}\delta \tag{5.36}$$

can be obtained. The first three terms on the right-hand side of the equals sign constitute the linear part of the evolution equation of E, while the expression involving the variable δ corresponds to the nonlinear part. Let us consider next the dynamics of solutions close to the fixed point $\mathbf{X}_{st}$. In this case δ, E, I correspond to small quantities and evolve according to linearized equations that can be obtained from Eqs. (5.35) and (5.36). More precisely, using the evolution equation of I in Eq. (5.35) and the linearized evolution equation for E given by Eq. (5.36) when neglecting the nonlinear part, leads to the evolution equations for E and I defined by

$$\frac{\mathrm{d}}{\mathrm{d}t}\begin{pmatrix} E \\ I \end{pmatrix} = L\begin{pmatrix} E \\ I \end{pmatrix} , \quad L = \begin{pmatrix} \beta_E - \alpha & \beta_I \\ \alpha & -\gamma \end{pmatrix} . \tag{5.37}$$

Importantly, the linearized evolution equations for E and I do not depend on δ. Since $\mathrm{d}S/\mathrm{d}t \leq 0$ holds in any case (see Eq. (5.35)), the objective is to study the dynamics of the E-I subsystem.

The trace $\hat{T}$ and the determinant $\hat{D}$ of the linearization matrix L read $\hat{T} = \beta_E - \alpha - \gamma$ and $\hat{D} = -[(\beta_E - \alpha)\gamma + \beta_I\alpha]$, respectively. From Eq. (5.27) it then follows that the eigenvalues of L are given by

$$\lambda_{1,2} = \frac{\beta_E - \alpha - \gamma}{2} \pm \sqrt{U} , \quad U = \frac{[\beta_E - \alpha - \gamma]^2}{4} + (\beta_E - \alpha)\gamma + \beta_I\alpha . \tag{5.38}$$

The eigenvalues are in any case real-valued because U can be cast into the form

$$U = \frac{[\beta_E - \alpha + \gamma]^2}{4} + \beta_I\alpha \geq 0 . \tag{5.39}$$

Consequently, the fixed point $(E, I) = (0, 0)$ in the E-I subsystem corresponds to a stable node, unstable node, or saddle under the assumption that the linearized system (5.37) determines the dynamics in the E-I subspace (i.e., the impact of the nonlinear δ term in Eq. (5.36) can be neglected). The following three cases can be studied.

Case I: $\hat{\mathbf{T}} > 0$

This case holds only if $\beta_E > \alpha + \gamma$ holds (i.e., β_E is large relative to α and γ) and it cannot occur in 1β models with $\beta_E = 0$. From $\beta_E > \alpha + \gamma$ it follows that $\beta_E > \alpha$ and, consequently, $U > \hat{T}^2/4$ (see Eq. (5.38)). Furthermore, from Eq. (5.38) and

$U > \hat{T}^2/4$ it follows that $\lambda_1 > 0$ and $\lambda_2 < 0$. The fixed point $(E, I) = (0, 0)$ is an unstable fixed point in terms of a saddle. Consequently, the SEIR model describes epidemic outbreaks in terms of wave-solutions. The case I corresponds to the case 3B described in Table 5.1 with $\hat{T} > 0$, $\hat{D} < 0$, and $U > \hat{T}^2/4$. Note that the classification of $(E, I) = (0, 0)$ as a saddle point only holds as long as $S(t) \approx N$ or $\delta \approx 0$. That is, $(E, I) = (0, 0)$ reflects a saddle point in the initial stage of the wave-solution under consideration. When time elapses the nonlinear term (i.e., the δ term) in Eq. (5.36) becomes relevant. This term makes that for $t \to \infty$ the dynamics converges to $(E, I) = (0, 0)$ in analogy to the discussion for the standard SEIR model presented in Sect. 3.7.

Case II: $\hat{\mathbf{T}} < 0$ and bifurcations leading to subsiding epidemics

Case II holds for $\beta_E < \alpha + \gamma$ and, in particular, for 1β models with $\beta_E = 0$. In this case, we deal either with case 2A or 3A of Table 5.1. The stability depends on the sign of the determinant $\hat{D}$. Therefore, let us rewrite the determinant like

$$\hat{D} = -[(\beta_E - \alpha)\gamma + \beta_I \alpha] = \alpha \left(\gamma - [\beta_I + \frac{\gamma}{\alpha} \beta_E] \right) . \tag{5.40}$$

Introducing the weighed effective contact rate [31, 32]

$$\beta_w = \beta_I + \frac{\gamma}{\alpha} \beta_E \tag{5.41}$$

with $\hat{D} = \alpha(\gamma - \beta_w)$, we see that $\beta_w > \gamma$ implies $\hat{D} < 0$. Provided that the additional constraint $\delta \approx 0$ holds, the inequality $\hat{D} < 0$, in turn, implies that $(E, I) = (0, 0)$ corresponds to a saddle and the SEIR model exhibits wave-solutions. In contrast, $\beta_w < \gamma$ implies $\hat{D} > 0$ and $(E, I) = (0, 0)$ corresponds to a stable node. In this case, the SEIR model describes subsiding epidemics.

The case II is of particular interest when discussing the subsiding of epidemics due to the impact of intervention measures. Let us assume a novel infectious disease such as COVID-19 leads to an epidemic outbreak with $\beta_w > \gamma$. Let us further assume that as a reaction to the outbreak intervention measures are implemented that reduce the effective contact rates β_I and β_E. For example, in the context of the COVID-19 pandemic physical distancing, home office, shutdown of businesses and schools may reduce the contact rates ν_I and ν_E and wearing face masks may reduce the probabilities p_I and p_E of susceptibles to get infected in such contacts. Decreasing the effective contact rates β_I and β_E implies that β_w decreases. If the parameter decreases below the critical value of $\beta_{w,crit} = \gamma$, then the disease-free fixed point becomes stable. The epidemic in a certain country or region begins to subside. This scenario can be interpreted as the backwards scenario of the scenario discussed at the end of Sect. 5.4.3. In the final paragraph of Sect. 5.4.3, a bifurcation is discussed that involves a stable node that turns into a unstable saddle and takes place when the value of $\hat{D}$ is decreased such that it switches from a positive to a negative value. In contrast, the intervention-induced bifurcation scenario describes a bifurcation in terms of a unstable disease-free state that turns into a stable state when intervention measures reduce β_w below the critical value γ such that $\hat{D} = \alpha(\gamma - \beta_w)$ switches from a negative to a positive value.

Case III: The 1β model with $\beta_\mathrm{E} = 0$

In the case $\beta_E = 0$ and $\beta_I = \beta$ the trace satisfies $\hat{T} = -(\alpha + \gamma) < 0$, just as in case II. Consequently, $\hat{D}$ determines again the stability of the disease-free fixed point. Importantly, $\beta_w = \beta$ holds. Consequently, for $\beta > \gamma$ (and assuming that $\delta \approx 0$ holds) the fixed point $(E, I) = (0, 0)$ corresponds to a saddle and the SEIR model exhibits wave-solutions describing epidemic outbreaks. For $\beta < \gamma$ the fixed point is stable and the model describes subsiding epidemics. Consequently, for 1β models, the effective contact rate β can be used as a bifurcation parameter that exhibits a critical value $\beta_{crit} = \gamma$. Alternatively, the dimensionless parameter

$$\kappa = \frac{\beta}{\gamma} \tag{5.42}$$

may be considered as bifurcation parameter with critical value $\kappa_{crit} = 1$. The bifurcation parameter κ corresponds to the stability parameter $\xi = \beta/\gamma$ (see Eq. (3.26)) that was derived by means of an ad-hoc approach in Sect. 3.7 for the standard SEIR model. In Sect. 3.7 the stability parameter $\xi = \beta/\gamma$ was derived under the assumption that the initial state $\mathbf{X}_0$ is close to the fixed point $\mathbf{X}_{st} = (N, 0, 0)$ (i.e., $S(0) \approx N$). This assumption holds for the stability analysis that yields Eq. (5.42). Therefore, the two approaches address the same circumstances and produce consistent results.

The fixed point $(E, I) = (0, 0)$ and the fixed point $\mathbf{X}_{st} = (N, 0, 0)$

In the previous discussion the focus was on the two-dimensional E-I subspace. It was shown that in this space the fixed point $(E, I) = (0, 0)$ corresponds to a stable node, unstable node, or saddle assuming that the state $\mathbf{X}$ in the three-dimensional state space is sufficiently close to the fixed point $\mathbf{X}_{st} = (N, 0, 0)$ (i.e., $(\delta, E, I) \approx (0, 0, 0)$ holds). This classification holds when ignoring the dynamics of the susceptibles. With respect to the three-dimensional state space spanned by the variables S, E, and I at issue is to account for the neutral direction along the S-axis. That is, when fixed points $\mathbf{X}_{st} = (S_{st}, 0, 0)$ of the SEIR model (5.35) are shifted along the axis describing the number of susceptibles of the population of interest, then they end up again on fixed points. Consequently, if the fixed point $(E, I) = (0, 0)$ for $\delta \approx 0$ and $\beta_w < \gamma$ corresponds to an asymptotically stable fixed point in terms of a stable node in the E-I subspace, then the corresponding fixed point $\mathbf{X}_{st} = (N, 0, 0)$ only corresponds to a neutrally stable fixed point. Perturbations out of $\mathbf{X}_{st} = (N, 0, 0)$ do not return to $\mathbf{X}_{st} = (N, 0, 0)$. They converge to one of the fixed points $\mathbf{X}_{st} = (S_{st}, 0, 0)$ with $S_{st} < N$.

The fixed points $\mathbf{X}_{st} = (\mathbf{S}_{st}, \mathbf{0,0})$ with $\mathbf{S}_{st} < \mathbf{N}$ and (E,I) = (0,0)

The stability of fixed points $\mathbf{X}_{st} = (S_{st}, 0, 0)$ with $S_{st} < N$ can be discussed by analogy to the discussion of the fixed point with $S_{st} = N$. The key step is to put

$S = S_{st} + \delta$ rather than $S = N + \delta$. This implies that all results obtained in this section have to be modified by replacing the effective contact rates β_I and β_E by reduced effective contact rates

$$\beta_{I,r} = \frac{S_{st}}{N}\beta_I \ , \ \beta_{E,r} = \frac{S_{st}}{N}\beta_E \tag{5.43}$$

that are reduced by a factor $S_{st}/N < 1$. That is, in Eqs. (5.36) to (5.42) the substitutions $\beta_I \rightarrow \beta_{I,r}$ and $\beta_E \rightarrow \beta_{E,r}$ need to be made. For example, β_w and κ as defined in Eqs. (5.41) and (5.42), respectively, become

$$\beta_{w,r} = \frac{S_{st}}{N}\left(\beta_I + \frac{\gamma}{\alpha}\beta_E\right) \ , \ \kappa_r = \frac{S_{st}}{N}\frac{\beta}{\gamma} \ . \tag{5.44}$$

For the 2β models it follows that if $\beta_{w,r} > \gamma$ ($\beta_{w,r} < \gamma$) holds, then $\mathbf{X}_{st} = (S_{st}, 0, 0)$ is an unstable (neutrally stable) fixed point and $(E, I) = (0, 0)$ corresponds to a unstable (asymptotically stable) fixed point. Likewise, for 1β models it follows that if $\kappa_r > 1$ ($\kappa_r < 1$) holds, then $\mathbf{X}_{st} = (S_{st}, 0, 0)$ is an unstable (neutrally stable) fixed point and $(E, I) = (0, 0)$ corresponds to an unstable (asymptotically stable) fixed point. These classifications hold under the constraint that the state $\mathbf{X}$ is sufficiently close to the fixed point $\mathbf{X}_{st} = (S_{st}, 0, 0)$. In particular, the constraint $\delta \approx 0$ must be satisfied.

Note that if the fixed point $\mathbf{X}_{st} = (N, 0, 0)$ is considered, we have $\delta \leq 0$. In contrast, if $\mathbf{X}_{st} = (S_{st}, 0, 0)$ is considered in the context of four-variable SEIR models with $R(t) = N - S(t) - E(t) - I(t)$, then $S \in [0, N]$ holds, which implies $\delta \in [-S_{st}, N - S_{st}]$. That is, δ may assume positive values.

5.5.2 EI Order Parameters of SEIR Models in E-I Subspaces

The eigenvectors of SEIR-type models in their E-I subspaces can be computed from Eq. (5.30). Let us consider 1β models with $\beta_I = \beta$ and $\beta_E = 0$. Then, substituting the matrix coefficients of L defined in Eq. (5.37) for $\beta_E = 0$ into Eq. (5.30), we obtain [13, 31]

$$\mathbf{v}_i = \begin{pmatrix} v_{i,E} \\ v_{i,I} \end{pmatrix} = \frac{1}{\sqrt{(\lambda_i - \alpha)^2 + \beta^2}}\begin{pmatrix} \beta \\ \lambda_i + \alpha \end{pmatrix} \tag{5.45}$$

for $i = 1, 2$ with λ_i defined by Eq. (5.38) and $\beta_E = 0$, again. In the case of an epidemic outbreak $\lambda_1 > 0$ and $\lambda_2 < 0$ holds such that $\mathbf{v}_1$ is an unstable eigenvector or order parameter, while $\mathbf{v}_2$ corresponds to a stable eigenvector (see Chap. 2). Consequently, the initial stage of the epidemic evolves along $\mathbf{v}_1$.

Let us illustrate the role of the eigenvectors $\mathbf{v}_1$ and $\mathbf{v}_2$ for a wave-solution of the standard the SEIR model defined by (5.35) and $k_0 = \beta I/N$. Let us use the model

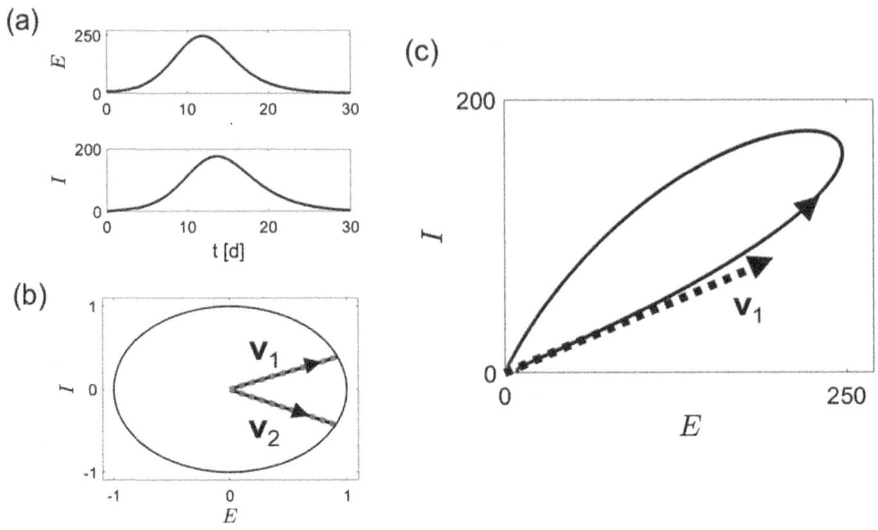

Fig. 5.2 Eigenvector analysis of the SEIR model (5.35) and EI order parameter (unstable eigenvector) in the E-I subspace. Panel **a**: Solutions $E(t)$ and $I(t)$ as functions of time computed from Eq. (5.35). Panel **b**: Eigenvectors $\mathbf{v}_1$ and $\mathbf{v}_2$ of the E-I subspace. Panel **c**: The phase curve $I(E)$ (solid line) computed from $E(t)$ and $I(t)$ and the unstable eigenvector $\mathbf{v}_1$ (dotted line) are shown. The unstable eigenvector was magnified by a factor for visualization purposes. $\mathbf{v}_1$ corresponds to the EI order parameter. Parameters and initial conditions: $N = 1,000$, $\beta = 2.0$/d, $\gamma = 0.5$/d, $\alpha = 0.4$/d, $E(0) = 10$, $I(0) = 0$, $S(0) = N - E(0)$

parameters $\beta = 2$/d, $\alpha = 0.4$/d, $\gamma = 0.5$/d and a small population with $N = 1000$ individuals. Since $\beta > \gamma$ holds, the fixed point $\mathbf{X}_{st} = (N, 0, 0)$ is unstable and the model solution describes an epidemic wave. Figure 5.2 illustrates the eigenvector analysis of the wave-solution. In order to produce Fig. 5.2, the SEIR model (5.35) was solved for an initial state $\mathbf{X}_0$ close to the stationary state $\mathbf{X}_{st} = (N, 0, 0)$. Panel (a) shows the trajectories E and I as function of time. In fact, the same parameters as for the simulation presented in panel (b) in Fig. 3.12 were used. Consequently, the curves $E(t)$ and $I(t)$ shown in panel (a) of Fig. 5.2 are identical with those presented in panel (b) of Fig. 3.12. The eigenvectors $\mathbf{v}_1$ and $\mathbf{v}_2$ were computed from Eq. (5.45). They are shown as solid lines in panel (b) of Fig. 5.2. The circle describes the unit circle and indicates that the eigenvectors are normalized with respect to 1. The two eigenvectors were also computed numerically from the matrix L shown in Eq. (5.37). The numerically obtained eigenvectors are shown as gray dotted lines. As expected, the numerical solutions were identical with the analytical ones.

Panel (c) shows the functions $I(t)$ versus $E(t)$ as phase curve $I(E)$ in the E-I subspace. Consistent with the explicit functions $I(t)$ and $E(t)$ shown in panel (a), in panel (c) when following the phase curve the following can be seen. The simulated epidemic under consideration evolves away from the unstable fixed point $(E, I) = (0, 0)$. Both E and I increase over time until E reaches a maximum value and starts to decay. For a short period I increases even further. Subsequently, I reaches its maximum and starts to decay as well. During the remaining course of the epidemic both the numbers of exposed and infectious individuals decrease monotonically. The loop-shaped trajectory shown in panel (c) illustrates the observation made in Sect. 3.7 that the wave of the exposed individuals is leading and reaches its peak earlier as compared to the wave of infectious individuals. Importantly, panel (c) of Fig. 5.2 shows the the unstable eigenvector $\mathbf{v}_1$ (dotted black line). The vector was magnified for illustration purposes. It can be seen that the wave-solution followed in the initial stage the direction of $\mathbf{v}_1$. Consequently, the initial stage of the simulated outbreak was determined by the unstable eigenvector and the exponential increase of the amplitude $A_1(t)$ (see Eqs. (5.31) and (5.32)). During that initial stage the direction $\mathbf{v}_2$ and the corresponding amplitude A_2 plays a negligible contribution to the overall dynamics. Subsequent to the initial stage, the solution $\mathbf{X}(t)$ (or the phase curve $I(E)$)) branches off from the direction $\mathbf{v}_1$, performs a loop and returns to the fixed point $(E, I) = (0, 0)$. Since the dynamics in the initial stage is dominated by $\mathbf{v}_1$, the unstable eigenvector $\mathbf{v}_1$ is the order parameter of the simulated epidemic (see Sect. 2.8).

5.6 Biorthogonal Vectors of Amplitude Spaces: 2D, 3D, and Beyond

Amplitude equations of SEIR-type models provide an amplitude space description of epidemics that goes beyond the initial stage. The whole course of an epidemic can be studied in amplitude space. In order to derive amplitude equations using the vector calculation method (see Sect. 2.9), biorthogonal vectors need to be determined (see also the road map in Sect. 5.4.1).

Case $n = 2$

To begin with, the explicit construction of biorthogonal vectors $\mathbf{w}_i$ for two-dimensional systems is discussed. The departure point is the dynamical system $\mathrm{d}\mathbf{X}/\mathrm{d}t = \mathbf{N}(\mathbf{X})$ with $\mathbf{X} = (X_1, X_2)$ discussed in Sect. 5.4.3 exhibiting at the fixed point $\mathbf{X}_{st}$ two (right) eigenvectors $\mathbf{v}_i$ defined by Eq. (5.30). Let us write the eigenvectors in terms of

$$\mathbf{v}_1 = \begin{pmatrix} v_{1,1} \\ v_{1,2} \end{pmatrix}, \quad \mathbf{v}_2 = \begin{pmatrix} v_{2,1} \\ v_{2,2} \end{pmatrix}. \tag{5.46}$$

It is assumed that the vectors are linearly independent such that they span the two-dimensional space (X_1, X_2). That is, any relative state (or perturbation) $\mathbf{u} = \mathbf{X} - \mathbf{X}_{st}$ can be expressed with the help of the amplitudes A_1 and A_2 like $\mathbf{u} = A_1\mathbf{v}_1 + A_2\mathbf{v}_2$ and any state can be expressed like $\mathbf{X} = \mathbf{X}_{st} + A_1\mathbf{v}_1 + A_2\mathbf{v}_2$ (see also Eq. (5.31)). These relations describe a mapping from the amplitude space to the state space. The biorthogonal eigenvectors (or left eigenvectors) $\mathbf{w}_1$ and $\mathbf{w}_2$ can be used to invert that mapping and, as mentioned above, to derive the evolution equations for the amplitudes A_1 and A_2.

The biorthogonal vectors $\mathbf{w}_1$ and and $\mathbf{w}_2$ may be constructed as follows. First, let us define the 2×2 eigenvector matrix M by

$$M = \begin{pmatrix} v_{1,1} & v_{2,1} \\ v_{1,2} & v_{2,2} \end{pmatrix} . \tag{5.47}$$

The matrix consists of columns defined by $\mathbf{v}_i$. Second, let us consider the inverse matrix M^{-1}. In this context, we note that the determinant $|M|$ of M does not vanish because $\mathbf{v}_i$ are assumed to be linearly independent. Explicitly, the determinant reads $|M| = v_{1,x}v_{2,y} - v_{2,x}v_{1,y}$. This implies that M^{-1} exists. Moreover, M^{-1} satisfies

$$M^{-1}M = \begin{pmatrix} 1 & 0 \\ 0 & 1 \end{pmatrix} . \tag{5.48}$$

Let

$$M^{-1} = \begin{pmatrix} a & b \\ c & d \end{pmatrix} . \tag{5.49}$$

Then from Eq. (5.48) the following vector multiplications (i.e., dot products) hold:

$$\begin{pmatrix} a \\ b \end{pmatrix} \mathbf{v}_1 = 1 , \begin{pmatrix} a \\ b \end{pmatrix} \mathbf{v}_2 = 0 , \begin{pmatrix} c \\ d \end{pmatrix} \mathbf{v}_1 = 0 , \begin{pmatrix} c \\ d \end{pmatrix} \mathbf{v}_2 = 1 . \tag{5.50}$$

Consequently, let us define the biorthogonal vectors as $\mathbf{w}_1 = (a, b)$ and $\mathbf{w}_2 = (c, d)$. Then, the dot product (or scalar product) between a vector $\mathbf{w}_i$ and an eigenvector $\mathbf{v}_k$ reads

$$\mathbf{w}_i\mathbf{v}_k = \delta_{ik} \tag{5.51}$$

as discussed in Sect. 2.6.2 (see Eq. (2.28)). That is, we have obtained a biorthogonal system (see Sect. 2.6.2 again). As mentioned in Sect. 2.6.2, the vectors $\mathbf{w}_i$ are also referred to as left eigenvectors. Explicitly, from Eqs. (5.47) and (5.48) it follows that

$$M^{-1} = \frac{1}{|M|} \begin{pmatrix} v_{2,2} & -v_{2,1} \\ -v_{1,2} & v_{1,1} \end{pmatrix} , \tag{5.52}$$

which can be seen by substituting Eqs. (5.47) and (5.52) into Eq. (5.48). Comparing Eqs. (5.49) and (5.52) and taking the result $\mathbf{w}_1 = (a, b)$ and $\mathbf{w}_2 = (c, d)$ obtained above into account, the explicit form of the biorthogonal vectors can be obtained as

$$\mathbf{w}_1 = \frac{1}{|M|} \begin{pmatrix} v_{2,2} \\ -v_{2,1} \end{pmatrix} , \quad \mathbf{w}_2 = \frac{1}{|M|} \begin{pmatrix} -v_{1,2} \\ v_{1,1} \end{pmatrix} . \tag{5.53}$$

Note that $|\mathbf{w}_i| = 1/|M|$ holds which implies that the vectors $\mathbf{w}_i$ are not necessarily normalized to 1.

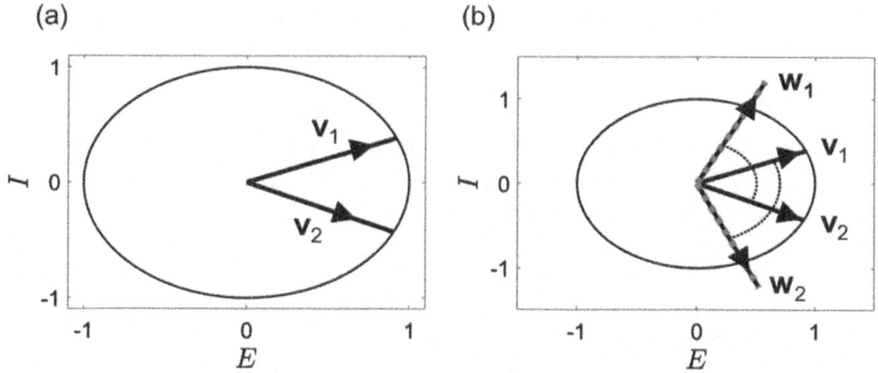

Fig. 5.3 Construction of biorthogonal vectors of the SEIR model in the E-I subspace. Panel (**a**) shows $\mathbf{v}_1$ and $\mathbf{v}_2$ as in panel (**b**) of Fig. 5.2. Panel **b** shows again $\mathbf{v}_1$ and $\mathbf{v}_2$. In addition, the biorthogonal vectors $\mathbf{w}_1$ and $\mathbf{w}_2$ are depicted. Dotted arcs indicate 90° angles. Parameters as in Fig. 5.2

Let us illustrate this construction for the eigenvectors $\mathbf{v}_i$ defined by Eq. (5.45) in the in E-I subspace of 1β SEIR-type models (5.3). More precisely, let us consider the eigenvectors shown in panel (b) of Fig. 5.2. For illustration purposes, these two eigenvectors $\mathbf{v}_1$ and $\mathbf{v}_2$ are shown again in panel (a) of Fig. 5.3. Equation (5.45) provides us with the coefficients $v_{i,E}$ and $v_{i,I}$ that corresponds to $v_{i,1}$ and $v_{i,2}$ given that $\mathbf{X} = (X_1, X_2) = (E, I)$. Substituting $v_{i,1} = v_{i,E}$ and $v_{i,2} = v_{i,I}$ as defined by Eq. (5.45) into Eq. (5.53) we arrive at the analytical expressions of the vectors $\mathbf{w}_1$ and $\mathbf{w}_2$ in terms of the model parameters β, α, and γ. Panel (b) of Fig. 5.3 presents $\mathbf{w}_1$ and $\mathbf{w}_2$ thus obtained as solid black lines. In addition, the biorthogonal vectors $\mathbf{w}_i$ were obtained numerically. To this end, the coefficients of the matrix M (see Eq. (5.47)) were computed from Eq. (5.45) again. Subsequently, M was numerically inverted to obtain M^{-1}. In a final step $\mathbf{w}_1 = (a, b)$ and $\mathbf{w}_2 = (c, d)$ were extracted as the row vectors of M^{-1} (see Eq. (4.52)). The numerically computed vectors $\mathbf{w}_i$ are shown in panel (b) as gray dotted lines. As can be seen, the numerically obtained results are identical with the results obtained via the analytical approach. Panel (b) also presents again the eigenvectors $\mathbf{v}_1$ and $\mathbf{v}_2$. The circle in panel (b) is the unit circle and indicates again that $\mathbf{v}_1$ and $\mathbf{v}_2$ are normalized to unity. In contrast, for the SEIR model parameters under consideration, the norm of the vectors $\mathbf{w}_i$ exceeds 1.

Panel (b) illustrates the meaning of the biorthogonal condition (5.51). Accordingly, $\mathbf{w}_2$ is orthogonal to $\mathbf{v}_1$. The angle between $\mathbf{v}_1$ and $\mathbf{w}_2$ is indicated by a dotted line and corresponds to a 90° angle. Likewise, $\mathbf{w}_1$ is orthogonal to $\mathbf{v}_2$. Again, the angle between $\mathbf{v}_2$ and $\mathbf{w}_1$ is indicated by a dotted line and corresponds to a 90° angle. Note that the reader probably will experience some difficulties to see these angles as 90° angles. The reason for this is that the angles only show up as "right angles" if physically the horizontal and vertical axes are printed in the same scales. That is, if the distances on the horizontal and vertical axis between the marks 0 and 1 are physically the same in the print out (or on the screen) on which Fig. 5.3 is presented, then the angles show up as 90° angles. If this is not the case, then it requires some imagination to see the angles as what they should indicate: 90° angles.

Case n = 3

For a three-dimensional problem, the biorthogonal vectors can be constructed in a similar way. The mapping from amplitude space (A_1, A_2, A_3) to state space (X_1, X_2, X_3) via the relative state $\mathbf{u}$ reads

$$\mathbf{u} = A_1\mathbf{v}_1 + A_2\mathbf{v}_2 + A_3\mathbf{v}_3 \,. \tag{5.54}$$

The 3×3 eigenvector matrix M is defined by

$$M = (\mathbf{v}_1\mathbf{v}_2\mathbf{v}_3) \,, \tag{5.55}$$

where the eigenvectors appear as columns in the matrix. That is, $\mathbf{v}_1$ describes the elements in the first column of M, $\mathbf{v}_2$ corresponds to the elements in the second column, and $\mathbf{v}_3$ constitutes the elements in the third column. Assuming that the eigenvectors are linearly independent, the inverse matrix M^{-1} exists. The inverse matrix M^{-1} formally reads

$$M^{-1} = \begin{pmatrix} a & b & c \\ d & e & f \\ g & h & j \end{pmatrix} \,. \tag{5.56}$$

Using the matrix coefficients defined in Eq. (5.56), the biorthogonal eigenvectors can be constructed like

$$\mathbf{w}_1 = \begin{pmatrix} a \\ b \\ c \end{pmatrix} \,, \quad \mathbf{w}_2 = \begin{pmatrix} d \\ e \\ f \end{pmatrix} \,, \quad \mathbf{w}_3 = \begin{pmatrix} g \\ h \\ j \end{pmatrix} \,. \tag{5.57}$$

Just as in the case $n = 2$, the biorthogonal vectors correspond to the rows of the inverse matrix. Since the matrix product of M^{-1} and M yields the identity matrix, from Eqs. (5.55), (5.56), and (5.57) it follows that the dot product (scalar product) between a pair $\mathbf{v}_i$ and $\mathbf{w}_k$ is zero for $i \neq k$ and equals 1 for $i = k$. That is, the construction yields for the three-dimensional case the required biorthogonal relation defined by Eq. (5.51).

As an alternative approach, the biorthogonal vectors may be defined via the vector cross-product. For example, $\mathbf{w}_1$ can be defined by

$$\mathbf{w}_1 = \frac{1}{Z}(\mathbf{v}_2 \times \mathbf{v}_3) \tag{5.58}$$

with $Z = \mathbf{v}_1(\mathbf{v}_2 \times \mathbf{v}_3)$, which measure the volume spanned by the three vectors. The cross-product of two vectors is orthogonal to the plane spanned by the two vectors. Consequently, $\mathbf{w}_1$ satisfies the required orthogonal relations

$$\mathbf{w}_1\mathbf{v}_2 = 0 \,, \ \mathbf{w}_1\mathbf{v}_3 = 0 \,. \tag{5.59}$$

By analogy, the vectors $\mathbf{w}_2$ and $\mathbf{w}_3$ can be defined by

$$\mathbf{w}_2 = \frac{1}{Z}(\mathbf{v}_3 \times \mathbf{v}_1) \,, \mathbf{w}_3 = \frac{1}{Z}(\mathbf{v}_1 \times \mathbf{v}_2) \,. \tag{5.60}$$

General case *n*

The general case was previous discussed in Sect. 2.6.2 and will be reviewed in what follows for sake of completeness. Let us consider a n-dimensional dynamical system exhibiting a fixed point $\mathbf{X}_{st}$. Then, the relative state $\mathbf{u}$ can be expressed in terms of the superposition $\mathbf{u} = \sum_{i=1}^{n} A_i \mathbf{v}_i$ involving n eigenvectors $\mathbf{v}_i$ (see Eq. (2.34)). Assuming that these eigenvectors are linearly independent, the biorthogonal vectors can be obtained by means of the $n \times n$ eigenvector matrix

$$M = (\mathbf{v}_1 \cdots \mathbf{v}_n) \,. \tag{5.61}$$

The inverse matrix

$$M^{-1} = \begin{pmatrix} \mathbf{w}_1 \\ \cdots \\ \mathbf{w}_n \end{pmatrix} \tag{5.62}$$

is composed of the biorthogonal vectors $\mathbf{w}_i$. As indicated in Eq. (5.62), they correspond to the rows of the matrix M^{-1}.

5.7 Applications and SEI Order Parameters

5.7.1 1β SEIR Model and Its 3D Autonomous Amplitude Description

Having discussed the explicit construction of biorthogonal vectors, the amplitude equations of SEIR models can be derived using the vector calculation method. To this end, let us first consider the SEIR model defined by Eq. (5.35) and $R(t) = N - E(t) - I(t)$ and for the sake of simplicity the 1β model with $k_0 = \beta I/N$. Using vector and matrix notations, the model for the autonomous three-variable system composed of S, E, I reads

$$\frac{\mathrm{d}}{\mathrm{d}t}\begin{pmatrix} S \\ E \\ I \end{pmatrix} = \begin{pmatrix} 0 & 0 & 0 \\ 0 & -\alpha & 0 \\ 0 & \alpha & -\gamma \end{pmatrix}\begin{pmatrix} S \\ E \\ I \end{pmatrix} + \frac{\beta}{N}SI\begin{pmatrix} -1 \\ 1 \\ 0 \end{pmatrix} \,. \tag{5.63}$$

Let us consider the fixed point $\mathbf{X}_{st} = (N, 0, 0)$. Substituting $S = N + \delta$ into Eq. (5.63), where δ is the first component of the relative state vector $\mathbf{u} = (\delta, E, I)$, leads to

$$\frac{\mathrm{d}}{\mathrm{d}t}\begin{pmatrix}\delta\\ E\\ I\end{pmatrix} = L\begin{pmatrix}\delta\\ E\\ I\end{pmatrix} + \frac{\beta}{N}\delta I\begin{pmatrix}-1\\ 1\\ 0\end{pmatrix} \tag{5.64}$$

with

$$L = \begin{pmatrix}0 & 0 & -\beta\\ 0 & -\alpha & \beta\\ 0 & \alpha & -\gamma\end{pmatrix} \tag{5.65}$$

(see also Eq. (2.65)). Comparing Eq. (5.64) with the general case $\mathrm{d}\mathbf{u}/\mathrm{d}t = L\mathbf{u} + \mathbf{R}$ (see Eq. (2.89)) discussed in Sect. 2.9.3, the remainder term $\mathbf{R}$ can be identified as

$$\mathbf{R}(\mathbf{u}, \mathbf{X}_{st}) = \frac{\beta}{N}\delta I\begin{pmatrix}-1\\ 1\\ 0\end{pmatrix} . \tag{5.66}$$

As previously discussed in Sect. 2.9.2, the matrix L defined by Eq. (5.65) exhibits the eigenvalue $\lambda_1 = 0$ and two additional eigenvalues $\lambda_{2,3}$ that can be computed from the 2×2 matrix shown in Eq. (2.67) (or in Eq. (5.37) for $\beta_E = 0$ and $\beta_I = \beta$). In particular, from Eq. (5.38), $\beta_E = 0$, and $\beta_I = \beta$ it follows that

$$\lambda_{2,3} = -\frac{\alpha+\gamma}{2} \pm \sqrt{\frac{(\alpha+\gamma)^2}{4} + \alpha(\beta-\gamma)} , \tag{5.67}$$

where the upper (plus) sign holds for λ_2 and the lower (minus) sign for λ_3. As discussed in Sect. 5.5.1, this implies that for $\beta > \gamma$ the fixed point $\mathbf{X}_{st}$ is unstable, while for $\beta < \gamma$ the fixed point $\mathbf{X}_{st}$ is neutrally stable.

The matrix L defined by Eq. (5.65) exhibits the Eigenvector $\mathbf{v}_1 = (1, 0, 0)$ associated with the eigenvalue $\lambda_1 = 0$. In order to determine the eigenvectors associated to $\lambda_{2,3}$, the eigenvectors (5.45) derived for the two-dimensional E-I subspace can be used as building blocks. Accordingly, let us put

$$\mathbf{v}_i \propto \begin{pmatrix}s\\ \beta\\ \lambda_i + \alpha\end{pmatrix} \tag{5.68}$$

with $i = 2, 3$ and s unknown. The coefficient s can be determined from $(L - \lambda_i E)\mathbf{v} = 0$ (see Eq. (5.21)). From the first row of this matrix equation it follows that

$$\lambda_i s + \beta(\lambda_i + \alpha) = 0 , \tag{5.69}$$

which implies $s = -\beta(\lambda_j + \alpha)/\lambda_j$ such that

$$\mathbf{v}_i \propto \begin{pmatrix} -\beta(\lambda_j+\alpha)/\lambda_j \\ \beta \\ \lambda_j+\alpha \end{pmatrix} = \frac{1}{\lambda_j}\begin{pmatrix} -\beta(\lambda_j+\alpha) \\ \beta\lambda_j \\ \lambda_j(\lambda_j+\alpha) \end{pmatrix} . \tag{5.70}$$

Taking the normalization of $\mathbf{v}_i$ into account, gives us the final result

$$\mathbf{v}_i = \frac{1}{Z_i}\begin{pmatrix} -\beta(\lambda_i+\alpha) \\ \beta\lambda_i \\ \lambda_i(\lambda_i+\alpha) \end{pmatrix} = \begin{pmatrix} v_{i,S} \\ v_{i,E} \\ v_{i,I} \end{pmatrix} \tag{5.71}$$

with $Z_i = \sqrt{\beta^2[(\lambda_i+\alpha)^2+\lambda_i^2]+\lambda_i^2(\lambda_i+\alpha)^2}$ and $i = 2, 3$ (see also Ref. [33] with $T_{st} = 1$). Note that the eigenvectors (5.71) differ from the SEIR model vectors $\mathbf{v}_2$ and $\mathbf{v}_3$ derived in Sect. 2.9.2. We will dwell on the two different perspectives in Chap. 6. In summary, relative states $\mathbf{u}$ of the SEIR model can be expressed like

$$\mathbf{u} = \begin{pmatrix} \delta \\ E \\ I \end{pmatrix} = A_1 \begin{pmatrix} 1 \\ 0 \\ 0 \end{pmatrix} + A_2\mathbf{v}_2 + A_3\mathbf{v}_3 \tag{5.72}$$

with $\mathbf{v}_i$ defined by Eq. (5.71). Equation (5.72) also describes the mapping from the amplitude space (A_1, A_2, A_3) to the state space $(X_1 = S, X_2 = E, X_3 = I)$ via $\mathbf{X} = \mathbf{X}_{st} + \mathbf{u}$.

The biorthogonal vectors $\mathbf{w}_i$ may be derived in a systematic manner as described in Sect. 5.6 by inverting the eigenvector matrix M defined by

$$M = \begin{pmatrix} 1 & v_{2,S} & v_{3,S} \\ 0 & v_{2,E} & v_{3,E} \\ 0 & v_{2,I} & v_{3,I} \end{pmatrix} . \tag{5.73}$$

A short-cut approach can be applied in view of the relatively simple first eigenvector $\mathbf{v}_1 = (1, 0, 0)$. Since $\mathbf{v}_1$ exhibits vanishing coefficients in the E-I subspace, the biorthogonal eigenvectors derived previous for two-dimensional spaces as defined by Eq. (5.53) can be used to construct the second and third coefficients of $\mathbf{w}_2$ and $\mathbf{w}_3$. The first coefficient can be put to zero such that

$$\mathbf{w}_2 = \frac{1}{|M|}\begin{pmatrix} 0 \\ v_{3,I} \\ -v_{3,E} \end{pmatrix} , \quad \mathbf{w}_3 = \frac{1}{|M|}\begin{pmatrix} 0 \\ -v_{2,I} \\ v_{2,E} \end{pmatrix} \tag{5.74}$$

with $|M| = v_{2,E}v_{3,I} - v_{3,E}v_{2,I}$. More explicitly, $|M| = \beta\lambda_2\lambda_3(\lambda_3-\lambda_2)/(Z_2Z_3)$. As a proof that this procedure yields the desired orthogonal relations, we note that from $\mathbf{v}_1 = (1, 0, 0)$ and Eqs. (5.71) and (5.74) it follows that $\mathbf{w}_2\mathbf{v}_1 = 0$, $\mathbf{w}_2\mathbf{v}_2 = 1$, $\mathbf{w}_2\mathbf{v}_3 = 0$, $\mathbf{w}_3\mathbf{v}_1 = 0$, $\mathbf{w}_3\mathbf{v}_2 = 0$, and $\mathbf{w}_3\mathbf{v}_3 = 1$ holds. Finally, $\mathbf{w}_1$ is given by $\mathbf{w}_1 = (1, a, b)$ such that $\mathbf{w}_1\mathbf{v}_1 = 1$ irrespective of the coefficients a and b. The coefficient

a and b are determined from the two requirements $\mathbf{w}_1\mathbf{v}_2 = 0$ and $\mathbf{w}_1\mathbf{v}_3 = 0$. The results reads

$$\mathbf{w}_1 = \frac{1}{\lambda_2\lambda_3}\begin{pmatrix} \lambda_2\lambda_3 \\ (\lambda_2+\alpha)(\lambda_3+\alpha) \\ -\alpha\beta \end{pmatrix} . \tag{5.75}$$

Having obtained the biorthogonal vectors $\mathbf{w}_i$, as described in Sect. 2.9.3, the amplitude equations can be obtained by multiplying $\mathrm{d}\mathbf{u}/\mathrm{d}t = L\mathbf{u} + \mathbf{R}$ by $\mathbf{w}_i$. Accordingly, multiplying Eq. (5.64) by $\mathbf{w}_i$, the intermediate result

$$\frac{\mathrm{d}}{\mathrm{d}t}A_i = \lambda_i A_j + G_i \ , \ G_i = \mathbf{w}_i\mathbf{R} = \frac{\beta}{N}\delta I \mathbf{w}_i \begin{pmatrix} -1 \\ 1 \\ 0 \end{pmatrix} = C_i\frac{\beta}{N}\delta I \tag{5.76}$$

can be obtained. The factors C_i introduced in Eq. (5.76) correspond to the vector products

$$C_i = \mathbf{w}_i \begin{pmatrix} -1 \\ 1 \\ 0 \end{pmatrix} = w_{i,E} - w_{i,S} \tag{5.77}$$

and can be explicitly obtained from Eqs. (5.74) and (5.75) as

$$C_1 = \alpha\frac{\lambda_2+\lambda_3+\alpha}{\lambda_2\lambda_3} \ , \ C_2 = \frac{v_{3,I}}{|M|} \ , \ C_3 = -\frac{v_{2,I}}{|M|} \tag{5.78}$$

(see also Ref. [33]). Finally, δ and I and, consequently, the product δI can be expressed in terms of amplitudes A_1, A_2, A_3 with the help of Eq. (5.72) like

$$\delta I = \left(\sum_{i=1}^{3} v_{i,S}A_i\right)\left(\sum_{i=2,3} v_{i,I}A_i\right) . \tag{5.79}$$

Substituting this result into Eq. (5.76), we obtain the SEIR amplitude equations in form of

$$\begin{aligned} \frac{\mathrm{d}}{\mathrm{d}t}A_i &= \lambda_i A_i + C_i\delta(A_1, A_2, A_3)k_0(A_2, A_3) \\ &= \lambda_i A_i + C_i p_2(A_1, A_2, A_3) \ , \\ p_2 &= \frac{\beta}{N}\left(\sum_{k=1}^{3} v_{k,S}A_k\right)\left(\sum_{k=2,3} v_{k,I}A_k\right) , \end{aligned} \tag{5.80}$$

$\delta = \sum_{k=1}^{3} v_{k,S}A_k$, $k_0 = \beta(v_{2,I}A_2 + v_{3,I}A_3)/N$, and p_2 is a multivariate polynomial of order 2 and does not contain any terms linear in the amplitudes. Moreover, let us reiterate that $\lambda_1 = 0$ holds. Furthermore, for $\beta > \gamma$ we have $\lambda_2 > 0$, while for

$\beta < \gamma$ we have $\lambda_2 < 0$. As discussed in Sect. 5.5.1, $\lambda_3 < 0$ holds for any parameters $\beta, \alpha, \gamma > 0$ (see also Eq. (5.67)).

Let T_L denote the initial period during which the linearized model (5.64) describes a good approximation. Variations in A_1 can be neglected due to $\lambda_1 = 0$. The amplitude A_3 decays during an intermediate period T_i towards zero because of $\lambda_3 < 0$. Assuming that λ_3 in the amount is sufficiently large such that $T_i < T_L$, it follows that for $t \in [T_i, T_L]$ the relative state $\mathbf{u}$ is given by $\mathbf{u} \approx A_1(t_0)\mathbf{v}_1 + A_2(t)\mathbf{v}_2$. Consequently, the state $\mathbf{X}(t)$ evolves (in the initial stage but subsequent to the decay of A_3) like

$$\mathbf{X}(t) \approx \mathbf{X}_{st} + A_1(t_0)\mathbf{v}_1 + \mathbf{v}_2 A_2(t) , \tag{5.81}$$

which can equivalently be expressed like

$$\mathbf{X}(t) \approx \mathbf{X}_0 - A_3(t_0)\mathbf{v}_3 + \mathbf{v}_2 \Delta A_2(t) , \tag{5.82}$$

where we have used $\mathbf{X}_0 = \mathbf{X}_{st} + A_1(t_0)\mathbf{v}_1 + A_2(t_0)\mathbf{v}_2 + A_3(t_0)\mathbf{v}_3$ and $\Delta A_2 = A_2(t) - A_2(t_0)$. Equation (5.82) can be written like

$$\Delta\mathbf{X} = \begin{pmatrix} \Delta S \\ \Delta E \\ \Delta I \end{pmatrix} = \mathbf{v}_2 \Delta A_2 - \mathbf{v}_3 A_3(t_0) \tag{5.83}$$

with $\Delta\mathbf{X} = \mathbf{X}(t) - \mathbf{X}(t_0)$, which states that when the vector $\mathbf{v}_2$ is plotted at the origin $\mathbf{X} = (0, 0, 0)$ then state changes $\Delta\mathbf{X}$ take place parallel to $\mathbf{v}_2$ with a state shift determined by the expression $-\mathbf{v}_3 A_3(t_0)$. Likewise, when when the vector $\mathbf{v}_2$ is plotted at the location $-\mathbf{v}_3 A_3(t_0)$, then $\mathbf{v}_2$ determines the direction of state changes $\Delta\mathbf{X}$ and ΔA_2 determines the corresponding dynamics. If the initial state $\mathbf{X}_0$ is sufficiently close to $\mathbf{X}_{st}$ such that $A_1(t_0)$ and $A_3(t_0)$ are negligibly small, then Eq. (5.83) simplifies to [32–34]

$$\Delta\mathbf{X} = \begin{pmatrix} \Delta S \\ \Delta E \\ \Delta I \end{pmatrix} = \mathbf{v}_2 \Delta A_2 \tag{5.84}$$

and Eq. (5.81) reduces to the fundamental relation [9, 20, 34]

$$\mathbf{X}(t) \approx \mathbf{X}_{st} + \mathbf{v}_2 A_2(t) \approx \mathbf{X}_{st} + \mathbf{v}_2 A_2(t_0) \exp\{\lambda_2(t - t_0)\} , \tag{5.85}$$

as discussed in Sect. 2.7 in the context of Eq. (2.47). The unstable eigenvector $\mathbf{v}_2$ and its amplitude A_2 dominate the infectious disease outbreak. They correspond to the order parameter and order parameter amplitude of the epidemic under consideration. Equations (5.84) and (5.85) are the counterpart relations to Eqs. (4.57) and (4.60) obtained in Sect. 4.2.6) for SIR-type models.

Let us consider the wave-solution of the SEIR model with parameters $\beta = 2$/d, $\alpha = 0.4$/d, $\gamma = 0.5$/d, $N = 1,000$ individuals that is shown in Fig. 5.2. While Fig. 5.2 presents a two-dimensional analysis in the E-I subspace, with the results derived

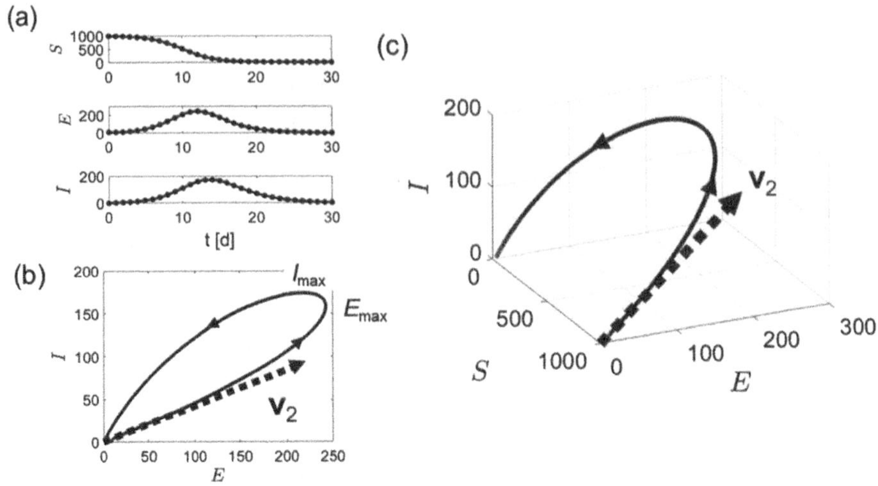

Fig. 5.4 Eigenvector analysis of the SEIR model (5.35) and SEI order parameter (unstable eigenvector) in the (S, E, I) subspace. Panel **a**: Solutions $S(t)$, $E(t)$, and $I(t)$ as functions of time computed from Eq. (5.35). Panel **b**: The phase curve $I(E)$ (solid line) computed from $E(t)$ and $I(t)$ and the (magnified) unstable eigenvector $\mathbf{v}_2$ (dotted line) are shown. Panel **c**: The phase curve (solid line) of the trajectory $\mathbf{X}(t)$ and $\mathbf{v}_2$ (dotted line) in the 3D (S, E, I) state space are shown. $\mathbf{v}_2$ was magnified and corresponds to the SEI order parameter. Parameters and initial conditions as in Fig. 5.2

above a full three-dimensional analysis can be presented. Figure 5.4 presents the analysis in state space. Figure 5.5 presents the corresponding analysis in amplitude space.

To begin with Fig. 5.4, panel (a) of Fig. 5.4 shows S, E, I as functions over time. The graphs E and I have been presented above in panel (a) of Fig. 5.2. The solution S shown in panel (a) of Fig. 5.4 describes a monotonically decaying population of susceptibles. For the selected parameters and initial conditions almost all susceptibles become eventually infected. Accordingly, the epidemic wave dies out when the number of susceptibles has decreased to a sufficiently small value. From Eq. (5.44) the interval of neutrally stable disease-free fixed points can be determined by requiring $\kappa_r < 1$, which implies $S_{st} < N\gamma/\beta$. For the selected model parameters the result reads $S_{st} < 250$ individuals. That is, the epidemic wave converges to a fixed point with less than 250 susceptible individuals. Panel (b) of Fig. 5.4 shows E and I in the two-dimensional E-I plane, just as in panel (c) of Fig. 5.2. Unlike panel (c) of Fig. 5.2, in panel (b) of Fig. 5.4 the projection of the unstable eigenvector $\mathbf{v}_2$ is plotted. Mathematically speaking, this projection of the 3D vector is identical to the 2D vector $\mathbf{v}_1$ of the E-I analysis except for a scaling factor. That is, the components $v_{2,E}$ and $v_{2,I}$ as defined by Eq. (5.70) are identical to the components $v_{1,E}$ and $v_{1,I}$ of $\mathbf{v}_1$ as defined by Eq. (5.45) except for a multiplication factor (note also that λ_2 of the 3D analysis corresponds to λ_1 of the 2D analysis). In short, the vectors point in the same direction in the E-I subspace. Therefore, panel (b) of Fig. 5.4 shows the same

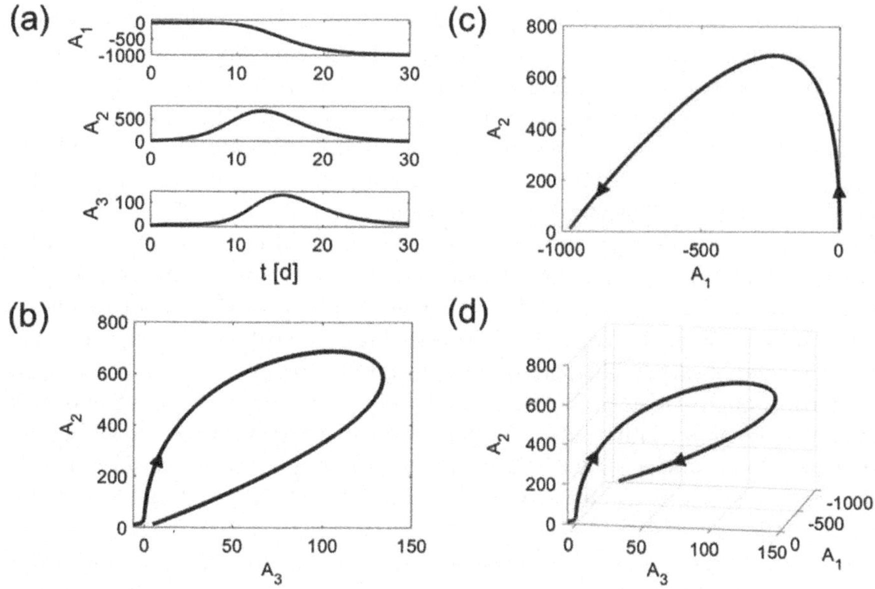

Fig. 5.5 SEIR model amplitude dynamics of the simulated epidemic shown in Figs. 5.2 to 5.4. Panel **a**: Amplitudes A_1, A_2, and A_3 as functions of time as computed from Eq. (5.80). Panels **b** and **c** show the phase curves $A_2(A_3)$ and $A_2(A_1)$ in their respective 2D amplitude subspaces. Panel **d** shows the amplitude trajectory $\mathbf{A}(t)$ as phase curve in the full 3D amplitude space

situation as presented above in panel (c) of Fig. 5.2: the simulated epidemic wave initially follows the unstable eigenvector and, only in a later stage, begins to branch off from the eigenvector. Panel (c) of Fig. 5.4 shows the wave-solution in the full three-dimensional state space. The unstable eigenvector $\mathbf{v}_2$ is plotted as well (and is located at the fixed point $\mathbf{X}_{st}$). It can be seen that the wave follows the eigenvector in the initial stage. That is, not only the changes in E and I over time are determined by $\mathbf{v}_2$ and its corresponding amplitude A_2, but also the decay of the susceptibles S is determined by $\mathbf{v}_2$ and A_2. Panel (c) illustrates that the simulated epidemic satisfies during its initial stage Eq. (5.85) (or alternatively Eq. (5.84)).

The dominant role of A_2 becomes also obvious when studying the outbreak in amplitude space. To this end, the amplitude equations (5.80) were solved numerically for the initial conditions $A_i(0) = \mathbf{w}_i(\mathbf{X}_0 - \mathbf{X}_{st})$ (see Eq. (2.38)). Panel (a) of Fig. 5.5 shows the amplitudes A_1, A_2, A_3 as functions of time thus obtained. It can be seen that A_2 increases during the initial stage, while the remaining amplitudes A_1 and A_3 do not vary considerably during that period. For example, in the simulated epidemic within the first 10 days A_2 increases to a number of 500 individuals, while A_1 during the same period only decreases by 50 individuals and A_3 increases to a level of 30 individuals. Having said that, when magnifying the graph of A_3 it can be seen that A_3 actually assumes a negative initial value $A_3(0) < 0$. $A_3(t)$ decays in magnitude quickly (within an intermediate period T_i of about 1 day for the simulation under consideration) and approaches zero, as expected from the linearized amplitude equation $dA_3/dt = \lambda_3 A_3$ with $\lambda_3 < 0$. Subsequently, as described above, A_3 increases slowly in the initial stage dominated by A_2. In other words, A_3 describes the dynamics in the stable direction defined by $\mathbf{v}_3$ of the saddle point $\mathbf{X}_{st}$, whereas A_2 describe the dynamics in the unstable direction of the saddle as defined by $\mathbf{v}_2$. Panel (b) shows the plot of A_2 versus A_3. The fast decay of A_3 from a negative value towards zero for $t < T_i$ can be seen. Subsequently, that is, for $t \in [T_i, T_L]$, the amplitude A_2 dominates the dynamics. As stated above, A_2 increases to a level of about 500 individuals while A_3 remains almost constant and eventually increases towards 30 individuals. This produces in the A_2 versus A_3 plot a graph that corresponds to a steeply increasing, almost vertical line. Panel (c) shows the plot of A_2 versus A_1. Again, the initial stage is characterized by an almost vertical line. A_2 increases while A_1 remains approximately constant. Combining panels (b) and (c), we obtain the amplitude dynamics in the full three-dimensional amplitude space shown in panel (d). Panel (d) illustrates again the dramatic increase of A_2 in the initial stage in terms of a graph that starts off almost vertically (after A_3 has decayed to zero), that is, parallel to the A_2 axis.

From Eq. (5.84) it follows that the initial changes over time of compartment sizes (or subpopulation size) are determined by the evolution of A_2 and the direction given by $\mathbf{v}_2$. In analogy to Eq. (4.61) that holds for SIR-type models, Eq. (5.84) implies that for SEIR models and SEIR-type models the changes of the subpopulation sizes relative to each other are determined by the order parameter $\mathbf{v}_2$. That is, we have [32]

$$\frac{\Delta S}{\Delta I} \approx \frac{v_{2,S}}{v_{2,I}}, \quad \frac{\Delta S}{\Delta E} \approx \frac{v_{2,S}}{v_{2,E}}, \quad \frac{\Delta I}{\Delta E} \approx \frac{v_{2,I}}{v_{2,E}}. \tag{5.86}$$

In closing these consideration on the 1β SEIR model, the equivalence of the amplitude space and state space descriptions should be demonstrated. To this end, the solutions $A_1(t)$, $A_2(t)$, $A_3(t)$ shown in panel (a) of Fig. 5.5 were used to determine the state variables $S(t)$, $E(t)$, $I(t)$. That is, the solutions $A_1(t)$, $A_2(t)$, $A_3(t)$ were substituted into Eq. (5.72) for the relative state $\mathbf{u}(t)$ and, subsequently, $\mathbf{X}(t) = (S(t), E(t), I(t))$ was computed from $\mathbf{X}(t) = \mathbf{X}_{st} + \mathbf{u}(t)$. The resulting trajectories are depicted in panel (a) of Fig. 5.4 as black solid circles. As can be seen in panel (a) the solutions obtained via the amplitudes were identical to those obtained directly by solving the state space equations of the SEIR model.

5.7.2 2β SEIR Model and Its 3D Autonomous Amplitude Description

The 2β SEIR model defined by Eq. (5.1) with $k_0 = (\beta_I I + \beta_E E)/N$ (see Eq. (5.3)) and $n = 4$ with $X_4(t) = R(t) = N - S(t) - E(t) - I(t)$ can be analyzed in a similar way as shown in Sect. 5.7.1. The linearization matrix L shown in Eq. (5.65) reads for the 2β model

$$L = \begin{pmatrix} 0 & -\beta_E & -\beta_I \\ 0 & \beta_E - \alpha & \beta_I \\ 0 & \alpha & -\gamma \end{pmatrix} . \tag{5.87}$$

The matrix exhibits the eigenvector $\mathbf{v}_1 = (1, 0, 0)$ with $\lambda_1 = 0$. The eigenvalues $\lambda_{2,3}$ have been discussed in Sect. 5.5.1 and are given by Eq. (5.38) when making the replacements $\lambda_1 \to \lambda_2$ and $\lambda_2 \to \lambda_3$. For sake of completeness, let us write them down explicitly as [13]

$$\lambda_{2,3} = \frac{\beta_E - \alpha - \gamma}{2} \pm \sqrt{U} \, , \; U = \frac{[\beta_E - \alpha - \gamma]^2}{4} + (\beta_E - \alpha)\gamma + \beta_I \alpha \, , \tag{5.88}$$

where the plus (minus) sign holds for λ_2 (λ_3). The prototype eigenvectors $\mathbf{v}_2$ and $\mathbf{v}_3$ with unknown coefficients $v_{i,s} = s$ can be obtained from the matrix (5.87) like [13, 32]

$$\mathbf{v}_i \propto \begin{pmatrix} s \\ \beta_I \\ \lambda_i + \alpha - \beta_E \end{pmatrix} . \tag{5.89}$$

The parameter s can be determined from L again such that $\mathbf{v}_i$ eventually reads

$$\mathbf{v}_i = \frac{1}{Z_i} \begin{pmatrix} -\beta_I(\lambda_i + \alpha) \\ \beta_I \lambda_i \\ \lambda_i(\lambda_i + \alpha - \beta_E) \end{pmatrix} = \begin{pmatrix} v_{i,S} \\ v_{i,E} \\ v_{i,I} \end{pmatrix} \tag{5.90}$$

with $Z_i = \sqrt{\beta_I^2[(\lambda_i + \alpha)^2 + \lambda_i^2] + \lambda_i^2(\lambda_i + \alpha - \beta_E)^2}$ and $i = 2, 3$. The eigenvector matrix M is given by Eq. (5.73) again and the biorthogonal vectors $\mathbf{w}_2$ and $\mathbf{w}_3$ satisfy Eq. (5.74) with $|M| = v_{2,E} v_{3,I} - v_{3,E} v_{2,I} \Rightarrow |M| = \beta_I \lambda_2 \lambda_3 (\lambda_3 - \lambda_2)/(Z_2 Z_3)$. For a matrix of the form (5.73) the biorthogonal vector $\mathbf{w}_1$ reads in general

$$\mathbf{w}_1 = \begin{pmatrix} 1 \\ a \\ b \end{pmatrix} , \; a = \frac{v_{3,S} v_{2,I} - v_{2,S} v_{3,I}}{|M|} , \; b = \frac{v_{2,S} v_{3,E} - v_{3,S} v_{2,E}}{|M|} \tag{5.91}$$

such that $\mathbf{w}_1$, $\mathbf{w}_2$, and $\mathbf{w}_3$ constitute the rows of the inverse matrix M^{-1} (see Eqs. (5.56) and (5.57)). Having obtained the biorthogonal vectors, the vector calculation method can be carried out to derive the amplitude equations of the 2β SEIR model. In analogy to the derivation of Eq. (5.80), the result reads

$$\frac{\mathrm{d}}{\mathrm{d}t} A_i = \lambda_i A_i + C_i \delta(\mathbf{A}) k_0(\mathbf{A}) = \lambda_i A_i + C_i p_2(A_1, A_2, A_3) \, ,$$
$$p_2 = \frac{1}{N} \left(\sum_{k=1}^{3} v_{k,S} A_k \right) \left[\beta_I \sum_{k=2,3} v_{k,I} A_k + \beta_E \sum_{k=2,3} v_{k,E} A_k \right] \tag{5.92}$$

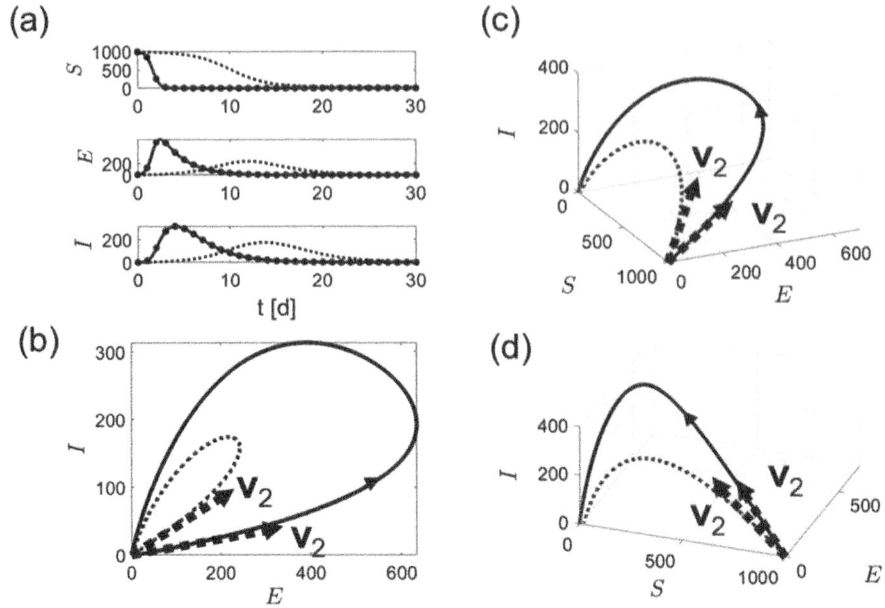

Fig. 5.6 Eigenvector analysis comparing solutions of the 1β and 2β SEIR models (5.35) and their SEI order parameters (unstable eigenvectors). Dotted (solid) lines show solutions of the 1β (2β model) with $\beta_E = 0$ ($\beta_E > 0$). The solutions of the 1β model (dotted lines) have been presented earlier in Fig. 5.4. Panel **a**: Solutions $S(t)$, $E(t)$, and $I(t)$ as functions of time computed from Eq. (5.35) for $\beta_E = 0$ (dotted lines) and $\beta_E > 0$ (solid lines). Panel **b**: The phase curves $I(E)$ (thin dotted and solid lines) computed from $E(t)$ and $I(t)$ and the (magnified) unstable eigenvectors $\mathbf{v}_2$ (thick dotted lines) for $\beta_E = 0$ and $\beta_E > 0$ are shown. Panel **c**: The phase curves (thin dotted and solid lines) of the trajectory $\mathbf{X}(t)$ and $\mathbf{v}_2$ (thick dotted lines) for $\beta_E = 0$ and $\beta_E > 0$ in the 3D (S, E, I) state space are shown. Panel **d** shows the same quantities as presented in panel (**c**) but uses a different viewing angle. Parameters and initial conditions for the 1β model simulation as in Fig. 5.2. For the 2β model simulation all parameters and initial conditions were the same except for $\beta_E = 3.0$/d

and C_i defined again as $C_i = w_{i,E} - w_{i,S}$. In Eq. (5.92) the relative state δ and the rate constant k_0 are given by $\delta(\mathbf{A}) = \sum_{k=1}^{3} v_{k,S} A_k$ and $k_0(\mathbf{A}) = [\beta_I \sum_{k=2,3} v_{k,I} A_k + \beta_E \sum_{k=2,3} v_{k,E} A_k]/N$, respectively. Explicitly, C_1 reads

$$C_1 = \frac{v_{3,S} v_{2,I} - v_{2,S} v_{3,I}}{|M|} - 1 \tag{5.93}$$

and C_2, C_3 are defined as shown in Eq. (5.78).

The impact of the second effective contact parameter β_E is illustrated in Fig. 5.6. Figure 5.6 shows as dotted lines the simulation results obtained for the 1β SEIR model with $\beta = \beta_I = 2$/d, $\alpha = 0.4$/d, $\gamma = 0.5$/d, and $N = 1000$ individuals as discussed above and presented in Fig. 5.4. In addition, Fig. 5.6 shows as solid lines the simulation results for a 2β SEIR model with $\beta = \beta_I = 2$/d, $\alpha = 0.4$/d, $\gamma = 0.5$/d, and $N = 1000$ individuals, again, but $\beta_E = 3$/d. To this end, Eq. (5.34) with k_0 given by Eq. (5.35) was solved numerically. The order parameter $\mathbf{v}_2$ was computed from Eq. (5.90). Panel (a) shows the graphs of S, I, and E for the two scenarios with $\beta_E = 0$ (1β scenario shown as dotted lines) and $\beta_E = 3$/d (2β scenario shown as solid lines). Comparing the two scenarios, it can be seen that due to the additional infections originating from the group of exposed and possibly asymptomatic infectious individuals in the 2β scenario, in the 2β scenario the epidemic evolves faster and reaches higher $E_{\max}$ and $I_{\max}$ values as compared to the 1β scenario. Panel (b) depicts the trajectories in the E-I subspace. The loop-like trajectory for the 2β scenario is increased in magnitude as compared to the 1β scenario, which reflects again that the maximal values $E_{\max}$ and $I_{\max}$ are increased under the 2β scenario. Importantly, the direction of the order parameter $\mathbf{v}_2$ in the E-I space is rotated such that the angle with the E axis becomes smaller. From the last relation in Eq. (5.86) it follows that the initial stages of the two scenarios are characterized by

$$\frac{\Delta I(\beta_E = 0)}{\Delta E(\beta_E = 0)} > \frac{\Delta I(\beta_E = 3)}{\Delta E(\beta_E = 3)} \Rightarrow \frac{\Delta E(\beta_E = 0)}{\Delta I(\beta_E = 0)} < \frac{\Delta E(\beta_E = 3)}{\Delta I(\beta_E = 3)} . \tag{5.94}$$

In words, due to the infections caused by the exposed and possibly asymptomatically infectious individuals, in the initial stage of the epidemics the increase ΔE when measured in units of the increase ΔI becomes larger. In general, substituting Eqs. (5.88) and (5.89) into the last relation of Eq. (5.86), we obtain

$$\frac{\Delta I}{\Delta E} \approx \frac{2\sqrt{U} + \alpha - \gamma - \beta_E}{2\beta_I} , \tag{5.95}$$

where U can (alternative to Eq. (5.88)) expressed like

$$U = \frac{[\beta_E + \gamma - \alpha]^2}{4} + \beta_I \alpha . \tag{5.96}$$

Introducing $e = \beta_E + \gamma - \alpha$, Eq. (5.95) can be written like

$$\frac{\Delta I}{\Delta E} \approx f(e) = \frac{-e + \sqrt{e^2 + 4\beta_I \alpha}}{2\beta_I} , \tag{5.97}$$

where $f(e)$ is a function with respect to e that decays monotonically from $\sqrt{\alpha/\beta_I}$ for $e = 0$ to zero for $e \to \infty$ (note: the first derivative of f with respect to e is negative). Consequently, in general, when increasing β_E while holding all other model parameters constant, the ratio $\Delta I / \Delta E$ characterizing the initial stage becomes smaller and, conversely, $\Delta E / \Delta I$ becomes larger. Returning to Fig. 5.6, in panel (c) of Fig. 5.6 the trajectories of the simulated 1β and 2β epidemics are shown in the full three-dimensional state space (dotted and solid lines hold for the 1β and 2β scenarios, respectively). Again, the epidemic in the case of the 2β scenario is described by a loop that is inflated relative to the 1β scenario (due to larger values of $E_{\max}$ and $I_{\max}$).

The order parameters $\mathbf{v}_2$ were computed from Eq. (5.90) and are shown for both scenarios. It can be seen that the simulated epidemics during their respective initial stages follow their respective order parameters, as expected. Comparing the order parameters, by visual inspection, the β_E parameter changes the orientation of $\mathbf{v}_2$ not only in the E-I subspace but also in the entire three-dimensional state space. Panel (d) shows the same graphs and order parameters as depicted in panel (c). However, panel (d) presents these graphs from a different viewing point that focuses more on the S-I plane. From this viewing point it can be seen that the ratio $\Delta S/\Delta I < 0$ that is determined by $\mathbf{v}_2$ (see Eq. (5.86)) is affected by β_E. Increasing β_E seems to make $\Delta S/\Delta I$ smaller in the amount and, consequently, $\Delta I/\Delta S$ larger in the amount. That is, ΔI when measured in units of $|\Delta S|$ becomes larger.

5.7.3 SEIR-Type Models: 3D Autonomous Amplitude Descriptions

Equation (5.5) formally describes epidemic models of the SEIR-type as defined by Eq. (5.3) and k_0 given by $k_0 = \beta I/N$ or $k_0 = (\beta_I I + \beta_E E)/N$ with the help of an amplitude space description that involves a three-dimensional autonomous amplitude space. From the results derived in the previous sections, it follows that in the absence of demographic terms (i.e., for $B = \mu = 0$) the amplitude space description is explicitly given by

$$\begin{aligned}
\frac{\mathrm{d}}{\mathrm{d}t} A_1 &= C_1 p_2(A_1, A_2, A_3) , \\
\frac{\mathrm{d}}{\mathrm{d}t} A_2 &= \lambda_2 A_2 + C_2 p_2(A_1, A_2, A_3) , \\
\frac{\mathrm{d}}{\mathrm{d}t} A_3 &= \lambda_3 A_3 + C_3 p_2(A_1, A_2, A_3) , \\
\frac{\mathrm{d}}{\mathrm{d}t} X_k &= N_k(S(\mathbf{A}), E(\mathbf{A}), I(\mathbf{A}), X_4, \ldots, X_n) , \text{ for } k = 4, \ldots, n
\end{aligned} \tag{5.98}$$

with $\mathbf{A} = (A_1, A_2, A_3)$ and

$$C_1 = \frac{v_{3,S} v_{2,I} - v_{2,S} v_{3,I}}{|M|} - 1 , \; C_2 = \frac{v_{3,I}}{|M|} , \; C_3 = -\frac{v_{2,I}}{|M|} ,$$

$$p_2 = \delta(\mathbf{A}) k_0(\mathbf{A}) = \left(\sum_{k=1}^{3} v_{k,S} A_k \right) \frac{1}{N} \left[\beta_I \sum_{k=2,3} v_{k,I} A_k + \beta_E \sum_{k=2,3} v_{k,E} A_k \right] , \tag{5.99}$$

where $v_{k,S}, v_{k,E}, v_{k,I}$ correspond to the eigenvector components defined by Eq. (5.90) for $k = 2, 3$ and $\mathbf{v}_1 = (1, 0, 0)$. Moreover, the eigenvalues λ_2 and λ_3 are defined by Eq. (5.88) and $\lambda_1 = 0$ holds. The amplitude description in terms of Eqs. (5.98) and (5.99) describes the epidemic under consideration from the perspective of the fixed point $\mathbf{X}_{st}$ with $X_{1,st} = S_{st} = N$, $X_{2,st} = E_{st} = 0$, $X_{3,st} = I_{st} = 0$. The initial conditions $A_1(t_0)$, $A_2(t_0)$, $A_3(t_0)$ are related to the initial state $\mathbf{X}_0$ by $A_i(t_0) = \mathbf{w}_i(\mathbf{X}_0^{(3)} - \mathbf{X}_{st}^{(3)})$ with $\mathbf{X}_0^{(3)} = (S_0, E_0, I_0)$ and $\mathbf{X}_{st}^{(3)} = (N, 0, 0)$. In this context, the biorthogonal vectors $\mathbf{w}_i$ are defined by Eqs. (5.74) and (5.91) with $|M| = v_{2,E}v_{3,I} - v_{3,E}v_{2,I} = \beta_I\lambda_2\lambda_3(\lambda_3 - \lambda_2)/(Z_2Z_3)$ and Z_i as shown below Eq. (5.90). The amplitude description by means of Eqs. (5.98) and (5.99) and the state space description by means of Eq. (5.3) are equivalent in the sense that $\mathbf{X}(t)$ can be obtained from solution $A_1(t)$, $A_2(t)$, $A_3(t)$ by computing $\mathbf{X}^{(3)}(t) = (X_1(t), X_2(t), X_3(t)) = \mathbf{X}_{st}^{(3)} + \sum_{i=1}^{3} A_i(t)\mathbf{v}_i$ and solving the evolution equations for X_k with $k = 4, \ldots, n$ listed in Eq. (5.98). For 1β models with $k_0 = \beta I/N$ we put $\beta_I = \beta$ and $\beta_E = 0$.

The following holds both for 1β and 2β SEIR-type models. For any parameter set $\lambda_3 < 0$ holds. Therefore, close to the fixed point $\mathbf{X}_{st}$ with $X_{1,st} = S_{st} = N$, $X_{2,st} = E_{st} = 0$, $X_{3,st} = I_{st} = 0$, the amplitude A_3 describes the evolution of the state along the stable eigenvector (direction) $\mathbf{v}_3$ with $dA_3/dt = \lambda_3 A_3$. The eigenvalue λ_2 is positive (negative) for $\beta_w > \gamma$ ($\beta_w < \gamma$) with $\beta_w = \beta_I + \gamma\beta_E/\alpha$ (see Eq. (5.41)) such that for $\beta_w > \gamma$ the fixed point $\mathbf{X}_{st}$ is unstable. For $\beta_w > \gamma$ the fixed point $\mathbf{X}_{st}$ in the subspace S, E, I corresponds to a saddle point with a neutrally stable direction $\mathbf{v}_1$, an unstable direction $\mathbf{v}_2$, and a stable direction $\mathbf{v}_3$. Close to the fixed point, the amplitude A_2 describes the epidemic outbreak of interest as an increase of the disease state along the unstable direction (or order parameter) $\mathbf{v}_2$. Equations (5.81) and (5.83) hold for $t \in [T_i, T_L]$ (i.e., after A_3 decayed to a sufficiently low value) when $\mathbf{X}$, $\mathbf{X}_{st}$, and $\mathbf{X}_0$ in Eqs. (5.81) and (5.83) are replaced by the corresponding subspace vectors $\mathbf{X}^{(3)} = (S, E, I)$, $\mathbf{X}_{st}^{(3)}$, and $\mathbf{X}_0^{(3)}$. Likewise, if $\mathbf{X}_0^{(3)}$ is sufficiently close to $\mathbf{X}_{st}^{(3)}$, then Eqs. (5.81) and (5.83) with $\mathbf{X}, \mathbf{X}_{st}, \mathbf{X}_0 \rightarrow \mathbf{X}^{(3)}, \mathbf{X}_{st}^{(3)}, \mathbf{X}_0^{(3)}$ for models of the SEIR-type reduce to the respective simplified expressions in terms of Eqs. (5.84) and (5.85) with $\mathbf{X}, \mathbf{X}_{st}, \mathbf{X}_0 \rightarrow \mathbf{X}^{(3)}, \mathbf{X}_{st}^{(3)}, \mathbf{X}_0^{(3)}$.

For $\beta_w < \gamma$ the fixed point $\mathbf{X}_{st}$ is neutrally stable provided that for vanishing amplitudes, that is, for $A_1 = A_2 = A_3 = 0$, the driven dynamical system defined by the variables X_k with $k = 4, \ldots, n$ converges in the $n - 3$ dimensional subspace $(X_4, \ldots, X_n)$ to a neutrally stable or asymptotically stable fixed point.

5.8 COVID-19 Outbreak in Wuhan city 2020 and its SEI Order Parameter

The events that took place at the end of the year 2019 in the city of Wuhan, located in the Hubei province of China, played a key role in the origin of the COVID-19 pandemic [35]. Different publications present different timelines about those events. Therefore, no attempt will be made to present here a clear report of the history of the beginning of the COVID-19 pandemic. Some timelines can be found in Refs. [10, 15, 18, 36–39]. In what follows the study by Pang et al. [10] and the timeline of events reported there will be reviewed. Pang et al. considered December 31, 2019, as the day on which the first COVID-19 case in Wuhan city was reported, which is consistent with an initial WHO report [35]. As such, Pang et al., considered the period from December 2019 to March 25, 2020 and decomposed the period into three phases. The first phase starts at a certain day in December 2019, on which the first individual living presumably in the area of Wuhan city was infected by SARS-CoV-2. The phase lasts until January 22, 2020. During this phase several COVID-19 cases were confirmed. On January 23, the city of Wuhan was put on lockdown [10, 15, 40, 41]. The second phase covers the 20 days period from January 23 to February 11. During the second phase people could not enter or leave the city. Diagnosed COVID-19 cases were quarantined. However, more rigorous quarantining procedures were not put in place. According to Pang et al., around February 11, 2020, such more rigorous quarantining procedures were imposed on the citizens of Wuhan city. In addition to diagnosed COVID-19 cases, all individuals with pneumonia-related fever [10] and all individuals who had contact with diagnosed COVID-19 cases were quarantined [42]. Consequently, Pang et al. considered the phase from February 12 to the end of their data analysis period, which was March 25, as the third phase.

Data about the number of confirmed COVID-19 individuals in Wuhan city can be found in Ref. [39] for the second phase from January 23 to February 11, 2020. Furthermore, for some selected days of the second phase some data can also be found in Refs. [10, 15].

The second phase was modeled by Pang et al. [10] and Frank [31] with a 2β model of SEIR-type. The model can be simplified such that it corresponds to the 2β SEIR-model defined by Eq. (5.34) and (5.35) with a fourth variable R that satisfies $\mathrm{d}R/\mathrm{d}t = bI$. The compartment R is interpreted as the number of cumulative COVID-19 cases (for a similar idea on the level of a SIR-type model, see the discussion of Eq. (4.82) in Sect. 4.4) and includes COVID-19 associated deaths. For sake of clarity, let us replace R in Eq. (5.35) by C in order to indicate that C denotes the diagnosed cases. The SEIR model (5.35) becomes the SEIC model

$$\frac{\mathrm{d}}{\mathrm{d}t}S = -k_0 S \ , \ \frac{\mathrm{d}}{\mathrm{d}t}E = k_0 S - \alpha E \ , \ \frac{\mathrm{d}}{\mathrm{d}t}I = \alpha E - \gamma I \ , \ \frac{\mathrm{d}}{\mathrm{d}t}C = bI \tag{5.100}$$

with k_0 defined by Eq. (5.34). Accordingly, the parameter γ denotes the removal rate of individuals in compartment I. In this context note that the model considers individuals in the compartment I as non-diagnosed symptomatic cases. The model assumes that such individuals either decease due to COVID-19 with a certain death rate a or they are diagnosed with a particular rate b and, subsequently, recover or decease due to COVID-19. The parameter $\gamma = a + b$ reflects both possibilities. Importantly, the model assumes that as soon as individuals are diagnosed they are quarantined (see the description of phase 2 above) such that they cannot infect others. Note that the model could be completed with the variable R_2 that denotes the deceased non-diagnosed cases and satisfies $\mathrm{d}R_2/\mathrm{d}t = aI$. If so, $N = S(t) + E(t) + I(t) + C(t) + R_2(t)$ holds. As far as the interpretation of N is concerned, N will be taken below from demographic records and is in the order of millions. Consequently, although the number of COVID-19 associated death during the first wave in Wuhan is a tragic number, it makes a negligibly small contribution to N. This implies that

N can be considered as the proper population size occurring in the factor β/N. For more details see Refs. [10, 13].

The best-fit parameters of the 2β model taken from Ref. [31] (which are related to those reported in Ref. [10]) are: $\beta_I = 0.25$/d, $\beta_E = 0.12$/d, $\alpha = 0.17$/d, $\gamma = 0.26$/d, $a = 0.04$/d, and $b = 0.22$/d. A population of $N = 9 \cdot 10^6$ was assumed based on demographic records [10, 31]. From the model parameters it follows $\beta_w = 0.45$/d, which is larger than the removal rate γ. Consequently, the model-based analysis suggests that during the phase 2 the disease-free fixed point for the population of Wuhan was unstable [13, 31]. In particular, from Eq. (5.88) and the aforementioned parameter estimates, it follows that the COVID-19 outbreak in Wuhan during phase 2 was characterized by the eigenvalues $\lambda_2 = 0.08$/d and $\lambda_3 = -0.39$/d [31]. The time constant $\tau = 1/\lambda_2$ was $\tau = 12.4$ days. Accordingly, the unstable amplitude A_2 increased in 12.4 days by a factor $e \approx 2.72$. Moreover, λ_3 was in the amount five times larger than λ_2. Therefore, the dynamics evolved quickly along the stable direction $\mathbf{v}_3$ of the saddle towards the unstable direction $\mathbf{v}_2$. The notion of A_2 as slowly evolving unstable amplitude and A_3 as fast evolving stable amplitudes as discussed in Sect. 2.10 was supported.

The 2β SEIR model defined by Eq. (5.34) and (5.100) was solved numerically [31]. January 23 was used as the initial time point and was referred to as day $t = 0$. The initial conditions were taken from Ref. [10] as $E(0) = 3251$, $I(0) = 2731$, $R(0) = 354$, and $S(0) = N - E(0) - I(0) - R(0)$. Figure 5.7 shows the simulation results obtained in Ref. [31] and the reported confirmed COVID-19 cases.

Panel (a) shows the trajectories $S(t)$ (top subpanel), $E(t)$ and $I(t)$ (bottom subpanel) obtained from the simulation. Accordingly, during phase 2 the number of susceptibles $S(t)$ decayed monotonically, while the number of exposed and possibly asymptomatic infectious E and the number of symptomatic infectious I increased monotonically. Panel (b) shows the model solution $C(t)$ as function of time and the data as reported in Ref. [39] assuming 496 COVID-19 cases on January 23 as reported in Ref. [10]. The model fits the data with moderate accuracy.

In panel (c) the trajectories $E(t)$ and $I(t)$ shown in panel (a) are plotted in the E-I plane as phase curve $I(E)$. The order parameter $\mathbf{v}_2$ computed from Eq. (5.90) located at $(E, I) = (0, 0)$ and projected into the E-I plane is depicted in panel (c) as well. As can be seen in panel (c), during an intermediate period T_i the trajectory approaches the direction defined by the order parameter. A detailed analysis based on the amplitude equations shows that during the intermediate period the amplitude A_3 converges from its finite initial value to a value close to zero, consistent with the linearized amplitude equation $\mathrm{d}A_3/\mathrm{d}t = \lambda_3 A_3$ with $\lambda_3 < 0$. Such an intermediate period during which A_3 decays in magnitude has been discussed in the context of Figs. 5.4 and 5.5 (in particular, see panel (b) of Fig. 5.4).

Panel (d) shows the trajectory of the SEIR-type model in the three-dimensional state space. The order parameter $\mathbf{v}_2$ is shown centered at three different locations. First, $\mathbf{v}_2$ is shown located at the fixed point $\mathbf{X}_{st}^{(3)} = (N, 0, 0)$ (dashed-dotted black line). By visual inspection, the trajectory $\mathbf{X}^{(3)}(t)$ evolves parallel to this vector after an intermediate period T_i. The parallel shift is quantitatively given by the term $A_1(t_0)\mathbf{v}_1$ occurring in Eq. (5.81). The order parameter $\mathbf{v}_2$ is also plotted in panel (d) at the location $\mathbf{X}_{st} + A_1(0)\mathbf{v}_1$ (see dotted black line), which accounts for the aforementioned shift. As can be seen, as expected, after the intermediate period T_i, the trajectory $\mathbf{X}^{(3)}(t)$ evolves along the direction specified by $(\mathbf{X}_{st} + A_1(0)\mathbf{v}_1) + A_2\mathbf{v}_2$, where A_2 plays the role of a coordinate. Simulation of the amplitude equations (5.92) reveals that the intermediate period T_i of the epidemic in Wuhan took about 5 days. Panel (d) shows a third option for plotting the order parameter $\mathbf{v}_2$. Accordingly, the vector $\mathbf{v}_2$ is shifted to the end state $\mathbf{X}^{(3)}(T_i)$ of the intermediate period (see dotted gray line). Panel (d) reveals that the trajectory $\mathbf{X}^{(3)}(t)$ followed the order parameter direction $\mathbf{v}_2$ depicted in this way. The second and third options to plot $\mathbf{v}_2$ suggest that the epidemic in Wuhan city as measured in terms of S, E, and I approximately followed the order parameter $\mathbf{v}_2$ after the amplitude A_3 converged to zero.

In summary, the model-based analysis of the COVID-19 outbreak in Wuhan city supports the notion that the epidemic during phase 2 evolved along the SEI order parameter $\mathbf{v}_2$ defined by Eq. (5.90) and shown in panels (c) and (d) of Fig. 5.7. The

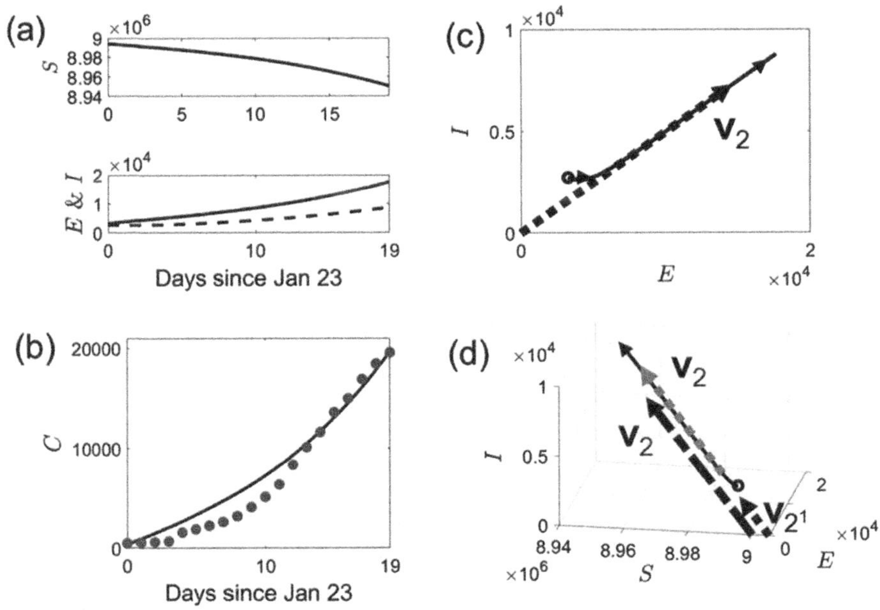

Fig. 5.7 Eigenvector analysis of the COVID-19 outbreak in Wuhan, China, based on the 2β SEIR model (5.100). Panel **a** shows the solutions $S(t)$ (top subpanel), $E(t)$ (solid line, bottom subpanel), and $I(t)$ (dashed line, bottom subpanel) of Eq. (5.100). Panel **b**: Confirmed COVID-19 cases during the observation period starting January 23, 2020 (gray circles) reconstructed from Refs. [10, 39] (see text) and $C(t)$ as computed from Eq. (5.100). Panel **c**: Phase curve $I(E)$ (solid line) obtained from $E(t)$ and $I(t)$ plotted in the E-I subspace. The projection of the order parameter $\mathbf{v}_2$ (magnified by a certain factor) in the E-I subspace is shown as well (dotted line). Panel **d**: The disease state $\mathbf{X}(t)$ is shown as phase curve in the 3D (S, E, I) state space together with $\mathbf{v}_2$ (dotted and dashed-dotted gray and black lines). $\mathbf{v}_2$ was magnified for visualization purposes and is shown at three different locations. See text for parameters and initial conditions

analysis also reveals that Eq. (5.84) only holds if $A_3(t_0)$ is sufficiently small as stated in Sect. 5.7.1. This was not the case for the COVID-19 outbreak in Wuhan – at least when analyzing the outbreak within a SEIR modeling framework. The epidemic outbreak in Wuhan followed the unstable direction of a saddle as described by Eq. (5.81) that takes the impact of $A_3(t_0)$ into account. Alternatively, it was shown that the saddle point characterizing the COVID-19 outbreak in Wuhan can be described with the help of $\Delta\mathbf{X}^{(3)} \approx \mathbf{v}_2 \Delta A_2$ (see Eq. (5.84)) when interpreting state changes $\Delta\mathbf{X}^{(3)}$ and amplitude changes ΔA_2 in a different way. Accordingly, $\Delta\mathbf{X}^{(3)} \approx \mathbf{v}_2 \Delta A_2$ reflects the dynamics of the epidemic along the unstable saddle direction subsequent to the intermediate period T_i with

$$\Delta\mathbf{X}^{(3)} = \mathbf{X}^{(3)}(t) - \mathbf{X}^{(3)}(T_i)\ , \ \Delta A_2 = A_2(t) - A_2(T_i)\ . \tag{5.101}$$

In this context, $\mathbf{X}^{(3)}(T_i)$ and $A_2(T_i)$ act as reference state and reference amplitude, respectively.

References

1. T.D. Hollingsworth, R.M. Anderson, C. Fraser, HIV-1 transmission, by stage of infection. J. Infect. Dis. **198**, 687–693 (2008)
2. Y. Bai, L. Yao, T. Wei, F. Tian, D.Y. Jin, L. Chen, M. Wang, Presumed asymptomatic carrier transmission of COVID-19. JAMA **323**, 1406–1407 (2020)
3. H. Nishiura, N.M. Linton, A.R. Akhmetzhanov, Serial interval of novel coronavirus (COVID-19) infections. Int. J. Infecti. Dis. **93**, 284–286 (2020)
4. Y. Pan, D. Zhang, P. Yang, L.M. Poon, Q. Wang, Viral load of SARS-CoV-2 in clinical samples. Lancet **20**, 411–412 (2020)
5. C. Rothe, M. Schunk, P. Sothmann et al., Transmission of 2019-nCoV infection from an asymptomatic contact in Germany. New Engl. J. Med. **382**:article 10 (2020)
6. P. Sah, M.C. Fitzpatrick, C.F. Zimmer, E. Abdollahi, L. Juden-Kelly, S.M. Moghadas, B.H. Singer, A.P. Galvani, Asymptomatic SARS-CoV-2 infection: a systematic review and meta-analysis. Proc. Natl. Acad. Sci. USA **118**:article e2109229118 (2021)
7. J. Li, N. Cui, Dynamic analysis of an SEIR model with distinct incidence for exposed and infectives. The Sci. World J. **2013**:article 871393 (2013)
8. B. Mukhopadhyay, R. Bhattacharyya, Analysis of a spatially extended nonlinear SEIS epidemic model with distinct incidences for exposed and infectives. Nonlinear Anal.: Real World Appl. **9**, 585–598 (2008)
9. T.D. Frank, S. Chiangga, SEIR order parameters and eigenvectors of the three stages of completed COVID-19 epidemics: with an illustration for Thailand January to May 2020. Phys. Biol. **18**:article 046002 (2021)
10. L. Pang, S. Liu, X. Zhang, T. Tian, Z. Zhao, Transmission dynamics and control strategies of COVID-19 in Wuhan, China. J. Biol. Syst. **28**, 543–561 (2020)
11. J.P. Arcede, R.L. Caga-Anan, C.Q Mentuda, Y. Mammeri. Accounting for symptomatic and asymptomatic in a SEIR-type model of COVID-19. Math. Model. Nat. Phenom **15**:article 34 (2020)
12. Y. Ding, L. Gao, An evaluation of COVID-19 in Italy: a data-driven modeling analysis. Infect. Dis. Model. **5**, 495–501 (2020)
13. T.D. Frank, COVID-19 order parameters and order parameter time constants of Italy and China: a modeling approach based on synergetics. J. Biol. Syst. **28**, 589–608 (2020)

14. B. Ivorra, M.R. Ferrandez, M. Vela-Perez, A.M. Ramos, Mathematical modeling of the spread of coronavirus disease 2019 (COVID-19) taking into account the undetected infections. the case of China. Commun. Nonlinear Sci. Numer. Simulat. **88**:article 105303 (2020)
15. M. Liu, J. Ning, Y. Du, J. Cao, D. Zhang, J. Wang, M. Chen, Modelling the evolution trajectory of COVID-19 in Wuhan, China: experience and suggestions. Public Health **183**, 76–80 (2020)
16. Z. Wang, Z. Yu, C. Tian, L.C. Rodriguez, L.S. Batista, B. Zhao, Differential equation analysis of COVID-19. Novel Tech. Nutr. Food Sci. **5**, 422–430 (2020)
17. C. Yang, J. Wang, A mathematical model for the novel coronavirus epidemic in Wuhan, China. Math. Biosci. Eng. **17**, 2708–2724 (2020)
18. Z. Wu, J.M. McGoogan, Characteristics of and important lessons from the coronavirus disease 2019 (COVID-19) outbreak in China. JAMA **323**, 1239–1242 (2020)
19. C. You, Y. Deng, W. Hu, J. Sun, Q. Lin, F. Zhou, C.H. Peng, Y. Zhang, Z. Chen, X.H. Zhou, Estimation of the time-varying reproduction number of COVID-19 outbreak in China. Int. J. Hygiene Environ. Health **228**:article 113555 (2020)
20. A. Das, A. Dhar, S. Goyal, A. Kundu, S. Pandey, COVID-19: Analytical results from a modified SEIR model and comparison of different intervention strategies. Chaos, Solitons and Fractals **144**:article 110595 (2021)
21. M. Gatto, E. Bertuzzo, L. Mari, S. Miccoli, L. Carraro, R. Casagrandi, A. Rinaldo, Spread and dynamics of the COVID-19 epidemic in Italy: effects of emergency containment measures. Proc. Natl. Acad. Sci. USA **117**, 10484–10491 (2020)
22. J. Arino, F. Bauer, P. van den Driessche, J. Watmough, J. Wu, Simple models for containment of a pandemic. J. Roy. Soc. Interf. **3**, 453–457 (2006)
23. M.A. Acuna-Zegarra, M. Santana-Cibrian, J.X. Velasco-Hernandez, Modeling behavioral change and COVID-19 containment in Mexico: a trade-off between lockdown and compliance. Math. Biosci. **325**:article 108370 (2020)
24. S.M. Garba, J.M.S. Lubuma, B. Tsanou, Modeling the transmission dynamics of the COVID-19 pandemic in South Africa. Math. Biosci. **328**:article 108441 (2020)
25. H.B. Taboe, K.V. Salako, J.M. Tison, C.N. Ngonghala, R.G. Kakai, Predicting COVID-19 spread in the face of control measures in West Africa. Math. Biosci. **328**:article 108431 (2020)
26. K.N. Nabi, Forecasting COVID-19 pandemic: a data-driven analysis. Chaos, Solitons, and Fractals **139**:article 110046 (2020)
27. H. Zhao, Z. Feng, Staggered release policiees for COVID-19 control: costs and benefits of relaxing restrictions by age and risk. Math. Biosci. **326**:article 108405 (2020)
28. J.M. Hyman, J. Li, E.A. Stanley, The differential infectivity and staged progression models for the transmission of HIV. Math. Biosci. **155**, 77–109 (1999)
29. P. van den Driessche, J. Watmough, Reproduction numbers and sub-threshold endemic equilibria for compartment models of disease transmission. Math. Biosci. **180**, 29–48 (2002)
30. K. Rock, S. Brand, J. Moir, M.J. Keeling, Dynamics of infectious diseases. Rep. Prog. Phys. **77**:article 026602 (2014)
31. T.D. Frank, Simplicity from complexity: on the simple amplitude dynamics underlying COVID-19 outbreaks in China. Adv. Complex Syst. **23**:article 2050022 (2020)
32. T.D. Frank, COVID-19 outbreaks follow narrow paths: a computational phase portrait approach based on nonlinear physics and synergetics. Int. J. Modern Phys. C **32**, 2150110 (2021)
33. T.D. Frank, SARS-Coronavirus-2 nonlinear dynamics in patients: three-dimensional state and amplitude state description. J. Phys. Soc. Jpn. **90**:article 073802 (2021)
34. T.D. Frank, A. Guevara-Flores, K. Nguyen, Fundamental properties of eigenvalues of dynamical systems and their applications: from fluid dynamics to human reactions and COVID-19 epidemics, in *Advances in Health and Disease, Vol. 38*, ed. by L.T. Duncan (Nova Publishers, New York, 2021), pp. 185–216
35. World Health Organization. Novel coronavirus (2019-nCoV) Situation Report-1, 21. Jan. 2020. *Novel coronavirus (2019-nCoV) Situation Report-1, 21. Jan. 2020*, https://www.who.int/emergencies/diseases/novel-coronavirus-2019/situation-reports (2020)

36. Z. Allam, *Surveying the COVID-19 Pandemic and Its Implications* (Elsevier, Amsterdam, 2020)
37. Y. Huang, Y. Wu, W. Zhang, Comprehensive identification and isolation policies have effectively suppressed the spread of COVID-19. Chaos, Solitons, and Fractals **139**, 110041 (2020)
38. A. Pan, L. Liu, C. Wang, H. Guo, X. Hao, Q. Wang, J. Huang, N. He, H. Yu, X. Lin, S. Wei, T. Wu, Association of public health interventions with the epidemiology of the COVID-19 outbreak in Wuhan, China. JAMA **323**, 1915–1923 (2020)
39. P. Shi, Y. Dong, H. Yan, C. Zhao, X. Li, W. Liu, M. He, S. Tang, S. Xi, Impact of temperature on the dynamics of the COVID-19 outbreak in China. Sci. Total Environ. **728**, 138890 (2020)
40. CNA Asia. China halts flights and trains out of Wuhan as WHO extends talks. *China halts flights and trains out of Wuhan as WHO extends talks*, http://www.channelnewsasia.com/news/asia/wuhan--virus--quarantine--city--flights--trains--china--12306684
41. New York Times. Scale of China's Wuhan shutdown is believed to be without precedent. *Scale of China's Wuhan shutdown is believed to be without precedent*, http://www.nytimes.com/202001/22/world/asia/coronavirus--quarantines--history.html
42. Globaltimes. Wuhan imposes strict quarantine measures amid fight against the novel coronavirus. *Wuhan imposes strict quarantine measures amid fight against the novel coronavirus*, http://www.globaltimes.cn/content/1178250.shtml

Chapter 6
Nonlinear Physics of Epidemics: Part C

This chapter addresses compartmental, epidemiological models, in general, and high-dimensional models featuring a relatively large number of variables, in particular. It begins with the introduction of the minimal subspace of variables leading to an autonomous amplitude space description in the linear domain. Such autonomous amplitude space descriptions are worked out in detail for the SIR and SEIR models. The benefit of this approach is demonstrated for high-dimensional models for which the autonomous amplitude space description in the linear domain allows for a stability analysis, on the one hand, but, on the other hand, involves a relatively small number of variables. Applications to the COVID-19 outbreaks during the year 2020 in Wuhan, China, and West Africa are also presented.

6.1 Higher-Dimensional Models and Non-autonomous Amplitude Equation Descriptions

In Chaps. 4 and 5 amplitude space descriptions of epidemics are presented that involve closed sets of amplitude equations. As explained in Sect. 5.1, in order to arrive at such descriptions, it is plausible to focus on the variables that describe infectious individuals who are actually in the position to infect others. In order to arrive at closed descriptions that address those variables, typically auxiliary variables need to be added. For example, for models of the SIR-type and SEIR-type the size of the susceptible population needs to be added to derive closed descriptions that address the evolution of the relevant infectious individuals. By adding auxiliary variables, the dimensionality of the problem at hand increases. In general, higher-dimensional problems are more difficult to analyze than lower-dimensional problems. Therefore, in order to study epidemics from high-dimensional perspectives but keep the respective amplitude spaces low-dimensional, an approach alternative to the approach discussed in Chaps. 4 and 5 may be considered.

T. D. Frank, *COVID-19 Epidemiology and Virus Dynamics*,
Understanding Complex Systems, https://doi.org/10.1007/978-3-030-97178-6_6

In such an alternative approach, the objective no longer is the derivation of amplitude equations that reflect autonomous systems. Instead, the objective is to derive amplitude equations that are autonomous in the linear domain. It is then accepted that in the nonlinear domain the amplitude space descriptions come in terms of non-autonomous systems. Following this idea, in what follows, amplitude equations that depend on state variables and, consequently, describe non-autonomous systems will be considered. More precisely, just as in Chaps. 4 and 5, amplitude descriptions will be considered that capture all compartment variables of infectious individuals who are in the position to infect others (i.e., belong to the class of actually infectious individuals). Unlike the approach used in Chaps. 4 and 5, non-infected variables are typically not addressed from the amplitude space perspective and additional variables describing infected but non-infectious individuals may be added (i) if they are of interest or (ii) if they are needed to arrive at linearized autonomous systems (see below).

Let D_0 denote the n-dimensional state space that describes all possible disease states of a population of interest. Let D^+ denote the subspace that involves the actually infectious individuals (i.e., those who are the sources of new infections) and some supplemental variables, if necessary. The remaining variables span the space D^-. The subspace D^+ is then transformed into an amplitude space, that is, described with the help of a new basis. The alternative approach has the benefit that the dimension of the amplitude space can be kept relatively small. The subspace D^+ has to be selected in such a way that the linearized system at the fixed point $\mathbf{X}_{st}$ of interest corresponds to an autonomous system. Therefore, the additional (i.e., supplemental) variables can be selected for two reasons: (i) they are of interest or (ii) they are needed to establish an autonomous linearized dynamical systems that describes accurately the dynamics in the subspace D^+ as long as the state $\mathbf{X} \in D_0$ is close to $\mathbf{X}_{st}$. If so, close to the fixed point the amplitude space description allows for a convenient discussion of the stability of the fixed point projection in D^+. That is, for the components $\mathbf{X}^+$ of $\mathbf{X}$ that belong to D^+ the dynamics close to $\mathbf{X}_{st}$ can be inferred from the amplitude dynamics. Typically, the supplemental variables reflect infected but non-infectious variables (see below).

6.1.1 Model Formulation and Decomposition of States

Our departure point is an epidemic that can be described in terms of a compartmental model of the form $d\mathbf{X}/dt = \mathbf{N}(\mathbf{X})$ (see Eq. (2.1)) with state vector $\mathbf{X} = (X_1, \ldots, X_n)$. The state space D_0 is given by $\mathbb{R}^n_+$. Without loss of generality, let the first m compartments with $m \geq 1$ and $m < n$ denote various types of infectious individuals who actually infect others and other individuals who are selected because they are needed to establish a linearized autonomous model or are of particular interest. That is, the first m variables span the space D^+ introduced above. Accordingly, $\mathbf{X}^+ = (X_1, \ldots, X_m)$ and $\mathbf{X}^- = (X_{m+1}, \ldots, X_n)$ denote vectors of the respective subspaces D^+ and D^- (with $D_0 = D^+ \cup D^-$) and $\mathbf{X} = (\mathbf{X}^+, \mathbf{X}^-)$.

Note that this approach to decompose the original state space includes as a special case the approach that will be discussed in Chap. 7 and has been briefly addressed in Sect. 5.1. In this case, the state space is decomposed into infected and non-infected compartments and the first m variables denote the variables that describe infected individuals [1, 2].

Let us return to the more general case that does not require that D^+ contains all infected compartments. The disease-free state $\mathbf{X}_{st}$ is defined by $\mathbf{X}_{st} = (\mathbf{X}_{st}^+, \mathbf{X}_{st}^-)$. If D^+ only contains infectious and infected variables, then $\mathbf{X}_{st}^+ = (0, \ldots, 0)$. If D^- contains infected or infectious variables (e.g., infectious hospitalized patients that are perfectly isolated and cannot infect others), then some components of $\mathbf{X}_{st}^-$ vanish for the disease-free state. As indicated, it is assumed that the disease-free state is a fixed point (or stationary state) of the dynamics $\mathrm{d}\mathbf{X}/\mathrm{d}t = \mathbf{N}(\mathbf{X})$. Let us split the right-hand side function $\mathbf{N}$ into $\mathbf{N} = (\mathbf{N}^+, \mathbf{N}^-)$. Then the dynamical system of interest can equivalently be expressed in terms of two coupled dynamical systems like

$$\frac{\mathrm{d}}{\mathrm{d}t}\mathbf{X}^+ = \mathbf{N}^+(\mathbf{X} = (\mathbf{X}^+, \mathbf{X}^-)) \, , \; \frac{\mathrm{d}}{\mathrm{d}t}\mathbf{X}^- = \mathbf{N}^-(\mathbf{X} = (\mathbf{X}^+, \mathbf{X}^-)). \tag{6.1}$$

6.1.2 Non-autonomous Amplitude Equation Descriptions

Using the vector and matrix calculation methods described in Sects. 2.9.3 and 2.9.4, let us map the m-dimensional subspace D^+ with state vector $\mathbf{X}$ to a m-dimensional amplitude space described by the amplitude vector $\mathbf{A} = (A_1, \ldots, A_m)$. In this context, it is frequently assumed that the linearized dynamics of the infected subsystems at the disease-free fixed point is independent of the the variables of the non-infected subsystem [1, 2].

Likewise, in Sect. 6.1.1 it has been explained that supplemental variables may be added to obtain an autonomous linearized model at $\mathbf{X}_{st}$, where those supplemental variables typically correspond to infected variables. In general, let $\mathbf{u} = \mathbf{X} - \mathbf{X}_{st}$ denote a perturbation in D_0 and $\mathrm{d}\mathbf{u}/\mathrm{d}t = L\mathbf{u}$ (see Eq. (2.16)) describe the linearized system at $\mathbf{X}_{st}$ with linearization matrix L composed of coefficients L_{ik}. Then the two aforementioned assumptions that (i) the linearized infected system at $\mathbf{X}_{st}$ is independent of the non-infected system and (ii) supplemental, infected variables are added to obtain an autonomous linearized system, implies that L can be expressed by means of the three matrices L^+, A, and B like

$$L = \begin{pmatrix} L^+ & 0 \\ A & B \end{pmatrix} . \tag{6.2}$$

In Eq. (6.2) the matrix L^+ is the upper, left-corner $m \times m$ matrix with coefficients L_{ik} and $i, k \leq m$. The zero entry in the upper, right corner of L stands for $L_{ik} = 0$ with $i \leq m$ and $k \geq m + 1$. From Eq. (6.2) the linearized equation of $\mathrm{d}\mathbf{X}^+/\mathrm{d}t = \mathbf{N}^+(\mathbf{X})$ at $\mathbf{X}_{st}$ is given by

$$\frac{\mathrm{d}}{\mathrm{d}t}\mathbf{X}^+ = L^+\mathbf{X}^+ \ , \ L^+_{ik} = \frac{\partial}{\partial X^+_k} N_i(\mathbf{X}_{st}), \tag{6.3}$$

where L^+ is the linearization matrix in D^+. It is fair to say that Eq. (6.3) should not be considered as a conclusion. Rather, Eq. (6.3) should be considered as starting point for decomposing D_0 into D^+ and D^-. That is, given an epidemic described in terms of n variables $X_1, \dots, X_n$, the first step is to select all variables describing (individuals who act as) sources of new infections (i.e., actually infectious individuals) and to add additional variables (which are typically variables of infected individuals) such that Eqs. (6.2) and (6.3) hold. In this sense, Eq. (6.3) describes a key assumption of the approach that will be presented below.

Irrespective whether Eq. (6.3) is regarded as conclusion or starting point, in what follows, it is further assumed that the matrix L^+ exhibits m linearly independent eigenvectors $\mathbf{v}_k$ associated with the eigenvalues λ_k. As pointed out in Sections 2.6.2 and 5.6, if the vectors $\mathbf{v}_k$ are taken as column vectors, then they constitute the matrix M_+ defined by

$$M_+ = (\mathbf{v}_1, \dots, \mathbf{v}_m) \ , \tag{6.4}$$

whose inverse matrix M_+^{-1} exists. Let $\mathbf{w}_k$ describe the rows of the inverse matrix like

$$M_+^{-1} = \begin{pmatrix} \mathbf{w}_1 \\ \dots \\ \mathbf{w}_m \end{pmatrix} . \tag{6.5}$$

Then, $\mathbf{w}_i\mathbf{v}_k = \delta_{ik}$ holds and $\mathbf{w}_i$ and $\mathbf{v}_k$ form a biorthogonal basis. In what follows, for sake of simplicity, it will be assumed that all supplemental variables are infected variables or they vanish at the fixed point of interest such that $\mathbf{X}^+_{st} = (0, \dots, 0)$. In this case, there is no need to distinguish between the relative state $\mathbf{u}^+$ in D^+ and the state $\mathbf{X}^+$ because $\mathbf{u}^+ = \mathbf{X}^+ - \mathbf{X}^+_{st} = \mathbf{X}^+$. The amplitude space spanned by the amplitude variables $A_1, \dots, A_m$ can be defined with the help of the mapping

$$\mathbf{X}^+ = \sum_{k=1}^{m} \mathbf{v}_k A_k \ \Leftrightarrow \mathbf{X}^+ = M_+\mathbf{A} \ , \tag{6.6}$$

which implies

$$A_k = \mathbf{w}_k\mathbf{X}^+ \ \Leftrightarrow \mathbf{A} = M_+^{-1}\mathbf{X}^+ \ . \tag{6.7}$$

Equation (6.6) describes the superposition of a state $\mathbf{X}^+ \in D^+$ in terms of the amplitude variables $A_1, \dots, A_m$. In other words, Eq. (6.6) defines the mapping of the amplitude space to the state space D^+. Vice versa, Eq. (6.7) describes the mapping of the state space D^+ to the amplitude space. In order to derive the amplitude equations defining the dynamics of $A_1, \dots, A_m$, the dynamical system $\mathrm{d}\mathbf{X}^+/\mathrm{d}t = \mathbf{N}^+(\mathbf{X})$ is decomposed into the linear part (6.3) and a remainder term $\mathbf{R}$ such that

$$\frac{\mathrm{d}}{\mathrm{d}t}\mathbf{X}^+ = \mathbf{N}^+(\mathbf{X}) = L^+\mathbf{X}^+ + \mathbf{R}(\mathbf{X}^+, \mathbf{X}^-) \tag{6.8}$$

(see Sect. 2.9.3). Let us multiply Eq. (6.8) by M_+^{-1}. Then, with the help of $M_+^{-1}\mathbf{X}^+ = \mathbf{A}$ we obtain

$$\frac{\mathrm{d}}{\mathrm{d}t}\mathbf{A} = M_+^{-1}L^+\mathbf{X}^+ + M_+^{-1}\mathbf{R}(\mathbf{X}^+, \mathbf{X}^-) . \tag{6.9}$$

Finally, $\mathbf{X}^+$ occurring in Eq. (6.9) can be expressed like $\mathbf{X}^+ = M_+\mathbf{A}$ and the identity $M_+^{-1}L^+M_+ = D$ can be exploited. From Eq. (6.9) it then follows

$$\frac{\mathrm{d}}{\mathrm{d}t}\mathbf{A} = D\mathbf{A} + M_+^{-1}\mathbf{R}(M_+\mathbf{A}, \mathbf{X}^-) . \tag{6.10}$$

As far as the dynamics in D^- is concerned, from Eq. (6.1) it follows that the evolution equation for $\mathbf{X}^-$ can be expressed like

$$\frac{\mathrm{d}}{\mathrm{d}t}\mathbf{X}^- = \mathbf{N}^-(\mathbf{X}^+, \mathbf{X}^-) = \mathbf{N}^-(M_+\mathbf{A}, \mathbf{X}^-) . \tag{6.11}$$

Equations (6.10) and (6.11) read in components

$$\frac{\mathrm{d}}{\mathrm{d}t}A_k = \lambda_k A_k + \mathbf{w}_k\mathbf{R}(M_+\mathbf{A}, \mathbf{X}^-) , \tag{6.12}$$

$$\frac{\mathrm{d}}{\mathrm{d}t}X_j^- = N_j^-(M_+\mathbf{A}, \mathbf{X}^-) \tag{6.13}$$

with $k = 1, \ldots, m$ and $j = m + 1, \ldots, n$, respectively. Equation (6.10) as a vector equation or Eqs. (6.12) in component form describe the epidemic of interest in terms of a m-dimensional amplitude space description that captures all actually infectious individuals. In addition, the amplitude space description may address additional individuals (e.g., infected but non-infectious individuals). The evolution of the entire state of the epidemic of interest is described by Eqs. (6.10) and (6.11) or, alternatively, by Eqs. (6.12) and (6.13).

In general, the amplitude space description does not correspond to an autonomous system. Due to the nonlinear terms $\mathbf{w}_k\mathbf{R}$ occurring in Eqs. (6.10) and (6.12), the amplitude dynamics depends in general on the variables $\mathbf{X}^- \in D^-$. However, the linearized system at $\mathbf{X}_{st}$ in the subspace D^+ is described in terms of the linear parts of the amplitude equation description. That is, the evolution equations $\mathrm{d}A_k/\mathrm{d}t = \lambda_k A_k$ (see Eq. (6.12)) capture the dynamics in D^+ close to the fixed point. They are sufficient to discuss the evolution of the state $\mathbf{X}^+$ in D^+ close to the fixed point. To this end, $\mathbf{X}^+ = \sum_{k=1}^{m} \mathbf{v}_k A_k$ can be used (see Eq. (6.6)).

6.1.3 Epidemic Outbreaks and Subsiding Epidemics

Let $\lambda_{\max}$ denote the eigenvalue of L^+ with the largest real part. Then, we can distinguish between two cases.

Epidemic outbreaks

If $\mathbb{R}\{\lambda_{\max}\} > 0$ holds, then the disease-free fixed point $\mathbf{X}_{st}^{+} = (0, \ldots, 0)$ in D^{+} is an unstable fixed point. Let $k(\max)$ denote the index of the eigenvalue with the largest real part: $\lambda_{k(\max)} = \lambda_{\max}$. In order to discuss the main aspects of the case $\mathbb{R}\{\lambda_{\max}\} > 0$, let us assume that there is only a single positive eigenvalue (for an example in this regard see the SEIR model below). In this case, the fixed point $\mathbf{X}_{st}^{+} \in D^{+}$ corresponds to a saddle point that exhibits one unstable direction given by $\mathbf{v}_{k(\max)}$. All other directions in D^{+} correspond to stable directions characterized by stable eigenvectors. Consequently, the dominant dynamics takes place along $\mathbf{v}_{k(\max)}$. The amplitude $A_{k(\max)}$ describing the dynamics along this direction increases in magnitude during the initial stage of an epidemic [3, 4]. The unstable eigenvector $\mathbf{v}_{k(\max)}$ corresponds to the order parameter of the disease outbreak and the amplitude $A_{k(\max)}$ denotes the order parameter amplitude [5, 6].

Subsiding epidemic driven by the linear system

If $\mathbb{R}\{\lambda_{\max}\} > 0$ holds, nevertheless, it is possible that due to the impact of the nonlinear terms the amplitudes $A_1, \ldots, A_m$ converge to zero and the epidemic subsides. In contrast to this kind of subsiding of an epidemic driven by nonlinear terms, let us consider the case $\lambda_{\max} < 0$ and let us assume that the linear terms in Eq. (6.12) dominate over the nonlinear terms. Furthermore, just as in the case of a single positive eigenvalue $\lambda_{\max}$ discussed above that is qualitatively different from all other (negative eigenvalues) λ_j with $j \neq k(\max)$, let us assume that there exist a gap in the eigenvalue spectrum between $\lambda_{\max}$ and the remaining eigenvalues λ_j with $j \neq k(\max)$. In order to describe this gap, we consider the time constants $\tau_{\max} = 1/\lambda_{\max}$ and $\tau_j = 1/\mathbb{R}(\lambda_j)$. If $|\lambda_{\max}| \ll |\mathbb{R}(\lambda_j)|$ then $\tau_j \gg \tau_{\max}$ holds. Consequently, the dynamics of $A_{k(\max)}$ along $\mathbf{v}_{k(\max)}$ is slow relative to the dynamics of the amplitudes in other directions. In other words, the amplitudes A_j with $j \neq k(\max)$ quickly decay to zero, while $A_{k(\max)}$ decreases slowly. If so, the subsiding of an epidemic is determined by the dynamics of $A_{k(\max)}$ and the direction $\mathbf{v}_{k(\max)}$.

6.2 SIR and SEIR Models: Non-autonomous Amplitude Equation Descriptions

6.2.1 SIR Model: Trivial Case $m = 1$

Let us briefly consider the trivial case $m = 1$. This case can be illustrated for the SIR model (3.16). In this case, we have $m = 1$, $n = 3$, and $\mathbf{X} = (I, S, R)$. The space D^{+} is one-dimensional with $X^{+} = I$. The space D^{-} is described by the state vector $\mathbf{X}^{-} = (S, R)$. In order to introduce an amplitude space, let us use the fixed point $\mathbf{X}_{st} = (0, N, 0)$ with $S_{st} = N$. Accordingly, the relative state $\mathbf{u} = (u_1, u_2, u_3)$ with respect to the fixed point $\mathbf{X}_{st} = (0, N, 0)$ exhibits the components $u_1 = I$,

$u_2 = S - N, u_3 = R$. The linearization of the evolution equation of I (see Eq. (3.16)) at $\mathbf{X}_{st} = (0, N, 0)$ is given by Eq. (6.3) for $m = 1$ and, consequently, yields a 1×1 matrix L^+ with a single coefficient: the eigenvalue $\lambda = \beta - \gamma$ [7]. Next, the subspace D^+ can be mapped to a one-dimensional amplitude space described by the variable A_1. In fact, Eq. (6.6) for $m = 1$ reduces to the identity $I = A_1$ because the one-dimensional eigenvector $\mathbf{v}_1$ corresponds to the scalar 1. Substituting $I = A_1$ and $S = N + u_2$ into the evolution equation of I (see Eq. (3.16) again), the amplitude equation

$$\frac{\mathrm{d}}{\mathrm{d}t}A_1 = \lambda A_1 + u_2 k_0(A_1) \tag{6.14}$$

can be obtained, where the rate constant (i.e., "force of infection") k_0 is given in terms of the linear function $k_0(A_1) = \beta A_1/N$. From the evolution equations of S and R (see Eq. (3.16)), it follows that

$$\begin{aligned}\frac{\mathrm{d}}{\mathrm{d}t}u_2 &= -(N + u_2)k_0(A_1) , \\ \frac{\mathrm{d}}{\mathrm{d}t}R &= \gamma A_1 .\end{aligned} \tag{6.15}$$

Equations (6.14) and (6.15) exemplify Eqs. (6.12) and (6.13), respectively, for the case of the SIR model.

6.2.2 *1β and 2β SEIR Models and m = 2*

Next, let us consider the case $m = 2$ for the SEIR model (3.43). This case was discussed in Refs. [3, 4]. In this case, we have $m = 2, n = 4, \mathbf{X} = (E, I, S, R), \mathbf{X}^+ = (E, I)$, and $\mathbf{X}^- = (S, R)$. That is, the system variables are split into the infected and non-infected variables. In what follows, let us consider the fixed point $\mathbf{X}_{st} = (0, 0, N, 0)$ with $S_{st} = N$. In this case, the linearization matrix L^+ reads

$$L^+ = \begin{pmatrix} -\alpha & \beta \\ \alpha & -\gamma \end{pmatrix} \tag{6.16}$$

and exhibits the eigenvalues

$$\lambda_{1,2} = -\frac{\alpha + \gamma}{2} \pm \sqrt{\frac{(\alpha + \gamma)^2}{4} + \alpha(\beta - \gamma)} \tag{6.17}$$

(see Eq. (5.67)). The plus (minus) sign holds for λ_1 (λ_2). Equation (6.17) implies that for arbitrary parameters values $\alpha, \beta, \gamma > 0$ we have $\lambda_2 < 0$ and $\lambda_1 = \lambda_{\max} > \lambda_2$. For $\beta > \gamma$ ($\beta < \gamma$) we have $\lambda_1 > 0$ ($\lambda_1 < 0$). The eigenvectors of L^+ read

$$\mathbf{v}_i = \begin{pmatrix} v_{i,E} \\ v_{i,I} \end{pmatrix} = \frac{1}{\sqrt{(\lambda_i + \alpha)^2 + \beta^2}} \begin{pmatrix} \beta \\ \lambda_i + \alpha \end{pmatrix} \tag{6.18}$$

and have been derived in Sect. 5.5.2 (see Eq. (5.45)). The relative state $\mathbf{u} = \mathbf{X} - \mathbf{X}_{st} = (E, I, u_3, R)$ involves $u_3 = S - N \leq 0$. The amplitude equations can be obtained as outlined in Sect. 6.1.2 and read [3, 4, 8]

$$\frac{\mathrm{d}}{\mathrm{d}t} A_1 = \lambda_1 A_1 + C_1 u_3 k_0(A_1, A_2) ,$$
$$\frac{\mathrm{d}}{\mathrm{d}t} A_2 = \lambda_2 A_2 + C_2 u_3 k_0(A_1, A_2) \tag{6.19}$$

with

$$C_1 = \frac{v_{2,I}}{|M_+|} , \; C_2 = -\frac{v_{1,I}}{|M_+|} , \; k_0 = \frac{\beta}{N}(v_{1,I} A_1 + v_{2,I} A_2) . \tag{6.20}$$

In Eq. (6.20) the determinant $|M_+| = v_{1,E} v_{2,I} - v_{2,E} v_{1,I}$ of M_+ reads explicity $|M_+| = \beta(\lambda_2 - \lambda_1)/(Z_1 Z_2)$ with $Z_i = \sqrt{(\lambda_i + \alpha)^2 + \beta^2}$. The dynamics of the non-infected system is given by

$$\frac{\mathrm{d}}{\mathrm{d}t} u_3 = -(N + u_3) k_0(A_1, A_2) ,$$
$$\frac{\mathrm{d}}{\mathrm{d}t} R = \gamma(v_{1,I} A_1 + v_{2,I} A_2) . \tag{6.21}$$

Furthermore, from Eqs. (6.6) and (6.18) the mappings $(A_1, A_2) \to (E, I)$ and $(E, I) \to (A_1, A_2)$ can be found as

$$E = v_{1,E} A_1 + v_{2,E} A_2 , \; I = v_{1,I} A_1 + v_{2,I} A_2 \tag{6.22}$$

and

$$A_1 = \frac{v_{2,I} E - v_{2,E} I}{|M_+|} , \; A_2 = \frac{v_{1,I} E - v_{1,I} I}{|M_+|} . \tag{6.23}$$

Equation (6.19) shows that the amplitude equations depend on u_3, that is, the state variable $S = N + u_3$. Consequently, the amplitude equations describe a non-autonomous system. However, u_3 only occurs in the nonlinear parts of the amplitude equations. Close to the fixed point, the amplitude equations are independent of u_3 (and R) and describe the dynamics of an autonomous system.

The SEIR amplitude equations given by Eqs. (6.19) and (6.21) have already been derived in Sect. 2.9.2 (see Eq. (2.84)). While in Sect. 2.9.2 the scalar calculation method was used, in the derivation above the vector calculation method was used. Importantly, in Sect. 2.9.2 the amplitude equations were derived in order to demonstrate the scalar calculation method. An explanation why only the variables E and I are transformed into amplitude space was not given. This motivation has been given in Sect. 6.1.

Let us briefly compare the autonomous systems approach with the non-autonomous systems approach in the context of the SEIR model. That is, let us compare the autonomous three-variable amplitude equation description (5.80) of the SEIR model with the non-autonomous two-variable amplitude equation description (6.19) of the SEIR model. In fact, the evolution equations of both descriptions exhibit the same structure: $\mathrm{d}A_i/\mathrm{d}t = \lambda_i A_1 + C_i y k_0$, where $y = \delta = S - N$ in Eq. (5.80) and $y = u_3 = S - N$ in Eq. (6.19) holds. That is, while the deviation $S - N$ is denoted by δ in Eq. (5.80), it is denoted by u_3 in Eq. (6.19). The key difference between the two descriptions is that Eq. (5.80) describes a three-variable autonomous system, whereas Eq. (6.19) describes a two-variable non-autonomous systems. The variable $y = \delta$ in Eq. (5.80) is expressed in terms of amplitudes like $\delta(\mathbf{A})$, which yields a closed description in terms of amplitudes. In contrast, the variable $y = u_3$ in Eq. (6.19) corresponds to a variable external to the amplitude space description. The amplitude dynamics depends on this external time-dependent variable. Of course, when taking A_1, A_2, and u_3 together, a closed set of evolution equations is obtained. That is, the triplet A_1, A_2, and u_3 that evolves according to Eqs. (6.19) and (6.21) describes an autonomous system. Importantly, the 3D eigenvectors $\mathbf{v}_2$ and $\mathbf{v}_3$ of the autonomous description are related to the 2D eigenvectors $\mathbf{v}_1$ and $\mathbf{v}_2$. The components of the 3D vectors $\mathbf{v}_2$ and $\mathbf{v}_3$ point in the E-I space in the same direction as their respective 2D eigenvectors $\mathbf{v}_1$ and $\mathbf{v}_2$. That is, $v_{2,E}(3D)/v_{2,I}(3D) = v_{1,E}(2D)/v_{1,I}(2D)$ and $v_{3,E}(3D)/v_{3,I}(3D) = v_{2,E}(2D)/v_{2,I}(2D)$ holds. However, since the 3D vectors have three components, while the 2D vectors only exhibit two components, the normalization of the vectors to unity implies that the components of $\mathbf{v}_2(3D)$ and $\mathbf{v}_3(3D)$ in the E-I space are smaller in magnitude than the respective components of $\mathbf{v}_1(2D)$ and $\mathbf{v}_2(2D)$. This also affects the coefficients C_i occurring in the amplitude equations of the form $\mathrm{d}A_i/\mathrm{d}t = \lambda_i A_1 + C_i y k_0$ and the weights $v_{i,E}$ and $v_{i,I}$ occurring in k_0. In particular, the autonomous approach yields a 3D unstable eigenvector $\mathbf{v}_2$ (or SEI order parameter) that determines in the three-dimensional state space spanned by S, E, and I the initial dynamics of an epidemic close to an unstable disease-free fixed point (see Sects. 5.7.1 and 5.7.2 and Figs. 5.4 and 5.6). In contrast, the non-autonomous approach yields a 2D unstable eigenvector $\mathbf{v}_1$ (or EI order parameter) that determines in the two-dimensional subspace spanned by E and I the initial dynamics of an epidemic emerging from the unstable disease-free fixed point $E = I = 0$. Both descriptions are consistent. As mentioned above, the 3D eigenvector $\mathbf{v}_2$ points in the E-I subspaces in the same direction as its 2D counterpart vector $\mathbf{v}_1$. This issue has been discussed and illustrated explicitly in Sect. 5.5 (e.g., compare panel (c) of Fig. 5.2) with panel (b) of Fig. 5.4).

The benefit of the non-autonomous approach to the SEIR model is that the approach produces an amplitude space description that is reduced by 1 dimension. Moreover, the closed description in terms of A_1, A_2, and u_3 involves the perturbation or relative state $u_3 = S - N$ that is directly linked to the number of susceptibles. u_3 describes the change of susceptibles with respect to the fixed point value $S_{st} = N$. The disadvantage is that the evolution equation of u_3 contains linear terms in A_1 and A_2 like $\mathrm{d}u_3/\mathrm{d}t = -Nk_0(A_1, A_2) +$ nonlinear terms. That is, the three-variable system A_1, A_2, u_3 does not exhibit diagonal form close to the fixed point $A_1 = A_2 = u_3 = 0$.

So far, the 1β SEIR model was discussed. The 2β SEIR model defined by Eqs. (5.34) and (5.35) with $\mathrm{d}R/\mathrm{d}t = \gamma I$ can be discussed by analogy. The following results can be obtained [3, 8]. The eigenvalues $\lambda_{1,2}$ are given by Eq. (5.38), which we copy here for sake of completeness as

$$\lambda_{1,2} = \frac{\beta_E - \alpha - \gamma}{2} \pm \sqrt{U}\ ,\ U = \frac{[\beta_E - \alpha - \gamma]^2}{4} + (\beta_E - \alpha)\gamma + \beta_I \alpha\ , \quad (6.24)$$

where the plus (minus) sign holds for λ_1 (λ_2). The eigenvectors $\mathbf{v}_1$ and $\mathbf{v}_2$ read

$$\mathbf{v}_i = \begin{pmatrix} v_{i,E} \\ v_{i,I} \end{pmatrix} = \frac{1}{Z_i} \begin{pmatrix} \beta_I \\ \lambda_i + \alpha - \beta_E \end{pmatrix} \quad (6.25)$$

with $Z_i = \sqrt{(\lambda_i + \alpha - \beta_E)^2 + \beta_I^2}$ (see also the E, I components in Eq. (5.90)). The evolution equations for the amplitudes A_1 and A_2 and the non-infected variables u_3 and R are given by Eqs. (6.19) and (6.21), respectively, with C_1 and C_2 as defined in Eq. (6.20), $|M_+| = \beta_I(\lambda_2 - \lambda_1)/(Z_1 Z_2)$, and

$$k_0(A_1, A_2) = \frac{1}{N}\left(\beta_I \sum_{k=1,2} v_{k,I} A_k + \beta_E \sum_{k=1,2} v_{k,E} A_k\right) = h_1 A_1 + h_2 A_2 \quad (6.26)$$

(see also Eq. (5.92)) with $h_1 = (\beta_I v_{1,I} + \beta_E v_{1,E})/N$ and $h_2 = (\beta_I v_{2,I} + \beta_E v_{2,E})/N$ [3, 8]. The mappings given by Eqs. (6.22) and (6.23) also hold for the 2β case provided that the eigenvectors components listed in Eq. (6.25) are used.

6.2.3 SEIR-Type Models and Their Non-autonomous $m = 2$ Amplitude Equation Descriptions

Model of SEIR-type as defined by Eq. (5.3) in the absence of demographic terms (i.e., for $B = \mu = 0$) can be studied using the 2D amplitude equation description. To this end, the compartment variables are listed like $\mathbf{X} = (E, I, S, X_4, \ldots, X_n)$. In this case, Eq. (5.3) reads

$$\frac{\mathrm{d}}{\mathrm{d}t}E = k_0 S - \alpha E\ ,$$

$$\frac{\mathrm{d}}{\mathrm{d}t}I = \alpha E - \gamma I\ ,$$

$$\frac{d}{dt}S = -k_0 S\ ,$$

$$\frac{d}{dt}X_k = N_k(E, I, S, X_4, \ldots, X_n) \text{ for } k = 4, \ldots, n \tag{6.27}$$

with $k_0 = \beta I/N$ or $k_0 = (\beta_I I + \beta_E E)/N$. Subsequently, the three state variable S, E, and I are transformed into the variables A_1, A_2, and u_3 by means of Eq. (6.22) and $u_3 = S - N$. In analogy to Eqs. (6.19) and (6.21), the corresponding evolution equations read

$$\frac{d}{dt}u_3 = -(N + u_3)k_0(A_1, A_2)\ ,$$

$$\frac{d}{dt}A_1 = \lambda_1 A_1 + C_1 u_3 k_0(A_1, A_2)\ ,$$

$$\frac{d}{dt}A_2 = \lambda_2 A_2 + C_2 u_3 k_0(A_1, A_2)\ . \tag{6.28}$$

For the 1β and 2β cases the coefficients C_1 and C_2 and the rate constant k_0 are defined as discussed in Sect. 6.2.2. The evolution equations for the remaining variables X_k become

$$\frac{d}{dt}X_k = N_k(E(\mathbf{A}), I(\mathbf{A}), S = N + u_3, X_4, \ldots, X_n) \tag{6.29}$$

with $k = 4, \ldots, n$. Equations (6.28) and (6.29) include the 1β and 2β SEIR models discussed in the previous section as special cases.

The three-variable model (6.28) has been used in Refs. [3, 4, 8] to discuss COVID-19 epidemics in China and Italy (see also Sect. 6.3 below). In particular, in Ref. [3] a detailed discussion of the properties of the coupled set of differential equations (6.28) can be found.

Accordingly, let us first discuss the signs of the variables A_1, A_2, and u_3. From $S = N + u_3$ it follows that $u_3 \le 0$. Furthermore, for any parameter set $\alpha, \beta_I, \beta_E, \gamma > 0$ the following inequalities hold:

$$\lambda_2 < 0\ ,\ k_0 \ge 0\ ,\ \lambda_1 > \lambda_2 \Rightarrow |M_+| < 0\ ,\ v_{i,E} > 0\ ,$$

$$\lambda_1 + \alpha - \beta_E > 0 \Rightarrow v_{1,I} > 0\ ,$$

$$\lambda_2 + \alpha - \beta_E < 0 \Rightarrow v_{2,I} < 0\ . \tag{6.30}$$

They also hold for 1β models (i.e., for $\beta_I = \beta$ and $\beta_E = 0$). In particular, from Eqs. (6.24) and (6.25) and the alternative way to express U shown in Eq. (5.39), it follows that

$$\lambda_i + \alpha - \beta_E = -\frac{e}{2} \pm \sqrt{\frac{e^2}{4} + \beta_I \alpha} \tag{6.31}$$

with $e = \beta_E + \gamma - \alpha$, where the plus (minus) sign holds for $i = 1$ ($i = 2$), which implies that $\lambda_1 + \alpha - \beta_E > 0$ and $\lambda_2 + \alpha - \beta_E < 0$ holds as listed in Eq. (6.30).

From Eq. (6.30), the mapping $A_1 = (v_{2,I}E - v_{2,E}I)/|M_+|$, and $E, I \geq 0$ it follows that $A_1(t) \geq 0$ at any time. In contrast, A_2 can assume both positive and negative values as can be shown by numerical calculations.

Let us study next the dynamics of a wave-solution of the SEIR model. Accordingly, let us consider the case $\beta > \gamma$ (1β model) or $\beta_w > \gamma$ (2β model) such that $\lambda_1 > 0$ and $\lambda_2 < 0$ holds. Furthermore, it is assumed that the disease (or health) state $\mathbf{X}$ of the population under consideration is sufficiently close to the fixed point $\mathbf{X}_{st}$ with $E = I = 0$ and $S_{st} = N$ (i.e., $u_3 = S - N$, A_1, A_2 are sufficiently small such that the nonlinear terms $G_i = C_i u_3 k_0$ of the amplitude equations can be neglected). Accordingly $\mathbf{X}_{st}^+ = (E, I) = (0, 0)$ is an unstable fixed point (a saddle point). From the linear parts of the amplitude equations listed in Eq. (6.28) it follows that initially A_1 increases exponentially like $A_1(t) = A_1(t_0)\exp\{\lambda_1(t - t_0)\}$. In contrast, the amplitude A_2 initially decays in the amount like $|A_2(t)| = |A_2(t_0)|\exp\{-|\lambda_2|(t - t_0)\}$. Consequently, the amplitude $A_1(t)$ describes the outbreak of the epidemic in the direction specified by $\mathbf{v}_1$. In other words, in the space spanned by E and I the dominant dynamics takes place in the direction $\mathbf{v}_1$. The vector $\mathbf{v}_1$ can be regarded as the EI order parameter of the disease outbreak and A_1 as its order parameter amplitude (see Sect. 2.8).

In particular, for bifurcations [5, 6] typically the inequality $|\lambda_2| \gg \lambda_1$ holds. In terms of the time constants $\tau_i = 1/|\lambda_i|$ related to the eigenvalues λ_i the condition $|\lambda_2| \gg \lambda_1$ implies that $\tau_2 \ll \tau_1$ holds. In this case A_2 decays in magnitude to zero faster than A_1 increases over time. For epidemic outbreaks that satisfy this kind of time scale separation, after an intermediate period T_i the contribution of A_2 to the dynamics in the E-I can be neglected. From Eq. (6.6) it then follows that

$$\frac{\mathrm{d}}{\mathrm{d}t}\mathbf{X}^+ \approx \mathbf{v}_1 \frac{\mathrm{d}}{\mathrm{d}t} A_1 \Rightarrow \Delta\mathbf{X}^+ \approx \mathbf{v}_1 \Delta A_1 \tag{6.32}$$

with $\Delta\mathbf{X}^+ = \mathbf{X}^+(t + \Delta t) - \mathbf{X}^+(t)$ and $\Delta A_1 = A_1(t + \Delta t) - A_1(t)$. Consequently,

$$\frac{\Delta E}{\Delta I} \approx \frac{v_{1,E}}{v_{1,I}} . \tag{6.33}$$

holds. Eq. (6.33) states that the order parameter describes the ratio of the changes in the size of the compartments E and I. In particular, using t_0 as reference point, from Eq. (6.32) and $\mathrm{d}A_1/\mathrm{d}t \approx \lambda_1 A_1$ it follows that

$$\mathbf{X}^+(t) = \mathbf{v}_1 A_1(t_0)\exp\{\lambda_1(t - t_0)\} \tag{6.34}$$

consistent with the general discussion in Sect. 2.7 (see also Eq. (2.47)). Equations (6.32) and (6.34) state that the precise initial conditions $E(t_0)$ and $I(t_0)$ of an epidemic do not matter [4, 9]. The dynamics is pushed towards the trajectory $\mathbf{X}^+(t)$ described by Eqs. (6.32) and (6.34). For similar considerations see also Sect. 5.7.1.

Having discussed the initial dynamics of a wave-solution close to the unstable disease-free fixed point, let us consider next the long term dynamics. In this case, the nonlinearities $G_i = C_i u_3 k_0$ play a role. In this context note that for any parameters $\alpha, \beta_I, \beta_E, \gamma > 0$ the following inequalities hold:

$$C_1 > 0\,, C_2 > 0\,,\ u_3 \leq 0\,, k_0 \geq 0\ \Rightarrow G_i = C_i u_3 k_0 \leq 0\,. \tag{6.35}$$

In particular, the inequalities $C_1 > 0$ and $C_2 > 0$ follow from Eq. (6.20) and the inequalities listed in Eq. (6.30). As stated in Eq. (6.35), the nonlinearities are negative functions $G_i = C_i u_3 k_0 < 0$ for $u_3 \neq 0$ and $k_0 \neq 0$ or vanish like $G_i = 0$ for $u_3 = 0$ or $k_0 = 0$. As mentioned above, $A_1 > 0$ increases initially like an exponential function, whereas A_2 decease in magnitude during the intermediate period T_i. However, A_2 may increase in magnitude at a later time point $t > T_i$. In the long time limit, A_1 decays in magnitude due to the impact of the nonlinearity $G_1 < 0$, whereas A_2 decays in magnitude due to the combined impacts of the linear term $\lambda_2 A_2$ and the nonlinear term $G_2 < 0$. Eventually, A_1 and A_2 vanish. That is, the amplitudes converge to the fixed point $A_1 = A_2 = 0$, which corresponds in state space to the location $\mathbf{X}_{st}^+ = (E, I) = (0, 0)$.

6.3 COVID-19 Outbreak in Wuhan City 2020 and Its EI Order Parameter

In Sect. 5.8 the COVID-19 outbreak in Wuhan city during January and February 2020 was discussed using a SEIR modeling perspective. The 3D amplitude space description of the SEIR model was used. In what follows, the COVID-19 outbreak will be discussed using the 2D amplitude space description as presented in Ref. [8]. Just as in Sect. 5.8, the 2β SEIR model equations (5.34) and (5.100) were solved numerically for the parameters and initial conditions reported in Sect. 5.8. Panel (a) of Fig. 6.1 shows the trajectories S, E, and I thus obtained. They are identical to those presented in panel (a) of Fig. 5.7. Unlike the steps taken in Sect. 5.8 for the 3D amplitude equation approach, for the 2D amplitude equation approach, the model equations (6.28) for the amplitudes A_1 and A_2 and the relative state u_3 were solved. The initial values $A_1(t_0)$ and $A_2(t_0)$ were obtained from Eq. (6.22) and $u_3(t_0)$ was computed from $u(t_0) = S_0 - N$. Panel (b) of Fig. 6.1 shows the numerical solutions $A_1(t)$ (solid line) and $A_2(t)$ (dashed line). As expected, A_1 increased in a more or less exponential manner. In contrast, A_2 decayed monotonically in magnitude. Panel (c) presents the eigenvectors $\mathbf{v}_1$ and $\mathbf{v}_2$ as computed from Eq. (6.25) in the E-I plane. In particular, panel (c) demonstrates that $\mathbf{v}_1$ points at an positive angle in the E-I plane such that changes along $\mathbf{v}_1$ imply an increase of both E and I. In contrast, $\mathbf{v}_2$ points in a negative angle such that changes along $\mathbf{v}_2$ imply that when E increases then I decreases or vice versa. Therefore, $\mathbf{v}_1$ can capture most of the COVID-19 outbreak in Wuhan city during which supposedly the population sizes of both compartments E and I increased over time. The dotted line in panel (c) is the unit circle. Panel (d) shows the trajectory $\mathbf{X}^+(t)$ as phase curve $I(E)$ in the E-I plane (solid thick

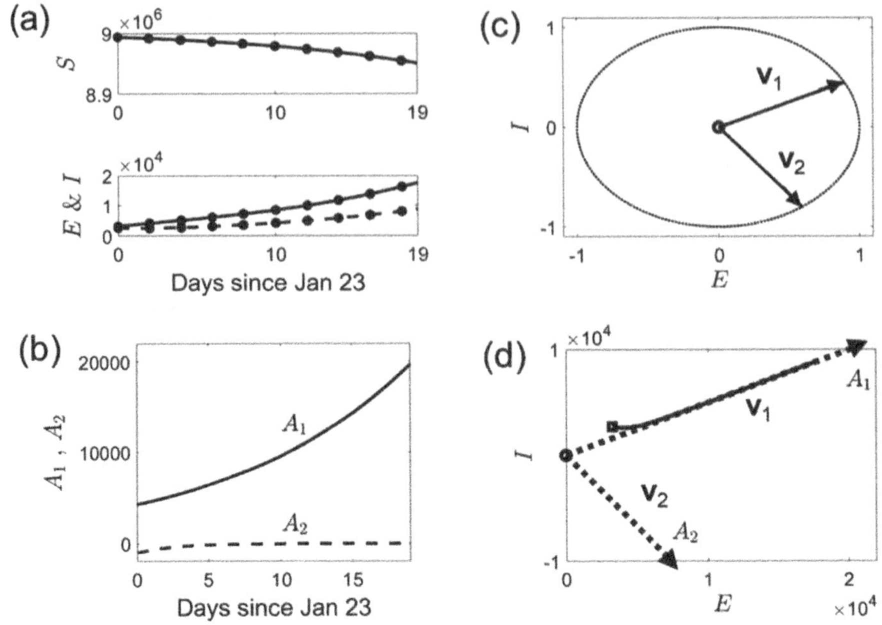

Fig. 6.1 State space and amplitude space description of the COVID-19 outbreak in Wuhan city, China, during 2020, using a reduced 2D amplitude space approach. Panel **a** shows solutions $S(t)$ (top subpanel), $E(t)$ (solid line, bottom subpanel), and $I(t)$ (dashed line, bottom subpanel) of the 2β SEIR model given by Eqs. (5.34) and (5.100). Panel **b** shows the amplitudes A_1 (solid line) and A_2 (dashed line) computed from Eq. (6.28). Panel **c** shows the eigenvectors $\mathbf{v}_1$ and $\mathbf{v}_2$ computed from Eq. (6.18) that constituted the 2D amplitude space basis of the outbreak in Wuhan city. Panel **d** shows the phase curve $I(E)$ (solid line) and eigenvectors $\mathbf{v}_1$ and $\mathbf{v}_2$ (dotted lines) magnified for visualization purposes. Panel **d** illustrates that $\mathbf{v}_1$ was the EI order parameter of the COVID-19 outbreak in Wuhan city

black line). The square indicates the initial value of the simulation. The eigenvectors $\mathbf{v}_1$ and $\mathbf{v}_2$ are shown there as well. It can be seen that the trajectory $\mathbf{X}^+(t)$ follows the direction specified by $\mathbf{v}_1$. Consequently, panel (d) illustrates that the COVID-19 outbreak in Wuhan city, when described from the perspective of a SEIR model, followed the EI order parameter $\mathbf{v}_1$ during the period from January 23 to February 11, 2020. As such, the eigenvector $\mathbf{v}_1(2D)$ shown in panels (c) and (d) points in the same direction as the eigenvector projection of $\mathbf{v}_2(3D)$ shown in panel (c) of Fig. 5.7. Moreover, since in panel (c) of Fig. 5.7 and in panel (d) of Fig. 6.1 the eigenvectors are increased in magnitude for the sake of visibility, panel (c) of Fig. 5.7 and panel (d) of Fig. 6.1 actually show the same results except for the fact that in panel (d) of Fig. 6.1 the eigenvector $\mathbf{v}_2$ is presented as well.

In order to demonstrate the equivalence of the amplitude description via Eq. (6.28) and the state space description given by Eq. (5.35) the variables S, E, and I were computed from the amplitudes A_1 and A_2 and the relative state u_3. That is, the solutions $A_1(t)$ and $A_2(t)$ of Eq. (6.28) as shown in panel (c) were substituted

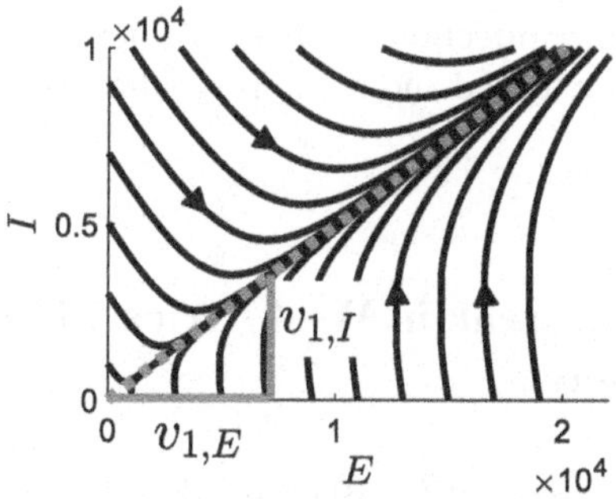

Fig. 6.2 Phase portrait of the dynamical system that determined the COVID-19 outbreak in the year of 2020 in Wuhan city, China, when taking a SEIR modeling perspective. The phase portrait illustrates that the outbreak followed a narrow path that was determined by the order parameter $\mathbf{v}_1$ (dotted gray line)

into Eq. (6.22) and $u_3(t)$ (not shown) obtained from Eq. (6.28) was used to compute $S(t) = N + u_3(t)$. In doing so, the time-dependent functions $E(A_1(t), A_2(t))$, $I(A_1(t), A_2(t))$ and $S(u_3(t))$ were obtained. They are plotted in the respective subpanels in panel (a) as full circles. As expected, the two approaches produced identical solutions.

Figure 6.2 present the phase portrait of the dynamical system that determined the disease state of the population of Wuhan city during January and February 2020 as it can be inferred from the analysis conducted so far. In order to produce the phase portrait, the SEIR model equations (5.34) and (5.35) were solved numerically for various initial conditions but for the fixed model parameters mentioned in Sect. 5.8. The order parameter $\mathbf{v}_1$ is shown as dotted gray line. The two components $\mathbf{v}_{1,E}$ and $\mathbf{v}_{1,I}$ (magnified by a factor 8000) are illustrated as well as gray bars. The phase portrait illustrates the statement made above, namely, that under appropriate circumstances epidemics approach their order parameters irrespective of the initial conditions at hand and, subsequently, evolve along those order parameters [4, 9]. The appropriate circumstances are the requirement that λ_2 is larger in the amount than λ_1. This was the case for the epidemic outbreak in Wuhan with $\lambda_1 = 0.08$/d and $\lambda_2 = -0.39$/d (see Sect. 5.8). In view of Fig. 6.2, it is fair to say that to some extent the initial numbers of infected individuals in the compartments E and I on the initial day January 23 considered in Fig. 6.1 did not play a crucial role. Irrespective of those numbers, the epidemic followed the “narrow path” [4] defined by the order parameter $\mathbf{v}_1$ shown in Fig. 6.2.

Finally, Fig. 6.2 can also be used to illustrate the application of Eq. (6.33) in a quantitative way. Substituting the model parameters listed in Sect. 5.8 into Eq. (6.25) for $\mathbf{v}_1$, the vector components are obtained as $v_{1,E} = 0.90$ and $v_{1,I} = 0.45$ [8]. From Eq. (6.33) and $v_{1,E} = 0.90$ and $v_{11,I} = 0.45$ it then follows that the outbreak evolved along $\mathbf{v}_1$ such that the compartment sizes of E and I varied like $\Delta E/\Delta I \approx 2$. Accordingly, when during the COVID-19 outbreak in Wuhan city in a certain period the population of symptomatic infectious individuals increased by 1, then in the same period the population of exposed and possibly asymptomatic infectious individuals increased by 2.

6.4 COVID-19 Outbreak in West Africa 2020 and Its EIA Order Parameter

Taboe et al. [10] studied the COVID-19 outbreak of the year 2020 in the region of West Africa. To this end, a SEIAR model [11] with compartments similar to the models that are discussed in Sect. 5.3.2 was used. The six-variable model by Taboe et al. [10] accounts for susceptible individuals (S), exposed individuals (E), asymptomatic infectious individuals (I_a), symptomatic infectious individuals (I_s), and recovered individuals (R). It also accounts for quarantined individuals or individuals in clinical treatment that are taken together in a single compartment (I_c). Exposed individuals are considered as individuals in their latent phase who cannot infect others. Individuals under treatment and quarantined individuals are assumed to be perfectly isolated such that they cannot infect others. Consequently, the model involves only two compartments of actually infectious individuals: I_a and I_s. Accordingly, the rate constant k_0 of reactions $S \to Y$, where Y are infected individuals, reads

$$k_0 = (1 - \Psi)\frac{\beta_a I_a + \beta_s I_s}{N} \, . \tag{6.36}$$

In Eq. (6.36) the variable N is given by $N = S + E + I_a + I_s + I_c + R$ and denotes the size of the total population. N is constant. In the original study by Taboe et al. [10] the parameter N in Eq. (6.36) is replaced by $N - I_c$. However, as argued in Ref. [4], $N \gg I_c$ holds such that the number of individuals I_c makes a negligibly small contribution to N. Therefore, in what follows, the contact rate as defined by Eq. (6.36) will be used with N being constant. The parameter $\Psi \in [0, 1]$ occurring in Eq. (6.36) measures the impact of intervention measures. The parameters β_a and β_s denote the effective contact rates for contacts between susceptibles and asymptomatic infectious individuals (β_a) and susceptibles and symptomatic infectious individuals (β_s), respectively. The evolution equations for the compartments read [10]

$$\begin{aligned}
&\frac{\mathrm{d}}{\mathrm{d}t}S = -k_0 S \, , \; \frac{\mathrm{d}}{\mathrm{d}t}E = k_0 S - k_1 E \, , \; \frac{\mathrm{d}}{\mathrm{d}t}I_a = p k_1 E - k_2 I_a \, , \\
&\frac{\mathrm{d}}{\mathrm{d}t}I_s = (1-p) k_1 E - k_3 I_s \, , \; \frac{\mathrm{d}}{\mathrm{d}t}I_c = \rho_a I_a + \rho_s I_s - k_4 I_c \, , \\
&\frac{\mathrm{d}}{\mathrm{d}t}R = \gamma_a I_a + \gamma_s I_s + \gamma_c I_c \, .
\end{aligned} \tag{6.37}$$

In Eq. (6.37) the model parameters are presented using a simplified notation suggested in Ref. [4]. The factor $p \in [0, 1]$ describes the proportion of exposed individuals for which COVID-19 develops in an asymptomatic manner. The parameters ρ_a and ρ_s denote diagnosis rates for asymptomatic and symptomatic individuals, respectively. The parameters γ_a, γ_s, and γ_c stand for recovery rates. Finally, the parameters $k_1, \ldots, k_4$ describe removal rates. In particular, $k_2, \ldots, k_4$ can be expressed in terms of the other model parameters like [10]

$$k_2 = \rho_a + \gamma_a \ , \ k_3 = \rho_s + \gamma_s + d_s \ , \ k_4 = \gamma_c + d_c \ , \tag{6.38}$$

where d_s and d_c are the death rates due to COVID-19 of non-diagnosed symptomatic individuals (I_s) and individuals of compartment (I_c). While COVID-19 associated deaths as such decrease the population size, the argument can again be made that the effect of such deaths on N can be neglected. In other words, N can be considered as constant despite the fact that the model addresses COVID-19 associated deaths. Note that a compartment of COVID-19 associated deaths could be added but would not change the following considerations.

In Ref. [4] the non-autonomous amplitude space description discussed in Sect. 6.1.1 was worked out for the model (6.37). To this end, the six-dimensional state space was decomposed into the three-variable subspace vector $\mathbf{X}^+ = (E, I_a, I_s)$ describing actually infectious individuals and exposed individuals and the three-variable subspace vector $\mathbf{X}^- = (S, I_s, R)$ describing the remaining variables. The rationale for this approach is that the variables I_c and R can be considered as variables driven by the remaining variables. They do not feed back and do not play a role for the instability of the epidemiological system described by Eq. (6.37). Consequently, when considering a linearized system of the four variables (S, E, I_a, I_s) they will not show up in the linearized evolution equation. As far as the variable S is concerned, as shown in Sect. 6.2.1 in the context of the 1β SEIR model, the susceptible variable S does not occur in the linearized evolution equations for E and I. Likewise, in the context of the epidemiological model (6.37), S or its corresponding deviation variable $u_S = S - N$ does not occur in the linearized evolution equations for E, I_a, and I_s. Consequently, the decomposition of $\mathbf{X}$ into $\mathbf{X} = (\mathbf{X}^+, \mathbf{X}^-)$ with $\mathbf{X}^+ = (E, I_a, I_s)$ and $\mathbf{X}^- = (S, I_s, R)$ leads to a 6×6 linearization matrix L that exhibits the required form (6.2).

Let $\mathbf{X} = (E, I_a, I_s, S, I_s, R)$ denote the state vector. Then, the disease-free fixed point of interest reads $\mathbf{X}_{st} = (0, 0, 0, N, 0, 0)$ and exhibits the projection $\mathbf{X}_{st}^+ = (0, 0, 0)$ in the subspace D^+. From Eq. (6.37) it follows that the 3×3 submatrix L^+ relevant for the dynamics in D^+ reads

$$L^+ = \begin{pmatrix} -k_1 & (1-\Psi)\beta_a & (1-\Psi)\beta_s \\ pk_1 & -k_2 & 0 \\ (1-p)k_1 & 0 & -k_3 \end{pmatrix} . \tag{6.39}$$

The eigenvalues λ_1, λ_2, and λ_3 of L^+ satisfy a cubic polynomial and for given model parameters can be obtained by numerical methods. In order to determine the eigenvectors $\mathbf{v}_i$ of L^+, the matrix $J = L^+ - \lambda E$ can be considered that assumes the form

$$J = \begin{pmatrix} J_{11} & J_{12} & J_{13} \\ J_{21} & J_{22} & 0 \\ J_{31} & 0 & J_{33} \end{pmatrix} \tag{6.40}$$

with off-diagonal elements $J_{ik} = L^+_{ik}$ and diagonal elements $J_{kk} = L^+_{kk} - \lambda$ that depend on λ. Note that by definition, the rows are linearly dependent. Therefore, it is sufficient to evaluate only two rows. Due to the special structure of J, it is convenient to evaluate the second and third rows of J. Then, the vectors $\mathbf{v}_i$ that satisfy $J\mathbf{v}_i = (0, 0, 0)$ can be written like

$$\mathbf{v}_i = x \begin{pmatrix} 1 \\ -J_{21}/J_{22} \\ -J_{31}/J_{33} \end{pmatrix} \Rightarrow \mathbf{v}_i = x' \begin{pmatrix} J_{22}J_{33} \\ -J_{21}J_{33} \\ -J_{31}J_{22} \end{pmatrix}, \tag{6.41}$$

where x or x' are determined by the normalization condition $|\mathbf{v}_i| = 1$. Let $\mathbf{v}_j = (v_{j,E}, v_{j,I_a}, v_{j,I_s})$ denote the eigenvectors of L^+ in D^+ with $j = 1, 2, 3$. Then, from Eqs. (6.39), (6.40), and (6.41) the analytical expressions [4]

$$\mathbf{v}_j = \frac{1}{W_j} \begin{pmatrix} (\lambda_j + k_2)(\lambda_j + k_3) \\ pk_1(\lambda_j + k_3) \\ (1-p)k_1(\lambda_j + k_2) \end{pmatrix} \tag{6.42}$$

for $j = 1, 2, 3$ can be obtained, where W_j are normalization constants. As a result, $\mathbf{X}^+$ can be expressed like

$$\mathbf{X}^+ = \sum_{j=1}^{3} A_j \mathbf{v}_j \tag{6.43}$$

(see Eq. (6.6)). Using the cross-product approach suggested in Sect. 5.6 for the case $n = 3$ (e.g., $\mathbf{w}_1 = (\mathbf{v}_2 \times \mathbf{v}_3)/Z$), the biorthogonal vectors may be obtained. Subsequently, following the procedure in Sect. 6.1.2, from Eqs. (6.37) and (6.43) the amplitude equations

$$\frac{\mathrm{d}}{\mathrm{d}t} A_j = \lambda_j A_j - C_j u_S k_0(A_1, A_2, A_3) \tag{6.44}$$

with $j = 1, 2, 3$ and the rate constant

$$k_0 = \frac{(1-\Psi)}{N} \sum_{k=1}^{3} (\beta_a v_{k,I_a} + \beta_s v_{k,I_s}) A_k \tag{6.45}$$

can be obtained [4]. In Eq. (6.44) the coefficients C_j are functions of the model parameters and do not depend on the amplitudes $A_1, \ldots, A_3$. They can be written like

$$C_j = \frac{(k_2 - k_3)W_j}{f_{j,1} + f_{j,2} + f_{j,3}} \tag{6.46}$$

with

$$\begin{aligned} f_{j,1} &= (k_2 - k_3)(\lambda_j + k_2)(\lambda_j + k_3) \, , \\ f_{j,2} &= -(\lambda_s + k_2)(\lambda_u + k_2)(\lambda_j + k_3) \, , \\ f_{j,3} &= (\lambda_s + k_3)(\lambda_u + k_2)(\lambda_j + k_2) \end{aligned} \tag{6.47}$$

and indices $(s, u) = (2, 3)$ for $j = 1$, $(s, u) = (1, 3)$ for $j = 2$, and $(s, u) = (1, 2)$ for $j = 3$. The evolution equation for $u_S = S - N$ reads

$$\frac{\mathrm{d}}{\mathrm{d}t} u_S = -(N + u_S) k_0(A_1, A_2, A_3) \, . \tag{6.48}$$

As expected, in the linear domain, the amplitude equations (6.44) constitute an autonomous system. However, in the full nonlinear domain, they describe a non-autonomous system that depends on u_S (whose dynamics, in turn, depends on the amplitude vector $\mathbf{A} = (A_1, A_2, A_3)$). The four-variable model defined Eqs. (6.44) and (6.48) provides a closed description of the disease dynamics in the subspace E, I_a, I_s, S of the original six-dimensional space. The dynamics of the two remaining variables I_c and R can then be obtained from the dynamics in the subspace E, I_a, I_s, S.

As mentioned at the beginning of this paragraph, Taboe et al. [10] applied the model defined by Eq. (6.37) to describe the outbreak of COVID-19 in West Africa. In this context, let us briefly review the discussion in Ref. [10] about the role of the intervention parameter Ψ. It was assumed that for $\Psi = 0$ the model describes the COVID-19 outbreak in the absence of intervention measures and, consequently, features an unstable fixed point. In contrast, for $\Psi = 1$ the rate constant k_0 vanishes and, consequently, the disease-free fixed point is stable. In Ref. [10] the existence of a critical parameter Ψ_c was shown such that for $\Psi < \Psi_c$ the disease-free fixed point is unstable, whereas for $\Psi > \Psi_c$ the fixed point is stable. This implies that for $\Psi < \Psi_c$ at least one eigenvalue with $j = 1, 2, 3$ is positive or has positive real part, whereas for $\Psi > \Psi_c$ all eigenvalues $j = 1, 2, 3$ are negative or have negative real part. In particular, Taboe et al. fitted the model to the data of the outbreak in the 16 countries region of West Africa for the period from February 28 to June 26, 2020, which is a 120 days period. They found a best-fit estimate of $\Psi = 0.261$ with $\Psi > 0$ and $\Psi < \Psi_c$ indicating that the disease-free fixed point during that period was unstable. Using the model parameters reported in Ref. [10], in Ref. [4] the eigenvalues λ_1, λ_2, and λ_3 were computed. They were found to be real-valued with $\lambda_1 = 0.031$/d, $\lambda_2 = -0.365$/d, and $\lambda_3 = -0.547$/d. The eigenvalues confirm that the disease-free fixed point was unstable.

Figure 6.3 shows simulation results obtained from the model (6.37) and confirmed COVID-19 cases [12]. In order to compute the graphs shown in Fig. 6.3, the model (6.37) was solved numerically for the period from February 28 to June 26, 2020 and

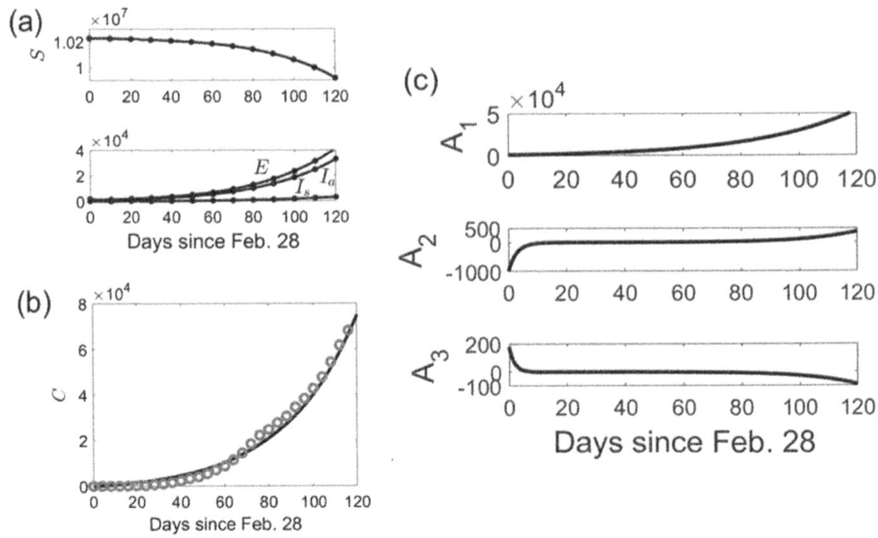

Fig. 6.3 State space and amplitude space descriptions of the COVID-19 outbreak in West Africa during the first half of the year 2020. Panel **a**: State space solutions of the SEIAR-like model (6.37). Panel **b**: Confirmed cases (gray circles) [12] and the model solution $C(t)$ (solid line) computed from Eq. (6.50). Panel **c**: Amplitudes A_1, A_2, and A_3 computed from Eqs. (6.44) and (6.48) indicating that A_1 after an intermediate period of less than 10 days was the dominant amplitude during the remaining observation period

the parameters reported in Ref. [10]. Panel (a) shows $S(t)$ (top subpanel) and $E(t)$, $I_s(t)$, and $I_a(t)$ (bottom subpanel) as functions of time. The graphs $E(t)$, $I_s(t)$, and $I_a(t)$ exhibit a monotonic increase consistent with the notion that they are determined by the exponentially increasing order parameter amplitude like

$$\mathbf{X}^+(t) \approx \mathbf{v}_1 A_1(t) = \mathbf{v}_1 A_1(t_0) \exp\{\lambda_1(t - t_0)\} \; . \tag{6.49}$$

Panel (b) shows the cumulative confirmed COVID-19 cases of the 16 countries region of West Africa [12]. With respect to the model (6.37) the diagnosed cumulative cases C satisfy

$$\frac{\mathrm{d}}{\mathrm{d}t} C = \rho_a I_a(t) + \rho_s I_s(t) \; . \tag{6.50}$$

That is, C corresponds to the cumulative cases in the compartment I_c (compare Eq. (6.50) with the evolution equation of I_c in Eq. (6.37)). The function $C(t)$ was computed numerically from Eq. (6.50) and $I_a(t)$ and $I_s(t)$ shown in panel (a). The result is plotted in panel (b). The model solution $C(t)$ fits the data with moderate accuracy. Panel (c) shows the amplitudes A_1, A_2, and A_3 as function of time as computed from the amplitude equation description given by Eqs. (6.44) and (6.48). As expected, the model-based analysis suggests that initially during the COVID-19 outbreak all amplitudes showed an exponential dynamics like $A_i = A_i(t_0) \exp\{\lambda_i(t - t_0)\}$. How-

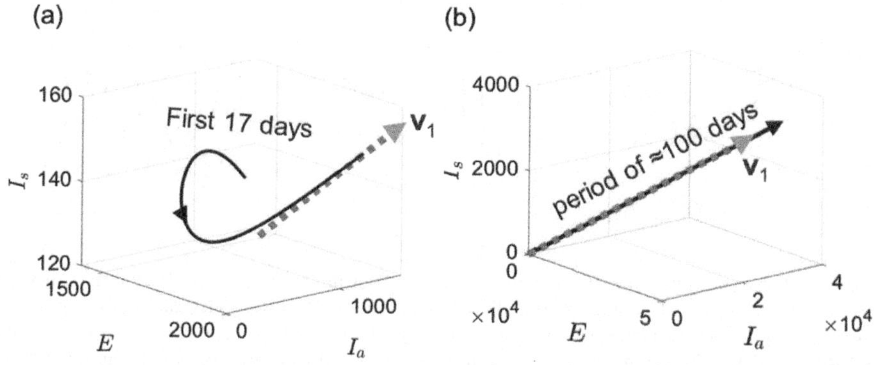

Fig. 6.4 Visualization of the amplitude dynamics of the COVID-19 outbreak in West Africa by means of the outbreak-specific EIA order parameter and phase curves. Panel (**a**): The trajectory $\mathbf{X}(t)$ (as shown in panel (**a**) of Fig. 6.3) is shown as phase curve in the (E, I_a, I_s) space for the first 17 days since February 28, 2020. Panel (**b**): $\mathbf{X}(t)$ is shown in the (E, I_a, I_s) space for the whole 120 days period. Panels (**a**) and (**b**): The EIA order parameter $\mathbf{v}_1$ (gray dotted line) computed from Eq. (6.42) is shown in a magnified scale

ever, A_1 increased monotonically during the whole 120 days observation period. In contrast, A_2 and A_3 decayed quickly during an intermediate period T_i of about 10 days. During the following 80 to 90 days, the amplitudes remained close to zero. Only in a late stage starting at about 90 to 100 days after February 28, the amplitudes started to deviate from zero. Finally, panel (a) also demonstrates the equivalence between amplitude space and state space descriptions. The variables S, E, I_a, I_s were computed from the solutions u_s and A_1, A_2, A_3 using $S = N + u_S$ and Eq. (6.43) and are shown as full circles. The solutions S, E, I_a, I_s as presented in panel (a) as solid lines are obtained directly from Eq. (6.37). As expected, both approaches produced identical results.

Let us illustrate the role of the order parameter $\mathbf{v}_1$ for the COVID-19 outbreak in West Africa. Figure 6.4 shows phase curves in the 3D subspace D^+ for the first 10 days following February 28 (panel (a)) and the full observation period of 120 days (panel (b)). The 3D EIA order parameter $\mathbf{v}_1$ computed from Eq. (6.42) is depicted as well (magnified in size for visualization purposes). As can be seen in panel (a), during the period $T_i = 10$d the dynamics was not solely determined by the direction $\mathbf{v}_1$. Rather, other directions as described by $\mathbf{v}_2$ and $\mathbf{v}_3$ played a role. As shown in panel (c) of Fig. 6.3, during this 10 days period A_2 and A_3 decayed in magnitude. In the 3D subspace D^+ this dynamics correspond to an approach of the trajectory $\mathbf{X}^+(t)$ (or its phase curve) towards the direction specified by $\mathbf{v}_1$. As shown in panel (b) of Fig. 6.4, after the intermediate period I_i, for the remaining observation period, the disease dynamics followed closely the direction determined by $\mathbf{v}_1$.

Let us illustrate graphically this approach towards the axis specified by $\mathbf{v}_1$. To this end, Eq. (6.37) was solved numerically for various initial conditions. In doing so, the phase portraits shown in Fig. 6.5 were obtained. In Fig. 6.5 two-dimensional phase portraits are presented in the E-I_a plane (panel (a)), E-I_s plane (panel (b)),

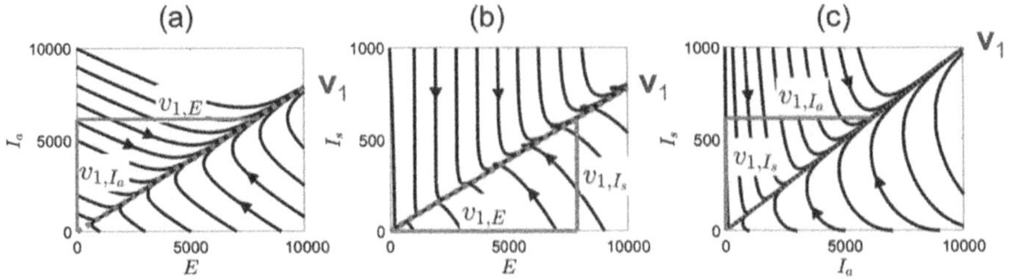

Fig. 6.5 Phase portrait of the dynamical system that determined the COVID-19 outbreak in West Africa, 2020, as seen in the context of the SEIAR model (6.37). Panels (**a**), (**b**), and (**c**) show possible disease dynamics phase curves in the planes (E, I_a), (E, I_s), and (I_a, I_s), respectively. The axis specified by the order parameter $\mathbf{v}_1$ is shown in all planes (dotted gray lines) as computed from Eq. (6.42)

and I_a-I_s plane (panel (c)). In order to interpret the phase portraits, let us return to the eigenvalues. As mentioned above, the disease-free fixed point was characterized by $\lambda_1 = 0.031$/d, $\lambda_2 = -0.365$/d, $\lambda_3 = -0.547$/d. That is, $\lambda_1 > 0$, $\lambda_2 < 0$, and $\lambda_3 < 0$. Consequently, the outbreak was consistent with the classical picture of a bifurcation exhibiting a single positive eigenvalue [5, 6]. The dynamics in D^+ was characterized by a saddle with one unstable direction $\mathbf{v}_1$ and two stable directions $\mathbf{v}_2$ and $\mathbf{v}_3$. The time constants of the eigenvalues were $\tau_1 = 32.1$d, $\tau_2 = 2.7$d, and $\tau_3 = 1.8$d. This implies that during the outbreak A_1 evolved slowly relative to A_2 and A_3. In other words, the amplitudes A_2 and A_3 evolved fast ($\tau_3 = 2.7$d, and $\tau_4 = 1.8$d) relative to A_1 ($\tau_2 = 32.1$d). Since A_2 and A_3 decayed ($\lambda_2 < 0, \lambda_3 < 0$) in the amount relatively quickly to zero all trajectories converged during the intermediate period T_i towards the $\mathbf{v}_1$-axis. As shown in the panels of Fig. 6.5, irrespective of the initial conditions, the disease dynamics converged relatively quickly towards the $\mathbf{v}_1$-axis (indicated in each panel by a dotted gray line) and, subsequently, evolved along the $\mathbf{v}_1$-axis. Figures 6.3, 6.4, and 6.5 demonstrates that after a short intermediate period T_i of about 10 days the disease state $\mathbf{X}^+(t)$ of the COVID-19 outbreak in West Africa evolved along the order parameter $\mathbf{v}_1$ such that the disease state satisfied Eq. (6.49). Alternative to Eq. (6.49), the dynamics for $t > T_i$ may be described by

$$\frac{\mathrm{d}}{\mathrm{d}t}\mathbf{X}^+ \approx \mathbf{v}_1 \frac{\mathrm{d}}{\mathrm{d}t} A_1 \Rightarrow \Delta\mathbf{X}^+ \approx \mathbf{v}_1 \Delta A_1 \; . \tag{6.51}$$

From Eq. (6.51) it follows that the compartment sizes changed relative to each other like

$$\begin{pmatrix} \Delta E \\ \Delta I_a \\ \Delta I_s \end{pmatrix} \approx \begin{pmatrix} v_{2,E} \\ v_{2,I_a} \\ v_{2,I_s} \end{pmatrix} \Delta A_1 \; . \tag{6.52}$$

For the model parameters reported in Ref. [10] and used throughout this section, the components of $\mathbf{v}_1$ read

$$\mathbf{v}_1 = \begin{pmatrix} 0.78 \\ 0.62 \\ 0.06 \end{pmatrix} \Rightarrow \begin{pmatrix} \Delta E \\ \Delta I_a \\ \Delta I_s \end{pmatrix} \approx \begin{pmatrix} 0.78 \\ 0.62 \\ 0.06 \end{pmatrix} \Delta A_1 \,. \tag{6.53}$$

The numerical values presented in Eq. (6.53) suggest that during the first-wave COVID-19 epidemic during Spring 2020 in West Africa when the number of symptomatic infectious individuals increased by 6, then in the same period the number of asymptomatic infectious individuals increased by 62 and the number of exposed individuals increased by 78. In other words, approximately for every symptomatic infectious person who appeared in the epidemic approximately ten asymptomatic infectious individuals and somewhat more than ten exposed individuals appeared as well.

Just as in the previous example presented in Sect. 6.3 and illustrated in Fig. 6.2, Fig. 6.5 demonstrates that the COVID-19 outbreak in West Africa exhibited a certain order. The dynamics followed a specific direction given by the unstable eigenvector $\mathbf{v}_1$. The outbreak in West Africa followed a "narrow path" just as the outbreak in Wuhan city (see Sect. 6.3). The order parameter concept, the notion of a "narrow path", and the phase portraits in Fig. 6.5 also suggest that the initial conditions are not necessarily crucial for understanding an epidemic. Under appropriate circumstances (e.g., a single positive eigenvalue), an epidemic evolves in a certain way or order irrespective of the precise initial conditions.

References

1. O. Diekmann, J.A.P. Heesterbeek, *Mathematical Epidemiology of Infectious Diseases* (Wiley, Chichester, 2000)
2. P. van den Driessche, J. Watmough, Reproduction numbers and sub-threshold endemic equilibria for compartment models of disease transmission. Math. Biosci. **180**, 29–48 (2002)
3. T.D. Frank, COVID-19 order parameters and order parameter time constants of Italy and China: a modeling approach based on synergetics. J. Biol. Syst. **28**, 589–608 (2020)
4. T.D. Frank, COVID-19 outbreaks follow narrow paths: a computational phase portrait approach based on nonlinear physics and synergetics. Int. J. Modern Phys. C **32**, 2150110 (2021)
5. T. Frank, *Determinism and Self-organization of Human Perception and Performance* (Springer, Berlin, 2019)
6. H. Haken, *Synergetics: An Introduction* (Springer, Berlin, 1977)
7. J. Ma, Estimating epidemic exponential growth rate and basic reproduction number. Infect. Dis. Model. **5**, 129–141 (2020)
8. T.D. Frank, Simplicity from complexity: on the simple amplitude dynamics underlying COVID-19 outbreaks in China. Adv. Complex Syst. **23**:article 2050022 (2020)

9. A. Das, A. Dhar, S. Goyal, A. Kundu, S. Pandey, COVID-19: analytical results from a modified SEIR model and comparison of different intervention strategies. Chaos, Solitons and Fractals **144**:article 110595 (2021)
10. H.B. Taboe, K.V. Salako, J.M. Tison, C.N. Ngonghala, R.G. Kakai. Predicting COVID-19 spread in the face of control measures in West Africa. Math. Biosci. **328**:article 108431 (2020)
11. J. Arino, F. Bauer, P. van den Driessche, J. Watmough, J. Wu, Simple models for containment of a pandemic. J. Roy. Soc. Interface **3**, 453–457 (2006)
12. COVID-19 tracker. Timeline data from Johns Hopkins Center for Systems Science and Engineering. COVID-19 tracker, https://vac-lshtml.shinyapps.io/ncov_tracker

Chapter 7
Model-Based Reproduction Numbers

This chapter introduces the concept of reproduction numbers in the context of compartmental models. Heuristic approaches to determine reproduction numbers of the SIR and SEIR models are presented. Subsequently, the next generation method is presented that allows to determine reproduction numbers for a variety of high-dimensional compartmental model. It is discussed that reproduction numbers can be regarded as bifurcation parameters that determine the stability of disease-free fixed points. Applications to the COVID-19 outbreaks during the year 2020 in Wuhan, China, and Pakistan are also presented.

7.1 Basic and Effective Reproduction Numbers

Reproduction numbers are frequently used quantities in epidemiology. The basic reproduction number R_0 is defined as the number of secondary infected cases in a completely susceptible population produced by a typically infectious individual [1–5]. Let N denote the size of a population. Then the basic reproduction number considers the situation $S = N$, which, for example, frequently holds when a novel infectious disease emerges in the population. In contrast, the effective reproduction number R_e refers to the expected number of secondary infected cases produced by a typically infectious individual when the epidemic is underway or when intervention measures have been put into place or both [5, 6]. In particular, when the epidemic is underway, part of the population has been infected by the virus under consideration such that the population is not completely susceptible. That is, $S < N$ holds. Let us consider an epidemic outbreak that has infected a non-negligible portion of the population such that the factor $x = S/N$ differs considerably from 1. However, no intervention measures have been put into place. Then x denotes the probability that an infectious individual in a contact meets with a susceptible individual. Likewise,

T. D. Frank, *COVID-19 Epidemiology and Virus Dynamics*,
Understanding Complex Systems, https://doi.org/10.1007/978-3-030-97178-6_7

$1 - x$ denotes the probability that the infectious individual meets with an individual that is infected or immune. Consequently, the effective reproduction number is related to the basic reproduction number like [6]

$$R_e = x R_0. \tag{7.1}$$

If intervention measures have been implemented, then typically $R_e < R_0$ holds even if $x \approx 1$. In this case, $R_e < R_0$ indicates that the spread of the infectious disease is reduced due to the impact of the intervention measures and not primarily due to a reduction of the number of susceptibles.

The basic reproduction number has also been called basic reproduction ratio, basic reproductive number, and basic reproductive ratio [1, 3]. As such reproduction numbers correspond to gain coefficients and amplification factors that have been extensively studied in physics (e.g., laser physics) and engineering (e.g., electronics). That is, a reproduction number describes how much the size of a certain generation i of infected individuals is amplified (or weakened) such that the amplified (or decreased) size corresponds to the size of the follow-up generation $i + 1$ of infected individuals.

Reproduction numbers have been used as threshold values [2]. Let us assume we could track the generations of infected individuals. Accordingly, the first infected individual describes the first generation. The infected individuals by that first infected individual correspond to the second generation. The infected individuals infected by the second generation correspond to the third generation individuals, and so on. Let c_i denote the size of the infected individuals in generation i. Then, let us assume $c_1 = 1$. Assuming that there are no intervention measures and that the decrease in susceptible is negligible (i.e., $x \approx 1$) during the first p generations, then $c_2 = R_0 c_1 = R_0$, $c_3 = R_0 c_2 = (R_0)^2 c_1 = (R_0)^2$ and $c_p = (R_0)^{p-1} c_1 = (R_0)^{p-1}$. Likewise, if interventions are implemented but do not vary during the epidemic that covers the first p generations, then we have $c_p = (R_e)^{p-1}$. This implies that for amplification factors $R_{0,e} > 1$, $R_{0,e} = 1$, and $R_{0,e} < 1$, respectively, generation sizes c_i increase, stay the same, and decrease over time. Consequently, if a single infected individual can infect more than one individual (i.e., if $R_{0,e} > 1$ holds), then the infectious disease spreads out. Conversely, if an infected individual can infect less than one other individual (i.e., $R_{0,e} < 1$ holds) or more realistically if an initial number of 100 infectious individuals can infect less than 100 other individuals, then the epidemic dies out. In other words, for $R_{0,e} > 1$ an infection can invade a population and there is an epidemic outbreak. For $R_{0,e} < 1$ the infectious disease subsides in the population. The threshold value of interest is $R_{0,e} = 1$.

In the context of this definition and utilization of reproduction numbers two questions arises. First, how can they be derived for epidemics that satisfy evolution equations of the form $\mathrm{d}\mathbf{X}/\mathrm{d}t = \mathbf{N}$ (see Eq. (2.1))? Second, how are the conditions $R_{0,e} > 1$ and $R_{0,e} < 1$ related to the nonlinear physics perspective according to which the stability of disease-free fixed points determines whether or not there is an epidemic wave or a subsiding epidemic? Let us address these questions in the subsequent Sects. 7.2, 7.3, and 7.4.

7.2 Case of a Single Infected Compartment

7.2.1 Heuristic Approach

Let us consider the spread of an infectious disease that can be described with the help of a single infected compartment. The infected compartment describes the infected and infectious individuals. For an example, let us consider the SIR model (3.16) with the compartment I of infectious individuals. As in Sect. 3.2, let ν denote the contact rate (i.e., contacts per day) of infectious individuals with other individuals. Let us consider the case in which all other individuals are susceptibles. Let p denote the probability of virus transmission per contact. Then $\beta = p\nu$ is the effective contact rate (see Eq. (3.3)). Consequently, if T is the duration of the infectious period, then the number of secondary infections caused by a single infectious individuals is given by [4, 6]

$$R_{0,e} = p\nu T = \beta T. \tag{7.2}$$

7.2.2 SIR Model: Heuristic Approach

Let us apply Eq. (7.2) to the SIR model (3.16). In the context of the SIR model, the infectious period T is given by the recovery period, which, in turn, is related to the rate of recovery γ like $T = 1/\gamma$ (see Sect. 3.4). Consequently, in the absence of interventions and for a completely susceptible population ($S = N$) the basic reproduction number reads

$$R_0 = \beta T = \frac{\beta}{\gamma}. \tag{7.3}$$

Interventions may affect the effective contact rate β or the recovery rate γ. If so, assuming that the epidemic under consideration has just started such that $S \approx N$, Eq. (7.3) holds again but the ratio β/γ should be interpreted as effective reproduction number

$$R_e = \frac{\beta}{\gamma}. \tag{7.4}$$

Finally, if a time point t is considered at which the number of susceptibles has decreased considerably such that $x = S/N$ is no longer close to 1, then the course of the epidemic is characterized by Eqs. (7.3) or (7.4) multiplied by x. Consequently, the reproduction number

$$R_e = \frac{S}{N}\frac{\beta}{\gamma} \tag{7.5}$$

is obtained. As indicated, this number is considered as an effective reproduction number regardless whether or not intervention measures are put in place. The reproduction numbers R_0 and R_e shown in Eqs. (7.3), (7.4), and (7.5) correspond to the stability parameters ξ of the SIR model derived in Sect. 3.5.1 (see Eqs. (3.24) and (3.26)) when replacing S by S_0. That is, the conditions $R_{0,e} > 1$ and $R_{0,e} < 1$ for an epidemic outbreak and subsiding epidemic, respectively, correspond to the conditions $\xi > 1$ and $\xi < 1$ that imply that the SIR model exhibits wave-solutions and monotonically decaying solutions, respectively. Likewise, Eqs. (7.3), (7.4), and (7.5) correspond to the bifurcation parameters α of the SIR model derived in Sect. 4.2.3 (see Eqs. (4.27) and (4.30)) when replacing S by S_{st}. That is, the conditions $R_{0,e} > 1$ and $R_{0,e} < 1$ correspond the conditions $\alpha > 1 \Rightarrow \lambda_2 > 0$ and $\alpha < 1 \Rightarrow \lambda_2 < 0$ that indicate the stability of the disease-free fixed point and again imply that the SIR model exhibits wave-solutions and monotonically decaying solutions, respectively. In other words, the amplification factor $R_{0,e}$ of the SIR model may be interpreted as a bifurcation parameter of the SIR model that exhibits a critical value 1 and determines the sign of the eigenvalue λ_2 of the SIR model.

7.2.3 SIR Model: Towards a Next Generation Approach

Let us determine $R_{0,e}$ in a more direct way from the SIR model defined by Eq. (3.16). The procedure that applies to the SIR model can then be generalized to more comprehensive epidemic models as will be shown in Sects. 7.3 and 7.4. To begin with, let $I_1(t)$ and $I_2(t)$ denote the numbers of infectious individuals of the first and second generations. In general, let $I_p(t)$ describe the size of the pth infectious generation over time. Furthermore, let us introduce the cumulative numbers of infectious individuals $I_{1,c}(t)$, $I_{2,c}(t)$, and $I_{p,c}(t)$ of the first, second, and pth generation as functions of time. The total number of infectious individuals in a generation p is denoted by i_p and corresponds to $i_p = I_{p,c}(\infty)$.

The key idea is to consider the dynamics of a disease state close to an appropriately defined fixed point such that the dynamics satisfies linearized evolution equations [1]. Accordingly, let us consider the disease-free fixed point with $S_{st} = N$. In order to simplify the presentation, in what follows, t_1 will denote the begin of the epidemic (rather than t_0 as used in other parts of this book). The following initial conditions hold

$$I_1(t_1) = I_{1,0},\ I_2(t_1) = I_{2,c}(t_1) = 0. \tag{7.6}$$

Furthermore, for the first generation $i_1 = I_{1,0}$ holds, which means, that the initially present number of infectious individuals corresponds to the cumulative number of first-generation infectious individuals. The evolution equation of I of the SIR model (3.16) reads $\mathrm{d}I/\mathrm{d}t = \beta SI/N - \gamma I$. The linearization at $S_{st} = N$ implies that $\beta SI/N \approx \beta I$. The term $-\gamma I$ describes the decrease of the first-generation infectious individuals like

$$\frac{\mathrm{d}}{\mathrm{d}t} I_1 = -\gamma I_1. \tag{7.7}$$

The evolution of the second-generation infectious individuals is determined by the production term βI that becomes βI_1 and the decay term $-\gamma I$ that becomes $-\gamma I_2$ such that

$$\frac{\mathrm{d}}{\mathrm{d}t} I_2 = \beta I_1 - \gamma I_2. \tag{7.8}$$

From Eq. (7.8) it follows that the number of cumulative second-generation infectious individuals increases like

$$\frac{\mathrm{d}}{\mathrm{d}t} I_{2,c} = \beta I_1(t). \tag{7.9}$$

Solving Eq. (7.7) gives us $I_1(t) = I_{1,0} \exp\{-\gamma(t - t_1)\} = i_1 \exp\{-\gamma(t - t_1)\}$. Substituting this result into Eq. (7.9) and integrating over time, Eq. (7.9) becomes

$$I_{2,c}(t) = \beta i_1 \int_{t_1}^{t} \exp\{-\gamma(t' - t_1)\} \mathrm{d}t', \tag{7.10}$$

where $I_{2,c}(t_1) = 0$ has been used (see Eq. (7.6)). Consequently, the second-generation infectious individuals $i_2 = I_{2,c}(\infty)$ can be computed from

$$i_2 = I_{2,c}(\infty) = \beta i_1 \int_{t_1}^{\infty} \exp\{-\gamma(t' - t_1)\} \mathrm{d}t' = \frac{\beta}{\gamma} i_1. \tag{7.11}$$

The infectious individuals of second generation occur in time at different time points $t \geq t_1$. In order to proceed with the derivation of $R_{0,e}$ within a deterministic, analytical framework, the following simplification can be made. It is assumed that all second-generation infectious individuals occur in the population of interest at the same time point $t_2 > t_1$. That is, just as the first generation individuals show up at the time point t_1 at once, all second generation individuals show up together at the time point t_2. This simplification allows us to repeat the procedure described above in order to determine the third generation infectious individuals. Accordingly, at t_2 the conditions

$$I_2(t_2) = I_{2,0} = i_2, \; I_3(t_2) = I_{3,c}(t_2) = 0 \tag{7.12}$$

hold. Equations (7.7) becomes $\mathrm{d}I_2/\mathrm{d}t = -\gamma I_2$, which leads to $I_2(t) = i_2 \exp\{-\gamma(t - t_2)\}$. Equation (7.9) becomes $\mathrm{d}I_{3,c}/\mathrm{d}t = \beta I_2(t)$ such that Eq. (7.10) reads

$$I_{3,c}(t) = \beta i_2 \int_{t_2}^{t} \exp\{-\gamma(t' - t_2)\} \mathrm{d}t', \tag{7.13}$$

which eventually leads to

$$i_3 = I_{3,c}(\infty) = \frac{\beta}{\gamma} i_2. \tag{7.14}$$

In general, when taking the simplified notion that the pth generation of infectious individuals occurs at time t_p with a size i_p, then epidemics described by the SIR model satisfy at later time points $t \geq t_p$ the following relations:

$$\begin{aligned} I_p(t) &= i_p \exp\{-\gamma(t - t_p)\}, \\ \frac{\mathrm{d}}{\mathrm{d}t} I_{p+1,c}(t) &= \beta I_p(t) = \beta i_p \exp\{-\gamma(t - t_p)\}, \\ i_{p+1} &= I_{p+1,c}(\infty) = \beta i_p \int_{t_p}^{\infty} \exp\{-\gamma(t' - t_p)\}\mathrm{d}t' = \frac{\beta}{\gamma} i_p. \end{aligned} \quad (7.15)$$

For the SIR model, the size c_p of the infected generation p corresponds to i_p (i.e., $c_p = i_p$) because the model exhibits only a single infected compartment. Consequently, the iterative equation $i_{p+1} = \beta i_p/\gamma$ occurring in Eq. (7.15) is equivalent to $c_{p+1} = \beta c_p/\gamma$, which implies that the ratio β/γ can be identified as amplification factors R_0 or R_e depending whether or not intervention measures are put into place that affect β and γ. As a by-product, the iterative equation for i_p can be written like

$$i_{p+1} = R_{0,e} i_p, \quad R_{0,e} = \frac{\beta}{\gamma}. \quad (7.16)$$

In summary, the more explicit approach re-produces Eq. (7.3) that was obtained above using a somewhat heuristic approach. The explicit procedure can also be applied to the disease-free fixed point with $S_{st} < N$. In this case, the iterative equation $i_{p+1} = S_{st}\beta i_p/(N\gamma)$ is obtained, which implies $R_e = \beta S_{st}/(N\gamma)$.

7.2.4 Next Generation Time Grid

The following considerations apply to all kind of epidemiological models (including the SIR model as a special case). Let us define a time grid $t_1, t_2, t_3, \ldots$ in terms of the occurrence time points t_p of the generations $p = 1, 2, 3, \ldots$. At t_p the infected individuals of the pth generation are present. For $t > t_p$ they make transitions through or out of infected compartments (for the SIR model, they only make transitions out of the compartment I). During that period, they also infect others and cause the build-up of the generation $p + 1$ of infected individuals. That is, the period $t > t_p$ describes the subsiding of the pth generation of infected individuals and the emergence of the $(p + 1)$th generation of infected individuals. The notion is that if we wait for a sufficiently long time, that is, if $t_{p+1} - t_p$ is sufficiently large, then the pth generation has disappeared and, consequently, all individuals of the $(p + 1)$th generation have emerged. The simplified point of view is that during the interval $[t_p, t_{p+1}]$ the individuals of the $(p + 1)$th generation do not make any transitions out or through compartments. The individuals are "frozen". At the time point t_{p+1} the procedure starts again with the $(p + 1)$th generation of individuals as those who make transitions and the $(p+2)$th generation of individuals as individuals who appear in the epidemic under consideration. In reality, the transition phases of two (or more than two) subsequent generations overlap and are not clearly separated. Likewise, the build-up phases overlap in time.

7.3 Two Infected Compartments

As an example of an epidemic model that involves two infected compartments let us consider the 1β SEIR model (3.43). However, let us take demographic terms into account. Then Eq. (3.43) reads [3]

$$\frac{\mathrm{d}}{\mathrm{d}t}S = -\frac{\beta}{N}IS + B - \mu S,\ \frac{\mathrm{d}}{\mathrm{d}t}E = \frac{\beta}{N}IS - (\alpha+\mu)E,$$
$$\frac{\mathrm{d}}{\mathrm{d}t}I = \alpha E - (\gamma+\mu)I,\ \frac{\mathrm{d}}{\mathrm{d}t}R = \gamma I - \mu R \tag{7.17}$$

(see also Eq. (5.3) with $n = 4$, $\beta_I = \beta$, and $X_4 = R$). For $\mu > 0$ and $B = \mu N$ the model exhibits the disease-free fixed point $\mathbf{X}_{st} = (N, 0, 0, 0)$. In contrast, for $\mu = B = 0$ (i.e., in the absence of demographic terms) fixed points are given by $\mathbf{X}_{st} = (S_{st}, 0, 0, R_{st})$ with $S_{st} + R_{st} = N$. Furthermore, the infected compartments of the SEIR model are E and I

Following [3], let us define the probability p of infected individuals to stay infected as the probability to make transition from E to I without deceasing as $p = \alpha/(\alpha+\mu)$. In the special case $\mu = 0$, we have $p = 1$. In the general case of $S < N$ susceptibles, the rate with which a single infectious individual infects susceptibles is given by the product of the effective contact rate β and the factor $x = S/N$ (see Sect. 7.1), which yields the product $x\beta$. Since the SEIR model with demographic terms takes death events into account, the infectious period becomes shorter than $T = 1/\gamma$. We obtain $T = 1/(\gamma+\mu)$. The reproduction number $R_{0,e}$ (which either corresponds to the basic or effective one) is given by the product of the probability to stay infected (p), the rate of infection $(x\beta)$ by a single infectious individual, and the infectious period (T). Consequently, we obtain [3]

$$R_{0,e} = \frac{\alpha}{\alpha+\mu}\frac{S}{N}\beta\frac{1}{\gamma+\mu} = \frac{S}{N}\frac{\alpha\beta}{(\alpha+\mu)(\gamma+\mu)}. \tag{7.18}$$

Let us consider a few special cases. In the case of the outbreak of a novel infectious disease, that is, if the entire population is susceptible and there are no intervention measures put into place, $S = N$ holds and $R_{0,e}$ reflects the basic reproduction number such that Eq. (7.18) becomes

$$R_0 = \frac{\alpha\beta}{(\alpha+\mu)(\gamma+\mu)}. \tag{7.19}$$

For a wave-like outbreak that occurs over a relatively short period demographic terms can be neglected ($\mu = 0$). If we assume that $S \approx N$ holds and no intervention measures have been implemented, then Eq. (7.18) yields the basic reproduction number

$$R_0 = \frac{\beta}{\gamma}. \tag{7.20}$$

In contrast, if we assume that the epidemic is underway (i.e., $S < N$ holds) and neglect again demographic terms ($\mu = 0$), then from Eq. (7.18) we obtain the effective reproduction number

$$R_e = \frac{S}{N}\frac{\beta}{\gamma}. \tag{7.21}$$

According to the discussion in Sect. 7.1, for $R_{0,e} > 1$ and $R_{0,e} < 1$ solutions of the SEIR model describe epidemic outbreaks and subsiding epidemics, respectively. In fact, in Sect. 3.7 the expressions (7.20) and (7.21) have been derived as stability parameters ξ. It has been shown in Sect. 3.7 that for $\xi > 1$ the SEIR model (7.17) without demographic terms exhibits wave-solutions, whereas for $\xi < 1$ the model describes subsiding epidemics. Likewise, using a nonlinear physics perspective, in Sect. 5.5.1 the expressions (7.20) and (7.21) have been derived as bifurcation parameters κ (see Eq. (5.42)) and κ_r (see Eq. (5.44)), respectively, when replacing S in Eq. (7.21) by S_{st}. Accordingly, the condition $R_{0,e} < 1$ implies that the fixed point $E = I = 0$ in the E-I subspace exhibits two negative eigenvalues and is asymptotically stable, which implies that solutions of the SEIR model describe subsiding epidemics. In contrast, for $R_{0,e} > 1$ the fixed point $E = I = 0$ exhibits a positive eigenvalue and corresponds to a saddle point, which implies that the SEIR model exhibits wave-solutions. The reproduction number $R_{0,e}$ corresponds to a bifurcation parameter with critical value 1.

Let us use the (next generation) approach presented in Sect. 7.3 to derive Eqs. (7.20) and (7.21) in a more explicit manner from the evolution equation (7.17) of the 1β SEIR model. For sake of simplicity, the analysis is conducted first for the fixed point $S_{st} = N$ and the case $\mu = 0$ is considered. In this case, the linearized evolution equations for E and I read

$$\frac{\mathrm{d}}{\mathrm{d}t}E = \beta I - \alpha E, \ \frac{\mathrm{d}}{\mathrm{d}t}I = \alpha E - \gamma I \tag{7.22}$$

(see also Eq. (5.37) with $\beta_E = 0$ and $\beta_I = \beta$). The SEIR model does not account for deaths due to the infectious diseases or due to other causes. This implies that eventually all exposed individuals become infectious. In analogy to the discussion in Sect. 7.2.3, let us define $E_p(t)$ and $I_p(t)$ as the numbers of the pth generations of exposed and infectious individuals, respectively. Let e_p and i_p denote the (final, cumulative) sizes of the pth generations of exposed and infectious individuals, respectively. Then, for any generation $p \geq 2$ we have $e_p = i_p$ because all infected individuals of the second and higher generations begin as exposed individuals and eventually turn into infectious individuals. As far as the first generation of infected individuals is concerned, if $I_1(t_1) = 0$ and $E_1(t_1) > 0$ holds, then we have $e_1 = i_1 = E_1(t_1)$. In contrast, if $I_1(t_1) > 0$ and $E_1(t_1) \geq 0$ holds, then we have $e_1 = E(t_1)$ and $i_1 = I_1(t_1) + e_1 > e_1$. Let $E_{p,c}(t)$ denote the cumulative number of pth-generation exposed individuals with $e_p = i_p = E_{p,c}(\infty)$ for $p \geq 2$. For the initial time points t_p the initial conditions are denoted by $E_p(t_p) = E_{p,0}$ and $I_p(t_p) = I_{p,0}$. For all $p \geq 2$ we have $E_{p,0} + I_{p,0} = e_p = i_p$. For the argument that will be developed in the remainder of this section, the precise decomposition (e.g., $E_{2,0} = 0.9e_2$, $I_{2,0} = 0.1e_2$, $E_{3,0} = 0.8e_3$, $I_{3,0} = 0.2e_3$, etc.) is irrelevant (as it will become clear in a moment). From Eq. (7.22) it follows that

$$\frac{\mathrm{d}}{\mathrm{d}t} E_p = -\alpha E_p \;\Rightarrow\; E_p(t) = E_{p,0} \exp\{-\alpha(t - t_p)\} \tag{7.23}$$

and

$$\frac{\mathrm{d}}{\mathrm{d}t} I_p = \alpha E_p - \gamma I_p \;\Rightarrow$$
$$I_p(t) = I_{p,0} \exp\{-\gamma(t - t_p)\} + \frac{\alpha E_{p,0}}{\gamma - \alpha} \left[\exp\{-\alpha(t - t_p)\} - \exp\{-\gamma(t - t_p)\}\right] \tag{7.24}$$

holds (assuming $\alpha \neq \gamma$). As in Sect. 7.2.3, the initial condition $E_{p,c}(t_p) = 0$ holds for the cumulative number $E_{p,c}$. From Eq. (7.22) it follows that the evolution equation of E_{p+1} reads $\mathrm{d}E_{p+1}/\mathrm{d}t = \beta I_p - \alpha E_{p+1}$. Consequently, the evolution of the corresponding cumulative number $E_{p+1,c}(t)$ can be determined like

$$\frac{\mathrm{d}}{\mathrm{d}t} E_{p+1,c} = \beta I_p(t) \;\Rightarrow\; E_{p+1,c} = \beta \int_{t_p}^{t} I_p(t')\,\mathrm{d}t'. \tag{7.25}$$

Solving the integral in Eq. (7.25) for $t \to \infty$, the cumulative final size of the $(p+1)$th-generation exposed individuals can be computed like

$$\begin{aligned} e_{p+1} &= E_{p+1,c}(\infty) = \beta \int_{t_p}^{\infty} I_p(t')\mathrm{d}t' \\ &= \beta \left[\frac{I_{p,0}}{\gamma} + \frac{\alpha E_{p,0}}{\gamma - \alpha}\left(\frac{1}{\alpha} - \frac{1}{\gamma}\right)\right] = \frac{\beta}{\gamma}(I_{p,0} + E_{p,0}). \end{aligned} \tag{7.26}$$

Consequently, for $p \geq 2$ the iterative equation

$$e_{p+1} = \frac{\beta}{\gamma} e_p \tag{7.27}$$

can be obtained. Since the size of the infected individuals c_p equals $e_p = i_p$ for $p \geq 2$, Eq. (7.27) implies $c_{p+1} = \beta c_p/\gamma$ for $p \geq 2$. For $p = 1$ the size c_1 of infected individuals is given by $c_1 = I_{1,0} + E_{1,0}$. Consequently, from Eq. (7.26) it follows

$c_2 = \beta c_1/\gamma$. In total, the iterative equation $c_{p+1} = \beta c_p/\gamma$ holds for any $p \geq 1$. The ratio β/γ can be identified as R_0 or R_e. The interpretation of β/γ as R_0 or R_e depends on whether an epidemic is considered under the impact of intervention measures (R_e) or not (R_0).

So far, the amplification factor $R_{0,e}$ has been derived for the disease-free fixed point with $S_{st} = N$. By analogy, the amplification factor z in $e_{p+1} = ze_p$ can be derived for a disease-free fixed point with $S_{st} < N$. If so, a factor of $z = \beta S_{st}/(\gamma N)$ is obtained, which implies that $R_e = \beta S_{st}/(\gamma N)$ holds in this case. In conclusions, the next generation approach that takes the SEIR model equations more explicitly into account re-produces the amplification factors or reproduction numbers that are defined by Eqs. (7.20) and (7.21) and have been derived above in a more heuristic manner.

7.4 *m* Infected Compartments

7.4.1 *Next Generation Approach*

In order to discuss the general case of m infected compartments, it is convenient to use a matrix approach [4, 7]. In what follows, this matrix approach is introduced by revisiting the example of the SEIR model discussed in Sect. 7.3.

The departure point is the set of coupled, linear differential equations defined in Eq. (7.22). From this set the evolution equations for E_p and I_p as listed in Eqs. (7.23) and (7.24) are obtained. They can be written in vector and matrix notation as

$$\frac{\mathrm{d}}{\mathrm{d}t}\begin{pmatrix} E_p \\ I_p \end{pmatrix} = -V\begin{pmatrix} E_p \\ I_p \end{pmatrix}, \quad V = \begin{pmatrix} \alpha & 0 \\ -\alpha & \gamma \end{pmatrix}, \tag{7.28}$$

where V will be referred to as transition matrix. Note the there is a minus sign in front of V. The solution of Eq. (7.28) reads

$$\begin{pmatrix} E_p(t) \\ I_p(t) \end{pmatrix} = \exp\{-V(t-t_p)\}\begin{pmatrix} E_{p,0} \\ I_{p,0} \end{pmatrix}. \tag{7.29}$$

The exponential function is used here as a function of a matrix. Next, the evolution equation for $E_{p+1,c}$ listed in Eq. (7.25) can be equivalently expressed by introducing the vector $\mathbf{F} = (0, \beta)$ like

$$\frac{\mathrm{d}}{\mathrm{d}t}E_{p+1,c} = \mathbf{F}\begin{pmatrix} E_p(t) \\ I_p(t) \end{pmatrix}. \tag{7.30}$$

By analogy, the integral solution shown in Eq. (7.25) can be written in terms of a dot product (scalar product) involving $\mathbf{F}$ like

$$E_{p+1,c}(t) = \mathbf{F} \int_{t_p}^{t} \begin{pmatrix} E_p(t') \\ I_p(t') \end{pmatrix} \mathrm{d}t', \tag{7.31}$$

which with the help of Eq. (7.29) leads to

$$E_{p+1,c}(t) = \mathbf{F} \left[\int_{t_p}^{t} \exp\{-V(t - t_p)\} \, \mathrm{d}t' \right] \begin{pmatrix} E_{p,0} \\ I_{p,0} \end{pmatrix}. \tag{7.32}$$

Consequently, $e_{p+1} = E_{p+1,c}(\infty)$ can be computed from

$$\begin{aligned} e_{p+1} &= E_{p+1,c}(\infty) = \mathbf{F} \left[\int_{t_p}^{\infty} \exp\{-V(t - t_p)\} \, \mathrm{d}t' \right] \begin{pmatrix} E_{p,0} \\ I_{p,0} \end{pmatrix} \\ &= \mathbf{F} V^{-1} \begin{pmatrix} E_{p,0} \\ I_{p,0} \end{pmatrix}, \end{aligned} \tag{7.33}$$

where V^{-1} is the inverse matrix of V. Note that just as the integral

$$\int_{t_p}^{\infty} \exp\{-\gamma(t' - t_p)\} \, \mathrm{d}t' = \frac{1}{\gamma} = \gamma^{-1} = T \tag{7.34}$$

used in Eq. (7.15) yields $1/\gamma$, as indicated, where γ is a scalar, the integral relation

$$\int_{t_p}^{\infty} \exp\{-V(t' - t_p)\} \, \mathrm{d}t' = V^{-1} = T_{mat} \tag{7.35}$$

holds and involves the inverse matrix V^{-1} of V. Furthermore, as indicated in Eq. (7.34), the fraction $1/\gamma$ can be interpretation as a duration T (see Sect. 3.4). Likewise, V^{-1} can be interpreted as a matrix T_{mat} of durations (see Eq. (7.35)).

The vector $\mathbf{F}$ can be generalized further to yield a matrix. In order to demonstrate this step let us generalize the SIR model (3.16) by assuming that there are two virus transmission mechanisms. One mechanism involves a non-infectious intermediate period. The second does not involve such a period. After being infected, susceptibles become immediately infectious. Let $f \in [0, 1]$ denote the probability that a susceptible gets infected by means of the first mechanism. Then, Eq. (3.16) reads

$$\frac{\mathrm{d}}{\mathrm{d}t} S = -\frac{\beta}{N} SI, \; \frac{\mathrm{d}}{\mathrm{d}t} E = f \frac{\beta}{N} SI - \alpha E, \; \frac{\mathrm{d}}{\mathrm{d}t} I = (1 - f) \frac{\beta}{N} SI + \alpha E - \gamma I \tag{7.36}$$

and $R(t) = N - S(t) - E(t) - I(t)$ (or $\mathrm{d}R/\mathrm{d}t = \gamma I$). Since for $f < 1$ there is a direct route from susceptibles to infectious individuals, we have $i_p < e_p$ for $p \geq 2$, which indicates that there are infectious individuals in the pth generation that have not passed through the stage of being exposed (i.e., being infected but not infectious). Only if $f = 1$ holds (i.e., if the model reduces to Eq. (3.16)) we have $i_p = e_p$.

Let $I_{p,c,N.E.}(t)$ denote the cumulative infectious individuals that have emerged via the second route, where $N.E.$ stands for "not exposed", that is, not having passed through the exposed stage. Likewise, let $i_{p,N.E.}$ denote the size of the pth generation of such infectious individuals with $i_{p,N.E.} = I_{p,c,N.E.}(\infty)$. Then the evolution equation $\mathrm{d}E_{p+1,c}/\mathrm{d}t = \beta I_p$ in Eq. (7.25) must be revised to take the factor f into account like $\mathrm{d}E_{p+1,c}/\mathrm{d}t = f\beta I_p$. Importantly, a second evolution equation should be added to describe the emergence of new infected individuals given in terms of infectious individuals that have skipped the exposed stage. That is, $\mathrm{d}I_{p+1,c,N.E.}/\mathrm{d}t = (1-f)\beta I_p$ should be added to our description of the epidemic under consideration. Consequently, Eq. (7.30) is generalized like

$$\frac{\mathrm{d}}{\mathrm{d}t}\begin{pmatrix} E_{p+1,c} \\ I_{p+1,c,N.E.} \end{pmatrix} = F\begin{pmatrix} E_p(t) \\ I_p(t) \end{pmatrix}, \quad F = \begin{pmatrix} 0 & f\beta \\ 0 & (1-f)\beta \end{pmatrix}. \tag{7.37}$$

Accordingly, the vector $\mathbf{F}$ in the general case becomes a matrix F. Note that Eqs. (7.28) and (7.29) still hold for the generalized SEIR model (7.36). Therefore, Eq. (7.32) in the general case when F denotes a matrix becomes

$$\begin{pmatrix} E_{p+1,c}(t) \\ I_{p+1,c,N.E.}(t) \end{pmatrix} = F\left[\int_{t_p}^{t} \exp\{-V(t-t_p)\}\,\mathrm{d}t'\right]\begin{pmatrix} E_{p,0} \\ I_{p,0} \end{pmatrix}. \tag{7.38}$$

From Eq. (7.38) it follows that

$$\begin{aligned}\begin{pmatrix} e_{p+1} \\ i_{p+1,N.E.} \end{pmatrix} &= \begin{pmatrix} E_{p+1,c}(\infty) \\ I_{p+1,c,N.E.}(\infty) \end{pmatrix} = F\left[\int_{t_p}^{\infty} \exp\{-V(t-t_p)\}\,\mathrm{d}t'\right]\begin{pmatrix} E_{p,0} \\ I_{p,0} \end{pmatrix} \\ &= FV^{-1}\begin{pmatrix} E_{p,0} \\ I_{p,0} \end{pmatrix}, \end{aligned} \tag{7.39}$$

which is the generalization of Eq. (7.33). Let us return to the inequality $i_p > e_p$ that holds for $f < 1$. For $f < 1$ we have $i_p = i_{p,N.E.} + e_p > e_p$. All exposed individuals eventually turn into infectious individuals. However, there are infectious individuals that have skipped the exposed stage. The size c_p of the infected pth generation is given by $c_p = i_p = i_{p,N.E.} + e_p$. Furthermore, at any initial time point t_p the number of exposed ($E_{p,0}$) and number of infectious ($I_{p,0}$) individuals add up to the total size c_p of infected individuals of the generation p (see also Sect. 7.2.4). That is, $c_p = E_{p,0} + I_{p,0}$ holds. In summary, the identity $c_p = i_{p,N.E.} + e_p = E_{p,0} + I_{p,0}$ holds. Importantly, Eqs. (7.28) reads in components $\mathrm{d}E_p/\mathrm{d}t = -\alpha E_p$ and $\mathrm{d}I_p/\mathrm{d}t = \alpha E_p - \gamma I_p$ (see also Eqs. (7.23) and (7.24)), which means that given a time t_p, an initial number $E_{p,0}$ of exposed individuals, and an initial number $I_{p,0}$ of infectious individuals, then E_p decays monotonically, whereas I_p may increase due to transitions of exposed individuals to infectious individuals. However, the notion pointed out in Sect. 7.2.4 is that new infectious individuals of the generation p do not show up. In particular, while I_p may increase during a transient period, the cumulative number of infectious individuals who skipped the exposed stage does not increase and is fixed by $i_{p,N.E.}$. These individuals emerged during the transition dynamics of the generation $p-1$ of infected individuals took place (see Sect. 7.2.4 again). In other words, it can be motivated to put $E_{p,0}=e_p$ and $I_{p,0}=i_{p,N.E.}$. In doing so, Eq. (7.39) becomes a closed iterative equation of the form

$$\begin{pmatrix} e_{p+1} \\ i_{p+1,N.E.} \end{pmatrix} = G \begin{pmatrix} e_p \\ i_{p,N.E.} \end{pmatrix}, \quad G = FV^{-1} \tag{7.40}$$

The matrix G has been referred to a next generation matrix [4, 7]. F is a matrix composed of rates that describe new infections [4] and may be called infection transmission matrix [8].

Let us consider the eigenvalues of G. Let us assume for sake of simplicity that G exhibits only real-valued, semi-positive eigenvalues. If so, let

$$\lambda_{\max}(G) = \text{largest eigenvalue of } G \tag{7.41}$$

denote the largest eigenvalue among all possible eigenvalues, as indicated. Note that in the context of the SEIR model, G is a 2×2 matrix and exhibits two eigenvalues. Therefore, in this context, the discussion is centered about two eigenvalues only. However, the concept of a next generation matrix G holds in higher dimensions as well (see below). Therefore, in general, G exhibits more than two eigenvalues.

If $\lambda_{\max}(G) < 1$ holds (but $\lambda_{\max}(G) > 0$), all solutions of the iterative equation (7.40) converge to $e_p = i_{p,N.E.} = 0$ for $p \to \infty$, that is, c_p satisfies $\lim_{p\to\infty} c_p = 0$. This indicates that the disease-free fixed point $E = I = 0$ is asymptotically stable. In contrast, if $\lambda_{\max}(G) > 1$ holds, then Eq. (7.40) exhibits solutions e_p and $i_{p,N.E.}$ that increase as a power law of p. This indicates that $E = I = 0$ is an unstable fixed point. In particular, for $p \to \infty$ it is known that the dynamics of iterative maps such as Eq. (7.40) is determined by the largest eigenvalue $\lambda_{\max}(G)$ and its corresponding eigenvector $\mathbf{v}_{\max}(G)$ such that

$$\begin{pmatrix} e_{p+1} \\ i_{p+1,N.E.} \end{pmatrix} \approx C\,\mathbf{v}_{\max}(G)[\lambda_{\max}(G)]^{p-1} \tag{7.42}$$

holds if p is sufficiently large, where C is a coefficient independent of p. Consequently, for sufficiently large p the iterative equation (7.42) that describes how the sizes of the relevant infected compartments evolve over generations reads

$$\begin{pmatrix} e_{p+1} \\ i_{p+1,N.E.} \end{pmatrix} \approx \lambda_{\max}(G) \begin{pmatrix} e_p \\ i_{p,N.E.} \end{pmatrix}. \tag{7.43}$$

Therefore, it has been suggested to identify the basic reproduction number R_0 as $\lambda_{\max}(G)$ [1, 4, 7] like

$$R_0 = \lambda_{\max}(G). \tag{7.44}$$

In addition, depending on the context in which an epidemic model is used, $\lambda_{\max}(G)$ may be used to compute the effective reproduction number. In fact, in the COVID-19 modeling literature, frequently, R_e has been computed as $R_e = \lambda_{\max}(G)$.

Let us return to the SEIR model defined by (7.36). The inverse matrix V^{-1} of V reported in Eq. (7.28) reads

$$V^{-1} = \frac{1}{\alpha\gamma}\begin{pmatrix} \gamma & 0 \\ \alpha & \alpha \end{pmatrix} \tag{7.45}$$

(see Eq. (5.52)). Consequently, $G = FV^{-1}$ with F given in Eq. (7.37) reads

$$G = \frac{\beta}{\gamma}\begin{pmatrix} f & f \\ 1-f & 1-f \end{pmatrix}. \tag{7.46}$$

For arbitrary $f \in [0, 1]$ the matrix G when putting the pre-factor $\beta/\gamma = 1$ exhibits the eigenvalues 0 and 1 (which may be shown by noticing that the rows of $G - \lambda E$ become linearly dependent for $\lambda = 0$ and $\lambda = 1$ or can be derived by using Eq. (5.27)). Consequently, when taking the pre-factor β/γ into account, the matrix G exhibits as largest eigenvalue $\lambda_{max}(G) = \beta/\gamma$. This implies that the SEIR model (7.36) exhibits the reproduction number

$$R_{0,e} = \frac{\beta}{\gamma}, \tag{7.47}$$

which is equivalent with the reproduction number (7.20) obtained in Sect. 7.3 for the special case $f = 1$ by means of a more heuristic approach.

We are now in the position to consider the general case of an epidemic that can be described in terms of a compartment model involving m infected compartments. The following procedure has been suggested [4, 7].

Epidemics are considered that satisfy evolution equations of the form $d\mathbf{X}/dt = \mathbf{N}(\mathbf{X})$ (see Eq. (2.1)), where $\mathbf{X} = (X_1, \ldots, X_n)$ is a n-dimensional state vector of suitably defined compartments (i.e., compartment sizes). The first step is to rearrange the variables occurring in the state vector such that the first m variables denote infected compartments. At this stage, let us return to the approach discussed in Sect. 6.1.1. Using the notation suggested there, let the upper indices $+$ and $-$ denote variables reflecting infected and non-infected compartments, respectively. Consequently, the state vector $\mathbf{X} = (X_1, \ldots, X_n)$ reads $\mathbf{X} = (\mathbf{X}^+, \mathbf{X}^-) = (X_1^+, \ldots, X_m^+, X_1^-, \ldots, X_r^-)$ with $r = n - m$ and $\mathbf{X}^+ = (X_1, \ldots, X_m) \Rightarrow X_i^+ = X_i$ for $i = 1, \ldots, m$, $\mathbf{X}^- = (X_{m+1}, \ldots, X_n) \Rightarrow X_i^+ = X_{m+i}$ for $i = 1, \ldots, r$. However, note that the approach presented in Sect. 6.1.1 allows for more flexibility. While in Sect. 6.1.1 it is suggested to separate state variables into m variables that contain all infectious variables describing individuals who actually infect others and add certain additional variables, the next generation procedure as such comes with a strict (less flexible) decomposition into infected and non-infected compartments.

Let us return to the derivation of R_0 by means of the next generation approach. To this end, the disease-free fixed point $\mathbf{X}_{st}$ with $X_1 = \cdots = X_m = 0$ is considered. As mentioned in Sect. 6.1.2, it is frequently assumed [4] (and can be tested explicitly for a given model) that linearizing the evolution equation $\mathrm{d}\mathbf{X}/\mathrm{d}t = \mathbf{N}(\mathbf{X})$ at $\mathbf{X}_{st}$ yields a linearization matrix of the form (6.2) such that the linearized dynamics in the subspace spanned by the infected variables is given by Eq. (6.3). For convenience, let us repeat equation (6.3) as

$$\frac{\mathrm{d}}{\mathrm{d}t}\begin{pmatrix} X_1^+ \\ \cdots \\ X_m^+ \end{pmatrix} = L^+ \begin{pmatrix} X_1^+ \\ \cdots \\ X_m^+ \end{pmatrix} . \tag{7.48}$$

In general, the linearization matrix L^+ depends on $X_{1,st}^-, \ldots, X_{r,st}^-$. Importantly, Eq. (7.48) corresponds to a closed set of coupled first-order differential equations. Following the ideas presented above in the context of the SEIR model, the matrix L^+ is decomposed into the two matrixes F and V like

$$L^+ = F - V . \tag{7.49}$$

The matrix F describes processes leading to new infections. In particular, it is required that the models $\mathrm{d}\mathbf{X}/\mathrm{d}t = \mathbf{N}(\mathbf{X})$ involve matrices F such that all elements of F are semi-positive (i.e., $F_{ij} \geq 0$) [4]. The matrix V describes transitions between compartments that are not related to new infections. In the absence of new infections $F = 0$ holds (e.g., put $\beta = 0$ in Eq. (7.37)) and the infected compartments evolve like

$$\frac{\mathrm{d}}{\mathrm{d}t}\mathbf{X}^+ = -V\mathbf{X}^+ . \tag{7.50}$$

It is required that in this case the dynamics in the subspace D^+ exhibits the asymptotically stable disease-free fixed point $\mathbf{X}_{st}^+ = (0, \ldots, 0)$ [4]. Consequently, $-V$ exhibits only eigenvalues that exhibit negative real parts and, conversely, V exhibits eigenvalues that only exhibit positive real parts. Moreover, V is assumed to be non-singular and invertible [4] such that V^{-1} exists. In addition, off-diagonal elements of $-V$ describe transitions into compartments and for this reason are semi-positive. This implies that off-diagonal elements of V are semi-negative (i.e., $V_{ij} \leq 0$ for $i \neq j$) [4]. An example in this regard is the coefficient $V_{21} = -\alpha$ in Eq. (7.28). In summary, the matrices F and V are assumed to satisfy the following properties. F is a matrix with semi-positive coefficients such that $F_{ik} \geq 0$ holds for all i, k. V exhibits eigenvalues with positive real parts, is invertible, and $V_{ik} \leq 0$ holds for $i \neq k$. For examples of F and V see Eqs. (7.28) and (7.37).

Let us illustrate that the requirement that V is invertible and exhibits only eigenvalues with positive real parts implies that $\mathbf{X}^+$ converges to the disease-free state defined by the zero vector. A matrix V for which V^{-1} exist and that exhibits only eigenvalues with positive real parts corresponds to a positive definite matrix V in the sense that $\mathbf{X}^+ V\mathbf{X}^+ > 0$ holds for any $\mathbf{X}^+$, which is not zero. Note again that positive definite does not mean that V has only positive elements (in fact, V exhibits negative off-diagonal elements, see above). Let us consider the evolution of $|\mathbf{X}^+(t)|^2$. From Eq. (7.50) it follows that $\mathrm{d}|\mathbf{X}^+|^2/\mathrm{d}t = -\mathbf{X}^+ V \mathbf{X}^+ < 0$. Therefore, $|\mathbf{X}|$ decays over time to zero. The convergence to the disease-free state can be explicitly demonstrated for the linearized equations (7.22) of the SEIR model that involves the matrix V defined in Eq. (7.28). Let us put $\beta = 0$ in Eq. (7.22), then the solutions $E(t)$ and $I(t)$ are given by $E_p(t)$ and $I_p(t)$ defined in Eqs. (7.23) and (7.24), respectively. Clearly, for any initial conditions $E_{p,0}$ and $I_{p,0}$ the functions E_p and I_p converge to zero over time.

Following the decomposition of L^+ into F and V, the next step is to define the inverse matrix V^{-1} of V as in Eq. (7.35) as a $m \times m$ matrix of durations

$$T_{mat} = V^{-1}. \tag{7.51}$$

The final step is to construct the $m \times m$ next generation matrix G from F and T_{mat} (or V^{-1}) like

$$G = FV^{-1} = FT_{mat}, \; G_{ik} = \sum_{j=1}^{m} F_{ij} T_{jk}. \tag{7.52}$$

The basic reproduction number R_0 can then be defined in line with Eq. (7.44). In order to simplify the definition, it is useful to consider the case in which all eigenvalues of G are real and semi-positive definite (i.e., they are allowed to be equal to zero). If so, R_0 is defined as the largest eigenvalue of G [1, 4, 7]. As mentioned in the context of Eq. (7.44), in the literature, this definition has also been used to determine effective reproduction numbers of epidemics that evolved under the impact of intervention measures and/or evolved in populations that were no longer completely susceptible. Consequently, in line with Eq. (7.44) the reproduction numbers $R_{0,e}$ may be defined as

$$R_{0,e} = \lambda_{\max}(G). \tag{7.53}$$

In the literature, R_0 has been defined as the spectral radius of G [1, 4, 7]. The spectral radius of G applies to eigenvalues that might be complex-valued and is defined as the absolute value of the largest eigenvalue of G. The largest eigenvalue, in turn, is defined as the eigenvalue that exhibits the largest absolute value. If the matrix G exhibits only real-valued, semi-positive eigenvalues (as it is the case in many applications) then the spectral radius is given by the largest eigenvalue. Consequently, the definition of R_0 in Eq. (7.53) is a special case of the more general definition of R_0 in terms of the spectral radius.

7.4.2 Theorems Involving Reproduction Numbers

In the literature, Theorem 6.13 of Ref. [7] and Theorem 2 of Ref. [4] make key statements about reproduction numbers. Accordingly, if $R_{0,e} > 1$ holds then $L^+ = F - V$ exhibits at least one eigenvalue with positive real part. Consequently, the disease-free fixed point $\mathbf{X}^+_{st} = (0, \ldots, 0)$ corresponds to an unstable fixed point. This also implies that under appropriate conditions the disease-free fixed point $\mathbf{X}_{st}$ referring to the full n-dimensional description of a population is an unstable fixed point.

In contrast, if $R_{0,e} < 1$ holds, then all eigenvalues of $L^+ = F - V$ exhibit negative real parts. This implies that $\mathbf{X}^+_{st} = (0, \ldots, 0)$ is an asymptotically stable fixed point (provided that $\mathbf{X}$ is sufficiently close to $\mathbf{X}_{st}$). Consequently, $\mathbf{X}_{st}$ either corresponds to a neutrally stable fixed point or an asymptotically stable fixed point.

The third statement concerns the critical case $R_{0,e} = 1$. If $R_{0,e} = 1$ holds then $L^+ = F - V$ exhibits at least one real-valued eigenvalue $\lambda = 0$. All other eigenvalues exhibit real parts zero or negative real parts. This third statement can be exploited for a particular application that will be discussed in Sect. 7.5.4

7.5 Applications

7.5.1 SIR Model and 1β SEIR Model

The SIR model (3.16) with $\mathrm{d}S/\mathrm{d}t = -\beta IS/N$, $\mathrm{d}I/\mathrm{d}t = -\beta IS/N - \gamma I$, and $\mathrm{d}R/\mathrm{d}t = \gamma I$ corresponds to a three variable model with a single infected compartment I. Using the decomposition discussed in Sect. 7.4.1, the three-dimensional state vector reads $\mathbf{X} = (I, R, S)$ with $X^+_1 = I$, $X^-_1 = R$, and $X^-_2 = S$. Fixed points of the model read $\mathbf{X}_{st} = (0, 0, S_{st})$ for $S_{st} \leq N$. Linearizing the nonlinear term SI like $SI = S_{st} I$, then the linearized equation for I can be expressed like

$$\frac{\mathrm{d}}{\mathrm{d}t} I = \left(\frac{S_{st}}{N} \beta - \gamma \right) I = L^+ I = (F - V) I. \tag{7.54}$$

Consequently, the expressions F and V read $F = \beta S_{st}/N$ and $V = \gamma$, respectively, and correspond to scalars rather than matrices. The dwelling time of an individual in the infectious state I is given by $T = V^{-1} = 1/\gamma$. The next generation matrix G reduces to the factor $G = FT = \beta S_{st}/(N\gamma)$. Accordingly, the reproduction number (basic or effective) equals the next generation factor and reads

$$R_{0,e} = \frac{S_{st}}{N} \frac{\beta}{\gamma}. \tag{7.55}$$

In doing so, the results presented in Eqs. (7.3) to (7.5) are re-obtained within the next generation approach.

The SEIR model (7.17) with $B = \mu N$ features two infected compartments E and I. The state vector can be written like $\mathbf{X} = (E, I, R, S)$ with $X_1^+ = E$, $X_2^+ = I$, $X_1^- = R$, and $X_2^- = S$. If $\mu = 0$ the model exhibits fixed points $\mathbf{X}_{st} = (0, 0, R_{st}, S_{st})$ with $R_{st} + S_{st} = N$. For $\mu > 0$ it is required that $S_{st} = N$ and $R_{st} = 0$. Let us consider both cases simultaneously. To this end, let us consider the disease-free fixed point with $S_{st} \leq N$. Linearizing Eq. (7.17) with $B = \mu N$ for $S = S_{st}$, the corresponding linearized equations in the subspace E-I read

$$\frac{\mathrm{d}}{\mathrm{d}t}E = \frac{\beta S_{st}}{N}I - (\alpha + \mu)E, \quad \frac{\mathrm{d}}{\mathrm{d}t}I = \alpha E - (\gamma + \mu)I \tag{7.56}$$

and generalize Eq. (7.22). The linearized model can equivalently be expressed in matrix notation like

$$\frac{\mathrm{d}}{\mathrm{d}t}\begin{pmatrix} E \\ I \end{pmatrix} = L^+ \begin{pmatrix} E \\ I \end{pmatrix} = \begin{pmatrix} -(\alpha + \mu) & \beta S_{st}/N \\ \alpha & -(\gamma + \mu) \end{pmatrix} \begin{pmatrix} E \\ I \end{pmatrix} \tag{7.57}$$

The matrix L^+ in Eq. (7.57) can be decomposed into the square matrices F and V like

$$\frac{\mathrm{d}}{\mathrm{d}t}\begin{pmatrix} E \\ I \end{pmatrix} = \begin{pmatrix} 0 & \beta S_{st}/N \\ 0 & 0 \end{pmatrix} \begin{pmatrix} E \\ I \end{pmatrix} - \begin{pmatrix} (\alpha + \mu) & 0 \\ -\alpha & (\gamma + \mu) \end{pmatrix} \begin{pmatrix} E \\ I \end{pmatrix} \tag{7.58}$$

with

$$F = \begin{pmatrix} 0 & \beta S_{st}/N \\ 0 & 0 \end{pmatrix}, \quad V = \begin{pmatrix} (\alpha + \mu) & 0 \\ -\alpha & (\gamma + \mu) \end{pmatrix}. \tag{7.59}$$

The durations square matrix can be obtained by inverting V (see Eq. 5.52)) and reads

$$T_{mat} = V^{-1} = \frac{1}{(\alpha + \mu)(\gamma + \mu)} \begin{pmatrix} (\gamma + \mu) & 0 \\ \alpha & (\alpha + \mu) \end{pmatrix}. \tag{7.60}$$

Consequently, the next generation matrix becomes

$$G = F T_{mat} = \frac{1}{(\alpha + \mu)(\gamma + \mu)} \begin{pmatrix} \alpha\beta S_{st}/N & \beta(\alpha + \mu)S_{st}/N \\ 0 & 0 \end{pmatrix}. \tag{7.61}$$

The matrix exhibits two real-valued eigenvalues: a zero eigenvalue and $\lambda_{max} = \alpha\beta S_{st}/[N(\alpha + \mu)(\gamma + \mu)]$. Consequently, the (basic or effective) reproduction number reads

$$R_{0,e} = \lambda_{max}(G) = \frac{S_{st}}{N} \frac{\alpha\beta}{(\alpha + \mu)(\gamma + \mu)}. \tag{7.62}$$

The result is equivalent with the reproduction number that is presented in Eq. (7.18) provided S in Eq. (7.18) is replaced by S_{st}.

7.5.2 2β SEIR Model and COVID-19 Outbreak in Wuhan City 2020

The 2β SEIR model defined by Eq. (5.35) and (5.36) can be treated in analogy to the 1β SEIR model discussed above. Let $\mathbf{X} = (E, I, R, S)$ denote the state vector and $\mathbf{X}_{st} = (0, 0, 0, N)$ the disease-free fixed point of a completely susceptible population. The relevant two-dimensional linearized system is given by Eq. (5.37), which is repeated here as

$$\frac{\mathrm{d}}{\mathrm{d}t}\begin{pmatrix} E \\ I \end{pmatrix} = L^{+}\begin{pmatrix} E \\ I \end{pmatrix},\ L^{+} = F - V = \begin{pmatrix} \beta_E - \alpha & \beta_I \\ \alpha & -\gamma \end{pmatrix},$$
$$\Rightarrow F = \begin{pmatrix} \beta_E & \beta_I \\ 0 & 0 \end{pmatrix},\ V = \begin{pmatrix} \alpha & 0 \\ -\alpha & \gamma \end{pmatrix}. \tag{7.63}$$

Consequently, $T_{mat} = V^{-1}$ is given by Eq. (7.60) when putting $\mu = 0$. Multiplying F by T_{mat} thus obtained, the next generation matrix reads

$$G = FT_{mat} = \frac{1}{\alpha\gamma}\begin{pmatrix} \alpha\beta_I + \gamma\beta_E & \alpha\beta_I \\ 0 & 0 \end{pmatrix}. \tag{7.64}$$

The matrix exhibit a zero eigenvalue and $\lambda_{max}(G) = \beta_w/\gamma$ with $\beta_w = \beta_I + \gamma\beta_E/\alpha$ (see Eq. (5.41)). Accordingly, R_0 and R_e can be obtained as

$$R_{0,e} = \lambda_{max}(G) = \frac{\beta_w}{\gamma}. \tag{7.65}$$

Equation (7.65) can be generalized to take $\mu > 0$ into account [9]. Moreover, the 2β model can be considered as a staged progression model (see Sect. 5.2.3). For such staged progression models Eq. (7.65) has been derived, again, for the more general case of $\mu > 0$ in Ref. [4].

In Sect. 5.5.1 the cases $\beta_w > \gamma$ and $\beta_w < \gamma$ were examined and it was shown that the matrix L^{+} exhibits a positive eigenvalue for $\beta_w > \gamma$ and two negative eigenvalues for $\beta_w < \gamma$. Consequently, the disease-free fixed point $E = I = 0$ is unstable for $\beta_w > \gamma$ and asymptotically stable for $\beta_w < \gamma$ (assuming $\mathbf{X}$ is sufficiently close to $\mathbf{X}_{st}$). These findings are consistent with the fact that for $\beta_w > \gamma$ and $\beta_w < \gamma$ we have $R_{0,e} > 1$ and $R_{0,e} < 1$, respectively, and the interpretation of $R_{0,e}$ as a threshold parameter that indicates whether there is an epidemic outbreak ($R_{0,e} > 1$) or a subsiding epidemic ($R_{0,e} < 1$).

The reproduction number (7.65) (in its generalized form that takes $\mu > 0$ into account) was used in the study by Pang et al. [9] that was reviewed in Sect. 5.8. Pang et al. studied the COVID-19 outbreak in Wuhan city, China, during the period from December 2019 to March 25, 2020. They distinguished between the first phase from December 2019 to January 22, the second phase from January 23 to February 11 and the third phase from February 12 to March 25. Across these three phases the intervention measures to stop to spread of COVID-19 were increased in severity. Consequently, one should observe a decay of the reproduction number from phase to phase. For all three phases it was assumed that the disease (or health) state of the population of Wuhan was relatively close to the disease-free fixed point. The first phase was assumed to reflect the spread of the disease under negligible impact of intervention measures and, consequently, was characterized by the basic reproduction number of COVID-19 in Wuhan city. In contrast, phases two and three were assumed to reflect the dynamics of the COVID-19 epidemic under the impact of intervention measures. Accordingly, they were characterized by effective reproduction numbers. Pang et al. [9] found $R_0 = 4.6$ for the first phase, $R_e = 1.8$ for the second phase, and $R_e = 0.2$ for the third phase. As expected, $R_{0,e}$ decreased across the phases. Importantly, while the first and second phases were characterized by reproduction numbers larger than 1, the third phase exhibited an effective reproduction number smaller than the threshold value of 1. Consequently, the results suggest that the intervention measures implemented during the third phase of the COVID-19 epidemic in Wuhan city were able to stabilized the disease-free fixed point. This stabilization phenomenon was investigated in a follow-up study [10]. In Ref. [10] the maximal eigenvalue $\lambda_{\max} = \lambda_2$ (see Eq. (5.88)) of the 2β SEIR model was computed for the three phases. Accordingly, the eigenvalue changed from $\lambda_2 = 0.19$/d in phase 1 to $\lambda_2 = 0.09$/d in phase 2 and $\lambda_2 = -0.14$/d in phase 3 [10]. Consistent with the analysis of the epidemic via the reproduction number, the eigenvalue analysis reveals that in phase 1 and 2 the disease-free fixed point was unstable but in phase 3 it became a stable fixed point. We will return to this issue in Sect. 8.2.3.

7.5.3 SIR- and SEIR-Type Models and Beyond

Models of SIR-type, SEIR-type and various other models that involve a single susceptible compartment S can be cast into the following form. Let $\mathbf{X} = (S, X_2, ..., X_n)$ denote the (original) state vector. For models of SIR-type $X_2 = I$ and $m = 1$ holds. For models of SEIR-type $X_2 = E$, $X_3 = I$, and $m = 2$ holds. Let us assume there are m infected compartments. Then, the original state vector can be re-arranged like $\mathbf{X} = (X_1^+, \ldots, X_m^+, X_1^-, \ldots, X_r^-)$ with $X_r^- = S$ and $r = n - m$. For models of SIR-type we have $X_1^+ = I$. For models of SEIR-type we have $X_1^+ = E$ and $X_2^+ = I$. Importantly, let us consider models for which the rate constant k_0 of transformation reactions form susceptibles to infected individuals (i.e., the "force of infection") reads like

$$k_0(\mathbf{X}^+) = \sum_{k=1}^{m} b_k X_k^+ = \mathbf{b}\,\mathbf{X}^+, \mathbf{b} = (b_1, \ldots, b_m). \tag{7.66}$$

The coefficients b_k may be written like $b_k = \beta_k/N$, where β_k denote the respective effective contact rates. Note that some of the coefficients b_k might vanishes (e.g., for 1β SEIR-type models $b_1 = 0$ and $b_2 = \beta/N$). In summary, the models under consideration read

$$\frac{\mathrm{d}}{\mathrm{d}t}\begin{pmatrix} X_1^+ \\ \cdots \\ X_m^+ \\ X_1^- \\ \cdots \\ X_r^- \end{pmatrix} = A \begin{pmatrix} X_1^+ \\ \cdots \\ X_m^+ \\ X_1^- \\ \cdots \\ X_r^- \end{pmatrix} + k_0(\mathbf{X}^+) X_r^- \begin{pmatrix} 1 \\ 0 \\ \cdot \\ \cdots \\ 0 \\ -1 \end{pmatrix} + \begin{pmatrix} 0 \\ 0 \\ \cdot \\ \cdots \\ 0 \\ B \end{pmatrix}, \tag{7.67}$$

where A is a matrix describing transitions between different compartments. The matrix coefficients of A do not describe the emergence of new infections. Furthermore, in what follows it is assumed that A exhibits the block structure

$$A = \begin{pmatrix} A^+ & 0 \\ C & D \end{pmatrix}, \tag{7.68}$$

where A^+ is the upper, left-corner $m \times m$ submatrix. In the absence of demographic terms (i.e., for $\mu = B = 0$) the model is assumed to exhibit the disease-free fixed points $\mathbf{X}_{st}$ with $\mathbf{X}^+ = (0, \ldots, 0)$ and $S_{st} = X_{r,st}^- \leq N$. If demographic terms with $B = \mu N$ are taken into consideration, it is assumed that the matrix A is such that $\mathbf{X}_{st}$ with $\mathbf{X}^+ = (0, \ldots, 0)$ and $S_{st} = X_{r,st}^- = N$ is a fixed point. In both cases, the linearized model in the subspace of the infected variables reads

$$\frac{\mathrm{d}}{\mathrm{d}t}\begin{pmatrix} X_1^+ \\ \cdots \\ X_m^+ \end{pmatrix} = (F - V)\begin{pmatrix} X_1^+ \\ \cdots \\ X_m^+ \end{pmatrix},$$

$$F = S_{st}\begin{pmatrix} b_1 & \cdots & b_m \\ 0 & \cdots & 0 \\ \cdots & \cdots & \cdots \\ 0 & \ldots & 0 \end{pmatrix}, \quad V = -A^+ \tag{7.69}$$

where A^+ is aforementioned $m \times m$ submatrix of A. It is assumed that A^+ is invertible such that V^{-1} exists. Let $\mathbf{c}_1, \mathbf{c}_2, \ldots, \mathbf{c}_m$ denote the columns of V^{-1}. Then

$$G = F T_{mat} = F V^{-1} = S_{st}\begin{pmatrix} \mathbf{b}\,\mathbf{c}_1 & \cdots & \mathbf{b}\,\mathbf{c}_m \\ 0 & \cdots & 0 \\ \cdots & \cdots & \cdots \\ 0 & \ldots & 0 \end{pmatrix} \tag{7.70}$$

holds. The matrix G exhibits $m - 1$ zero eigenvalues and one non-vanishing eigenvalue $\lambda_{max}(G)$ with $\lambda_{max}(G) = S_{st}\mathbf{b}\,\mathbf{c}_1$. Consequently, for models of SIR-type and SEIR-type and for all other n-dimensional models defined by Eqs. (7.66) and (7.67), the (basic or effective) reproduction number reads

$$R_{0,e} = S_{st}\mathbf{b}\,\mathbf{c}_1 . \tag{7.71}$$

7.5.4 Determining Critical Effective Contact Rates

In Theorem 2 of Ref. [4] the third statement listed in Sect. 7.4.2 can be found. Accordingly, if $R_{0,e} = 1$ holds, then L^+ exhibits at least one real-valued vanishing eigenvalue and all other eigenvalues exhibit negative real parts. In what follows, a proof for the first part of this statement will be presented. Let us define the matrix P like

$$P = FV^{-1} - E \Rightarrow L^+ = F - V = PV , \tag{7.72}$$

where E is the identity matrix. Next, we consider the case $R_{e,0} = 1$, which means that the largest eigenvalue of $G = FV^{-1}$ is real-valued and equal to 1. Then, it follows that $P = G - E$ has at least one real-valued zero eigenvalue. The reason for this is that from $\lambda(G) = 1$ is follows that the determinant of $G - \lambda E$ equals zero for $\lambda = 1$: $\det(G - E) = 0$. Consequently, $\det(G - E - \lambda E) = \det(P - \lambda E) = 0$ has the solution $\lambda = 0$, which means that P exhibits a zero eigenvalue. Let $\mathbf{v}_0$ denote the eigenvector of P with $\lambda = 0$ such that $P\mathbf{v}_0 = 0$. Consequently, let us put $\mathbf{X}^+ = V^{-1}\mathbf{v}_0$. Then, the following calculation holds: $L^+\mathbf{X}^+ = (F - V)V^{-1}\mathbf{v}_0 = PVV^{-1}\mathbf{v}_0 = P\mathbf{v}_0 = 0$. This is tantamount to say that L^+ exhibits a zero eigenvalue. The corresponding eigenvector is $V^{-1}\mathbf{v}_0$.

As discussed in Sects. 2.7 and 2.9.1, when an eigenvalue changes its sign, then under appropriate condition the stability of a fixed point changes (from stable to unstable or vice versa) and a bifurcation occurs. Consequently, the critical condition at which the bifurcation occurs is characterized by a zero eigenvalue. The critical value $R_{0,e} = 1$ of the amplification factor $R_{0,e}$ describes such a critical condition.

Let us assume that the matrices L and L^+ depend on a parameter of interest for which it is plausible to assume that it is affected by intervention measures. For example, let us consider the effective contact rate β. Intervention measures are likely to affect and lower the effective contact rate β. In the context of the COVID-19 pandemic, measures that have been used to lower the effective contact rate were business shutdowns, school closures, and face mask wearing mandates. If so, the question arises how the critical parameter value β_{crit} can be determined at which a bifurcation takes place such that the epidemic starts to subside? That is, it is assumed that the initial stage of an epidemic exhibits an effective contact rate $\beta > \beta_{crit}$ such that the disease-free state is unstable. The epidemic evolves. Intervention measures are put into effect that decrease β. In order to find the critical β_{crit} at which the disease-free state becomes stable, the condition $R_{0,e} = 1$ can be used. Since $R_{0,e}$

depends on L^+, L^+ depends on k_0, and k_0 depends on β, the reproduction number $R_{0,e}$ can be regarded as a function of β. In a first step, an analytical expression for $R_{0,e}(\beta)$ is determined. In a second step, $R_{0,e}(\beta)$ is set to 1 like

$$R_{0,e}(\beta) = 1 \Rightarrow \beta = \beta_{crit}. \tag{7.73}$$

As indicted in Eq. (7.73), solving $R_{0,e}(\beta) = 1$ for β yields the to-be-determined critical value of β.

7.5.5 COVID-19 Epidemic in Pakistan 2020

Ullah and Khan (2020) investigated the COVID-19 epidemic in Pakistan from March 1 to May 28, 2020 [11]. During that period, confirmed COVID-19 cases increased monotonically in a more or less exponential manner. In order to describe the spread of the disease in terms of the observed increase of confirmed COVID-19 cases, Ullah and Khan [11] used an epidemic model with $n = 8$ compartments involving $m = 6$ infected compartments. The model satisfies the general model structure defined by Eqs. (7.66) and (7.67). Ullah and Khan estimated model parameters by fitting the model solution to the confirmed COVID-19 cases. Using the next generation method, they derived an analytical expression for R_0. They obtained a value of $R_0 = 1.9$ suggesting that the disease-free state was unstable during that period. The model used by Ullah and Khan involves an overall effective contact rate parameter β. The best fit estimate for β was found to be $\beta = 0.66$/d. The authors suggested in their study that intervention measures would be able to lower β and simulated various possible scenarios.

The analytical expression for R_0 derived by Ullah and Khan [11] can be cast into the form

$$R_0 = \beta f, \tag{7.74}$$

where f is an expression that involves other model parameters but not β. In a subsequent study [12], model solutions were fitted to a longer observation period from March 1 to September 30, 2020. COVID-19 case data from Pakistan suggest that the first-wave epidemic started to subside during the summer of 2020 and, in particular, at the end of the observation period (i.e., during August and September 2020). Following the suggestion by Ullah and Khan [11], the overall effective contact rate β was varied in three steps to fit the entire period from March 1 to September 30. To this end, it was assumed that in the second step intervention measures were able to decrease the overall effective contact rate such that it reached its critical value β_{crit}. In order to determine β_{crit}, R_0 in Eq. (7.74) was interpreted as R_e for the circumstances during the summer 2020. Subsequently, Eq. (7.74) was solved like

$$R_e = 1 \Rightarrow \beta_{crit} = 1/f. \tag{7.75}$$

A critical value of $\beta_{crit} = 0.35/\text{d}$ was obtained. In contrast, the best-fit parameter of β for the first step capturing the exponent increase of COVID-19 cases was $\beta = 0.60/\text{d}$. This value was close to the value estimated by Khan and Ullah [11]. Importantly, as expected it was larger than the critical value. In doing so, the first step parameter β with $\beta > \beta_{crit}$ indicates again that the disease-free fixed point was unstable during the first few months of the COVID-19 epidemic in Pakistan. In fact, the eigenvalues of L^+ were determined for that initial period of time and a positive eigenvalue was found [12]. The first COVID-19 wave in Pakistan and the eigenvalue analysis will be discussed in Sect. 8.7.2.

References

1. O. Diekmann, J.A.P. Heesterbeek, J.A.J. Metz, On the definition and the computation of the basic reproduction ratio R0 in models for infectious diseases in heterogeneous populations. Math. Biol. **28**, 365–382 (1990)
2. H.W. Hethcote, The mathematics of infectious diseases. SIAM Rev. **42**, 599–653 (2000)
3. K. Rock, S. Brand, J. Moir, M.J. Keeling. Dynamics of infectious diseases. Rep. Prog. Phys. 77:article 026602 (2014)
4. P. van den Driessche, J. Watmough, Reproduction numbers and sub-threshold endemic equilibria for compartment models of disease transmission. Math. Biosci. **180**, 29–48 (2002)
5. J. Wallinga, P. Teunis, Different epidemic curves for severe acute respiratory syndrome reveal similar impacts of control measures. Am. J. Epidemiol. **160**, 509–516 (2004)
6. M. Lipsitch, T. Cohen, B. Cooper, J.M. Robins, S. Ma, L. James, G. Gopalakrishna, S.K. Chew, C.C. Tan, M.H. Samore, D. Fisman, M. Murray, Transmission dynamics and control of severe acute respiration syndrome. Science **300**, 1966–1970 (2003)
7. O. Diekmann, J.A.P. Heesterbeek, *Mathematical epidemiology of infectious diseases* (Wiley, Chichester, 2000)
8. M. Gatto, E. Bertuzzo, L. Mari, S. Miccoli, L. Carraro, R. Casagrandi, A. Rinaldo, Spread and dynamics of the COVID-19 epidemic in Italy: effects of emergency containment measures. Proc. Natl. Acad. Sci. USA **117**, 10484–10491 (2020)
9. L. Pang, S. Liu, X. Zhang, T. Tian, Z. Zhao, Transmission dynamics and control strategies of COVID-19 in Wuhan China. J. Biol. Syst. **28**, 543–561 (2020)
10. T.D. Frank, COVID-19 order parameters and order parameter time constants of Italy and China: a modeling approach based on synergetics. J. Biol. Syst. **28**, 589–608 (2020)
11. S. Ullah, M.A. Khan. Modeling the impact of non-pharmaceutical interventions on the dynamics of novel coronavirus with optimal control analysis with a case study. Chaos, Solitons, and Fractals, 139:article 110075 (2020)
12. T.D. Frank, Emergence and subsiding of the first-wave COVID-19 pandemic in Pakistan 2020: an eigenvalue analysis based on synergetics. Proc. Pak. Acad. Sci. B **57**, 1–7 (2020)

Chapter 8
Modeling Interventions

This chapter discusses the impact of intervention measures on the course of epidemics from a nonlinear physics perspective. It is discussed how intervention measures can affect structural properties of populations and, in doing so, induce bifurcations from unstable disease-free states towards stable disease-free states. As a result of such bifurcations, epidemics begin to subside. The sign switching phenomenon of maximal eigenvalues is explored that indicates the occurrence of a bifurcation. The bifurcation scenario and sign switching phenomenon is demonstrated for several COVID-19 waves from Europe, Thailand, Pakistan, and the USA that took place during the year 2020 and exhibited a subsiding stage. A three-stage model of epidemic waves is discussed in this context.

8.1 Motivation

Intervention measures or protective measures against an infectious disease typically have three goals. First, they should lower the total number of infections. Second, if an infectious disease can cause the death of infected individuals (e.g., as it is the case for AIDS and COVID-19) the goal is to reduce the number of deaths caused by the disease. Third, intervention measures may be put into place to lower the peak of an emerging epidemic wave.

Let us illustrate these goals from a mechanistic perspective. In particular, let us illustrate that lowering the number of virus infections is one way to reduce the number of deaths caused by a virus disease. Let us consider an epidemic described by a 1β SEIR-type model (5.3) that captures both recovered individuals (R) and individuals deceased due to the disease (D). It is assumed that infectious individuals of the compartment I either recover with a rate a or decease with a rate b. In this case, Eq. (5.3) becomes

T. D. Frank, *COVID-19 Epidemiology and Virus Dynamics*,
Understanding Complex Systems, https://doi.org/10.1007/978-3-030-97178-6_8

$$\frac{\mathrm{d}}{\mathrm{d}t}S = -\frac{\beta I}{N}S \ , \ \frac{\mathrm{d}}{\mathrm{d}t}E = \frac{\beta I}{N}S - \alpha E \ , \ \frac{\mathrm{d}}{\mathrm{d}t}I = \alpha E - (a+b)I \ ,$$
$$\frac{\mathrm{d}}{\mathrm{d}t}R = aI \ , \ \frac{\mathrm{d}}{\mathrm{d}t}D = bI \ . \tag{8.1}$$

Let us consider an infectious disease that results in a given population in a effective contact rate β_{80} that requires that the number of susceptibles S drops by 20%, that is, reaches 80% of the initial total population before the disease-free fixed point becomes stable. From the critical value $\kappa_r = 1$, the definition of κ_r as $\kappa_r = S_{st}\beta/(\gamma N)$ (see Eq. (5.44)), and $S_{st}/N = 0.8$ it then follows that $\beta_{80} = \gamma/0.8$ with $\gamma = a + b$. Figure 8.1 shows a simulation of Eq. (8.1) for this scenario. For this simulation the parameters $a = 0.4$/d, $b = 0.1$/d ($\Rightarrow \gamma = 0.5$/d), $\alpha = 0.5$/d, and $N = 100{,}000$ and initial conditions $I(0) = 10$, $E(0) = 0$, and $S(0) = N - I(0)$ have been used. Panel (a) shows the decrease of S (top subpanel) from 100,000 (more precisely: 99,990) to the critical value 80,000 and eventually to a stationary value slightly larger than 60,000. Panel (a) also shows the number of infected individuals $C(t) = E(t) + I(t)$. The wave of infected individuals reaches its peak at the moment when S reaches the critical value of 80%. The reason for this is that the disease-free fixed point under that critical condition $S/N = 80\%$ becomes stable. C begins to decrease (see Sects. 3.7 and 5.5.1). In panel (b) E and I as functions of time are shown. The waves in E and I approximately reach their maxima at the time point when the disease-free fixed point becomes stable. A close inspection shows that the wave of exposed individuals E is slightly leading the wave of infectious individuals I. In summary, $E(t)$ reaches its maximum first, $C(t)$ follows, and $I(t)$ reaches its maximum at last. Panel (c) presents the cumulative infectious cases $I_c(t)$ as computed from $\mathrm{d}I_c/\mathrm{d}t = \alpha E$ with $I_c(0) = I(0)$. Panel (c) also presents the recovered and simulated individuals deceased due to the virus disease. Panel (c) illustrates that the curve of the deaths follows qualitatively the curve of the cumulative infectious individuals. In other words, when the cumulative number of infectious cases increases, then the number of deaths caused by the disease increases as well. Consequently, the number of deaths can be decreased by decreasing the total (or cumulative) number of infections.

Figure 8.1 illustrates a no-intervention scenario. The epidemic eventually subsides because the number of susceptibles S decreases below the critical value S_{crit} for which the disease-free fixed point with $S_{st} = S_{crit}$ and all fixed points with $S_{st} < S_{crit}$ are neutrally stable. Let us assume at a time t intervention measures are implemented that reduce the effective contact rate β (e.g., in the context of a COVID-19 epidemic taking place in a certain region by mask mandates or lockdown/shutdown orders). If β is decreased below its critical value γ (assuming $S \approx N$) then the disease-free fixed point becomes stable and the epidemic under consideration subsides. Figure 8.2 shows results of a simulation of this intervention scenario. The same parameters and initial conditions as for the simulation presented in Fig. 8.1 have been used. However, at $t = 60$ days, the effective contact rate was lowered to 95% of the value of γ. Consequently, at $t = 60$ days the disease-free state becomes stable in the simulated epidemic. Panel (a) of Fig. 8.2 shows E and I as functions of time. When β is switched

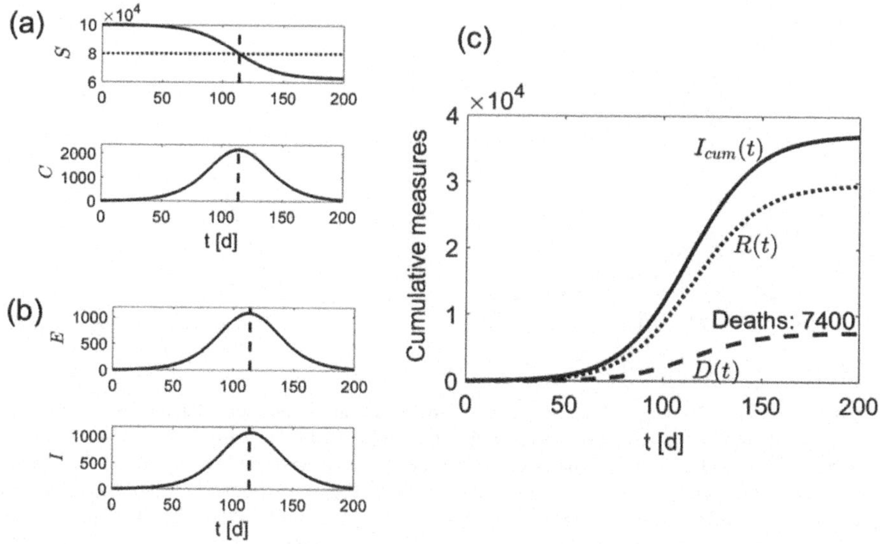

Fig. 8.1 Simulated epidemic in the absence of interventions to stop the epidemic. Panel (**a**): Solutions $S(t)$ and $C(t) = E(t) + I(t)$ of the SEIR-type model (8.1) over a 200 days period. The dotted horizontal line indicates the required decay in S for the disease-free fixed point to become stable. The vertical dashed line indicates the corresponding time point. Panel (**b**): Solutions $E(t)$ and $I(t)$ of Eq. (8.1). Panel (**c**): Cumulative measures $I_{cum}(t)$, $R(t)$, and $D(t)$ of the simulated epidemics as computed from Eq. (8.1) and $dI_{cum}/dt = \alpha E$. See text for parameters and initial conditions

from $\beta = \gamma/0.8 > \gamma$ to $\beta = 0.95\gamma < \gamma$, then the number of exposed individuals $E(t)$ immediately decays. After a short delay, $I(t)$ decays as well. Panel (b) shows I_c, R, and D for the intervention scenario. Again D mimics the time course of I_c. Importantly, the final death toll is lower as compared to the no-intervention scenarios (compare $D(\infty) \approx 1050$ in panel (b) of Fig. 8.2 with $D(\infty) \approx 7400$ in panel (c) of Fig. 8.1).

As mentioned above, a third goal of intervention measures is to lower the peak of an emerging epidemic wave of infections. The reason for this is that typically there is a proportion of the infectious individuals that requires intensive treatment. In general, infectious individuals who develop severe clinical symptoms need to be hospitalized. Hospitalization capacities have limits. A peak in infections may cause a situation in which hospitals reach their limits and health care systems partially break down.

For example, the COVID-19 pandemic has caused repeatedly breakdowns of health care systems all around the globe. When the first wave of COVID-19 hit Italy in March 2020, hospitals ran out of ventilators. Some COVID-19 patients in need for ventilators could not receive one [1]. In general, chances of such patients to survive without ventilation are very low [2]. During the first-wave COVID-19 epidemic in New York city in the Spring of 2020 the number of intensive care units (ICU) could not meet the demand. Hospitals in New York city had to convert non-

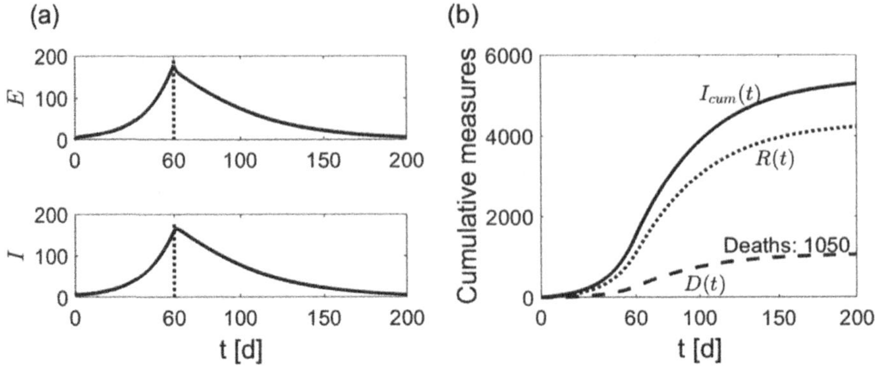

Fig. 8.2 Simulated epidemic when interventions measures are implemented and become effective at time $t = 60$ days. Panel (**a**): Solutions $E(t)$ and $I(t)$ of Eq. (8.1). The vertical dotted line indicates the time point $t = 60$ days when the effective contact rate β is lowered, which reflects the impact of intervention measures. Panel (**b**): Cumulative measures $I_{cum}(t)$, $R(t)$, and $D(t)$ of the simulated epidemics as computed from Eq. (8.1). Parameters and initial conditions as for the simulation in Fig. 8.1 but with β lowered at $t = 60$ days to a below-critical value

ICU rooms into ICU rooms [3]. Moreover, the ICU nursing staff was no longer sufficient. For example, before the beginning of the COVID-19 pandemic, the New York city hospital NYU-Tisch had one ICU team. During the first-wave of COVID-19 the hospital increased the number of teams to 11 in order to be able to deal with the dramatically increasing number of COVID-19 ICU patients [3]. The Brazilian health care system collapse both during the first COVID-19 wave during Spring of 2020 [4] and the second COVID-19 wave during Spring of 2021 [5]. During Spring of 2021 the so-called delta-variant of SARS-CoV-2 was driving an unprecedented second COVID-19 wave in India. This second wave caused a collapse of India's health care system. In particular, hospitals were running out of oxygen needed by COVID-19 patients and patients were dying as a result of that shortage [6].

When health care systems become overburdened and intensive care units become unavailable to patients, then chances are high that patients, who can be saved with proper treatment, lose their lives.

Figure 8.3 illustrates the goal to lower the peak of an epidemic wave by means of intervention measures. Figure 8.3 shows the rise and decay of the number of infectious individuals $I(t)$ as obtained in the no-intervention scenario (solid line) and the intervention scenario (dashed line). Let us assume that the health care system capacity can be measured on the basis of infectious individuals. That is, on the one hand, the burden of a typical infectious individual on the health care system can be determined and, on the other hand, it can be determined for how many of such typical infectious individuals a given health care system can provide appropriate service and treatment. This capacity measure corresponds to a threshold. For illustration purposes let us assume this threshold is at a level of 500 infectious individuals. In this case, the demand created by the simulated epidemic for the no-intervention scenario would

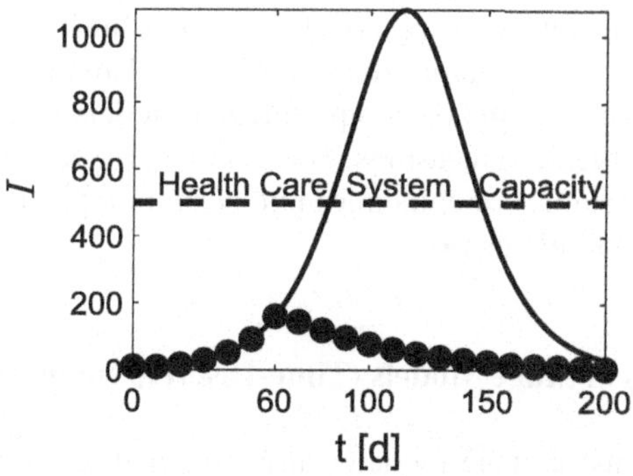

Fig. 8.3 Comparison of the solutions $I(t)$ (solid line) of the no-intervention scenario shown in Fig. 8.1 and $I(t)$ (full circles) of the intervention scenario shown in Fig. 8.2. The dashed line describes the assumed health care system capacity as measured in infectious individuals. In the no-intervention scenario the health care system is overburdened during the peak of the epidemic

exceed the health care system capacity—at least during a certain peak period. In contrast, for the simulated intervention scenario the peak of infectious individuals falls below the health care system capacity. Comparing the two curves with respect to the health care system capacity, the hypothesized intervention measures that decrease the effective contact rate below its critical value are such that they push the peak of the epidemic wave below the capacity threshold. In this context, frequently the phrase is used that during an epidemic intervention measures should "flatten the epidemic curve" such that hospitals can continue to provide sufficient care for their patients.

8.2 Types of Intervention Models

8.2.1 Overview

8.2.1.1 Discrete Versus Continuous

Interventions may be described with the help of stage models. To this end, the course of an epidemic may be decomposed into a number of stages. That is, the observation period may be partitioned into several intervals reflecting stages. Within each stage the impact of interventions is constant. Parameters describing the impact of interventions when plotted over the whole observation period are then given in terms of step-functions (i.e., piecewise-constant functions). Stage models of interventions may be regarded as time-discrete intervention models.

Alternatively, time-continuous intervention models may be used. In such approaches model parameters become explicitly time-dependent and are given in terms of time-continuous functions. Stage modes may be considered as approximation of time-continuous intervention models (i.e., step-functions may be considered as approximation of time-continuous functions) [7, 8]. Time-continuous intervention models can also come in terms of compartmental models that introduce compartments specific to intervention measures. For example, models that take the effect of vaccination on the dynamics of epidemics into account typically feature a compartment of vaccinated individuals [9].

8.2.1.2 Two Types of Stage Models (Time-Discrete Intervention Models)

The stages or intervals of stage models can be defined in two qualitatively different ways. They can be defined based on the history of events that presumably have affected the course of an epidemic. That is, they can be event-based. Alternatively, intervals may be defined based on the data describing an epidemic under consideration. For example, the time course of diagnosed infected individuals may be used to define a sequence of stages. This approach can be regarded as a data-driven approach.

8.2.1.3 Two Types of Time-Continuous Intervention Models

As mentioned above, there are two types of time-continuous intervention models. On the one hand, it might be assumed that model parameters such as the effective contact rate change in a continuous way due to the impact of intervention measures. On the other hand, compartments may be introduced that account for the effect of interventions.

In this context, the analysis of epidemiological models requires an additional level of caution. When model parameters become explicitly time-dependent, the models do no longer belong to the class of autonomous models. The analysis of fixed points described in previous chapters and the model-based derivation of reproduction numbers discussed in this chapter was carried out for autonomous epidemiological models (i.e., models that do not explicitly depend on time). Models of the second type that describe an sheltering of susceptibles from the infectious disease (e.g., due to vaccination) as in Eq. (4.85) or in some of the studies that will be reviewed below typically do not exhibit a disease-free fixed point with $S_{st} > 0$. Consequently, it might be not possible to apply the concepts and approaches discussed in the previous chapters that assume the existence of a fixed point $\mathbf{X}_{st}$ with $S_{st} > 0$.

8.2.1.4 Special Case: Vaccination Model SIRV

The rate constant k_0 that describes the rate with which susceptible individuals get infected (see Sect. 3.2) plays a crucial role for the stability of the disease-free fixed point. For the SIR model (3.16), the rate constant reads $k_0 = \beta S/N$. Intervention measures may be designed to decrease k_0. In Sect. 8.1 it has been exemplified that lowering the contact rate β can lead to the desired result of stabilizing the disease free-fixed point. An alternative approach is to decrease the size S of the susceptible population. Vaccination is one approach in this regard.

Vaccination can reduce the probability of individuals to get infected. In the ideal case, the infection probability becomes zero. Susceptible individuals who get vaccinated against a certain virus fall out of the group of susceptible and enter the group of vaccinated individuals. Epidemiological model may be supplemented by a compartment V of vaccinated individuals. For example, the SIR model that takes vaccination into account reads [9]

$$\frac{\mathrm{d}}{\mathrm{d}t}S = -\beta\frac{IS}{N} - \nu\frac{S}{N} \, , \; \frac{\mathrm{d}}{\mathrm{d}t}I = \beta\frac{IS}{N} - \gamma I \, , \; \frac{\mathrm{d}}{\mathrm{d}t}R = \gamma I \, , \; \frac{\mathrm{d}}{\mathrm{d}t}V = -\nu\frac{S}{N} \, . \quad (8.2)$$

Epidemiological models may be used to calculate the degree of vaccination of a population such that the disease free-fixed point becomes stable [10]. In this context, the phrase herd-immunity has been coined. Herd-immunity begins when S is sufficiently low such that the disease-free fixed point for an infectious disease in a given population with particular parameters becomes (neutrally) stable. For example, in Sect. 8.1, the case of a virus infection with a contact rate β_{80} has been considered. In this case, the fixed point becomes stable when the susceptible population decreases to 80% of the total population such that $S/N = 0.8$. Consequently, an infectious disease in a population with β_{80} requires that 20% of the population becomes vaccinated. In this example, herd-immunity begins at 20%. Worked out examples for diseases ranging from measles to chickenpox can be found in Ref. [10]. In those examples, vaccination between 80% and 95% of the whole populations is required for herd-immunity.

8.2.2 *SIR-Type Models Used in Studies Examining the Impact of Interventions*

Several studies used SIR-type models as defined by Eq. (4.1) to study the impact of interventions (see also Table 3.1).

8.2.2.1 Event-Based SIR-Type Stage Models

Wangping et al. [11] used an event-based three-stage model to describe the first-wave COVID-19 outbreak in the Hunan district, China. The Hunan district is the neighborhood district to the Hubei district which includes Wuhan city. They studied the period from January to March 2020. Based on the effective dates of intervention measures they distinguished between the following three stages. The stages were given by stage 1 from January 23 to February 4 during which Chinese provinces were put on lockdown, stage 2 from February 4 to February 8 during which enhanced quarantine measures were implemented, and stage 3 from February 8 to March

17 during which even more strict quarantine rules were put into effect in Hunan. Wangping et al. [11] assumed that during the stages the effective contact rate of the SIR model was constant with $\beta = x\beta_0$, where β_0 denotes the baseline effective contact rate in the absence of any containment measures. Fitting SIR-model solutions to COVID-19 case data, they obtained the following estimates for x: $x = 0.9$ for stage 1, $x = 0.5$ for stage 2, and $x = 0.1$ for stage 3. Their analysis suggests that the effective contact rate decayed due to the impact of intervention measures. In the same study, Wangping et al. [11] proposed a three stage description of the first-wave COVID-19 epidemic in Italy for the period from January 31 to March 31. Again, they identified stages on basis of the dates when specific intervention measures came into effect. The stages were given by stage 1 from January 31 to March10 during which a number of Italian cities were blocked and isolated, stage 2 from March 10 to March 22 during which a nationwide blockade of all cities and regions was implemented, and stage 3 from March 22 to March 31 during which all non-essential business were shut down. For those stages they estimated the parameter x as follows: $x = 0.95$ for stage 1, $x = 0.90$ for stage 2, and $x = 0.1$ for stage 3. Again, the factor x decayed in time across the stages suggesting that increasing the gravity of the intervention measures led during that period to a decrease of the effective contact rate.

8.2.2.2 Data-Driven SIR-Type Stage Models

Pedersen and Meneghini [12] used a data-driven two-stage model to describe first-wave COVID-19 epidemics in various countries. To this end, the SIQR model (4.82) (which belongs to the class of SIR-type models, see Sect. 4.4) was used. The study will be reviewed in Sects. 8.3.1 and 8.4.4.

8.2.2.3 SIR-Type Time-Continuous Intervention Models With Time-Dependent Contact Rates

As reviewed in Sects. 3.6.2 and 4.5, Fanelli and Piazza used the SIRD model (4.3) to analyze first COVID-19 waves in China and Italy [13]. As reviewed in Sect. 3.6.2, for the COVID-19 outbreak in Italy the period from February 11 to March 15, 2020 was examined in order to estimate SIRD model parameters. In addition, several simulations were conducted in order to predict possible impacts of intervention measure on the COVID-19 epidemic in Italy. To this end, it was assumed the intervention measures result in an exponential decrease of the effective contact rate like $\beta \propto \exp(-a(t - t_0))$ with $a > 0$, where t_0 denotes the starting point of the implementation of the measures interventions. Since this time-dependency results in an unrealistic long time dynamics like $\beta \to 0$, a more sophisticate relationship was suggested. Accordingly, it was assumed that intervention measures may result in a decay of β like $\beta = \beta_0(1 - x)\exp(-a(t - t_0)) + x\beta_0$ with x in $[0, 1]$ such that $t \to \infty \Rightarrow \beta = x\beta_0$. In other words, β was assumed to decay from β_0 at t_0 to $x\beta_0$ for $t \to \infty$.

A similar suggestion was made by Willis et al. [14]. In the context of a SIR modeling framework it was suggested that containment measures could decrease the effective contact rate like $\beta = \beta_0 - K(1 - \exp(-a(t - t_0)))$ with $K, a > 0$ such that $t \to \infty \Rightarrow \beta = \beta_0 - K$.

Zareie et al. [15] conducted a data driven approach to determine the presumed decay of the effective contact rate β due to intervention measures. To this end, a time-discrete version of the SIR model was used. Since N was large and the situation of a COVID-19 outbreak in a completely susceptible population was considered, it was assumed that variation in S can be neglected (see also Sect. 8.4.3). For $S = N$, the SIR model reduces to a single variable model for I that reads $I(t+1) = I(t) + \beta(t)I(t) - k(t)I(t)$ with $\beta(t)$ and $k(t)$ denote the time-dependent effective contact rate and the time-dependent removal rate, respectively. For the first-wave COVID-19 epidemic in China from January 22 to March 23, 2020 Zareie et al. [15] computed $\beta(t)$ and $k(t)$. They found that $\beta(t)$ as function of time approximately exhibited the shape of an exponential decaying function like $\beta = \beta_0 \exp(-a(t - t_0))$.

8.2.2.4 SIR-Type Time-Continuous Intervention Models With Vaccination-Like Sheltering

Barmparis and Tsironis [16] fitted SIR model solutions to first-wave COVID-19 epidemics observed in the USA and seven European countries. Rather than computing S from the evolution equation $\mathrm{d}S/\mathrm{d}t = -\beta IS/N$ it was assumed that S decayed linearly during those COVID-19 waves as a result of intervention measures that protected (i.e., sheltered) more and more susceptible individuals from becoming infected. That is, it was assumed that S evolved like $S = a - bt$ with $a, b > 0$. Using this linear ansatz, the SIR model yields an analytical solution for $I(t)$ in terms of a Gaussian function like $I(t) = I_0 \exp\{-c(t - t_0)^2 + d(t - t_0)\}$ with $c, d > 0$.

As mentioned in Sect. 4.4, an SIQR model defined by Eq. (4.85) was used to study the COVID-19 outbreak in Brazil in early 2020 [17], Accordingly, the evolution equation for S involves a term $-wS$ as used to describe vaccination [9] describing a sheltering effect of individuals from getting infected by COVID-19. Explicitly, the evolution equation reads $\mathrm{d}S/\mathrm{d}t = -\beta SI/N - wS$ (see Eq. (4.85)). As argued in Ref. [17], it is plausible to assume that during the initial stage of the COVID-19 epidemic in Brazil the approximations $S \approx N$ and $I \approx 0$ were satisfied. This implies that the evolution of S was determined by the linearized equations is $\mathrm{d}S/\mathrm{d}t = -\beta I - wS$. Neglect the impact of the infection term βI relative to the impact of the intervention wS, it follows that the population of susceptibles S decays exponentially like $S = S_0 \exp(-wt)$ for $t_0 = 0$. For short time periods, the exponential law implies a linear decay of S like $S = S_0(1 - wt)$ as assumed by Barmparis and Tsironis [16] (see above). In general, substituting the ansatz $S = S_0 \exp(-wt)$ into the linearized evolution equation $\mathrm{d}S/\mathrm{d}t = -\beta I - wS$ and analytical solution for $I(t)$ can be found [17].

8.2.3 *Modeling COVID-19 Interventions Beyond SIR Models*

In this section let us briefly review a few studies that are about the impact of COVID-19 intervention measures and based on epidemiological models with two or more than two infectious compartments.

8.2.3.1 Event-Based Stage Models

In Sect. 5.8, the study by Pang et al. [18] was introduced. Pang et al. investigated the epidemic in Wuhan city at the beginning of the year 2020. They used a 2β SEIR model. As described in Sect. 5.8, they conducted an event-based stage model approach and introduced three stages of the epidemic. These stages were stage 1 from December 2019 to January 22 (pre-lockdown stage), stage 2 from January 23 to February 11 (lockdown stage), and stage 3 from February 12 to the end of their data analysis period, which was March 25 (lockdown stage with strict quarantine measures). As part of their comprehensive study, they determined the effective contact rates and the basic and effective reproduction numbers of the stages. Table 8.1 summarizes their results. They found that the effective contact rates decayed across the stages. As discussed in Sect. 7.5.2, Pang et al. also found that the reproduction number decayed from an above-threshold value to a below-threshold value. The maximal eigenvalue of the three stages shown in Table 8.1 was computed in Ref. [19]. The eigenvalue switched from a positive to a negative value across the three stages, indicating that the disease-free fixed point was stabilized over the course of the observation period.

Gatto et al. [20] used an extended SEIAR model as discussed in Sect. 5.3.2 to study the emerging COVID-19 epidemic in Italy. They used an event-based three-stage approach. Based on measures implemented by the Italian government that restricted the mobility of individuals and the contacts between individuals, Gatto et al. defined the following three stages: stage 1 (February 21 to February 24/26), stage 2 (February 24/26 to March 8/10), and stage 3 (March 8/10 to March 25). From stage to stage the severity of the intervention measures increased [20]. The model uses the effective contact rates β_P, β_A, and β_I (see Eq. (5.17)). The parameters β_A and β_I were expressed as certain fixed multipliers of β_P like $\beta_A = x\beta_P$ and $\beta_I = y\beta_P$, where x and y did not change across the stages. The parameter β_P was estimated for the three

Table 8.1 Parameters of the 2β SEIR model describing the first COVID-19 wave in Wuhan city, China, as obtained in Refs. [18, 19]

Stage	β_I [1/d]	β_E [1/d]	$R_{0,e}$	λ_{max} [1/d]
1	0.29	0.17	4.64	0.19
2	0.26	0.13	1.75	0.09
3	0.02	0.02	0.16	−0.14

stages. In particular, the change of β_P across the stages was determined. Gatto et al. found that β_P decayed across the stages like $\beta_P(\text{Stage 2})/\beta_P(\text{Stage 1}) = 0.82$ and $\beta_P(\text{Stage 3})/\beta_P(\text{Stage 2}) = 0.66$.

Sun et al. [21] discussed the inverse problem. They studied events that took place in the Heilongjiang province, China, during Spring 2020. As part of the general COVID-19 related lockdown that was implemented in several regions of China at the beginning of the year 2020, during Spring 2020 initially strict intervention measure were implemented in the Heilongjiang province. These measures were subsequently relaxed. Sun et al. [21] studied the effect of relaxing intervention measures. More precisely, they used an event-based two-stage approach. Stage 1 was defined as the period from January 23 to March 25, 2020, during which of few local COVID-19 outbreaks occurred and subsided relative quickly due to the strict intervention measures. Stage 2 was defined as the period from April 9 to April 29, 2020 that was characterized by relaxed intervention measures. During that period another local COVID-19 outbreak took place. Sun et al. used a five-variable compartmental model involving three actually infectious compartments and, consequently, three effective contact rate parameters β_1, β_2, and β_3. They determined the effective contact rate parameters for stage 1. Subsequently, they assumed that in stage 2 all three parameters were increased by a factor F like $\beta_k(\text{Stage 2}) = F\beta_k(\text{Stage 1})$. They estimated that the factor F was in the range from 6 to 8.

Zhao et al. [22] conducted a theoretical study using a SEIAR-like compartmental model. Their considerations focused on an event-based stage description of the COVID-19 pandemic when intervention measures are implemented during certain periods and change across periods. Importantly, they discussed the possibility of implementing different intervention measures during the same period for different subgroups of a population.

8.2.3.2 Data-Driven Stage Models

Dickman [7] used a data-driven three-stage SEIAR-like model to study the COVID-19 epidemics during the first half of the year 2020 in Brazil, USA, and more than 100 regions and countries around the globe. The model involves three actually infectious compartments with effective contact rates β_1, β_2, and β_3 that are related to a reference effective contact rate β like $\beta_i = x_i\beta$, where x_i were fixed across the stages. The reference contact rate β was varied from stage to stage. That is, β was modeled in terms of a piecewise-constant step-function. Dickman pointed out that such step-functions may be considered as approximation of smooth and continuously varying functions as mentioned in Sect. 8.2.1. The time points t_1 and t_2 of the stage boundaries were estimated from the data together with the β(Stage 1), β(Stage 2), and β(Stage 3) to obtain an optimal fit between of the respective model solutions. Consequently, a data-driven stage approach was used.

Oliveira et al. [23] used an eight-variable compartment model to study the COVID-19 epidemic in the 14 million population of the state of Bahia, Brazil, during the period from March to September 2020. The model involves two actually infectious compartments with effective contact rates β and $\zeta\beta$, respectively. In order to account for the impact of intervention measures that were implemented during their observation period, Oliveira et al. [23] conducted a data-driven three-stage analysis. The factor ζ was fixed across the stages. In contrast, β varied and was estimated for the three stages together with the stage boundaries t_1 and t_2 to obtain an optimal fit of the model to the available data. Accordingly, the first stage $[t_0, t_1]$ was from March 6 to April 3, the second stage $[t_1, t_2]$ was the period from April 3 to June 11, and the third stage $[t_2, t_3]$ was the period from June 11 to the end of the observation period, which was September 13, 2020. They found that the effective contact rate β decreased during the three stages from $\beta(\text{Stage 1}) = 1.40/\text{d}$ to $\beta(\text{Stage 2}) = 0.96/\text{d}$ and $\beta(\text{Stage 3}) = 0.66/\text{d}$. This decrease was presumably due to the implemented intervention measures.

Serhani and Labbardi [24] used an SIAR-like model to study the first three months of the COVID-19 epidemic in Morocco of the year 2020. The model involves two actually infectious compartments with effective contact rates β and $\zeta\beta$. Serhani and Labbardi first determined a baseline effective contact rate β_0 in the absence of intervention measures. Using $\beta_0 = \nu p$ (see Eq. (3.3)), they assumed that individuals had $\nu = 40$ contacts per day and that the probability of infection was $p = 3\%$. The parameter ζ was taken from the literature. The effective contact rate β that accounts for intervention measures was then assumed to be proportional to β_0 like $\beta = h\beta_0$. Serhani and Labbardi [24] used a data-driven three-stage approach to fit the solution of their epidemiological model to the observed COVID-19 data from Morocco. In particular, they were interested to explain a seemingly secondary peak in the COVID-19 cases. According to their analysis, in stage 1 the factor h was $h = 0.844$ such that after the number of susceptibles decreased to a sufficiently low level the disease-free fixed point became stable and the epidemic started to subside. However, in the subsequent stage (stage 2), intervention measures were relaxed or ignored by the population such that h increased to $h = 0.846$. In stage 3, this relaxation or ignorance was even more pronounced such that h increased again to $h = 0.851$. According to the simulation and analysis conducted by Serhani and Labbardi, in stage 3 the disease-free fixed point became unstable again, which was the reason why a secondary peak in the epidemic curve of COVID-19 cases occurred.

A series of data-driven three-stage studies were conducted for several European countries [25], Thailand [8], Pakistan [26], and the USA [27]. They will be reviewed in more detail in Sects. 8.4 and 8.5.

8.2.3.3 Time-Continuous Intervention Models with Time-Dependent Contact Rates

Garba et al. [28] used a SEIAR-like model to examine the spread of COVID-19 in South Africa during the first half of the year 2020. The model involves a time-dependent effective contact rate $\beta(t)$. It was assumed that during the first 25 days since March 5, when the first COVID-19 case was confirmed in South Africa, the contact rate was constant at $\beta(t) = \beta_0$. However, on March 26 a nationwide lockdown was ordered [28]. Therefore, Garba et al. assumed that after that initial $t_0 = 25$ days period intervention measures showed an effect such that $\beta(t)$ decayed like $\beta(t) = \beta_1 + (\beta_0 - \beta_1)/[1 + \omega(t - t_0)]$ with $\beta_1 < \beta_0$ and $\omega > 0$. The function describes a monotonic decay from β_0 at $t = t_0$ towards β_1 for $t \to \infty$.

Ivorra et al. [29] used a similar approach to describe the first wave of the COVID-19 epidemic in China during the period from December 2019 to the end of March 2020. A comprehensive multi-variable compartmental model was used that involved three actually infectious compartments. The respective effective contact rates were assumed to be constant during an initial period. Subsequent to that period, they were assumed to vary in a smooth (i.e., time-continuous) manner. In particular, all three effective contact rates were assumed to be proportional to an intervention factor $m(t)$ that was modeled with the help of two stages. The first stage was given by the period from December 1, 2019 to January 23, 2020. During that period m was fixed at $m = 1$. The second stage was given by the period from January 23 to the end of the observation period (i.e., end of March 2020). It was assumed that m decayed exponentially during that period like $m(t) = \exp(-\kappa(t - t_0))$ with $\kappa > 0$ and $t_0 = 23$ January. This decay of the factor m should reflect the impact of the lockdown of the region around Wuhan and several neighboring regions that was ordered on January 23, 2020, by the Chinese government.

8.2.3.4 Time-Continuous Intervention Models with Vaccination-Like Sheltering

Zhao et al. [30] used a 1β SEIR model with a vaccination-like sheltering term such that the evolution equation of S was given as shown in Eq. (8.2) like $\mathrm{d}S/\mathrm{d}t = -\beta SI/N - \nu S$. The initial COVID-19 epidemics of six African countries during March 21 to April 13 were examined. To this end, the sheltering factor ν was varied and observed infections were compared with model solutions for various values of ν. In doing so, the African countries could be compared to each other on the basis of the parameter ν. For example, the data from South Africa was consistent with all larger sheltering factor ν as compared to the data from Egypt. This result could be interpreted to assume that during that initial period of the COVID-19 pandemic in Africa, the government of South Africa was more successful in implementing intervention measures than the government of Egypt.

Mandal et al. [31] fitted model solutions of a SEIQR model to COVID-19 data from several regions of India as observed during March/April 2020. An objective of the study was to predict the epidemics in those regions when assuming that intervention measures could shelter and protect people such that S evolves like $\mathrm{d}S/\mathrm{d}t = -k_0 S - \nu S$. That is, a vaccination-like term was used to model the impacts of measures to prevent the spread of COVID-19.

Nabi [32] used an eight-variable compartmental model to describe early COVID-19 epidemics in Russia, Brazil, India, Bangladesh, and UK. In order to address actions taken by the populations and governments to protect against the spread of the disease, in the model a fixed quarantine term $-\nu S$ was introduced that effectively reads like the vaccination-term in Eq. (8.2). That is, it was assumed that the susceptibles were quarantined at a fixed rate ν. After they finished quarantine, they returned to the group of susceptibles provided they were not infected. From a modeling point of view, such as linear term $-\nu S$ in the evolution equation of S makes that the epidemiological model no longer exhibits a disease-free fixed point with $S_{st} = N$, where N is the population size. An alternative approach to describe the impact of quarantining will be presented in Sect. 8.7.1 on the basis of the study by Ngonghala et al. [33]. In this alternative approach, quarantining only takes place when a population exhibits a non-zero number of infected individuals. Consequently, the disease-free state with $S_{st} = N$ remains a fixed point of the model.

8.3 Models with Analytical Solutions

Analytical solutions may be obtained under the assumption that the disease dynamics evolves close to the disease-free fixed point. In this case, the evolution is determined by linearized equations for which analytical solutions can often be obtained. Models exhibiting analytical solutions can be conveniently used as stage models. The reason for this is that it is often less tedious to estimate model parameters with the help of analytical solutions as compared to numerical solutions. For models exhibiting analytical solutions, the solutions can be fitted conveniently for each stage of an observed trajectory or data set under consideration.

8.3.1 SIR-Type Models

Analytical solutions for the SIQR models (4.82) and (4.85) have been derived and used to describe COVID-19 epidemics in Italy and Brazil in studies by Pedersen and Meneghini [12] and Crokidakis [17, 34], respectively. Pedersen and Meneghini [12] also studied 2020 COVID-19 outbreaks in Europe, Canada, Australia, and the USA with the help of similar analytical SIQR model solutions. The SIQR models are models of the SIR type. In particular, the model (4.82) has the benefit that individuals in the compartments Q and R taken together correspond to the cumulative reported (i.e., diagnosed) COVID-19 cases such that the model can be conveniently fitted to data (see Sect. 4.4). The analytical solutions of the SIQR model (4.82) will be presented in Sect. 8.4.3.

8.3.2 SEIR-Type Models

Let us consider the 2β SEIR-type model suggested in Refs. [8, 25]. Figure 8.4 shows a schematic of the model (top part). Accordingly, the model involves the populations

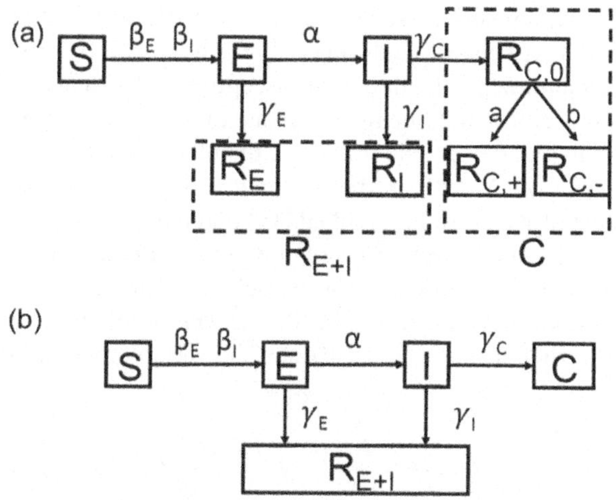

Fig. 8.4 Panel (**a**): Compartments and flow chart of the 2β SEIR-type model described by Eq. (8.3). Panel (**b**): Compartments and flow chart of the simplified version of the model as described by Eq. (8.4)

of susceptible (S), exposed and possibly asymptomatic infectious individuals (E), and symptomatic infectious but non-diagnosed individuals (I) and several types of removed individuals. These removed individuals are recovered individuals (R_E) from the class E, recovered individuals (R_I) from the class I, individuals who are diagnosed with COVID-19 and removed (e.g., hospitalized or quarantined) but not recovered or deceased ($R_{C,0}$), diagnosed recovered individuals ($R_{C,+}$), and diagnosed deceased individuals ($R_{C,-}$). The individuals in the compartment $R_{C,0}$ cannot infect others (e.g., due to perfect isolation). The total population is given by $N = S + E + I + R_E + R_I + R_{C,0} + R_{C,+}$. Moreover, $N(t) + R_{C,-}(t) = N_0$, where N_0 is a constant. The model equation reads [25]

$$\begin{aligned}
\frac{\mathrm{d}}{\mathrm{d}t} S &= -\beta_E \frac{ES}{N} - \beta_I \frac{IS}{N} , \\
\frac{\mathrm{d}}{\mathrm{d}t} E &= \beta_E \frac{ES}{N} + \beta_I \frac{IS}{N} - (\alpha + \gamma_E) E , \\
\frac{\mathrm{d}}{\mathrm{d}t} I &= \alpha E - (\gamma_I + \gamma_C) I , \\
\frac{\mathrm{d}}{\mathrm{d}t} R_E &= \gamma_E E , \\
\frac{\mathrm{d}}{\mathrm{d}t} R_I &= \gamma_I I , \\
\frac{\mathrm{d}}{\mathrm{d}t} R_{C,0} &= \gamma_C I - (a + b) R_{C,0} , \\
\frac{\mathrm{d}}{\mathrm{d}t} R_{C,+} &= a R_{C,0} , \\
\frac{\mathrm{d}}{\mathrm{d}t} R_{C,-} &= b R_{C,0} .
\end{aligned} \tag{8.3}$$

All parameters are semi-positive. They denote the effective contact rate between individuals of class E and susceptible individuals (β_E), the effective contact rate between symptomatic infectious and susceptible individuals (β_I), the rate of progression from class E to I (α), the recovery rate of exposed and possibly asymptomatic infectious individuals (γ_E) and, likewise, the recovery rate of symptomatic infectious individuals (γ_I). Moreover, the parameters describe the diagnoses rate (γ_C), the recovery rate of diagnosed individuals (a) and the death rate of diagnosed individuals (b).

The total population of individuals diagnosed with COVID-19 is given $C = R_{C,0} + R_{C,+} + R_{C,-}$. The model (8.3) can be simplified as shown in Fig. 8.4 (see bottom part). To this end, the recovered individuals from class E and I are taken together $R_{E+I} = R_E + R_I$ and the variable C is used. In this case, Eq. (8.3) reads [8]

$$\frac{\mathrm{d}}{\mathrm{d}t}S = -\beta_E \frac{ES}{N} - \beta_I \frac{IS}{N} , \quad \frac{\mathrm{d}}{\mathrm{d}t}E = \beta_E \frac{ES}{N} + \beta_I \frac{IS}{N} - (\alpha + \gamma_E)E ,$$
$$\frac{\mathrm{d}}{\mathrm{d}t}I = \alpha E - (\gamma_I + \gamma_C)I , \quad \frac{\mathrm{d}}{\mathrm{d}t}R_{E+I} = \gamma_E E + \gamma_I I , \quad \frac{\mathrm{d}}{\mathrm{d}t}C = \gamma_C I . \tag{8.4}$$

Consequently, C evolves like

$$C(t) = C(t') + \gamma_C \int_{t'}^{t} I(s)\,\mathrm{d}s . \tag{8.5}$$

The key assumptions that allow to use linearized equations, is to assume that N is approximately constant and the $S = N$ holds in good approximation [25] (see also Refs. [12, 15, 17]). Let us motivate these assumptions. First, while the number of COVID-19 associated deaths is a tragic number, it is typically a negligibly small number relative to the total population under consideration. For example, in Germany, as a result of the first-wave of COVID-19, in July 2020, about $R_{C,-} = 9{,}000$ deaths were reported. Note that $R_{C,-}$ does not describe daily deaths or "new deaths" but the total number of deaths until the time point under consideration. The population of Germany in 2020 was about $N = 83{,}000{,}000$ [35]. Consequently, $R_{C,-} \ll N$, which implies that $N \approx N_0$. The variable N occurring in Eqs. (8.3) and (8.4) is constant. Second, the approximation $S = N$ holds for first-waves of COVID-19 assuming that such waves are relative short in time. This implies that the diagnosed cases are relatively small ($C \ll N$). Moreover, it can be assumed that the unknown non-diagnosed infected cases E and I and the corresponding recovered cases R_E and R_I correspond to relatively small numbers as well ($E + I + R_E + R_I \ll N$). For example, in July 2020, in Germany a cumulative number of confirmed cases $C \approx 200{,}000$ was observed, which is 0.24% of the population with $N = 83{,}000{,}000$ such that the assumption $C \ll N$ is satisfied.

If we put $S = N$ then the rate constant $k_0 = (\beta_E E + \beta_I I)S/N$ reads $k_0 = \beta_E E + \beta_I I$ and the models (8.3) and (8.4) becomes linear. The linear evolution equations for E and I read

$$\frac{\mathrm{d}}{\mathrm{d}t}E = \beta_E^* E + \beta_I I \ , \ \frac{\mathrm{d}}{\mathrm{d}t}I = \alpha E - \gamma_I^* I \tag{8.6}$$

with $\beta_E^* = \beta_E - (\alpha + \gamma_E)$ and $\gamma_I^* = \gamma_I + \gamma_C$.

In summary, the compartmental model sketched in Fig. 8.4 and described by Eq. (8.3) or alternatively by Eq. (8.4) reduces under the circumstances described above to a three-variable model for E, I, and C defined by Eqs. (8.5) and (8.6), which is linear and can be solved analytically. The linear model (8.6) has been discussed in Sect. 5.5.1. In particular, Eq. (8.6) corresponds to Eq. (5.37) if in Eq. (5.37) γ is substituted like $\gamma = \gamma_I^*$ and $\beta_E - \alpha$ is replaced by β_E^*, that is, by $\beta_E - (\alpha + \gamma_E)$. In analogy to Eq. (5.38), the eigenvalues of Eq. (8.6) read

$$\lambda_{1,2} = \frac{\beta_E^* - \gamma_I^*}{2} \pm \sqrt{\frac{(\beta_E^* - \gamma_I^*)^2}{4} + \beta_E^* \gamma_I^* + \beta_I \alpha} \ , \tag{8.7}$$

where the upper (lower) sign holds for λ_1 (λ_2). Both eigenvalues are real-valued. The stability analysis can be conducted as in Sect. 5.5.1. In particular, in analogy to β_w as defined in Eq. (5.41), the weighted parameter

$$\beta_w = \beta_I + \frac{\gamma_I^*}{\alpha}(\beta_E - \gamma_E) \tag{8.8}$$

can be defined. If so, the determinant $\hat{D} = -(\beta_E^* \gamma_I^* + \beta_I \alpha)$ occurring in Eq. (8.7) reads $\hat{D} = \alpha(\gamma_I^* - \beta_w)$ such that for

$$\beta_w > \gamma_I^* \Rightarrow \lambda_1 > 0 \ , \ \lambda_2 < 0 \tag{8.9}$$

the fixed point $E_{st} = I_{st} = 0$ corresponds to a saddle point (as long as the linear approximation is valid, i.e., $S \approx N$), for

$$\beta_w = \gamma_I^* \Rightarrow \lambda_1 = 0 \ , \ \lambda_2 < 0 \tag{8.10}$$

the disease is at its bifurcation point, and for

$$\beta_w < \gamma_I^* \Rightarrow \lambda_1 < 0 \ , \ \lambda_2 < 0 \tag{8.11}$$

$E_{st} = I_{st} = 0$ corresponds to a stable node. Alternatively, the bifurcation parameter [8, 25]

$$\beta_{tot} = \beta_E + \frac{\alpha}{\gamma_I^*}\beta_I \tag{8.12}$$

with $\hat{D} = -(\beta_E^* \gamma_I^* + \beta_I \alpha) = \gamma_I^*(\alpha + \gamma_E - \beta_{tot})$ can be introduced such that for

$$\beta_{tot} > \alpha + \gamma_E \Rightarrow \lambda_1 > 0 \ , \ \lambda_2 < 0 \tag{8.13}$$

the fixed point $E_{st} = I_{st} = 0$ is a saddle point, for

$$\beta_{tot} = \alpha + \gamma_E \Rightarrow \lambda_1 = 0\ ,\ \lambda_2 < 0 \tag{8.14}$$

the disease is at its bifurcation point, and for

$$\beta_{tot} < \alpha + \gamma_E \Rightarrow \lambda_1 < 0\ ,\ \lambda_2 < 0 \tag{8.15}$$

$E_{st} = I_{st} = 0$ corresponds to a stable node. The basic and effective reproduction numbers of the SEIR model can be computed using the techniques described in Chap. 7. With the help of β_{tot} the reproduction numbers can be expressed like [8]

$$R_{0,e} = \frac{\beta_{tot}}{\beta_{tot,c}}\ , \tag{8.16}$$

where $\beta_{tot,c} = \alpha + \gamma_E$ is the critical value of the bifurcation parameter β_{tot}. Consequently, $R_{0,e} > 1$ and $R_{0,e} < 1$ corresponds to the two cases described by Eqs. (8.13) to (8.15), respectively. This implies that for $R_{0,e} > 1$ ($R_{0,e} < 1$) the fixed point $E_{st} = I_{st} = 0$ corresponds to an unstable (asymptotically stable) state, which is consistent with the general considerations made in Chap. 7 about reproduction numbers.

Let $\mathbf{X}^+ = (E, I)$ denote the two-dimensional subspace vector of the E-I plane D^+. Then, the amplitude space description of the subsystem dynamics (8.7) in D^+ reads

$$\mathbf{X}^+(t) = \begin{pmatrix} E(t) \\ I(t) \end{pmatrix} = A_1(t)\mathbf{v}_1 + A_2(t)\mathbf{v}_2\ , \tag{8.17}$$

where $\mathbf{v}_1$ and $\mathbf{v}_2$ denote the eigenvectors related to λ_1 and λ_2, respectively. The eigenvectors will be determined explicitly below. The amplitudes themselves satisfy the linear equations

$$\frac{\mathrm{d}}{\mathrm{d}t} A_j = \lambda_j A_j \tag{8.18}$$

and, consequently, evolve like $A_j(t) = A_j(t_0)\exp\{-\lambda_j(t - t_0)\}$. Finally, if the overall dynamics is dominated by the term $A_1\mathbf{v}_1$ of the amplitude and eigenvector related to the maximal eigenvalue λ_1, then Eq. (8.17) reduces to

$$\mathbf{X}^+(t) = \begin{pmatrix} E(t) \\ I(t) \end{pmatrix} \approx A_1(t)\mathbf{v}_1\ . \tag{8.19}$$

For $\lambda_1 > 0$ the vector $\mathbf{v}_1$ is the EI order parameter that determines in combination with the order parameter amplitude A_1 the disease subsystem dynamics $\mathbf{X}^+$ in the infected E-I subspace. For $\lambda_1 < 0$ the vector $\mathbf{v}_1$ is the remnant of the order parameter.

8.3.2.1 Single λ Approach

From Eq. (8.7) it follows that $\lambda_1 > \lambda_2$ and that $\lambda_2 < 0$ holds in any case. Consequently, the amplitude A_1 can be considered as the dominant amplitude. In particular, as discussed in Sect. 6.1.3, for $\lambda_1 > 0$ the amplitude A_1 corresponds to the order parameter amplitude and increases over time, while A_2 decays in magnitude. Moreover, for $\lambda_1 < 0$ there might be a time-scale separation as discussed in Sect. 6.1.3 such that A_1 slowly decays in magnitude, whereas A_2 decays relatively quickly. If so, A_1 determines the subsiding dynamics of an epidemic wave under consideration. Taken these considerations together, it has been suggested to neglect the evolution of the amplitude A_2 [25]. This leads to a single λ approach or single λ approximation. Let $\mathbf{v}_1$ be given in components like $\mathbf{v}_1 = (v_{1,E}, v_{1,I})$, then from Eq. (8.19) it follows that $I/E = g$ with $g = v_{1,I}/v_{1,E} > 0$. Consequently, the SEIR-type model (8.4) in the linear approximation reduces to

$$S = N_0 \ , \ \frac{\mathrm{d}}{\mathrm{d}t} E = \lambda_1 E \ , \ I = gE \ , \ \frac{\mathrm{d}}{\mathrm{d}t} C = \gamma_C I = g^* E \tag{8.20}$$

with $g^* = g\gamma_C$. In Eq. (8.20) the dynamics of R_{E+I} that is described in Eq. (8.4) has been neglected because it will be of no concern in what follows. For $\lambda_1 \neq 0$ Eq. (8.20) exhibits the analytical solutions [25]

$$\begin{aligned} E(t) = E(t_0)\exp\{\lambda_1(t-t_0)\} \ &\Rightarrow \ I(t) = I(t_0)\exp\{\lambda_1(t-t_0)\} \\ \Rightarrow \ C(t) &= C(t_0) + V[\exp\{\lambda_1(t-t_0)\} - 1] \end{aligned} \tag{8.21}$$

with $V = \gamma_C I(t_0)/\lambda_1$. The advantage of the single λ approach is that the unknown parameters V and λ_1 occurring in Eq. (8.21) can be estimated from data given in terms of $C(t)$ conveniently and with high accuracy. The disadvantage of the approach is that $\mathbf{v}_1$ (and $\mathbf{v}_2$) cannot be estimated solely on the basis of data $C(t)$.

8.3.2.2 Two λ Approach

In order to estimate the eigenvectors $\mathbf{v}_1$ and $\mathbf{v}_2$ the following procedure has been suggested [8]. Let us assume that all model parameters are known except for the effective contact rates β_E and β_I. In other words, literature values for α, γ_I, γ_E, and γ_C are taken. In this case, if λ_1 and λ_2 can be estimated from data, then β_E and β_I can be computed from λ_1 and λ_2 by solving Eq. (8.7). Consequently, the eigenvectors $\mathbf{v}_1$ and $\mathbf{v}_2$ can be determined. This two λ approach works explicitly as follows. First, Eq. (8.6) for the subsystem $\mathbf{X}^+ = (E, I)$ can be written like

$$\frac{\mathrm{d}}{\mathrm{d}t}\begin{pmatrix} E \\ I \end{pmatrix} = L^+ \begin{pmatrix} E \\ I \end{pmatrix} \ , \ L^+ = \begin{pmatrix} \beta_E^* & \beta_I \\ \alpha & -\gamma_I^* \end{pmatrix} . \tag{8.22}$$

The matrix equation can equivalently be expressed as the second order differential equation

$$\frac{d2}{dt^2} I = \hat{T}(L^+) \frac{d}{dt} I - \hat{D}(L^+) I \; , \tag{8.23}$$

where $\hat{T}$ and $\hat{D}$ denote the trace and determinant of L^+, respectively. For $\lambda_1, \lambda_2 \neq 0$ the analytical solution of Eq. (8.23) is given by [8]

$$I(t) = B_1 \exp\{\lambda_1(t - t_0)\} + B_2 \exp\{\lambda_2(t - t_0)\} \; , \tag{8.24}$$

where B_1 and B_2 are parameters related to initial conditions. Likewise, for $\lambda_1, \lambda_2 \neq 0$ using Eqs. (8.5) and (8.6) the analytical solutions for E and C can be obtained and read [8]

$$\begin{aligned} E(t) &= \frac{1}{\alpha} \left[(\lambda_1 + \gamma_I^*) B_1 \exp\{\lambda_1(t - t_0)\} + (\lambda_2 + \gamma_I^*) B_2 \exp\{\lambda_2(t - t_0)\} \right] \; , \\ C(t) &= C(t_0) + V_1[\exp\{\lambda_1(t - t_0)\} - 1] + V_2[1 - \exp\{\lambda_2(t - t_0)\}] \; , \end{aligned} \tag{8.25}$$

where V_1 and V_2 are certain parameters. Finally, from the matrix L^+ the eigenvectors can be obtained (see Eq. (5.30)) and read

$$\mathbf{v}_1 = \frac{1}{Z_1} \begin{pmatrix} \beta_I \\ \lambda_1 - \beta_E^* \end{pmatrix} , \mathbf{v}_2 = \frac{1}{Z_2} \begin{pmatrix} \beta_I \\ \lambda_2 - \beta_E^* \end{pmatrix} \tag{8.26}$$

with $Z_j = \sqrt{\beta_I^2 + (\lambda_j - \beta_E^*)^2}$ for $j = 1, 2$.

The two λ approach has been used in the study by Frank and Chiangga [8] to discuss the first-wave COVID-19 epidemic in Thailand during the year 2020 (see Sect. 8.4.5). In Frank and Chiangga [8] only λ_1 could be estimated with satisfactory accuracy based on case data $C(t)$. That is, the data of the first-wave epidemic in Thailand were such that the analytical solution for $C(t)$ involving two eigenvalues λ_1 and λ_2 could not be used to estimate λ_2. Nevertheless, the two λ approach could be used to determine the eigenvectors $\mathbf{v}_1$ and $\mathbf{v}_2$. To this end, a theoretical assumption about the relationship between β_E and β_I was made (see again Sect. 8.4.5 below). This example illustrates that, as such, the double-exponential function for $C(t)$ shown in Eq. (8.25) allows to estimate the eigenvalues λ_1 and λ_2 independently from each other from given cumulative COVID-19 case data $C(t)$. However, for any given data set $C(t)$ it should be checked whether or not the data is such that both parameters can indeed be estimated in a reliable way.

8.4 Three-Stage Models and the Bifurcation Scenario Underlying Epidemic Waves

8.4.1 Bifurcation Scenario of Epidemic Waves

Let us continue the discussion about interventions that was presented in Sect. 8.1. Accordingly, intervention measures may change structural population parameters such that unstable disease-free fixed points change their stability and become stable. Such a switch in stability describes a bifurcation (see Sect. 2.5). In this context, Fig. 8.5 illustrates two scenarios in this context. On the left, the disease-free state is shown. In what follows the case is considered in which the disease-free state in the presence of a virus corresponds to an unstable state (see Chaps. 1, 3–7) which is typically the case if a novel virus such as COVID-19 invades a population. In the very bottom part of Fig. 8.5, the unstable state is characterized in terms of eigenvalues. For sake of simplicity, the case is illustrated in which there is a single eigenvalue that is positive. In fact, this case applies to the SIR and SEIR compartmental models (see Chaps. 4 and 5) and this case holds for all other applications that have been discussed

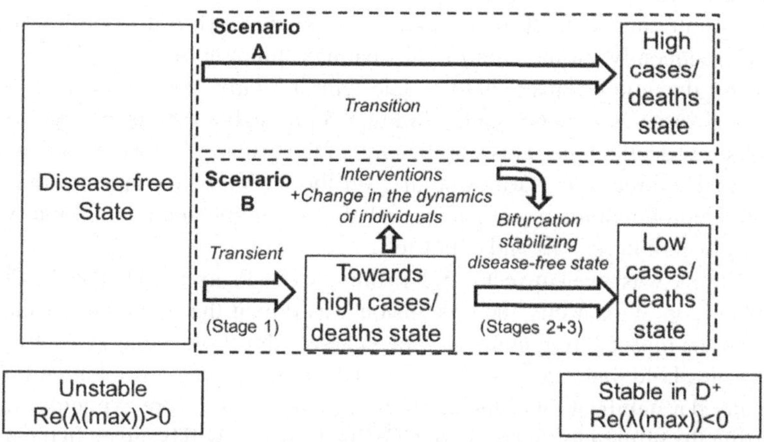

Fig. 8.5 No-intervention and intervention scenarios of an epidemic. In both scenarios the disease state of a population evolves initially away from an unstable diseases-free fixed point $\mathbf{X}_{st}$ towards a high cases/high deaths state. In the no-intervention scenario (scenario A) this state is reached. In the intervention scenario (scenario B) this state is not reached due to a change in the dynamics of the individuals that triggers a bifurcation and stabilizes the disease-free fixed point. Consequently, a low cases/low deaths state is reached. The real part of the maximal eigenvalue of the diseases-free state at the beginning of the epidemic is positive (i.e., $\text{Re}(\lambda(\text{max})) > 0$) indicating that the state is unstable. In contrast, the final states in the two scenarios when seen in a subspace D^+ (e.g., of infected individuals) exhibit a maximal eigenvalue that is negative or exhibit a negative real part (i.e., $\text{Re}(\lambda(\text{max})) < 0$). The final state is asymptotically stable in D^+ and typically neutrally stable when considering the entire state space. Note that the dynamics of the intervention scenario is an example of a dynamics that takes place, in general, in D1 systems [36] (compare with Fig. 8.6)

in Chap. 6. The remaining eigenvalues are assumed to exhibit negative real parts or real parts equal to zero. On the far right in Fig. 8.5, the final (stationary) states of the two scenarios are shown. These states are assumed to be stable (just as the stationary states shown in Fig. 8.1 and 8.2). They are again characterized by eigenvalues. For the stable states all eigenvalues exhibit real parts that are negative or zero.

The top part of Fig. 8.5 illustrates the scenario, in which no interventions are put into place, which is called the scenario A. Due to the instability of the disease-free fixed point, the disease (or health) state of a population evolves away from the unstable fixed point. The infectious disease invades the population and spreads out. Since no interventions are implemented, a relative large portion of the population becomes infected. Without an intervention, not only a relatively large number of individuals become infected but the disease also causes a relatively large number of disease-associated deaths (e.g., compare Figs. 8.1 and 8.2). Therefore, the stable fixed point of the dynamics of the disease under the no-intervention scenario A is referred to as the high cases/high deaths state.

In contrast, when intervention measures are implemented that reduce the spread of an infectious disease, the intervention scenario (scenario B) shown in the lower part of Fig. 8.5 is likely to take place. While initially the disease state evolves towards the high cases/high deaths state, after a certain period intervention measures are implemented and show an effect. In particular, the bifurcation scenario assumes that intervention measures affect the positive eigenvalue and shift it towards a negative value. Consequently, an intervention or treatment bifurcation takes place in which the disease dynamics converges to a state with a relative low number of infected cases and disease-associated deaths. In Fig. 8.5 this stable state is referred to as the low cases/low deaths state. This stabilization of the disease-free fixed point has been documented indirectly in various studies on the COVID-19 pandemic that determined the reproduction number before and after the implementation of intervention measures in certain regions and countries.

An example was given in Sect. 7.5.2. In the study by Pang et al. [18] on the COVID-19 epidemic in Wuhan city the conclusion was drawn that intervention measures decreased the reproduction number from an above-threshold value $R_0 = 4.6 > 1$ to a below-threshold value $R_e = 0.2 < 1$, which indicates that the intervention measures caused a stabilization of the disease-free fixed point. More explicity examples supporting the bifurcation scenario of COVID-19 waves will be given in the remainder of this chapter.

8.4.2 *Bigger Picture: Dynamical Diseases and D1-Systems*

8.4.2.1 Dynamical Diseases

The notion that diseases on the level of individuals emerge via bifurcations is at least 40 years old. In a seminal study, Mackey and Glass [37] argued that increasing the travel durations of feedback signals in humans can induce bifurcations in which states that are considered as healthy states become unstable and states that are considered as diseases states emerge. Clinical tremor [38], epileptic seizures [39, 40], Parkinson disease [41], falls of elderly individuals [42], and certain respiratory and hematological disorders [37, 43–45] may arise from bifurcations. Certain conditions that are considered in clinical psychology as disorders such as mood disorders [46–51] and, in particular, schizophrenia [52] may emerge via bifurcations as well (for short reviews see, e.g., Refs. [36, 53]).

In general, the possible health (or disease) states of an individual may be described by means of an appropriately defined n-dimensional state space. The state of an healthy individual may correspond to an attractor located in the n-dimensional state space [36]. If a bifurcation occurs such that the attractor becomes unstable and an another attractor emerges that corresponds to a condition less favorable for the individual, we may talk about a disease emerging in the individual [36]. Importantly, if such a bifurcation involves a single eigenvalue that changes its sign from a negative to a positive value then there is a single unstable eigenvector $\mathbf{v}_k$ in the n-dimensional state space that dominates at least initially the disease dynamics (see Chap. 2). This order parameter $\mathbf{v}_k$ and its amplitude A_k characterize the initial-stage progression of the disease under consideration.

It has been suggested that this bifurcation perspective does not only apply to the emergence of a disease but also to the decline of a disease due to treatment [36, 49, 54–56]. Accordingly, a treatment (or therapy) bifurcation takes place that brings the individual from the disease state either back to the original healthy state or to a new kind of healthy state [36]. Several clinical observations have been reported consistent with this bifurcation perspective of treatment and therapy of patients [57–61].

The instability-induced outbreaks of infectious diseases described in the previous section that lead to epidemics in populations may be seen as counterparts to bifurcations of health or disease states that lead to the emergence of diseases in individuals. Likewise, the intervention bifurcations described in the previous section that stabilize unstable disease-free fixed points and lead to the subsiding of epidemics in populations may be seen as counterparts to treatment and therapy bifurcations that lead to a disease decline in patients.

8.4.2.2 D1 Systems

In Ref. [36] a classification of humans and animals into five system classes with certain properties has been worked out. This five-system classes scheme allows to put phenomena in a systematic order. The classification scheme is not restricted to individual humans and animals. It applies to all kind of biological and physical system. In particular, it might be applied to epidemics spreading in populations. In the context of the five-system classes scheme, the intervention scenario (scenario B, see Fig. 8.5) can be considered as a phenomenon that typical occurs in D1 systems. In general, D1 systems are composed of structure, state, and environment. Figure 8.6 illustrates these components in the context of the intervention scenario. In this context, structure can be described in terms of parameters such as the maximal

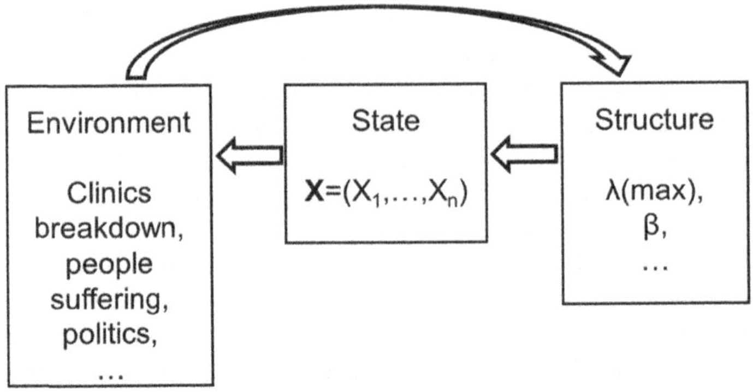

Fig. 8.6 Components of D1 systems and interactions between those components when applied to describe intervention scenarios of epidemics

eigenvalue or effective contact rate parameters. State is given in terms of the disease (or health) state of a population. Environment covers elements that do not occur in the state vector. For example, environment addresses the breakdown of clinics and the health care system due to dramatically rising numbers of COVID-19 patients, the suffering of the population due to the deaths associated with the infectious disease under consideration, and the reaction of politicians and lawmakers to those events. For a D1 system the evolution of the state is determined by the structure (as indicated by the arrow). Furthermore, the state affects the environment (as indicated again by the arrow). Importantly, the environmental circumstances affect the structure. In the context of the intervention scenario, this implies that the implementation of intervention measures affects parameters such as the maximal eigenvalue or effective contact rate parameters. Due to the such structural changes, the evolution of the disease (or health) state changes. D1 systems exhibit a circular causality loop. More sophisticated intervention scenarios may be discussed using higher system classes such as D2 and D3 systems or E systems (for details see Ref. [36]). Finally note that the no intervention scenario (scenario A, see Fig. 8.5) corresponds within the five-system classes scheme to a phenomenon occurring in an A1 system. Such A1 systems form the lowest (or most basic) class of systems. In classical mechanics, an example of an A1 system is a ball that rolls on flat surface.

8.4.3 Three-Stage Epidemic Waves

Interventions measures are assumed to affect virus transmission parameters. For example, with respect to the SEIR model (8.3) such measures may lower the effective contact rates β_E and β_I like $\beta_E = \beta_{E,B} < \beta_{E,A}$, $\beta_I = \beta_{I,B} < \beta_{I,A}$ and increase the diagnosis rate γ_C like $\gamma_C = \gamma_{C,B} > \gamma_{C,A}$, where the subindices "A" and "B" refer to the parameter values in the absence (scenario A) and presence (scenario B) of inter-

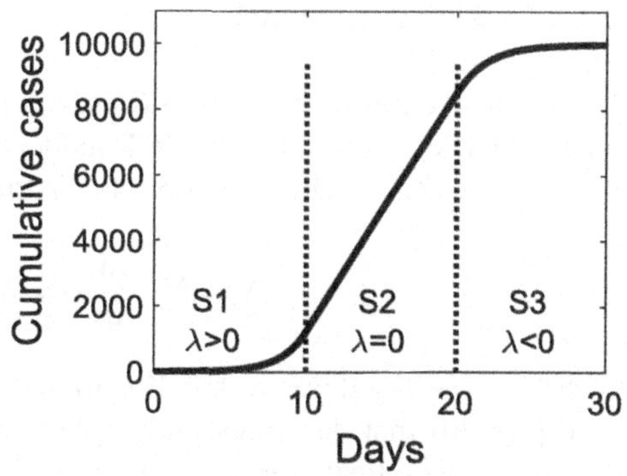

Fig. 8.7 Three-stage model of an epidemic wave illustrated for the evolution of cumulative diagnosed infected cases over time (as suggested, e.g., in Refs. [8, 25])

ventions, respectively. In the context of the COVID-19 pandemic, effective contact rates may decay, for example, due to measures like physical distancing and wearing face masks and the diagnoses rate may increase, for example, due to increased COVID-19 testing.

Let us pursue a parsimony approach. Accordingly, let us focus on the fact that the stability of a fixed point is determined by its eigenvalues. Therefore, while it is worth while to study how interventions affect specific model parameters such as β_E, β_I, and γ_C the key issue is that in the end they change eigenvalues. This can be shown explicitly. For example, Eq. (8.7) illustrates that the eigenvalues of the SEIR model (8.3) depend on the model parameters β_E, β_I, and γ_C. Consequently, intervention measures that change the model parameters β_E, β_I, and γ_C also change the eigenvalues λ_1 and λ_2 (except for rare situations in which the changes in β_E, β_I, and γ_C cancel out on the level of the eigenvalues).

When focusing on eigenvalues, then the evolution of an epidemic wave involves three stages that can be defined using a combined nonlinear physics and data-driven approach. Figure 8.7 illustrates the stages thus obtained. From a nonlinear physics perspective, stage 1 is characterized by an unstable fixed point with an positive eigenvalue ($\lambda > 0$). Stage 2 is the bifurcation stage at which the eigenvalue becomes zero ($\lambda = 0$). Stage 3 is the stage for which the disease-free fixed point has been stabilized. That is, the eigenvalue λ that was positive in stage 1, becomes negative in stage 3 ($\lambda < 0$) at least when considering a suitably defined subspace D^+ (see Chaps. 5, 6, and 7 for examples of D^+). From a data-driven perspective, stage 1 is defined by an approximately exponential increase (i.e., a first bend) of the cumulative diagnosed cases C as function of time t. Stage 2 is defined as an approximately linear increase of C, which is related to the fact that $\lambda = 0$ holds (see Sect. 4.2.4 and the worked out examples below). Finally, stage 3 is defined as the convergence towards the low cases/low deaths stationary state. Consequently, in stage 3 the cumulative cases as a function of time exhibit a second bend with $\mathrm{d}C/\mathrm{d}t > 0$ but $\mathrm{d}^2C/\mathrm{d}t^2 < 0$. In short, the stages of an epidemic wave from a data-driven perspective are given by a stage of exponential increase (S1), a stage of linear increase (S2), and a stage of de-accelerating increases (S3).

8.4.3.1 SIR-Type Model

In order to exemplify the three-stage scenario of epidemic waves by means of a SIR-type model, let us consider the SIQR model defined by Eq. (4.82) with the cumulative disease cases $C(t)$ defined by Eq. (4.83). Putting $S = N$, the linearized model reads

$$S = N \ , \ \frac{\mathrm{d}}{\mathrm{d}t} I = \lambda I \ , \ \frac{\mathrm{d}}{\mathrm{d}t} C = bI \tag{8.27}$$

with $\lambda = \beta - \gamma$ and $\gamma = a + b$. Equation (8.27) exhibits basically the same structure as Eq. (8.20) that describes the single λ approach of the SEIR model (8.3). For $\lambda \neq 0$ the analytical solutions of Eq. (8.27) read $I(t) = I(t_0)\exp\{\lambda(t - t_0)\}$ and $C(t) = C(t_0) + V[\exp\{\lambda(t - t_0)\} - 1]$ with the slope parameter $V = bI(t_0)/\lambda$, (see Eq. (4.86)). Let $[t_0, t_1]$, $[t_1, t_2]$, and $[t_2, t_3]$ denote the consecutive time intervals of the three stages S1, S2, and S3 under consideration. Then, from Eq. (8.27) and the requirements $\lambda > 0$, $\lambda = 0$, $\lambda < 0$ for the three stages it follows that (see also Ref. [12])

$$\begin{aligned}
&\mathrm{S1} \ , \ t \in [t_0, t_1] \ , \ \lambda = \lambda_{S1} > 0 \ : \\
&\quad C(t) = C(t_0) + V_{S1} \left(\exp\{\lambda_{S1}(t - t_0)\} - 1\right) \ , && (8.28)\\
&\mathrm{S2} \ , \ t \in [t_1, t_2] \ , \ \lambda = \lambda_{S2} = 0 \ : \ C(t) = C(t_1) + V_{S2}(t - t_1) \ , && (8.29)\\
&\mathrm{S3}, \ t \in [t_2, t_3] \ , \ \lambda = \lambda_{S3} < 0 \ : \\
&\quad C(t) = C(t_2) + V_{S3} \left(1 - \exp\{\lambda_{S3}(t - t_2)\}\right) && (8.30)
\end{aligned}$$

with $V_{S1} = b_{S1} I(t_0)/\lambda_{S1}$, $V_{S2} = b_{S2} I(t_1)$, and $V_{S3} = b_{S3} I(t_2)/|\lambda_{S3}|$. The notation is such that $\lambda_{S1}, \lambda_{S2}, \lambda_{S3}$ describe the eigenvalue λ in the stages S1, S2, S3. Likewise, b_{S1}, b_{S2}, b_{S3} and V_{S1}, V_{S2}, V_{S3} describe the parameters b and V, respectively, in the stages S1, S2, S3. In particular, Eq. (8.29) demonstrates the linear increase of cases in stage 2 as illustrated in Fig. 8.7. Equations (8.28) and (8.30) have been used in the context of a two-stage approach (skipping the stage $\lambda = 0$) to describe COVID-19 outbreaks in the year 2020 in Canada, Australia, USA, and several European countries [12]. Some results of the study will be presented in Sect. 8.4.4.

8.4.3.2 Single λ Approach and SEIR Model

As discussed in Sect. 8.3.2, the single λ approach of the SEIR model (8.3) yields the evolution equation (8.20) that is equivalent to Eq. (8.27) for E and I and, in particular, for the cumulative cases $C(t)$. Consequently, for the three stages with $\lambda_1 > 0$, $\lambda_1 = 0$, and $\lambda_1 < 0$ of an epidemic wave the cumulative cases $C(t)$ evolve like [25]

$$\text{S1}\ ,\ t \in [t_0, t_1]\ ,\ \lambda_1 = \lambda_{1,S1} > 0\ :$$
$$C(t) = C(t_0) + V_{S1}\left(\exp\{\lambda_{1,S1}(t - t_0)\} - 1\right)\ , \tag{8.31}$$
$$\text{S2}\ ,\ t \in [t_1, t_2]\ ,\ \lambda_1 = \lambda_{1,S2} = 0\ :\ C(t) = C(t_1) + V_{S2}(t - t_1)\ , \tag{8.32}$$
$$\text{S3},\ t \in [t_2, t_3]\ ,\ \lambda_1 = \lambda_{1,S3} < 0\ :$$
$$C(t) = C(t_2) + V_{S3}\left(1 - \exp\{\lambda_{1,S3}(t - t_2)\}\right)\ . \tag{8.33}$$

with $V_{S1} = \gamma_{C,S1} I(t_0)/\lambda_{1,S1}$, $V_{S2} = \gamma_{C,S2} I(t_1)$, and $V_{S3} = \gamma_{C,S3} I(t_2)/|\lambda_{1,S3}|$. The stage equations (8.31)–(8.33) describe the sigmoid pattern shown in Fig. 8.7. The eigenvalue λ shown in Fig. 8.7 corresponds to λ_1. Equations (8.31)–(8.33) have been used in a study on 2020 COVID-19 first-waves in Europe that will be reviewed in Sect. 8.4.4.

8.4.3.3 Two λ Approach and SEIR Model

As discussed in Sect. 8.3.2, the two λ approach of the SEIR model (8.3) yields the evolution equation for $I(t)$ given by Eq. (8.23) and the elementary solutions of I, E, and C in terms of exponential functions as listed in Eqs. (8.24) and (8.25) assuming that $\lambda_1 \neq 0$ and $\lambda_2 \neq 0$ holds. In analogy to those elementary solutions, analytical expressions for $C(t)$ can be obtained for all three stages. In particular, for stage 2 that exhibits $\lambda_1 = 0$ but $\lambda_2 < 0$ an analytical solution can be obtained. The solutions read [8]

$$\text{S1}\ ,\ t \in [t_0, t_1]\ ,\ \lambda_1 > 0 \wedge \lambda_2 < 0\ :$$
$$C(t) = C(t_0) + V_{1,S1}[\exp\{\lambda_1(t - t_0)\} - 1] + V_{2,S1}[1 - \exp\{\lambda_2(t - t_0)\}]\ , \tag{8.34}$$
$$\text{S2}\ ,\ t \in [t_1, t_2]\ ,\ \lambda_1 = 0 \wedge \lambda_2 < 0\ :$$
$$C(t) = C(t_1) + V_{1,S2}(t - t_1) + V_{2,S2}[1 - \exp\{\lambda_2(t - t_1)\}]\ , \tag{8.35}$$
$$\text{S3}\ ,\ t \in [t_1, t_2]\ ,\ \lambda_1 < 0 \wedge \lambda_2 < 0\ :$$
$$C(t) = C(t_2) + V_{1,S3}[1 - \exp\{\lambda_1(t - t_2)\}] + V_{2,S3}[1 - \exp\{\lambda_2(t - t_2)\}]\ . \tag{8.36}$$

The stage-1 equation (8.34) states that $C(t)$ increases approximately exponential due to the λ_1 term. The exponential term $U_2 = V_{2,S1}[1 - \exp\{\lambda_2(t - t_0)\}]$ involving λ_2 increase in magnitude over time from $U_2 = 0$ at $t = t_0$ to $U_2 = V_{2,S1}$ for $t \to \infty$. That is, if $t - t_0$ becomes larger relative to the time constant $\tau_2 = 1/\lambda_2$, then the second term becomes constant and the first term $U_1 = V_{1,S1}[\exp\{\lambda_1(t - t_0)\} - 1]$ describes an exponential increase (assuming $V_{1,S1} > 0$). Equation (8.34) describes the first bend or exponential increase of $C(t)$ shown in Fig. 8.7. Consistent with Fig. 8.7, the stage-2 equation (8.35) states that the increase of C is approximately linear provided that the linear term dominates the exponential term, that is, if $V_{1,S2} \gg |V_{2,S2}|$ holds, where it is assumed that $V_{1,S2}$ is positive while $V_{2,S2}$ can be positive or negative. Finally, Eq. (8.36) describes that $C(t)$ converges to the stationary value $C(\infty) = C(t_2) + V_{1,S3} + V_{2,S3}$. For the SEIR model $\lambda_2 > \lambda_1$ holds in general. If, in addition, there is a time-scale separation such that $|\lambda_1|$ is much smaller than $|\lambda_2|$, then (when assuming $V_{1,S3} > 0$) the eigenvalue λ_1 determines the speed or time scale of the subsiding of the epidemic (see also Sect. 6.1.3). As far as Fig. 8.7 is concerned, under appropriate circumstances, Eqs. (8.34)–(8.36) reproduce the sigmoid pattern shown in Fig. 8.7. In this context, the eigenvalue λ listed in Fig. 8.7 corresponds to λ_1. Equations (8.34)–(8.36) have been used in a study on the COVID-19 first-wave of 2020 in Thailand that will be reviewed in Sect. 8.4.5.

8.4.4 *COVID-19 First-Waves of 2020 in Europe: Stabilization Bifurcations and the Sign Switching Phenomenon*

8.4.4.1 Three-Stage Approach

In what follows, a study on the COVID-19 epidemics in 20 European countries [25] will be reviewed. The study considered the following countries: Austria, Belgium, Bulgaria, Czech Republic, Denmark, Finland, France, Germany, Greece, Hungary, Italy, Ireland, Netherlands, Poland, Portugal, Romania, Slovakia, Spain, Sweden, and United Kingdom. Data from January 1, 2020, to June 15, 2020 were analyzed. During the first-wave epidemic from January to June 2020 in Europe, countries of the European Union closed their borders in order to reduce the spread of SARS-CoV-2 infections. The end date, June 15, of the study period was chosen because around June 15, 2020 the countries of the European Union started to open their borders within the European Union [62]. Daily, cumulative reported COVID-19 cases were used as reported by the Johns Hopkins University and listed on the COVID-19 tracker website [63]. The reported COVID-19 cumulative cases were fitted to Eqs. (8.31)–(8.33) using a standard nonlinear fitting algorithm (for stages 1 and 3) and linear regression analysis (for stage 2). In doing so, the eigenvalues λ_{S1} and λ_{S3} were estimated and corresponding confidence intervals (CIs) were obtained. If a confidence interval of a stage 3 eigenvalue λ_{S3} included zero, then the eigenvalue λ_{S3} was not significantly different from zero. In such as case, the country was not able to stabilize in stage 3 the disease-free low cases/low deaths state (at least up to June 15, 2020). For one country, Bulgaria, a negative value for λ_{S3} could not be determined because the COVID-19 case trajectory did not exhibit the sigmoid pattern shown in Fig. 8.7. Rather, the trajectory followed a three-stage pattern with an exponential increase (S1), linear increase (S2), and another exponential or at least nonlinear increase (S3). Therefore, for this country, Eq. (8.33) was replaced by: $\lambda_1 = \lambda_{S3} > 0 : C(t) = C(t_2) + V_{S3}(\exp\{\lambda_{S3}(t - t_2)\} - 1)$.

For all countries, the time points t_0, t_1, t_2 of the stages were determined as follows. The beginning of stage 1 (t_0) was defined as the time point for which there was at least 1 new reported infection on every day of the two weeks period following t_0. That is, t_0 was the first time point for which COVID-19 cases increased monotonically on every day at least for a two weeks period. The stage boundaries t_1 and t_2 were varied

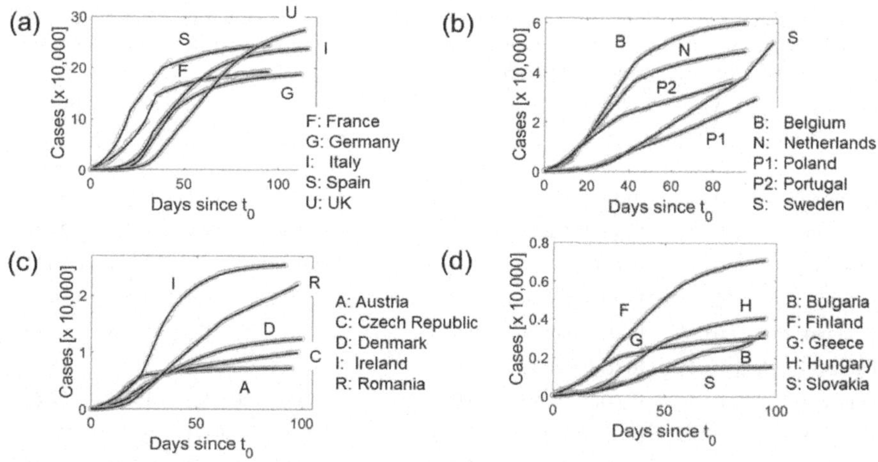

Fig. 8.8 Cumulative confirmed COVID-19 cases observed in 20 European countries during the period from January 1, 2020 to June 15, 2020 (gray circles) and three-stage model solutions $C(t)$ (solid black lines) defined by Eqs. (8.31), (8.32), and (8.33) of the single-λ 2β SEIR modeling approach. Panels (**a**), (**b**), (**c**), and (**d**) show groups of five countries. Within each group countries showed similar numbers of cases

under the constraint $t_0 < t_1 < t_2$. For each pair t_1, t_2 the model was fitted to the data of the country under consideration. The boundaries values t_1 and t_2 were selected as best-fit parameters that produced the best fit to the data as measured in terms of the root-mean-square error.

Figure 8.8 present the cumulative COVID-19 cases as functions of time for the 20 European countries (gray circles) and the model fits $C(t)$ (solid black lines) as obtained from Eqs. (8.33)–(8.35). As can be seen in Fig. 8.8, while for all countries the study period was January 1 to June 15, the case trajectories differ in length. The reason for this is that in general the parameter t_0 varied across countries. In order to easy the presentation, countries that showed similar numbers of COVID-19 cases as of June 15, 2020 were placed into the same group. In total, four country groups were formed.

Panel (a) shows the five countries with the highest numbers of confirmed COVID-19 infections up to June 15, 2020. The total confirmed infections ranged from 200,000 (France and Germany) to 270,000 (UK). Panel (b) presents the second country group for which diagnosed COVID-19 cases ranged from 30,000 (Poland) to 60,000 (Finland) cases. Panel (c) presents the cumulative case trajectories for the third group of countries which observed on June 15 between 8,000 (Austria) to 25,000 (Ireland) cases. Finally, panel (d) shows the remaining five countries that reported on June 15 between 2,000 (Slovakia) and 7,000 (Finland) total infections. By visual inspection, the model fits $C(t)$ (solid lines) were able to captured the characteristic sigmoid patterns of observed COVID-19 cases in all countries. Importantly, the epidemics of 15 of the 20 investigated countries exhibited the third stages with clearly de-accelerating

Table 8.2 Results of the three-stage analysis conducted in Ref. [25]: countries, eigenvalues, confidence intervals, and time constants for stages 1 and 3 are shown ("pos." and "n.s." stand for positive and not statistically significant, respectively)

Country	λ_{max} [1/d]		λ_{max} [1/d]		τ [d]	
	Stage 1	CI	Stage 3	CI	Stage 1	Stage 3
Austria	0.15	[0.14, 0.17]	−0.035	[−0.038, −0.032]	6.62	28.46
Belgium	0.12	[0.11, 0.14]	−0.041	[−0.042, −0.039]	8.10	24.66
Bulgaria	0.014	[0.010, 0.018]	0.10 pos.	[0.09, 0.11]	72.41	10.27
Czech R.	0.14	[0.13, 0.15]	−0.012	[−0.013, −0.010]	7.28	86.95
Denmark	0.054	[0.050, 0.059]	−0.040	[−0.041, −0.039]	18.38	24.95
Finland	0.046	[0.042, 0.051]	−0.046	[−0.048, −0.043]	21.51	21.83
France	0.051	[0.043, 0.059]	−0.035	[−0.038, −0.032]	19.75	28.67
Germany	0.18	[0.17, 0.19]	−0.045	[−0.046, −0.044]	5.47	22.31
Greece	0.04	[0.03, 0.05]	−0.027	[−0.029, −0.024]	25.21	37.65
Hungary	0.082	[0.075, 0.089]	−0.034	[−0.036, −0.033]	12.23	29.16
Ireland	0.064	[0.058, 0.069]	−0.063	[−0.064, −0.061]	15.69	15.98
Italy	0.124	[0.117, 0.131]	−0.045	[−0.046, −0.044]	8.06	22.19
Netherlands	0.128	[0.123, 0.133]	−0.028	[−0.030, −0.026]	7.83	35.21
Poland	0.115	[0.109, 0.120]	−0.0006 n.s.	[−0.005, 0.003]	8.72	N.A.
Portugal	0.086	[0.075, 0.097]	−0.0002 n.s.	[−0.002, 0.002]	11.64	N.A.
Romania	0.15	[0.14, 0.16]	−0.0007 n.s.	[−0.003, 0.002]	6.79	N.A.
Slovakia	0.045	[0.040, 0.049]	−0.032	[−0.036, −0.028]	22.42	31.52
Spain	0.089	[0.078, 0.099]	−0.030	[−0.033, −0.027]	11.29	33.43
Sweden	0.081	[0.077, 0.084]	−0.011 n.s.	[−0.034, 0.011]	12.41	N.A.
UK	0.15	[0.14, 0.16]	−0.0365	[−0.0371, 0.0358]	6.66	27.41

increasing cumulative cases. The remaining five countries did not show a third stage with a clearly subsiding epidemic. These countries were Poland, Sweden, and Portugal (panel (b)), Romania (panel (c)), and Bulgaria (panel (d)). The four countries Poland, Sweden, Portugal, and Romania exhibited a more or less linear increase of the cumulative COVID-19 cases in the weeks preceding June 15 (panels (b) and (c)). As mentioned above, Bulgaria showed an exponential or nonlinear increase in this period (panel (d)).

Table 8.2 presents for all countries the eigenvalues $\lambda_{max} = \lambda_1$ occurring in Eqs. (8.31) and (8.33) for stage 1 and 3, respectively, and their confidence intervals. As mentioned above, if for a stage 3 eigenvalue the confidence interval included zero, then the eigenvalue was regarded to be not different from zero in a statistically significant sense. This is indicated "n.s." in Table 8.2. Table 8.2 also reports the time constants $\tau = 1/\lambda_1$ of the eigenvalues. A small time constant τ (i.e., large eigenvalue) in stage 1 implies a fast exponential increase of infections. A small time constant τ in stage 3 implies a relatively strong stabilization of the desirable disease-free low cases/low deaths state and a relatively fast disease dynamics toward that fixed point. Note that the stage-3 time constants τ are not shown for Poland, Portugal, Romania, and Sweden because (according to the model-based analysis) these countries exhibited in stage 3 linearly increasing epidemics rather than subsiding epidemics (see the discussion below).

Table 8.2 illustrates for 15 countries the stage 2 bifurcations that were presumably induced by intervention measures and caused the epidemics in those countries to subside. These bifurcations can be revealed by comparing the signs of the eigenvalues of stages 1 and 3 as shown in columns 2 and 4 of Table 8.2. It was found that for all countries the eigenvalues λ_1 were positive in stage 1, as expected. For 15 countries λ_1 became negative in stage 3, consistent with the notion that in those countries intervention measures caused a stabilization of the low cases/low deaths state with $I_{st} = E_{st} = 0$. These countries were Austria, Belgium, Czech Republic, Denmark, Finland, France, Germany, Greece, Hungary, Italy, Ireland, Netherlands, Slovakia, Spain, and the United Kingdom. Table 8.2 illustrates explicitly that the sign of the leading (or maximal) eigenvalue λ_1 switched during the first-wave epidemics for those countries. In doing so, the results reported in Table 8.2 illustrate the bifurcation scenario and sign switching phenomenon that underlies under certain circumstances epidemic waves as discussed in Sect. 8.4.1 and 8.4.3. The eigenvalues of those countries reported in Table 8.2 and the corresponding COVID-19 trajectories shown Fig. 8.8 exemplify how the schematic Fig. 8.7 looks like in applications.

As far as the five remaining countries are concerned, Bulgaria showed a statistically significant positive eigenvalue λ_1 in stage 3. Furthermore, Poland, Portugal, and Romania exhibited negative eigenvalues λ_1 in stage 3 that were much smaller in the amount than those of the remaining countries. The confidence intervals of those eigenvalues indicate that they were not statistically significant different from zero. Accordingly, during the days and weeks around June 15 the COVID-19 pandemic still followed a linear increase in those countries. The populations of those countries were still in the bifurcation stage 2. Sweden exhibited a negative stage 3 eigenvalue $\lambda_{1,S3}$ that was small but in the same range as the eigenvalue $\lambda_{1,S3}$ of the Czech Republic (-0.011/d versus -0.012/d). However, while the eigenvalue $\lambda_{1,S3}$ of the epidemic in the Czech Republic was statistically significant different from zero, this was not the case for the eigenvalue $\lambda_{1,S3}$ of the COVID-19 epidemic in Sweden.

8.4.4.2 Two-Stage Approach by Pedersen and Meneghini

Pedersen and Meneghini [12] used a two-stage approach based on the SIQR model (4.82) to describe early COVID-19 outbreaks of 2020 in Australia, Canada, USA, and several European countries. To this end, cumulative COVID-19 cases reported in the period from February 1 to April 2, 2020 were analyzed. The data were fitted to Eqs. (8.28) and (8.30) like

$$\begin{aligned} &\text{S1}\ , t \in [t_0, t_1]\ ,\ \lambda = \lambda_{S1} > 0\ : \\ &\quad C(t) = C(t_0) + V_{S1}\,(\exp\{\lambda_{S1}(t - t_0)\} - 1)\ , \end{aligned} \tag{8.37}$$

$$\begin{aligned} &\text{S2}\ , t \in [t_1, t_2]\ ,\ \lambda = \lambda_{S2}\ : \\ &\quad C(t) = C(t_1) + W\,(\exp\{\lambda_{S2}(t - t_1)\} - 1) \\ &\quad = C(t_1) + V_{S2}\,(1 - \exp\{\lambda_{S2}(t - t_1)\}) \end{aligned} \tag{8.38}$$

with $W = -V_{S2}$. For each country or region, the initial day t_0 was given by the first day in February 2020 when the number of confirmed COVID-19 cases exceeded 50 in the respective country or region. The end day t_2 was April 2, 2020, for all countries and regions. For every given country or region the stage boundary t_1 was varied to find the optimal time point t_1 for which Eqs. (8.37) and (8.38) produced the best fit to the country or regional data. The overall aim of the study was to show that the eigenvalue λ decayed from the first stage to the second stage due to the impact of intervention measures. No specific hypothesis about the sign of λ_{S2} was made. Table 8.3 reports the eigenvalues obtained in the study. It was observed that the eigenvalues of all countries and regions either decreased and stayed positive or switch their signs and became negative. For Austria, Switzerland, and the region of British Columbia (Canada) the sign of the eigenvalue switched and values of λ_{S2} in the range of -0.03/d to -0.01/d was found. For the region of New South Wales (Australia) the eigenvalue switched the sign and a negative stage-2 eigenvalue of $\lambda_{S2} = -0.08$/d that was relatively large in the amount was observed. In view of the relatively short data set in stage 2 (i.e., short intervals $[t_1, t_2]$), in Ref. [12] it was questioned whether the negative values in the range of -0.03/d to -0.01/d were statistically different from zero. In contrast the eigenvalue of $\lambda_{S2} = -0.08$/d was identified as statistically different from zero. In summary, the eigenvalue analysis suggests that for those four countries and regions the disease-free fixed point was about to become stable (i.e., the populations were at their bifurcation points with $\lambda_{S2} = 0$) or was stabilized ($\lambda_{S2} < 0$) at the beginning of April, 2020.

It is difficult to compare the results of the three-stage study [25] discussed above and the two-stage study by Pedersen and Meneghini [12] because they used a different number of stages. Importantly, the three-stage study examined a longer period, namely, from the beginning of the pandemic to June 15. That is, the three-stage study looked at a two-months longer period. Comparing the results reported in Tables 8.2 and 8.3, the study with the longer observation period suggests that the COVID-19 dynamics in France, Germany, Italy, Spain, and UK exhibited a stable disease-free fixed point by the end of June 2020, while the study with the shorter observation period suggests that in those countries this stabilization was not yet achieved by April 2, 2020. Both studies suggest that in some countries and regions of the world the subsiding of first-wave COVID-19 epidemics was due to bifurcations that changed positive eigenvalues of disease-free fixed points towards critical values of zero and, eventually, switched them to negative ones. These bifurcations presumably were caused by the implementation of intervention measures.

Table 8.3 Results of the two-stage analysis conducted in Ref. [12]: regions under consideration and eigenvalues λ_{S1} and λ_{S2} of stages 1 and 2 are listed

Country/Region/City	λ_{S1} [1/d]	λ_{S2} [1/d]
Austria	0.250	−0.013
France	0.252	0.108
Germany	0.248	0.043
Italy	0.210	0.032
Spain	0.316	0.082
Switzerland	0.225	−0.022
UK	0.213	0.107
British Columbia (Canada)	0.347	−0.032
New York City (USA)	0.505	0.061
New South Wales (Australia)	0.217	−0.080

8.4.5 *First-Wave COVID-19 Epidemic in Thailand, 2020: EI Order Parameter and Its Remnant*

When COVID-19 cases were first reported in China in early 2020, Thailand was one of the first countries that reported patients with COVID-19 outside of China [64]. Despite being one of the first countries outside China that was confronted with the disease, in contrast to many countries around the globe in which COVID-19 spread out quickly during the year 2020 and caused high infections among their populations and a large number of COVID-19 associated deaths, the Thai government and the Thai population managed to stop the COVID-19 epidemic entirely in their country during the first three quarters of 2020. That is, Thailand belongs to the few countries in which the COVID-19 epidemic subsided completely in the year 2020. During the summer months June, July, August, and September no local transmission of COVID-19 were observed in Thailand [65, 66].

The COVID-19 first-wave in Thailand was studied by means of the SEIR model (8.3) using the two λ approach [8]. In what follows the study will be briefly reviewed. As such Fig. 8.7 captures the basic notion of the three-stage model used in Ref. [8]. However, the two λ approach allows to determine the pair of eigenvectors $\mathbf{v}_1$ and $\mathbf{v}_2$. In particular, the approach allows to determine the eigenvector $\mathbf{v}_1$ related to the maximal eigenvalue λ_1 that by theoretical reasonings should play the key role for rise and decay of an epidemic wave. Taking the role of $\mathbf{v}_1$ into account, Fig. 8.7 may be revised as shown in Fig. 8.9. Figure 8.9 presents schematically the eigenvector $\mathbf{v}_1$ throughout the assumed three stages of the epidemic wave. λ_1 is shown as well and the sigmoid shape of the wave is presented. The vertical axis shows cumulative cases as percentage values of the cumulative cases observed at the end of the wave. Just as in Fig. 8.7, the stages S1, S2, and S3 are characterized by an exponential increase, a linear increase, and a de-accelerating increases of cases. The eigenvalues

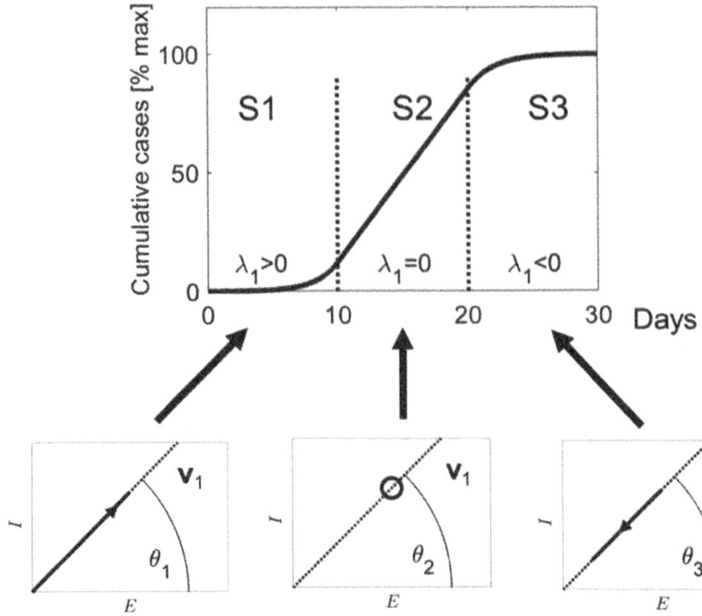

Fig. 8.9 Three-stage model of an epidemic wave in the context of a two-λ approach that addresses the eigenvector $\mathbf{v}_1$ of the maximal eigenvalue λ_1. The top part shows the evolution of the cumulative confirmed (or diagnosed) cases across the three stages. The bottom part shows $\mathbf{v}_1$ in the E-I plane and the disease state dynamics along $\mathbf{v}_1$ for the three stages

λ_1 that produce such three-stage pattern are shown as $\lambda_1 > 0$, $\lambda_1 = 0$, and $\lambda_1 < 0$, respectively.

The inserts on the bottom of Fig. 8.9 describe schematically the eigenvector $\mathbf{v}_1$ in the E-I plane for the three stages. The eigenvector forms an angle θ with respect to the E axis. Since $\mathbf{v}_1$ is the unstable eigenvector and should dominate the disease dynamics in stage 1, during stage 1 the dynamics in the E-I plane should evolve away from the disease-free fixed point along the eigenvector $\mathbf{v}_1$. The vector $\mathbf{v}_1$ is the EI order parameter of the disease-free state. In stage 2, the eigenvector $\mathbf{v}_1$ points towards a certain location in the E-I plane around which the state is almost constant (because $\mathrm{d}A_1/\mathrm{d}t \approx 0$). In stage 3, under the assumption that $\mathbf{v}_1$ (the remnant of the order parameter) makes the major contribution such that Eq. (8.19) holds, the dynamics $\mathbf{X}^+(t)$ should approach the disease-free state along $\mathbf{v}_1$ (as indicated by the arrow).

Let us compare stage 1 and 3. For stage 1, since $\lambda_1 > 0$ and $\lambda_2 < 0$ holds, it is expected that the direction $\mathbf{v}_1$ captures most of the dynamics (at least after the transient period with time constant $\tau_2 = 1/\lambda_2$) such that $\mathbf{X}^+(t) \approx A_1(t)\mathbf{v}_1$ holds and $C(t) \approx C(t_0) + V_{1,S1}[\exp\{\lambda_1(t - t_0) - 1]$ (see the discussion in Sect. 8.4.3 about Eq. (8.36))). That is, the schematic shown in the insert of stage 1 should hold in general. In contrast, as discussed in Sect. 6.1.3, the schematic shown in the insert of stage 3 holds only under the assumption that there is a time scale separation with $\tau_2 = 1/|\lambda_2|$ much shorter than $\tau_1 = 1/|\lambda_1|$. Therefore, for the COVID-19 first-wave in Thailand it must be checked to what extent the stage-3 dynamics in the E-I subspace evolves along the remnant vector $\mathbf{v}_1$ of the order parameter.

In Ref. [8] the cumulative COVID-19 infections as reported from Thailand in the data repository of the Johns Hopkins University and published on the COVID-19 tracker website [63] were analyzed. The period from January 1 to May 31, 2020 was considered. The beginning of stage 1 (t_0) was defined as the first day that showed new daily infections and for which in the consecutive seven days daily new infections were reported. This date t_0 fell on March 10, 2020. The final time point t_3 was fixed as May 31. This end day was selected because the last infection related to the Spring 2020 COVID-19 outbreak due to community transmission was reported on May 26. All COVID-19 cases that were reported after May 26, 2020, until end of September 2020 were imported cases from outside of Thailand [65, 66]. The time points t_1 and t_2 were varied (just as in the study reviewed in Sect. 8.4.4) in order to find optimal time points for the beginning of stages 2 and 3.

The model parameters α, γ_E, γ_I, γ_d were fixed. As suggested in Refs. [18, 67], the parameter α was taken as the reciprocal of the incubation time T: $\alpha = 1/T$. Incubation times T for COVID-19 in the range from 5 to 9 days are typically reported [18, 20, 30, 67, 68]. As in Pang et al. [18] $T_I = 5.9$ days ($\Rightarrow \alpha = 0.17$/d) was used. COVID-19 recovery rates vary considerably in the literature (i.e., values from 0.03/d to 0.3/d can be found in Refs. [13, 18, 28, 32, 69]). Again, followed Pang et al. [18] $\gamma_E = \gamma_I = 0.1$/d was used. Finally, again motivated by the study by Pang et al. [18], a diagnoses rate of $\gamma_d = 0.5$/d was used (for details see Ref. [8]).

An explorative analysis of the COVID-19 data from Thailand showed that the single λ approach given in terms of Eqs. (8.31) to (8.33) produced reliable estimates for λ_1 (just as in the study reviewed in Sect. 8.4.4). In contrast, when applying the two λ approach given in terms of Eqs. (8.34) to (8.36) reliable estimates for λ_2 could not be obtained. Consequently, the approach to compute β_E and β_I from independent estimates of λ_1 and λ_2 could not be pursued.

Therefore, an ordinary two-step fitting process was used. In a first step, λ_1 and V_1 were estimated using Eqs. (8.31) and (8.33) for stages 1 and 3. $V_{1,S2}$ was estimated using Eq. (8.32). β_E, β_I were then computed from the estimates of λ_1 using an addition theoretical assumption about the relationship between β_E and β_I. To this end, an overall reference value for β_E was taken from the literature. More precisely, the stage-2 effective contact rate $\beta_{E,S2} = 0.2$/d was used in line with effective contact rates reported in Refs. [18, 28, 32, 70, 71] (for details see Ref. [8] again). Given $\beta_{E,S2}$ the corresponding parameter $\beta_{I,S2}$ was obtained by solving $\beta_{tot} = \beta_{tot,crit}$ (see Eq. (8.14)). It was then assumed that the effective contact rates approximately showed a fixed ratio in all three stages. That is, it was assumed that the ratio $\beta_I/\beta_E = r$ remained constant across the epidemic wave. The precise value for r was computed from $r = \beta_{I,S2}/\beta_{E,S2}$. Given λ_1 in all three stages and the fixed parameters α, γ_I, γ_E, γ_C, and r, the effective contact rates β_E and $\beta_I = r\beta_E$ were computed from Eq. (8.7) by solving Eq. (8.7) numerically for β_E. As a by-product, the eigenvalue λ_2 could be computed as well from Eq. (8.7). In the second step of the two-step fitting

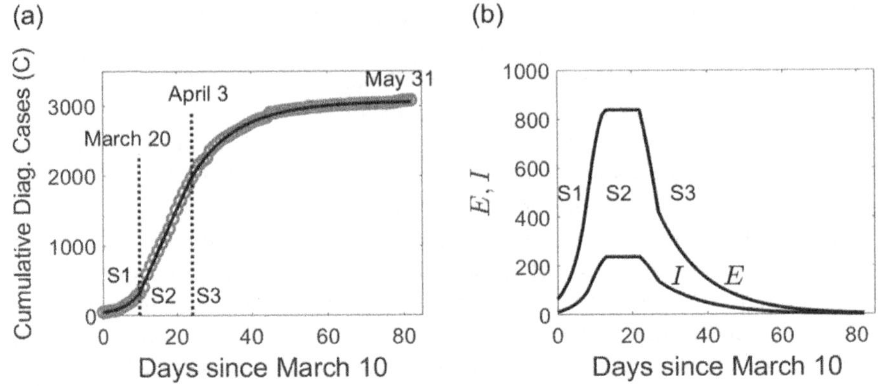

Fig. 8.10 Panel (**a**): Cumulative confirmed COVID-19 cases (gray circles) observed in Thailand during the period from March to May, 2020, and the three-stage model solution $C(t)$ (solid line) computed from Eqs. (8.34), (8.35), and (8.36). Panel (**b**): The corresponding trajectories of $E(t)$ and $I(t)$ are shown across the three stages as computed from Eqs. (8.24) and (8.25)

procedure V_2 was estimated for all three stages using Eqs. (8.34) to (8.36) given the previously obtained values for λ_1 and λ_2.

In Ref. [8] the cumulative cases C were computed from Eqs. (8.34) to (8.36). Panel (a) of Fig. 8.10 shows the cumulative COVID-19 cases (gray circles) reported in the database [63] and the solution $C(t)$ (solid black line) of the SEIR model (8.3) obtained from the analytical expressions (8.34) to (8.36) for the period from t_0 (March 10) to t_3 (May 31). Overall, the three-stage SEIR model captured the sigmoid pattern of the observed COVID-19 cumulative cases with moderate accuracy. The graph of cumulative COVID-19 cases observed in Thailand shows the typical first "up-swinging bend" in the days after March 10 that is captured by the stage-1 equation (8.34). The middle part of the graph can be interpreted as a linear increase that is captured by the stage-2 equation (8.35). Following the linear increase, the graph of observed cumulative COVID-19 cases shows a second "de-accelerating bend" that describes the convergence of the disease (or health) state of the Thai population to an almost stationary plateau at the end of March 2020. This part is captured by the stage-3 equation (8.36). The stage boundaries t_1 and t_2 were found as March 20 and April 3, respectively. As expected, the estimated maximal eigenvalue λ_1 was positive in stage 1 with $\lambda_1 = 0.24$ and negative in stage 3 with $\lambda_1 = -0.08$. That is, the sign switching phenomenon was observed. The epidemic wave was consistent with the bifurcation scenario according to which intervention measures performed by the Thai population triggered a bifurcation and switched the sign of λ_1 such that the disease-free fixed point $I_{st} = E_{st} = 0$ turned from an unstable fixed point to a stable one.

Panel (b) of Fig. 8.10 shows the functions E and I as computed from Eqs. (8.24) and (8.25). Panel (b) demonstrates that in stage 1 the number of exposed and infected individuals followed an exponential increase. This is consistent with the amplitude

Table 8.4 Results of the three-stage analysis of the first-wave COVID-19 epidemic in Thailand from January to May 2020. Eigenvalues λ_1 and λ_2, eigenvectors angles θ_1 and θ_2, and effective contact rates β_E and β_I are shown for the three stages. The ratio $\Delta E/\Delta I$ is reported as well

Stage	Eigenvalues		Eigenvectors		Contact rates		$\Delta E/\Delta I$
	λ_1	λ_2	θ_1	θ_2	β_E	β_I	
Stage 1	0.24/d	−0.70/d	11.4°	−59.0°	0.41/d	0.51/d	5:1
Stage 2	0/d	−0.67/d	15.8°	−67.6°	0.20/d	0.25/d	N.A.
Stage 3	−0.08/d	−0.65/d	18.1°	−72.2°	0.13/d	0.17/d	3:1

equation (8.19) which states that the term $A_1(t)\mathbf{v}_1$ makes the dominant contribution. In other words, it is consistent with the assumption that the λ_1 terms play the dominant role in Eqs. (8.24) and (8.25). In stage 2 the numbers of both populations remained constant, which is again consistent with Eq. (8.19) and assuming that $A_1(t) = A_1(t_1)$ is a constant value because of $\lambda_1 = 0$. That is, for the 2020 first-wave epidemic in Thailand the time-varying dynamics along the eigenvector $\mathbf{v}_2$ during the bifurcation stage 2 did not make an essential contribution. Finally, in stage 3 the E and I functions shown in panel (b) correspond to exponential decaying functions consistent with Eq. (8.19) and $A_1(t) = A_1(t_2)\exp\{\lambda_1(t - t_2)\}$.

In Ref. [8] the eigenvectors $\mathbf{v}_1$ and $\mathbf{v}_2$ were computed from Eq. (8.26) and the corresponding angles θ_1 and θ_2 with the horizontal axis (i.e., the E axis) were determined. Table 8.4 reports the angles θ_1, θ_2 for the stages 1, 2, and 3. The contact rates β_E and β_I are listed there as well. The ratios $\Delta E/\Delta I$ shown in Table 8.4 will be discussed below.

The angle θ_1 of $\mathbf{v}_1$ was positive, while θ_2 was negative. This indicates that $\mathbf{v}_1$ captures segments of the Thai 2020 COVID-19 wave during which E and I increased or decreased simultaneously. In contrast, the direction by $\mathbf{v}_2$ describes situation in which one of the two variables E and I increased while the other decreased. In view of these angles, it follows that $\mathbf{v}_1$ specifies the direction in the E-I subspace that was most relevant for the initial and subsiding stages of the 2020 COVID-19 wave in Thailand. In order to illustrate this issue, panel (a) of Fig. 8.11 shows $\mathbf{v}_1$ in the E-I plane for the three stages 1, 2, 3. As anticipated from the values shown in Table 8.4, the eigenvector $\mathbf{v}_1$ of the first-wave COVID-19 epidemic in Thailand gradually rotated from smaller to larger angles during the course of the epidemic. Panel (b) of Fig. 8.11 presents the trajectory $\mathbf{X}^+(t) = (E(t), I(t))$ (shown in panel (b) of Fig. 8.10) as phase curve $I(E)$ in the E-I plane (thick solid line). In addition, the eigenvector $\mathbf{v}_1$ (dotted lines) magnified for illustration purposes is shown for the three stages. Panel (b) of Fig. 8.11 reveals that in stage 1 of the first-wave COVID-19 epidemic in Thailand the disease state $\mathbf{X}^+(t)$ closely evolved in the direction specified by the unstable eigenvector $\mathbf{v}_1$. Accordingly, $\mathbf{v}_1$ was the EI order parameter that determined in combination with the corresponding amplitude $A_1(t)$ the initial evolution of the first-wave COVID-19 epidemic in Thailand. From panel (b) of Fig. 8.11 it also follows that in stage 2 the trajectory $\mathbf{X}^+(t)$ branched off from the direction specified by $\mathbf{v}_1(S1)$ and evolved towards the location labeled S2. That

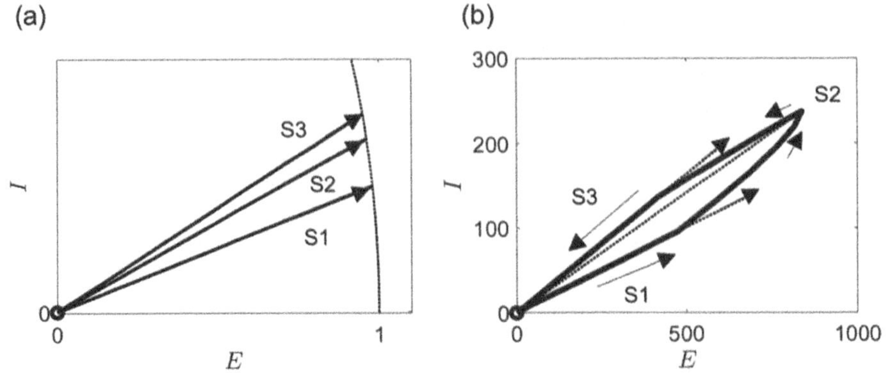

Fig. 8.11 Panel (**a**): Eigenvector $\mathbf{v}_1$ in stages 1, 2, 3. Panel (**b**): Dynamics of the disease state $\mathbf{X}^+(t)$ in the E-I plane across the three stages of the first COVID-19 wave of Thailand. The phase curve $I(E)$ (solid line) and the eigenvector $\mathbf{v}_1$ (dotted line) for the three stages 1, 2, 3 are shown. $\mathbf{v}_1$ is shown in a magnified scale

location corresponds to the plateaus of the $E(t)$ and $I(t)$ functions shown in panel (b) of Fig. 8.10. The vector $\mathbf{v}_1$ of stage 2 points to that location. At the end of stage 2 the trajectory $\mathbf{X}^+(t)$ abandoned the S2 location and evolved towards a point on the direction specified by the stage-3 eigenvector $\mathbf{v}_1$. Subsequently, the disease dynamics $\mathbf{X}^+(t)$ evolved along $\mathbf{v}_1(S3)$ towards the disease-free state with $E_{st} = I_{st} = 0$. Just as for stage 1, for stage 3, the dynamics in the direction of the second eigenvector $\mathbf{v}_2$ was negligible. In doing so, panel (b) of Fig. 8.11 illustrates graphically that not only for the initial stage but also for the final stage of the first-wave COVID-19 epidemic in Thailand Eq. (8.19) is a useful approximative description.

Quantitatively, the eigenvalues λ_1 and λ_2 of stage 3 may be compared. Table 8.4 shows the eigenvalues in all three stages. For stage 3 λ_2 was in magnitude 10 times larger than λ_1. Consequently, the dynamics underlying the subsiding epidemic in Thailand showed a time-scale separation (as discussed in Sect. 6.1.3)) in the sense that it exhibited a fast subsiding component in terms of A_2 and a slowly subsiding component in terms of A_1. The slowly evolving component determined the overall subsiding stage of the first-wave COVID-19 epidemic.

As argued above, the numerical and graphical analysis suggests that Eq. (8.19) accurately describes stages 1 and 3 of the first COVID-19 wave in Thailand. Practical insights based on Eq. (8.19) can be obtained as pointed out earlier in Chaps. 4, 5, and 6 the context of Eqs. (4.61), (5.86), (6.33), and (6.53), respectively. The eigenvector $\mathbf{v}_1$ defined by Eq. (8.26) can be expressed like $\mathbf{v}_1 = (\cos(\theta_1), \sin(\theta_1))$ with the help of the angle θ_1. From Eq. (8.19) it then follows that

$$\begin{pmatrix} \Delta E \\ \Delta I \end{pmatrix} \approx \mathbf{v}_1 \Delta A_1 = \begin{pmatrix} \cos(\theta_1) \\ \sin(\theta_1) \end{pmatrix} \Delta A_1 \Rightarrow \frac{\Delta I}{\Delta E} = \tan(\theta_1) \tag{8.39}$$

holds in states 1 and 3 with $\Delta E = E(t + \Delta t) - E(t)$, $\Delta I = I(t + \Delta t) - I(t)$, and $\Delta A = A(t + \Delta t) - A(t)$, where $\Delta t > 0$ is a time interval. Equation (8.39) states that $\mathbf{v}_1$ determines the relationship between the changes of the compartment sizes E and I. Using the angles θ_1 reported in Table 8.4, for stage 1 it follows that $\theta_1 = 11.4° \Rightarrow \Delta I/\Delta E = 0.20 \Rightarrow \Delta E/\Delta I = 5$. Accordingly, the analysis suggests that when during stage 1 of the 2020 epidemic in Thailand the number of individuals in class I increased by 1, then in the same period the number of individuals in the class E increased by 5. Graphically, this 5:1 ratio is illustrated by the stage-1 order parameter $\mathbf{v}_1(S1)$ shown in panel (b) of Fig. 8.11. For stage 3 the analogous calculation reads $\theta_1 = 18.1° \Rightarrow \Delta I/\Delta E = 0.32 \Rightarrow \Delta E/\Delta I = 3.05$. That is, when during the subsiding period of the 2020 COVID-19 wave in Thailand the class I decreased by 1 individual, then during the same period the class E decreased by approximately 3 individuals. Again, graphically, this 3:1 relationship is illustrated by the stage-3-remnant $\mathbf{v}_1(S3)$ shown in panel (b) of Fig. 8.11. These 5:1 and 3:1 ratios have also been reported in Table 8.4.

In closing this discussion of the first-wave COVID-19 epidemic in Thailand, let us return to the effective contact rates shown in Table 8.4. Both parameter β_E and β_I decreased from stage 1 to stage 3. This decrease is assumed to be due to the change in the daily routines of the Thai population in order to reduce the spread of COVID-19 as a result of intervention measures. As discussed in Sect. 8.2.3, similar observations were made by Pang et al. [18] for the COVID-19 epidemic in Wuhan city. Effective contact rate parameters decreased from $\beta_I = 0.30$/d and $\beta_E = 0.17$/d to $\beta_I = \beta_E = 0.02$/d presumably due to the implementation of intervention measures (see Table 8.1). Likewise, the study by Gatto et al. [20] concludes that the intervention measures imposed by the Italian government during the first COVID-19 wave in Italy in Spring 2020 caused effective contact rate parameters to decrease to lower levels. Further examples of model-based studies that have suggested that intervention or containment measures have been successful in reducing effective contact rates can be found in Sects. 8.2.2 and 8.2.3.

8.5 Three-Stage Models and the Bifurcation Scenario in Higher Dimensions

The bifurcation scenario and three-stage approach can be applied to address the nonlinear physics of epidemic waves in arbitrary dimensions. Figure 8.12 illustrates the bifurcation scenario and three-stage approach when applied to observed disease cases (panel (a)) and observed disease-associated deaths (panel (b)). Our departure point is the fundamental equation $\mathrm{d}\mathbf{X}/\mathrm{d}t = \mathbf{N}$ (see Eq. (2.1)) describing the evolution of the n-dimensional disease (or health) state of a population. As in Chaps. 6 and 7 the state is split into two subsystems using different criteria. In what follows, a data-focused view will be taken. The objective is to relate model solutions to cumulative confirmed infected cases or deaths that show patterns as illustrated in Fig. 8.12. Confirmed (or diagnosed) cases only count infected individuals (when neglecting false positives and other diagnoses issues). Consequently, we either follow the idea of Chap. 7 and decompose the state $\mathbf{X}$ such that the first m variables $X_1, \ldots, X_m$ denote infected compartments, whereas remaining variables $X_{m+1}, \ldots, X_n$ denote non-infected variables. Alternatively, the first m variables $X_1, \ldots, X_m$ denote infected variables of those individuals that are diagnosed with the infectious disease of interest and form at least in linear approximation an autonomous subsystem, whereas $X_{m+1}, \ldots, X_n$ denote the remaining variables. The alternative approach leads to a smaller number m of variables in the first group (and a lower dimension m of D^+). For example, if models distinguish between hospitalized cases and ICU cases (as the models that will be reviewed in Sect. 8.7) and all hospitalized cases are confirmed cases and only hospitalized cases can become ICU cases, then the compartment of ICU cases is not needed for counting the confirmed cases because the confirmed infected cases that receive ICU treatment are a subset of the confirmed hospitalized cases. Decomposing the disease (or health) state of a population into two classes implies that on the level of the amplitude space description $\mathbf{X}^+ = (X_1, \ldots, X_m)$ with $\mathbf{X}^+(t) = \sum_{k=1}^{m} A_k(t)\mathbf{v}_k$ evolves like $\mathrm{d}A_k/\mathrm{d}t = \lambda_k A_k + \mathbf{w}_k \mathbf{R}(MA, \mathbf{X}^-)$ and the remaining variables evolve like $\mathrm{d}X_j^-/\mathrm{d}t = N_j^-(MA, \mathbf{X}^-)$ (see Eqs. (6.12) and (6.13)).

Panel (a) of Fig. 8.12 shows schematically the time course of the cumulative confirmed infected cases (here: over a period of one month) during the three model stages of an epidemic wave under consideration. The stages are denoted by S1, S2, and S3. The stages are characterized in terms of the eigenvalues of the disease-free fixed point $\mathbf{X}^+ = (0, \ldots, 0)$ as discussed in Sect. 8.4.1. Stage 1 is characterized by an unstable fixed point. For sake of simplicity, it is assumed that the fixed point is

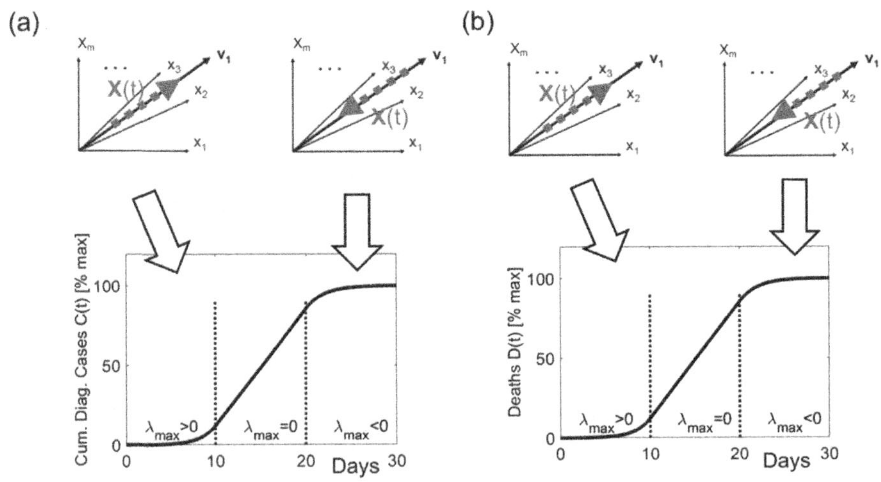

Fig. 8.12 Three-stage descriptions of epidemic waves in subspaces D^+ of arbitrary dimensions. Panels (**a**) and (**b**) refer to descriptions based on cumulative confirmed (or diagnosed) cases and disease-associated deaths, respectively

characterized by a single (real-valued) positive eigenvalue $\lambda_{\max} > 0$. Without loss of generality, the eigenvalues $\lambda_1, \ldots, \lambda_m$ are arranged in descending order with respect to the values of their real parts (i.e., $\lambda_1 > \mathbb{R}\{\lambda_2\} \geq \cdots \geq \mathbb{R}\{\lambda_m\}$) such that $\lambda_1 = \lambda_{\max}$. Stage 2 is the bifurcation stage defined by $\lambda_{\max} = \lambda_1 = 0$. Finally, stage 3 is the subsiding stage defined by $\lambda_{\max} = \lambda_1 < 0$ for which the disease-free fixed point $\mathbf{X}^+ = (0, \ldots, 0)$ is asymptotically stable in D^+ (provided the linearized dynamics holds). As far as the disease-free fixed point in the entire state space is concerned, in stage 3 the state $\mathbf{X}_{st} = (\mathbf{X}^+, \mathbf{X}^-)$ with $\mathbf{X}^+ = (0, \ldots, 0)$ and $X_{i,st} \geq 0$ for $i \geq m+1$ does typically not correspond to an asymptotically stable fixed point. Rather, when neglecting the impact of demographic terms, it typically corresponds to a neutrally stable fixed point.

The two inserts on the top of panel (a) of Fig. 8.12 illustrate schematically the dynamics in the outbreak (S1) and subsiding (S3) stages as seen in the infected m-dimensional subspace D^+. As far as stage 1 is concerned, the eigenvector $\mathbf{v}_1$ corresponding to $\lambda_1 = \lambda_{\max} > 0$ describes the unstable direction of the saddle point located at $\mathbf{X}^+ = (0, \ldots, 0)$. As indicated, the dynamics in D^+ is assumed to converge towards the direction specified by $\mathbf{v}_1$ and, subsequently, evolves along $\mathbf{v}_1$ away from the disease-free fixed point. In contrast, in stage 3 assuming there is a gap in the eigenvalue spectrum such that $|\lambda_{\max}| = |\lambda_1|$ is small relative to $|\mathbb{R}\{\lambda_j\}|$ for $j = 2, \ldots, m$ (i.e., there is a time scale separation, see Sect. 6.1.3), then the subsiding dynamics in D^+ quickly converges to the direction specified by $\mathbf{v}_1$ and, subsequently, evolves along $\mathbf{v}_1$ towards the disease-free fixed point. This is indicated in panel (a) in the insert related to the stage 3. If the epidemic wave under consideration evolves sufficiently close to the disease-free fixed point such that it can be described in terms of linearized evolution equations, then the amplitude A_1 related to the maximal eigenvalue $\lambda_{\max} = \lambda_1$ exhibits in the three stages the following dynamics:

$$\text{S1}\ ,\ t \in [t_0, t_1]\ ,\ \lambda_1 = \lambda_{1,S1} > 0\ :$$

$$A_1(t) = A_1(t_0) \exp\{\lambda_{1,S1}(t - t_0)\} \Rightarrow \frac{\mathrm{d}}{\mathrm{d}t}|A_1| > 0\ , \tag{8.40}$$

$$\text{S2}\ ,\ t \in [t_1, t_2]\ ,\ \lambda_1 = \lambda_{1,S2} = 0\ :$$

$$A_1(t) = A_1(t_1) = \text{const} \Rightarrow \frac{\mathrm{d}}{\mathrm{d}t} A_1 = 0\ , \tag{8.41}$$

$$\text{S3}\ ,\ t \in [t_2, t_3]\ ,\ \lambda_1 = \lambda_{1,S3} < 0\ :$$

$$A_1(t) = A_1(t_2) \exp\{\lambda_{1,S3}(t - t_2)\} \Rightarrow \frac{\mathrm{d}}{\mathrm{d}t}|A_1| < 0\ . \tag{8.42}$$

This three-stage sequence in which the maximal eigenvalue $\lambda_1 = \lambda_{\max}$ changes across the stages like $\lambda_{1,S1} > \lambda_{1,S2} = 0 > \lambda_{1,S3}$ and the dominant amplitude A_1 exhibits a dynamics as shown in Eqs. (8.40) to (8.42) produces under appropriate circumstances a sigmoid shape of the cumulative confirmed disease cases $C(t)$ sketched in panel (a) of Fig. 8.12: the function $C(t)$ shows an exponential increase in stage 1, a linear increase in stage 2, and a de-accelerating increases in stage 3. Let us derive this sigmoid pattern in a more quantitative way.

Let $C(t)$ denote the number of cumulative diagnosed (or confirmed) disease cases. Assuming linear diagnoses mechanisms, infected individuals of the compartment j are diagnosed at rates s_j, which implies that $C(t)$ evolves like

$$\frac{\mathrm{d}}{\mathrm{d}t}C = \sum_{k=1}^{m} s_k X_k(t) = \mathbf{s}\mathbf{X}^{+}(t) \,. \tag{8.43}$$

Special cases of Eq. (8.43) have been addressed previously in form of Eqs. (4.83), (5.100), (6.50), (8.4), (8.20), and (8.27). Assuming that $\mathbf{v}_{\max} = \mathbf{v}_1$ is dominant in all three stages, the superposition $\mathbf{X}^{+}(t) = \sum_{k=1}^{m} A_k(t)\mathbf{v}_k$ reduces to $\mathbf{X}^{+} \approx A_1\mathbf{v}_1$. Consequently, from Eq. (8.43) it follows that

$$\frac{\mathrm{d}}{\mathrm{d}t}C = \mathbf{s}\mathbf{v}_1 A_1(t) \,. \tag{8.44}$$

An example of Eq. (8.44) was discussed in Sect. 4.5 and is given by Eq. (4.87). In analogy to Eqs. (8.28) to (8.30), from Eq. (8.44) and Eqs. (8.40) to (8.42) it follows that

$$\begin{aligned}
&\mathrm{S1}\ ,\ t \in [t_0, t_1]\ ,\ \lambda_1 = \lambda_{1,S1} > 0\ : \\
&\qquad C(t) = C(t_0) + \frac{\mathbf{s}\mathbf{v}_{1,S1}}{\lambda_{1,S1}} A_1(t_0)\left(\exp\{\lambda_{1,S1}(t - t_0)\} - 1\right) \,, && (8.45) \\
&\mathrm{S2}\ ,\ t \in [t_1, t_2]\ ,\ \lambda_1 = \lambda_{1,S2} = 0\ : \\
&\qquad C(t) = C(t_1) + \mathbf{s}\mathbf{v}_{1,S2} A_1(t_1)(t - t_1) \,, && (8.46) \\
&\mathrm{S3}\ ,\ t \in [t_2, t_3]\ ,\ \lambda_1 = \lambda_{1,S3} < 0\ : \\
&\qquad C(t) = C(t_2) + \frac{\mathbf{s}\mathbf{v}_{1,S3}}{|\lambda_{1,S3}|} A_1(t_2)\left(1 - \exp\{\lambda_{1,S3}(t - t_2)\}\right) \,, && (8.47)
\end{aligned}$$

where $\lambda_{1,S1}, \lambda_{1,S2}, \lambda_{1,S3}$ and $\mathbf{v}_{1,S1}, \mathbf{v}_{1,S2}, \mathbf{v}_{1,S3}$ denote the eigenvalues λ_1 and eigenvectors $\mathbf{v}_1$ in the respective stages. The evolution equations describe an exponential increase (Eq. (8.45)) for stage 1, a linear increase (Eq. (8.46)) for stage 2, and a de-accelerating increase (Eq. (8.47)) for stage 3 as shown schematically in panel (a) of Fig. 8.12. Equation (8.45) can equivalently be expressed as

$$\frac{C(t)}{z} = A_1(t) + h \tag{8.48}$$

for $t \in [t_0, t_1]$ with $z = \mathbf{s}\mathbf{v}_{1,S1}/\lambda_{1,S1}$ and $h = C(t_0)/z - A_1(t_0)$. Accordingly, the order parameter amplitude $A_1(t)$ determines quantitatively up to a additive constant h the confirmed infected cases $C(t)$ in the stage 1 of an epidemic provided the confirmed cases are rescaled appropriately by the parameter z. Equation (4.89) discussed in Sect. 4.5 and illustrated in panel (d) of Fig. 4.6 is a special case of Eq. (8.48).

The same argumentation can be made when considering disease-related deaths as addressed in panel (b) of Fig. 8.12. Again, linear transmission mechanisms between the various types of infected individuals and deceased individuals are assumed. It is assumed that infected individuals decease due to the infectious disease at certain rates m_j. Consequently, the number of deaths $D(t)$ evolves like

$$\frac{\mathrm{d}}{\mathrm{d}t}D = \sum_{k=1}^{m} m_k X_k(t) = \mathbf{mX}^{+}(t) = \mathbf{mv}_1 A_1(t) \ , \tag{8.49}$$

where it has been assumed again that $\mathbf{X}^{+}(t) = A_1(t)\mathbf{v}_1$ is a useful approximation. It then follows, that $D(t)$ in the three stages 1, 2, 3 is given by

$$\begin{aligned}
&\mathrm{S1}\ , t \in [t_0, t_1]\ ,\ \lambda_1 = \lambda_{1,S1} > 0\ : \\
&\qquad D(t) = D(t_0) + \frac{\mathbf{mv}_{1,S1}}{\lambda_{1,S1}} A_1(t_0)\left(\exp\{\lambda_{1,S1}(t-t_0)\} - 1\right)\ , && (8.50)\\
&\mathrm{S2}\ , t \in [t_1, t_2]\ ,\ \lambda_1 = \lambda_{1,S2} = 0\ : \\
&\qquad D(t) = D(t_1) + \mathbf{mv}_{1,S2} A_1(t_1)(t-t_1)\ , && (8.51)\\
&\mathrm{S3}\ , t \in [t_2, t_3]\ ,\ \lambda_1 = \lambda_{1,S3} < 0\ : \\
&\qquad D(t) = D(t_2) + \frac{\mathbf{mv}_{1,S3}}{|\lambda_{1,S3}|} A_1(t_2)\left(1 - \exp\{\lambda_{1,S3}(t-t_2)\}\right)\ . && (8.52)
\end{aligned}$$

Consistent with the sigmoid pattern shown in panel (b) of Fig. 8.12, the evolution equations describe an exponential increase (Eq. (8.50)) for stage 1, a linear increase (Eq. (8.51)) for stage 2, and a de-accelerating increase (Eq. (8.52)) for stage 3. Equation (8.50) may be expressed as

$$\frac{D(t)}{z} = A_1(t) + h \tag{8.53}$$

for $t \in [t_0, t_1]$ with $z = \mathbf{mv}_{1,S1}/\lambda_{1,S1}$ and $h = D(t_0)/z - A_1(t_0)$. Accordingly, the order parameter amplitude $A_1(t)$ determines quantitatively up to a additive constant h the number $D(t)$ of individuals deceased from the infectious disease in the stage 1 of an epidemic provided $D(t)$ is rescaled appropriately by the parameter z.

8.6 Sequences of Stages in Amplitude Space

8.6.1 Semi-analytical Approach

The objective in this section is to follow an epidemic that can be described in terms of s stages through all stages $i = 1, \ldots, s$ using the amplitude space description. Let us consider an epidemiological model $\mathrm{d}\mathbf{X}/\mathrm{d}t = \mathbf{N}$ in terms of the two-classes amplitude space description given by Eqs. (6.12) and (6.13), which read $\mathrm{d}A_k/\mathrm{d}t = \lambda_k A_k + \mathbf{w}_k \mathbf{R}(M\mathbf{A}, \mathbf{X}^-)$ and $\mathrm{d}X_j^-/\mathrm{d}t = N_j^-(M\mathbf{A}, \mathbf{X}^-)$. For sake of simplicity these basic equations are repeated as

$$\frac{\mathrm{d}}{\mathrm{d}t}\mathbf{X} = \mathbf{N}(\mathbf{X}) \tag{8.54}$$

and

$$\begin{aligned} \frac{\mathrm{d}}{\mathrm{d}t} A_k &= \lambda_k A_k + \mathbf{w}_k \mathbf{R}(M\mathbf{A}, \mathbf{X}^-) , \\ \frac{\mathrm{d}}{\mathrm{d}t} X_j^- &= N_j^-(M\mathbf{A}, \mathbf{X}^-) . \end{aligned} \tag{8.55}$$

The following considerations hold in general but for illustration purposes the SEIR model discussed in Sect. 6.2.2 will be used that comes with the state space description (3.43) and the amplitude space description in form of Eqs. (6.19) and (6.21).

For stage 1 from the initial state $\mathbf{X}(t_0)$ and the corresponding state $\mathbf{X}^+(t_0)$ in the subspace D^+ (e.g., $\mathbf{X}^+(t_0) = (E(t_0), I(t_0)))$ the initial amplitude vector $\mathbf{A}(t_0)$ can be computed from

$$\mathbf{A}(t_0, S1) = M_+^{-1}(\text{Stage 1})\mathbf{X}^+(t_0) \tag{8.56}$$

(which for the SEIR model is given by Eq. (6.23)). Subsequently, the state space equation (8.54) and the evolution equations (8.55) of the amplitude space description can be solved numerically for stage 1. At the end of stage 1, that is, at the time point t_1, it is assumed that at least one of the model parameters changes (e.g., in case of the SEIR model, one of the parameters α, β, γ varies). The new initial state of stage 2 is defined by the final state $\mathbf{X}^+(t_1)$ of stage 1. That is, the final state $\mathbf{X}^+(t_1)$ serves as initial state of stage 2. Consequently, let $\mathbf{A}(t_1, S2)$ denote the initial amplitudes of stage 2, then

$$\mathbf{A}(t_1, S2) = M_+^{-1}(\text{Stage 2})\mathbf{X}^+(t_1) . \tag{8.57}$$

In general, this implies that $\mathbf{A}(t)$ exhibits a discontinuity at the stage boundary t_1. That is, let $\mathbf{A}(t, S1)$ denote the vector function $\mathbf{A}(t)$ computed in stage 1, then typically

$$\mathbf{A}(t_1, S2) \neq \mathbf{A}(t_1, S1) . \tag{8.58}$$

This discontinuity is due to the fact that the basis of the amplitude space (e.g., the eigenvectors $\mathbf{v}_1$ and $\mathbf{v}_2$ in the case of the SEIR model) in stage 2 in general differs from the basis of the amplitude space used in stage 1.

Having determined the initial states $\mathbf{X}^+(t_1)$ and $\mathbf{A}(t_1, S2)$ for stage 2, Eqs. (8.54) and (8.55) can be solved again numerically. In doing so, the state dynamics and amplitude dynamics for stage 2 is obtained. For example, in the context of the SEIR model, Eq. (3.43) and Eqs. (6.19) and (6.23) are solved numerically for stage 2. If $s = 2$, then the analysis is completed after the second stage. If $s > 2$, then the procedure is repeated until all stages have been processed.

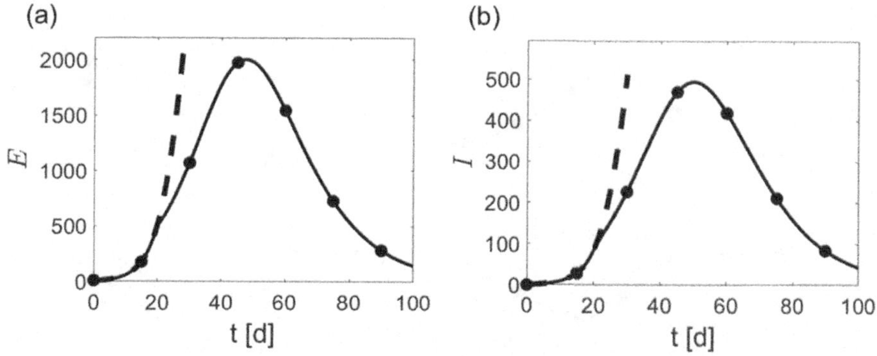

Fig. 8.13 Illustration of a simulated two-stage epidemic in state space. Panels (**a**) and (**b**) show $E(t)$ and $I(t)$ (solid lines) as computed from the 1β SEIR model (3.43), respectively. In the simulation, at $t = 20$ days the contact rate parameter β is decreased but still assumes a above-critical value. The dashed lines indicate the evolution of $E(t)$ and $I(t)$ when ignoring the change in β. See text for model parameters and initial conditions

Let us illustrate these considerations by means of a simulated two-stage epidemic that involves two stages S1 and S2 characterized by positive eigenvalues using the SEIR model (3.43). Let us assume that $\lambda_1(S1) > \lambda_1(S2) > 0$ holds. For example, at the end of stage 1, the effective contact rate β is reduced (which mimics the impact of intervention measures) such that λ_1 decreases but is still positive in stage 2. As far as α and γ are concerned, we assume that they are constant across S1 and S2. For the following simulation the parameters $\alpha = 0.1$/d, $\gamma = 0.4$/d, $N = 10000$, β(S1) $= 2.0$/d for the first 20 days (S1) and β(S2) $= 1.0$/d after day 20 (S2) were used. In this case, from Eq. (6.17) it follows that the maximal eigenvalue of the SEIR model is λ_1(S1) $= 0.22$/d and λ_1(S2) $= 0.10$/d in S1 and S2, respectively. The entire simulation period was 100 days and $t_0 = 0$. Furthermore, the initial conditions $E(0) = 10$, $I(0) = 0$ and $S(0) = N - E(0) - I(0)$ were used.

Different aspects of the simulation are illustrated in Figs. 8.13 and 8.14. Panels (a) and (b) of Fig. 8.13 show the trajectories $E(t)$ and $I(t)$, respectively, as computed from Eq. (3.43) as solid lines. The dashed lines indicate how $E(t)$ and $I(t)$ evolve if β would be kept constant for the whole simulation period. Comparing the solutions for constant β and an effective contact rates that decreases in stage 2, the impact of the decrease of the effective contact rate can be seen. Accordingly, the increase of the exposed and infectious individuals becomes less dramatic when β is changed from β(S1) to β(S2) $<$ β(S1). As in Sect. 8.1, the scenario for which β is kept constant corresponds to a no-intervention scenario, whereas the two-stage scenario with $\beta(S2) < \beta(S1)$ exemplifies an intervention scenario.

Panel (a) of Fig. 8.14 presents the amplitudes A_1 and A_2 computed from Eqs. (6.19) and (6.21) as functions of time. At $t_1 = 20$ days the amplitudes have been put to their respective new initial values $A_1(t_1, S2)$ and $A_2(t_1, S2)$ of stage 2. Accordingly, A_1 exhibits a discontinuity at t_1. However, on the relatively large scale of 2000 individuals this discontinuity is hardly visible (see top subpanel). In contrast, the

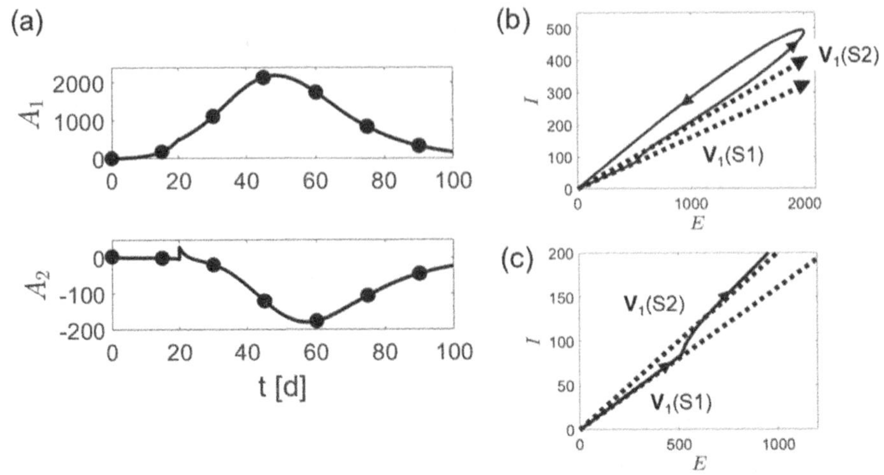

Fig. 8.14 Amplitude and state space description of the simulated two-stage epidemic that was introduced in Fig. 8.13 and exhibits a sudden change in β at $t = 20$ days. Panel (**a**) shows the amplitudes A_1 and A_2 as functions of time. Panel (**b**) presents the phase curve $I(E)$ in the E-I subspace. The unstable eigenvector $\mathbf{v}_1$ is presented for stages 1 and 2 (dotted lines). Panel (**c**) shows a detail of panel (**b**) in which the evolution of the disease state $\mathbf{X}^+(t)$ quickly switches from a dynamics along the axis specified by $\mathbf{v}_1(S1)$ to a dynamics along the axis specified by $\mathbf{v}_1(S2)$

discontinuity of A_2 at the stage boundary t_1 is visible as a small spike in the time course of A_2 at $t = 20$ days (see bottom subpanel). Panel (b) of Fig. 8.14 presents the trajectory $\mathbf{X}^+(t) = (E(t), I(t))$ as phase curve $I(E)$ in the E-I subspace. The curve describes a loop that starts close to the disease-free state $E = I = 0$ and after an excursion returns to $E = I = 0$. The axes defined by unstable eigenvector $\mathbf{v}_1$ in stages 1 and 2 are depicted in panel (b) as well (dotted lines). Panel (b) demonstrates that in stage 1 the trajectory $\mathbf{X}^+(t)$ follows the axis specified by $\mathbf{v}_1$ of stage 1. When β is switched to a lower value at t_1, then the trajectory quickly converges towards the new axis as defined by the stage-2 eigenvector $\mathbf{v}_1$. Subsequently, the trajectory branches off that axis, completes the excursion, and returns to the disease-free fixed point. Panel (c) highlights the transition dynamics at the beginning of stage 2. The unstable eigenvector $\mathbf{v}_1$ and its amplitude A_1 primarily determine the evolution of the state $\mathbf{X}^+$ in stage 1 and part of the dynamics in stage 2. The stable eigenvector $\mathbf{v}_2$ plays a negligible role during these periods. In particular, as can be seen in panel (a), the amplitude A_1 during the first 20 to 30 days is large as compared to $|A_2|$ and in this sense dominates the dynamics during that initial period. A_2 makes an essential contribution to the dynamics when A_1 reaches its maximum value at about 50 days (see panel (a)) and the $E(I)$ phase curve is close to its turning point (see panel (b)).

The equivalence of the description (8.54) and (8.55) throughout the different stages of an epidemic can be illustrated in two ways.

First, the trajectory $\mathbf{X}^+(t)$ of the subspace state vector $\mathbf{X}^+$ can be computed from the amplitude vector $\mathbf{A}(t)$ in each stage j like

$$\mathbf{X}^+(t) = \sum_{k=1}^{m} \mathbf{v}_k(\text{Stage } j) A_k(t) = M_+(\text{Stage } j)\mathbf{A}(t) . \tag{8.59}$$

For example, the state variables $E(t)$ and $I(t)$ can be computed from the amplitude functions $A_1(t)$ and $A_2(t)$ of the SEIR model. Substituting the functions $A_1(t)$ and $A_2(t)$ shown in panel (a) of Fig. 8.14 into Eq. (8.59), the graphs $E(t)$ and $I(t)$ indicated by full circles in panels (a) and (b) of Fig. 8.13 have been obtained. As expected, the solutions computed via Eq. (8.59) (full circles) and obtained directly by solving the SEIR model Eq. (3.43) (solid lines) are identical.

A second way to demonstrate the equivalence of the state space and amplitude space descriptions is to derive the amplitude vector $\mathbf{A}$ from the state vector $\mathbf{X}^+$ without solving the amplitude equations (8.55). Accordingly, $\mathbf{A}$ in each stage j can be obtained from

$$\mathbf{A}(t) = M_+^{-1}(\text{Stage } j)\mathbf{X}^+(t) \Leftrightarrow A_k(t) = \mathbf{w}_k(\text{Stage } j)\mathbf{X}^+(t) . \tag{8.60}$$

When applying this procedure to the SEIR model simulations presented in Figs. 8.13 and 8.14, the state functions $E(t)$ and $I(t)$ shown Fig. 8.13 are substituted in Eq. (8.60). In doing so, the amplitude functions shown in Fig. 8.14 as full circles can be obtained. As expected, computing A_1 and A_2 from Eq. (8.60) (full circles) or computing A_1 and A_2 directly from Eqs. (6.19) and (6.21) (solid lines) yields identical results.

As a second example let us consider a two-stage epidemic for which the effective contact rate drops in the second stage to a sufficiently low level such that a bifurcation occurs and the disease-free fixed point becomes stable. That is, let us consider the case $\lambda_{max} = \lambda_1(S1) > 0$ and $\lambda_{max} = \lambda_1(S2) < 0$. For the following simulation the parameters $\alpha = 0.1$/d, $\gamma = 0.4$/d, β(S1) $= 2.0$/d, and β(S2) $= 0.3$/d were used. A small population of $N = 10,000$ individuals was considered. Stage 1 was again 20 days long and the total simulation period was 100 days. From Eq. (6.17) it then follows that λ_1(S1) $= 0.22$/d and λ_1(S2) $= -0.02$/d. The same initial conditions were used as in the previous simulation. The simulation results are presented in Fig. 8.15. Panels (a) shows $E(t)$ and $I(t)$ as functions of time as obtained from Eq. (3.43). At day 20 due to change of β the increasing trend of the number of exposed individuals $E(t)$ turns into a decreasing one. The function $I(t)$ exhibits the same qualitative pattern as $E(t)$. However, $I(t)$ switches from an increasing to a decreasing function with a short delay. Panel (b) shows the amplitudes A_1 and A_2 as computed from Eqs. (6.19) and (6.21) as functions of time. As expected, A_1 dominates the amplitude dynamics and increases exponentially in stage 1, while A_2 remains almost constant. At day 20, both amplitude functions exhibit discontinuities. In stage 2 the fixed point $E = I = 0$ exhibits two negative eigenvalues. The simulation reveals that for the selected model parameters the linearized model in state space $\mathrm{d}E/\mathrm{d}t = -\alpha E + \beta I$,

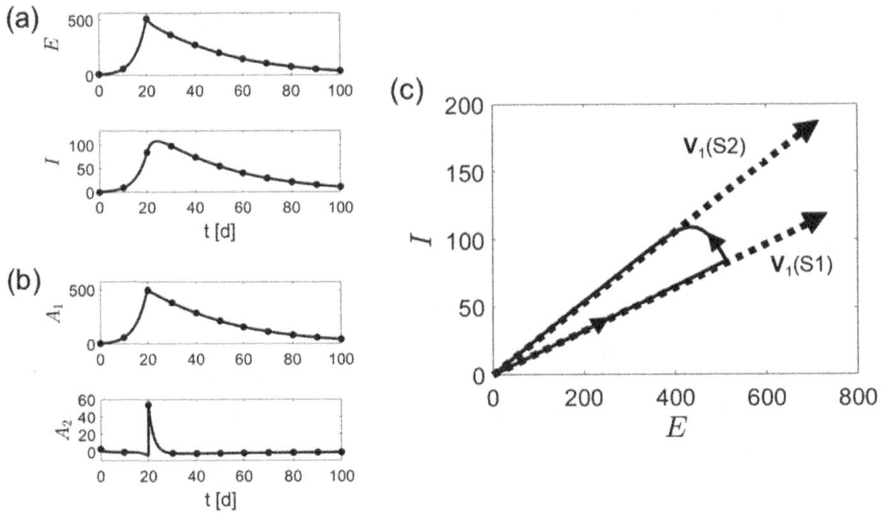

Fig. 8.15 Illustration of a simulated two-stage epidemic involving a relatively strong intervention measures. In this simulation, at time $t = 20$ days a sudden change in β to a sufficiently low below-critical value stabilize the disease-free fixed point. Panels (**a**): State space trajectories as computed from Eq. (3.43). Panel (**b**): Amplitude space trajectories as computed from Eqs. (6.19) and (6.21). Panel (**c**) shows the phase curve $I(E)$ (solid line) in the E-I plane and the eigenvector $\mathbf{v}_1$ (magnified for visualization purposes) for stages 1 and 2 (dotted lines)

$\mathrm{d}I/\mathrm{d}t = \alpha E - \gamma I$ (see Eq. (6.16)) and amplitude space $\mathrm{d}A_j/\mathrm{d}t = \lambda_j A_j$ captures the essential dynamics in stage 2. Accordingly, both amplitudes approach zero in a more or less exponential manner (see panel (b)) indicating that the epidemic is vanishing.

Panel (c) presents the trajectory $\mathbf{X}^+(t)$ as phase curve $I(E)$ in the E-I subspace. The axes given by the eigenvector $\mathbf{v}_1$ for stages 1 and 2 are shown as well. As can be seen in panel (c), during stage 1, $\mathbf{X}^+(t)$ follows the axis of the unstable eigenvector $\mathbf{v}_1$. When $\lambda_{\max}$ becomes negative at $t_1 = 20$ days, the trajectory quickly converges to the new axis specified by $\mathbf{v}_1$ in stage 2. The fast approach towards this direction specified by $\mathbf{v}_1$ indicates that there is gap in the eigenvalue spectrum as discussed in Sect. 6.1.3. In fact, for stage 2 the eigenvalues read $\lambda_1 = \lambda_{\max} = -0.02$/d and $\lambda_2 = -0.48$/d and the corresponding time constants $\tau_j = 1/|\lambda_j|$ read $\tau_1 = 47.9$ days and $\tau_2 = 2.1$ days, respectively. That is, A_2 exhibits a time constants that is by a factor 24 smaller than the time constant of A_1. In this sense, A_2 describes a fast dynamics, while A_1 describes a slow dynamics, which can also seen in panel (b) when comparing the dynamics of amplitudes for $t > 20$ days. As can be seen in panels (b) and (c), the slow dynamics of A_1, which is the amplitude of the remnant of the order parameter, determines the vanishing of the simulated epidemic.

Note that E and I are continuous functions at the stage boundary t_1, whereas A_1 and A_2 exhibit jumps at that boundary. Nevertheless, A_1 captures not only qualitatively but also quantitatively most of the dynamics of E. This relation between

E and A_1 follows from the selected simulation parameters. From the simulation parameters it follows that $v_{1,E} = 0.99$ holds in stage 1 and $v_{1,E} = 0.97$ in stage 2. Since $E = v_{1,E}A_1 + v_{2,E}A_2$ holds (see Eq. (6.22)) and $|A_2| \ll A_1$, we have $E(t) \approx v_{1,E}A_1(t) \Rightarrow E(t) \approx A_1(t)$.

Finally, in order to demonstrate the equivalence of the state space and amplitude space descriptions, just as for the previous example, we computed states from Eq. (8.59) and amplitudes from Eq. (8.60). The states and amplitudes thus obtained are shown as filled circles in panels (a) and (b) of Fig. 8.15. As can be seen, the solutions obtained in an indirect way (full circles) from Eqs. (8.59) and (8.60) were identical to those obtained directly (solid lines) from the respective model equations (3.43), (6.19), and (6.21).

8.6.2 *Numerical Stage Analysis*

From the discussion in the previous section it follows that a completely numerical stage analysis can be conducted for any given epidemiological model of the form (2.1) [72]. This numerical approach is based on Eqs. (2.1), (6.3), (6.4), and (8.60). For sake of conveniency, let us rewrite them as

$$\begin{aligned}
\frac{\mathrm{d}}{\mathrm{d}t}\mathbf{X} &= \mathbf{N}(\mathbf{X}) \, , \\
L^+_{ik} &= \frac{\partial}{\partial X^+_k} N_i(\mathbf{X}_{st}) \, , \\
M_+ &= (\mathbf{v}_1, \ldots, \mathbf{v}_m) \, , \\
\mathbf{A}(t) &= M_+^{-1}(\text{Stage } j)\mathbf{X}^+(t) \, . \qquad (8.61)
\end{aligned}$$

Accordingly, a given epidemiologic model in terms of $\mathrm{d}\mathbf{X}/\mathrm{d}t = \mathbf{N}(\mathbf{X})$ is solved numerically. The linearization matrix L^+ of the (infected) compartments $\mathbf{X}^+$ and the eigenvector matrix M_+ are determined numerically for each stage j of the epidemic under consideration. Subsequently, the amplitudes $A_1, \ldots, A_m$ in terms of the vector $\mathbf{A}$ are computed for each stage j with the help of the state vector $\mathbf{X}^+(t)$ and the inverse matrix M_+^{-1}. This numerical approach will be illustrated by means of two examples in Sect. 8.7.

8.7 Examples of Three-Stage COVID-19 Waves and 5D Order Parameters

Let us consider two examples of COVID-19 first-waves that have been analyzed in relative high-dimensional state spaces using a three-stage approach. Both example involve five-dimensional unstable stage-1 eigenvectors.

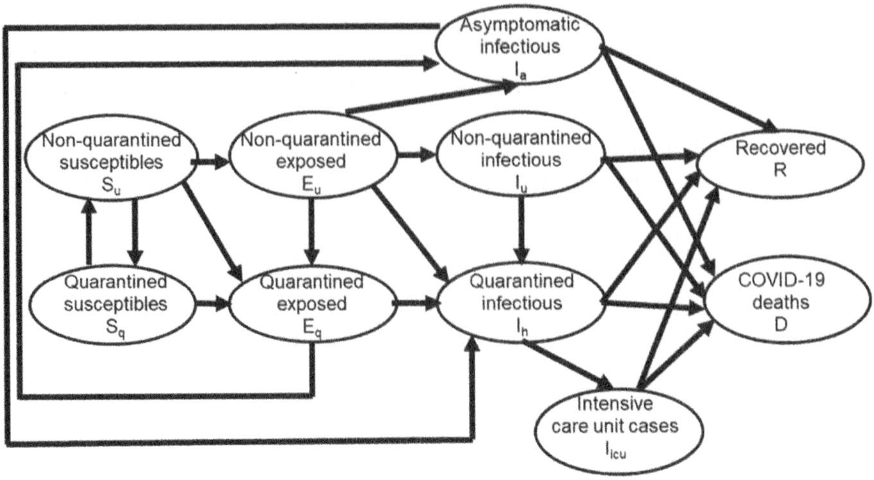

Fig. 8.16 Compartments and flow chart of the epidemiological model described by Eq. (8.62)

8.7.1 First COVID-19 Wave of 2020 in the State of New York

The study by Ngonghala et al. [33] used a generalized SEIAR model to describe the first few months of the COVID-19 outbreak in the state of New York, USA. More precisely, in the study, the period from March 1 to April 7, 2020 was considered. Figure 8.16 shows a flow chart of the model. The model makes a general distinction between non-quarantined and quarantined individuals. As far as non-quarantined individuals are concerned, the model describes susceptible (S_u), exposed (E_u), and symptomatic infectious (I_u) non-quarantined individuals. With respect to the quarantined individuals, the model considers susceptible (S_q), exposed (E_q), and symptomatic infectious (I_h) quarantined individuals. Quarantined symptomatic infectious individuals may be hospitalized, whence the subindex h of the compartment variable I_h. In addition, the model accounts for isolated symptomatic individuals that are hospitalized and require intensive care I_{icu}. They do not belong to the compartment I_h. Asymptomatic infectious COVID-19 cases (I_a) form a compartment of their own. Finally, the model involves the compartment of individuals recovered from COVID-19 (R) and the compartment of individuals deceased due to COVID-19 (D). Exposed individuals are in their latent phase and cannot infect others. Individuals staying in intensive care units are perfectly isolated and cannot infect others. The infectious compartments that describe individuals who can infect others (i.e., the compartments of actually infectious individuals) are I_u, I_a, and I_h.

As indicated in Fig. 8.16, non-quarantine susceptibles (S_u) can get quarantined ($S_u \rightarrow S_q$) or they can get infected and become exposed quarantined individuals (E_q) or exposed non-quarantined individuals (E_u). The quarantined susceptible individuals (S_q) either complete quarantine and return to the class of non-quarantined susceptibles (S_u) or they become infected during quarantine and become quarantined exposed individuals (E_q). Non-quarantined exposed individuals (E_u) either become quarantined exposed individuals (E_q), non-quarantined symptomatic individuals (I_u), quarantined symptomatic individuals (I_h), or asymptomatic infectious individuals (I_a). In contrast, quarantined exposed individuals (E_q) either become quarantined symptomatic individuals (I_h) or asymptomatic infectious individuals (I_a). As far as non-quarantined symptomatic individuals (I_u) are concerned, they become quarantined cases (I_h) or they recover ($I_u \to R$), or decease due to COVID-19 ($I_u \to D$). Likewise, quarantined symptomatic individuals (I_h) become ICU cases (I_{icu}), recover (R), or decease due to COVID-19 (D). ICU cases either recover (R) or decease due to COVID-19 (D). Finally, asymptomatic infectious individuals (I_a) either recover (R), decease due to COVID-19 (D), or develop symptoms and, consequently, become quarantined symptomatic cases (I_h). The model equations read [27, 33]

$$
\begin{aligned}
\frac{\mathrm{d}}{\mathrm{d}t} S_u &= -k_0 S_u + k_1 S_q \ , \\
\frac{\mathrm{d}}{\mathrm{d}t} S_q &= (1-p) k_0 S_u - (\theta_j k_0 + k_1) S_q \ , \\
\frac{\mathrm{d}}{\mathrm{d}t} E_u &= (1-q) p k_0 S_u - k_2 E_u \ , \\
\frac{\mathrm{d}}{\mathrm{d}t} E_q &= q p k_0 S_u + \alpha E_u + \theta_j k_0 S_q - k_3 E_q \ , \\
\frac{\mathrm{d}}{\mathrm{d}t} I_u &= f_1 \sigma_u E_u - k_4 I_u \ , \\
\frac{\mathrm{d}}{\mathrm{d}t} I_h &= f_2 \sigma_u E_u + r \sigma_q E_q + \phi I_u + \sigma_a I_a - k_5 I_h \ , \\
\frac{\mathrm{d}}{\mathrm{d}t} I_a &= (1 - f_1 - f_2) \sigma_u E_u + (1-r) \sigma_q E_q - k_6 I_a \ , \\
\frac{\mathrm{d}}{\mathrm{d}t} I_{icu} &= = \nu I_h - k_7 I_{icu} \ , \\
\frac{\mathrm{d}}{\mathrm{d}t} D &= m_u I_u + m_h I_h + m_a I_a + m_{icu} I_{icu} \ ,
\end{aligned}
\tag{8.62}
$$

where k_0 denotes the rate constant

$$
k_0 = \beta \frac{I_u + \eta_a I_a + \eta_h I_h}{N - \theta_q (E_q + I_h + I_{icu})} \ . \tag{8.63}
$$

The model involves the parameters p, q, f_1, f_2, θ_j, θ_q, α, σ_u, σ_q, ϕ, r, ν, δ_u, δ_h, δ_a, δ_{icu}, η_a, η_h, $k_1, \ldots, k_7$, and β, which are semi-positive. The parameters denote the probability of an infection per contact (p), proportion of being quarantined (q), proportion of exposed E_u who transition to I_u (f_1), proportion of exposed E_u who transition to I_h (f_2), efficacy of quarantine (θ_j), general efficacy of quarantine (θ_q), quarantine rate of exposed E_u (α), latent period of exposed E_u ($1/\sigma_u$), latent period of exposed E_q ($1/\sigma_q$), quarantine rate of I_u (ϕ), proportion of exposed E_q who transition to I_h (r), rate of progression to ICU case (ν), and two reduced infectiousness parameters for asymptomatic cases (η_a) and quarantined symptomatic cases I_h (η_h). The model involves several death rate parameters m_j: death rate of non-quarantined infectious I_u (m_u), death rate of quarantined infectious I_h (m_h), death rate of asymptomatic cases I_a (m_a), death rate of ICU patients (m_{icu}). The model involves several removal rate parameters: removal rate of S_q (k_1), removal rate of E_u (k_2), removal rate of E_q (k_3), removal rate of I_u (k_4), removal rate of I_h (k_5), removal rate of I_a (k_6), and the removal rate of I_{icu} (k_7). Note that the model parameters are not independent from each other. For example, some of the removal rate parameters include deaths parameters. Note also that the model equation for the recovered is not presented above because it does not play a role for the subsequent discussion.

Ngonghala et al. [33] determined the effective reproduction number R_e using the next generation method described in Chap. 7. For the model (8.62), R_e assumes the form $R_e = \beta f$, where f depends on other model parameters but is independent of β. As discussed in Sect. 7.5.4, by putting $R_e = 1$ and solving $R_e = \beta f$ for β, the critical effective contact rate $\beta_{crit} = 1/f$ can be obtained for which $\lambda_{\max} = 0$ holds and the population of the state of New York is at its bifurcation point. This procedure yields [27]

$$\beta_{crit} = \frac{k_2 k_4 k_5 k_6}{p(B_u + \eta_a B_a + \eta_h B_h)} \tag{8.64}$$

where B_u, B_a, and B_h are expressions of the model parameters that can be found in Ref. [33]. Finally, in the study by Ngonghala et al. [33], COVID-19 associated deaths were considered in order to fit the model parameters. As mentioned above, to this end, the period from March 1 to April 7, 2020, was used. Subsequently, the study examined possible impacts of intervention measures that among other things would affect the effective contact rate β.

In a study [27] subsequent to the original work by Ngonghala et al. [33], the three-stage scheme shown in panel (b) of Fig. 8.12 was applied to study the entire COVID-19 first-wave in the state of New York during the four months period from March 1 to June 30. All parameters were fixed as in the study by Ngonghala et al. [33]. The effective contact rate β was varied across the three stages like β_{S1}, β_{S2}, and β_{S3}. Accordingly, $\beta_{S1} > \beta_{S2} = \beta_{crit} > \beta_{S3}$ was assumed for the stages 1, 2, 3, respectively. In order to obtain β_{S1} and β_{S3} a standard nonlinear fitting procedure was used to minimize the error between the numerical model solutions $D(t)$ obtained from Eq. (8.62) and the observed deaths. For stage 2 the parameter β_{S2} was computed from Eq. (8.64). While t_0 with March 1 and t_3 with June 30 were fixed, the stage boundaries t_1 and t_2 were varied to find optimal time intervals $[t_0, t_1]$, $[t_1, t_2]$, $[t_2, t_3]$ that produced the best fit between the model solution $D(t)$ and the observed deaths. Data were taken from the website [63].

Figure 8.17 shows the COVID-19 deaths data of the state of New York and the model solution $D(t)$. Panel (a) shows the cumulative confirmed deaths (gray circles) in the 4 months period from March 1 to June 30 and the model solution $D(t)$ (solid black line) obtained via the three-stage approach. $D(t)$ fits the data with moderate

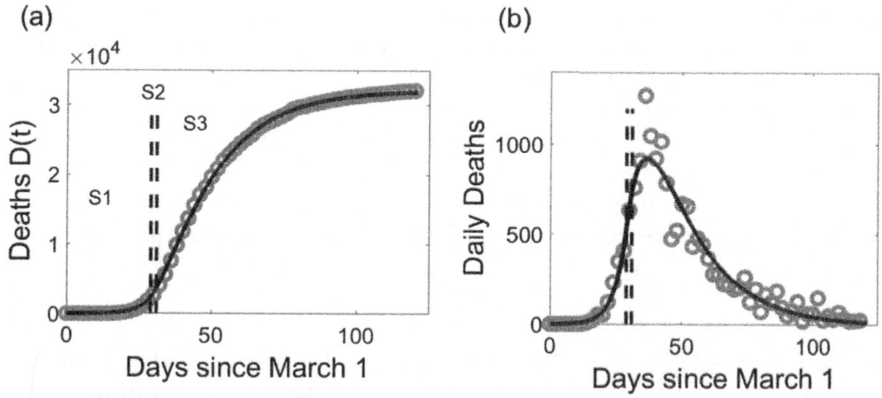

Fig. 8.17 Model-based analysis of the first COVID-19 wave of the state of New York during 2020. Panel (**a**): Cumulative deaths that occurred during the period from March 1 to June 30 (gray circles) and solution $D(t)$ (solid line) of Eq. (8.62) using a three-stage approach. Vertical dashed lines indicate stage boundaries. Panel (**b**): Daily death data (gray circles) during the same period and model fit (solid line)

accuracy and reflects the characteristic sigmoid shape of the trajectory. The two vertical lines indicate stage boundaries t_1 and t_2 at $t_1 = 29$ days (March 30) and $t_2 = 31$ days (April 1) since March 1st. Panel (b) presents the daily new deaths as reported (gray circles) and obtained from the model solution $D(t)$ (solid black line). The stage boundaries are indicated again by vertical lines. The results presented in Fig. 8.17 suggest that the bifurcation stage (i.e., stage 2) was relatively short. Overall, COVID-19 emerged in the population of the state of New York during March 2020 due to an instability (see also the eigenvalue analysis below). COVID-19 associated deaths increased dramatically. However, the model-based analysis suggests that the dynamics of the epidemic changed within the month of March such that at the end of March the dynamics entered stage 2 (i.e., the bifurcation stage) and the unstable disease-free fixed point was about to become stable. This bifurcation stage 2 was relatively short. Beginning of April, the dynamics of the spread of COVID-19 entered stage 3. The disease-free fixed point was stabilized. Panel (b) illustrates that the stabilization effect on the trajectory of daily deaths occurred somewhat later. The tragic peak of about 1000 new deaths per day occurred around days 35 to 40 (April 5 to April 10), that is, a few days after the epidemic entered stage 3.

In order to conduct an eigenvalue analysis and identify the relevant unstable eigenvector of the COVID-19 outbreak the five dimensional state $\mathbf{X}^+ = (E_u, E_q, I_u, I_u, I_h)$ of infected individuals was considered. The variable I_{icu} was neglected in those considerations because according to the model (8.62) and (8.63) it did not affect the stability of the disease-free fixed point. From Eq. (8.62) and (8.63) it follows that the linearization matrix L^+ of the dynamics in D^+ reads

Table 8.5 Results of the three-stage analysis of the first COVID-19 wave in the state of New York, 2020, based on an infected five-dimensional subspace $(E_u, E_q, I_u, I_u, I_h)$: the eigenvalues $\lambda_1, \ldots, \lambda_5$ and the effective contact rate β are shown

Stage	Eigenvalues [1/d]					β [1/d]
	λ_1	λ_2	λ_3	λ_4	λ_5	
1	0.02	−0.28	−0.32	−0.42	−0.64	1.81
2	0	−0.28	−0.33	$-0.42 + i0.04$/d	$-0.42 - i0.04$	0.43
3	−0.06	−0.27	−0.33	$-0.39 + i0.04$	$-0.39 - i0.04$	0.23

$$L^+ = \begin{pmatrix} -k_2 & 0 & (1-q)p\beta & (1-q)p\beta\eta_h & (1-q)p\eta_a \\ \alpha & -k_3 & qp\beta & qp\beta\eta_h & qp\beta\eta_a \\ f_1\sigma_u & 0 & -k_4 & 0 & 0 \\ f_2\sigma_u & r\sigma_q & \phi & -k_5 & \sigma_a \\ (1-f_1-f_2)\sigma_u & (1-r)\sigma_q & 0 & 0 & -k_6 \end{pmatrix} . \tag{8.65}$$

As far as the eigenvalues are concerned, in stage 1, from $\beta > \beta_{crit}$ it follows that the 5×5 matrix exhibits at least one eigenvalue with positive real part, whereas in stage 3, for $\beta < \beta_{crit}$ it follows that all eigenvalues exhibit eigenvalues with negative real parts. In the special case of $\beta = \beta_{crit}$, there is at least one zero eigenvalue parameter. All remaining eigenvalues exhibit real parts equal to zero or negative real parts.

Table 8.5 shows the eigenvalues of L^+ for the stages 1 (outbreak stage), 2 (bifurcation stage), and 3 (subsiding stage). As expected, stage 1 was characterized by a positive eigenvalue, whereas stage 3 showed a set of eigenvalues that were either real-valued and negative or had negative real parts. This illustrates that (as expected) the disease-free fixed point was unstable in stage 1 and asymptotically stable (assuming $\mathbf{X} \approx \mathbf{X}_{st}$) in stage 3. Qualitatively, Table 8.5 demonstrates the sign-switching phenomenon of the bifurcation scenario of epidemic waves: the positive eigenvalue $\lambda_1 = 0.02$/d indicating that during stage 1 the disease-free state of the state of New York's population was unstable turned into a negative one.

The estimated effective contact rates are reported in Table 8.5 as well. For stages 1 and 3 the parameters $\beta(S1) = 1.81$/d and $\beta(S3) = 0.23$/d, respectively, were found. The critical value, which was the stage-2 value, was $\beta(S2) = \beta_{crit} = 0.43$/d.

The eigenvectors $\mathbf{v}_1(S1)$ and $\mathbf{v}_1(S3)$ were computed from the matrix L^+ (see Eq. (8.65)). Figure 8.18 shows the phase curves of the COVID-19 epidemic in terms of the solutions E_u, E_q, I_u, I_h, I_a of Eq. (8.62) in the two-dimensional subspaces E_u-E_q, E_u-I_u, E_u-I_h, and $E_u - I_a$. The projections of $\mathbf{v}_1(S1)$ and $\mathbf{v}_1(S3)$ are plotted in the respective two-dimensional subspaces as well. As can be seen, during stage 1 the disease state $\mathbf{X}^+(t)$ in D^+ of the population of the state of New York followed closely the unstable eigenvector $\mathbf{v}_1(S1)$. Accordingly, $\mathbf{v}_1$ in stage 1 was the 5D order parameter of the COVID-19 outbreak of the state of New York. Likewise, during stage 3 the disease state followed $\mathbf{v}_1(S3)$ (i.e., the remnant of the order parameter).

The amplitude description of the first COVID-19 wave of the state of New York was derived using the numerical approach discussed in Sect. 8.6.2. To this end, the

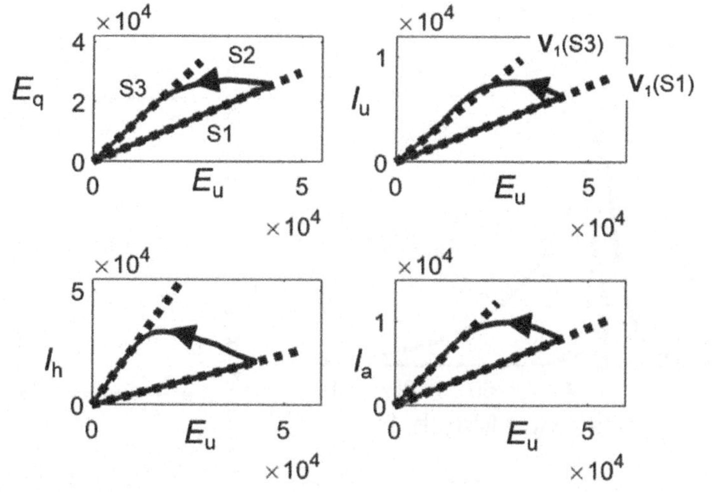

Fig. 8.18 State of New York phase curves (solid lines) in two-dimensional subspaces and the directions specified by the dominant eigenvector $\mathbf{v}_1$ in stage 1 (i.e., $\mathbf{v}_1(S1)$) and stage 3 (i.e., $\mathbf{v}_1(S3)$) (dotted lines). In all subpanels, the lower and upper directions refer to $\mathbf{v}_1(S1)$ and $\mathbf{v}_1(S3)$, respectively

trajectories E_u, E_q, I_u, I_h, I_a were substituted into Eq. (8.60) to obtain the amplitudes $A_1, \ldots, A_5$ as functions of time over the three stages. In this context, A_j was defined as the amplitude related to the eigenvalue λ_j listed in Table 8.5.

Figure 8.19 presents the amplitudes $A_1, \ldots, A_5$ as functions of time. Panel (a) shows the real-valued amplitudes A_1, A_2, A_3 related to the eigenvalues λ_1, λ_2, λ_3 that were real-valued in all three stages and exhibited the largest real parts (see Table 8.5). Panels (b) and (c) show the real- and imaginary-parts of the two remaining amplitudes A_4 and A_5. The corresponding eigenvalues λ_4 and λ_5 were real-valued in stage 1 but assumed complex numbers in stages 2 and 3 (see Table 8.5 again). Consequently, A_4 and A_5 were real-valued in stage 1 but formed a complex-valued pair in stages 2 and 3. Comparing A_1 with the remaining amplitudes, it can be seen that A_1 was by several order of magnitudes larger in stages 1 and 3 than the remaining amplitudes. That is, A_1 played the dominant role. This observation is consistent with the phase curves shown in Fig. 8.18 that suggest that the epidemic evolved primarily along the eigenvector $\mathbf{v}_1$ in stages 1 and 3. Returning to Fig. 8.19, in stage 2, A_1 was still the largest amplitude. However, A_1 remaining almost constant over time (due to its eigenvalue $\lambda_1 = 0$). During stage 2, the amplitudes A_2 and A_3 (panel (a)) and the imaginary parts of A_4 and A_5 (panel (c)) decayed in magnitude. Interestingly, the real parts of A_4 and A_5 (panel (b)) increased in magnitude during the very short bifurcation period of stage 2. Overall, changes of the state $\mathbf{X}^+(t)$ in stage 2 were determined by variations in A_2, A_3, A_4, and A_5. In other words, the stage 2 dynamics shown in the phase curves in Fig. 8.18 was primarily due to changes of the A_2, A_3, A_4, and A_5 related to eigenvalues with non-zero real parts.

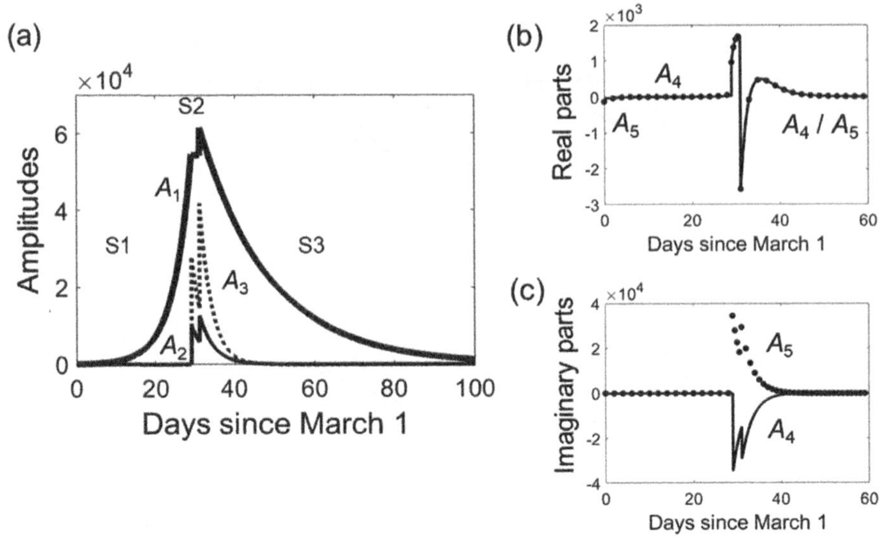

Fig. 8.19 Three-stage amplitude description of the first COVID-19 wave of the state of New York during 2020. Panel (**a**): Amplitudes $A_1(t)$ (thick solid line), $A_2(t)$ (thin solid line), and $A_3(t)$ (dotted line) are shown across the stages S1, S2, and S3. Panels (**b**) and (**c**): The real parts (panel (**b**)) and imaginary parts (panel (**c**)) of $A_4(t)$ (solid line) and $A_5(t)$ (full circles) are shown for stages S1, S2, and S3

In closing these considerations on the first COVID-19 wave in the state of New York, let us return to the analytical solutions (8.50) of $D(t)$. For the model (8.62) the variable I_{icu} makes a contribution to $D(t)$ via the rate constant m_{icu}. Let use denote L^+ as defined in Eq. (8.65) by $L^+_{5\times 5}$. When taking the variable I_{icu} into consideration, the linearization matrix becomes a 6×6 square matrix defined by

$$L^+_{6\times 6} = \begin{pmatrix} & & & & & 0 \\ & & & & & 0 \\ & & L^+_{5\times 5} & & & 0 \\ & & & & & 0 \\ & & & & & 0 \\ 0 & 0 & 0 & 0 & \nu & -k_7 \end{pmatrix} . \tag{8.66}$$

The matrix $L^+_{6\times 6}$ exhibits the same eigenvalues as listed in Table 8.5 and an additional eigenvalue $\lambda_6 = -k_7$. Moreover, let $\mathbf{v}_j^{(5)}$ denote the five-dimensional eigenvectors of $L^+_{5\times 5}$ and $\mathbf{w}_j^{(5)}$ the corresponding biorthogonal vectors. Then, the matrix $L^+_{6\times 6}$ exhibits the six-dimensional eigenvectors $\mathbf{v}_j^{(6)}$ defined by

$$\mathbf{v}_j^{(6)} = \frac{1}{Z_j} \begin{pmatrix} \mathbf{v}_j^{(5)} \\ c_j \end{pmatrix} \tag{8.67}$$

for $j = 1, \ldots, 5$ with c_j determined by the sixth row of $L^{+}_{6\times 6}$ like $\nu v^{(5)}_{j,5} - (k_7 + \lambda_1)c_j = 0$, which implies $c_j = \nu v^{(5)}_{j,5}/(k_7 + \lambda_1)$ and $Z_j = \sqrt{1 + c_j^2}$. The sixth eigenvector reads $\mathbf{v}^{(6)}_6 = (0, 0, 0, 0, 0, 1)$. Importantly, since the $\mathbf{w}^{(5)}_j \mathbf{v}^{(5)}_k = \delta_{jk}$ holds for $j = 1, \ldots, 5$ and $k = 1, \ldots, 5$, it follows that the biorthogonal vectors $\mathbf{w}^{(6)}_j$ for $j = 1, \ldots, 5$ are given by

$$\mathbf{w}^{(6)}_j = Z_j \begin{pmatrix} \mathbf{w}^{(5)}_j \\ 0 \end{pmatrix} . \tag{8.68}$$

For $j = 6$ we have $\mathbf{w}^{(6)}_6 = \mathbf{v}^{(6)}_6 = (0, 0, 0, 0, 0, 1)$. That is, from Eqs. (8.67) and (8.68) and the biorthogonality of the respective five-dimensional vectors, it follows that $\mathbf{w}^{(6)}_j \mathbf{v}^{(6)}_k = \delta_{jk}$ holds for $j = 1, \ldots, 6$ and $k = 1, \ldots, 6$. Consequently, let $\mathbf{X}^{(6,+)} = (E_u, E_q, I_u, I_u, I_h, I_{icu})$ and $A^{(6)}_1$ denote six-dimensional state vector and the order parameter amplitude related to λ_1 in stage 1, then

$$A^{(6)}_1 = \mathbf{w}^{(6)}_1 \mathbf{X}^{(6,+)}(t) = Z_1 \mathbf{w}^{(5)}_1 \mathbf{X}^{+}(t) = Z_1 A_1 \tag{8.69}$$

with the five-dimensional vector $\mathbf{X}^{+} = (E_u, E_q, I_u, I_h, I_a)$ and A_1 shown in Fig. 8.19. Equation (8.49) becomes

$$\frac{\mathrm{d}}{\mathrm{d}t} D = \mathbf{m}\mathbf{v}^{(6)}_1 Z_1 A_1(t) , \tag{8.70}$$

where again A_1 is shown in Fig. 8.19. Likewise, Eq. (8.50) becomes

$$\begin{aligned} &\mathrm{S1} \ , t \in [t_0, t_1] \ , \ \lambda_1 = \lambda_{1,S1} > 0 \ : \\ &D(t) = D(t_0) + \frac{\mathbf{m}\mathbf{v}^{(6)}_{1,S1}}{\lambda_{1,S1}} Z_1 A_1(t_0) \left(\exp\{\lambda_{1,S1}(t - t_0)\} - 1\right) . \\ &= D(t_0) + \frac{\mathbf{m}\mathbf{v}^{(6)}_{1,S1}}{\lambda_{1,S1}} Z_1 \left(A_1(t) - A_1(t_0)\right) . \end{aligned} \tag{8.71}$$

Accordingly, the exponential increase of $A_1(t)$ shown in Fig. 8.19 during the first 29 days (stage 1) of the epidemic in the state of New York determined the increase of deaths shown in panel (a) of Fig. 8.18 in that first 29 days period.

8.7.2 First COVID-19 Wave of 2020 in Pakistan

Ullah and Khan [71] developed an epidemiological model to describe the increasing COVID-19 cases during the first few months of the COVID-19 epidemic in Pakistan of the year 2020. In a subsequent study, the model was used to work out a three-stage description of the increasing and subsiding infection numbers of the first COVID-19 wave in Pakistan during the period March to September 2020 [26]. The model for the COVID-19 epidemic in Pakistan involves eight compartments, which are the compartments of susceptible individuals (S), exposed individuals (E), symptomatic infectious individuals (I_s), asymptomatic infectious individuals (I_a), hospitalized infectious individuals (I_h), and infectious individuals who require treatment in intensive care units (I_{icu}). The model also addresses infected quarantined individuals (Q) as well as recovered individuals (R). The model reads [26, 71]

$$\begin{aligned}
\frac{\mathrm{d}}{\mathrm{d}t}S &= -k_0 S \,, \\
\frac{\mathrm{d}}{\mathrm{d}t}E &= k_0 S - k_1 E \,, \\
\frac{\mathrm{d}}{\mathrm{d}t}I_s &= \rho\omega E - k_2 I_s \,, \\
\frac{\mathrm{d}}{\mathrm{d}t}I_a &= (1-\rho)\omega E - k_3 I_a \,, \\
\frac{\mathrm{d}}{\mathrm{d}t}I_h &= \eta I_s + \delta Q - k_5 I_h \,, \\
\frac{\mathrm{d}}{\mathrm{d}t}Q &= \kappa E - k_4 Q \,, \\
\frac{\mathrm{d}}{\mathrm{d}t}I_{icu} &= \phi I_h - k_6 I_{icu}
\end{aligned} \tag{8.72}$$

with the rate constant k_0 defined by

$$k_0 = \beta \frac{I_s + \psi I_a + \nu I_h}{N} \,. \tag{8.73}$$

The model parameters describe the latent period of exposed individuals ($1/\omega$), the hospitalization rate of symptomatic infectious individuals (η), the hospitalization rate of quarantined individuals (δ), the quarantine rate of exposed individuals (κ), the rate at which hospitalized individuals require ICU care (ϕ), and the proportion of asymptomatic infections (ρ). The parameters $k_1, \ldots, k_6$ denote removal rates of individuals out of certain compartments. The parameters occurring in Eq. (8.73) are the effective contact rate (β), the relative transmissibility of the asymptomatic individuals (ψ), and the relative transmissibility of hospitalized individuals (ν). In Eq. (8.72) the evolution equation of the recovered individuals is not listed because it will be of no concern in the following considerations. Furthermore, note that the original study by Ullah and Khan [71] takes demographic terms into account. In contrast, in Ref. [26] and in Eq. (8.72) those terms have been neglected because on the relative short period (March to September) they do not affect the disease dynamics [73].

Model parameters were fitted by introducing the variable C of cumulative COVID-19 cases. To this end, the assumption was made that symptomatic individuals were diagnosed quickly by health authorities as being infected with COVID-19. However, an expression like $\mathrm{d}C/\mathrm{d}t = gI_s$ with g large was not used. Rather, it was assumed that C increased when exposed individuals became symptomatic infectious like $\mathrm{d}C/\mathrm{d}t = \rho\omega E$, where $\rho\omega$ denotes the transition rate of $E \to I_s$ transitions (see Eq. (8.72)). Moreover, it was assumed that quarantined individuals who became hospitalized because they developed severe COVID-19 symptoms increased the number of confirmed cumulative COVID-19 cases. From these considerations it follows that $C(t)$ satisfies [26]

$$\frac{\mathrm{d}}{\mathrm{d}t}C = s_E E + s_Q Q\,. \tag{8.74}$$

with $s_E = \rho\omega$ and $s_Q = \delta$ (see Eq. (8.44)).

The key parameter of the model defined by Eqs. (8.72) and (8.73) is the effective contact rate β. As discussed in Sect. 7.5.5, the critical value β_{crit} at which according to the model (8.72) a bifurcation from an unstable to a stable disease-free fixed point takes place can be determined with the help of Eq. (7.75). The result reads [26]

$$\beta_{crit} = \frac{k_1 k_2 k_3 k_4 k_5}{Z}\,,\quad Z = k_4\omega[\rho k_3(\eta\nu + k_5) + (1-\rho)k_2 k_5 \psi] + \kappa\delta k_2 k_3 \nu\,. \tag{8.75}$$

The three-stage model shown in panel (a) of Fig. 8.12 was applied. To this end, it was assumed that β changed across the three stages $\beta_{S1} > \beta_{S2} = \beta_{crit} > \beta_{S3}$ (see also the previous example in Sect. 8.7.1). For all three stages the model parameters as determined by Ullah and Khan [71] were used except for the values of β. Solutions of Eqs. (8.72) and (8.74) were then fitted in the three-stage study [26] to observed cumulative COVID-19 case from Pakistan [63]. The time points t_1 and t_2 of the beginnings of stages 2 and 3 were varied. The parameters β_{S1} and β_{S3} were estimated for stages 1 and 3 as the optimal parameters that produced the best model fit given a the time points (t_1, t_2). For stage 2 the parameter $\beta = \beta(crit)$ was used and computed from Eq. (8.75). In summary, in Ref. [26] the four model parameters t_1, t_2, β_{S1}, and β_{S3} were estimated using COVID-19 case data from Pakistan.

Figure 8.20 present the confirmed COVID-19 cases in Pakistan and the best-fit model solution $C(t)$. Panel (a) shows the cumulative COVID-19 cases (gray circles) during the 7-months period (about 210 days) from March 1 to September 30 as reported on the website [63]. The data showed the typically sigmoid, three-stage pattern: a first (accelerating) bend, a linear stage, and a second (de-accelerating) bend. The optimal stage boundaries were t_1 as June 10 for the beginning of stage 2 and t_2 as July 1 for the beginning of stage 3. The solid black line in panel (a) shows the model solution $C(t)$ computed from Eqs. (8.72) and (8.74). The model captures the sigmoid shape of the first COVID-19 wave in Pakistan with moderate accuracy. Panel (b) of Fig. 8.20 shows daily new COVID-19 cases (gray circles) as reported in the database [63] and the model fit (solid black line). The three stages of the model are clearly visible. In stage 1 there is a rapid increase of daily new infections. In stage 2 there is a relative fast decrease of daily new infections to a plateau. In stage 3 a decay of daily new infections can be seen from the aforementioned plateau towards a low level of about 500 confirmed daily new infections.

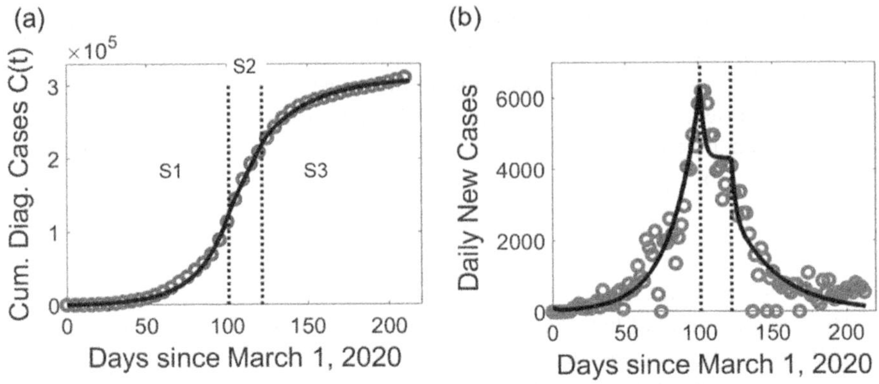

Fig. 8.20 Model-based analysis of the first COVID-19 wave in Pakistan, 2020. Panel (**a**): Cumulative COVID-19 cases confirmed during the period from March 1 to September 30 (gray circles) and solution $C(t)$ (solid line) computed from Eqs. (8.72) and (8.74) using a three-stage approach. Vertical dotted lines indicate stage boundaries. Panel (**b**): Daily new confirmed cases (gray circles) during the same period and model fit (solid line)

In Ref. [26], in order to determine the eigenvalues and eigenvectors that determined the three stages of the 2020 COVID-19 first-wave in Pakistan, the infected compartments E, I_s, I_a, I_h, and Q were considered. They span the subspace D^+. Individuals in intensive care units were neglected in Ref. [26] because according to the model (8.72) they did affect the stability of the disease-free fixed point. In the five-dimensional space D^+ the linearization matrix of Eq. (8.72) reads

$$L^+ = \begin{pmatrix} -k_1 & \beta & \beta\psi & \beta\nu & 0 \\ \rho\omega & -k_2 & 0 & 0 & 0 \\ (1-\rho)\omega & 0 & -k_3 & 0 & 0 \\ 0 & \eta & 0 & -k_5 & \delta \\ \kappa & 0 & 0 & 0 & -k_4 \end{pmatrix} . \tag{8.76}$$

The eigenvalues of L^+ are listed in Table 8.6 for all three stages. As can be seen in Table 8.6, the largest eigenvalue of stage 1 was positive with 0.05/d, indicating that the disease-free fixed point was unstable. In contrast, the largest eigenvalue was negative in stage 3 with −0.03/d, indicating that the fixed point was stabilized presumably due to the implementation of intervention measures. For sake of completeness, Table 8.6 also presents the effective contact rates β. As expected, the order $\beta(S1) > \beta(S2) = \beta(crit) > \beta(S3)$ was found.

In order to demonstrate the role of the eigenvector $\mathbf{v}_1$ related to the maximal eigenvalue λ_1, the vector $\mathbf{v}_1$ was computed numerically from L^+ for stages 1 and 3. Subsequently, phase curves were plotted and compared with $\mathbf{v}_1(S1)$ and $\mathbf{v}_1(S3)$. Figure 8.21 shows the phase curves of the COVID-19 epidemic as obtained from the solutions E, I_s, I_a, I_h, Q of Eq. (8.72) in the two-dimensional subspaces $E - I_a$, $I_s - I_a$, $I_h - I_a$, and $Q - I_a$. The eigenvector $\mathbf{v}_1$ for stages 1 and 3 is shown

Table 8.6 Results of the three-stage analysis of the first COVID-19 wave in Pakistan, 2020, as obtained for the infected five-dimensional subspace (E, I_s, I_a, I_h, Q): the eigenvalues $\lambda_1, \ldots, \lambda_5$ and the effective contact rate β are shown

Stage	Eigenvalues [1/d]					β [1/d]
	λ_1	λ_2	λ_3	λ_4	λ_5	
1	0.05	−0.49	−0.81	$-1.31 + i0.18$	$-1.31 - i0.18$	0.60
2	0	−0.51	−0.78	$-1.10 + i0.14$	$-1.10 - i0.14$	0.35
3	−0.03	−0.53	−0.75	$-1.10 + i0.10$	$-1.10 - i0.10$	0.21

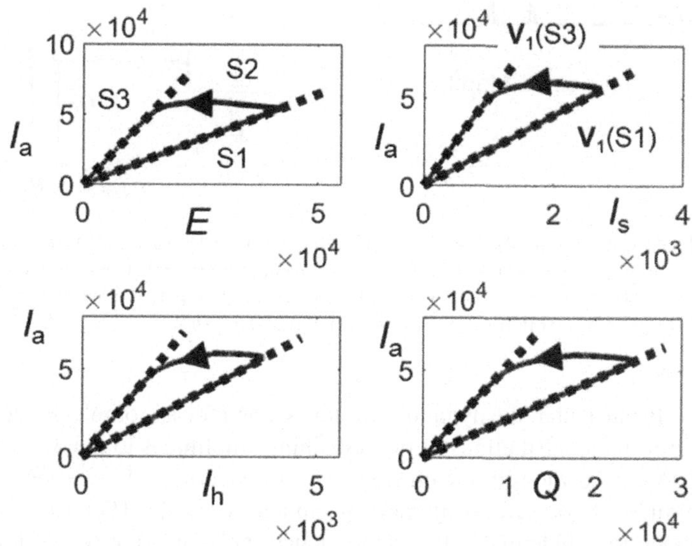

Fig. 8.21 Phase curves (solid lines) of the COVID-19 epidemic (March 1 to September 30, 2020) in Pakistan shown in two-dimensional subspaces. The directions (dotted lines) defined by the 5D order parameter $\mathbf{v}_1(S1)$ of stage 1 and its stage-3 remnant $\mathbf{v}_1(S3)$ are shown as well. In all four subpanels, the lower and upper directions refer to $\mathbf{v}_1(S1)$ and $\mathbf{v}_1(S3)$, respectively

there as well. As can be seen, the epidemic followed closely $\mathbf{v}_1$ in both stages 1 and 3. Accordingly, $\mathbf{v}_1$ was the five-dimensional order parameter in the first stage of the 2020 COVID-19 outbreak in Pakistan. Moreover, the remnant $\mathbf{v}_1(S3)$ of the order parameter $\mathbf{v}_1(S1)$ determined the way the first COVID-19 wave subsided in Pakistan.

Following the numerical approach discussed in Sect. 8.6.2, the amplitudes $A_1, \ldots, A_5$ related to the eigenvalues listed in Table 8.5 were computed from Eq. (8.60) for all three stages using the numerical solutions $E(t)$, $I_s(t)$, $I_a(t)$, $I_h(t)$, $Q(t)$ of Eq. (8.72). Figure 8.22 presents the amplitudes as functions of time. Panel (a) presents the three real-valued amplitudes A_1, A_2, A_3. Panels (b) and (c) show the real- and imaginary-parts of the amplitudes A_4 and A_5 related to the two remaining eigenvalues that were complex-valued in all three stages. Figure 8.22 reveals that at any point in time A_1 was at least by a factor 10 larger in magnitude than the other

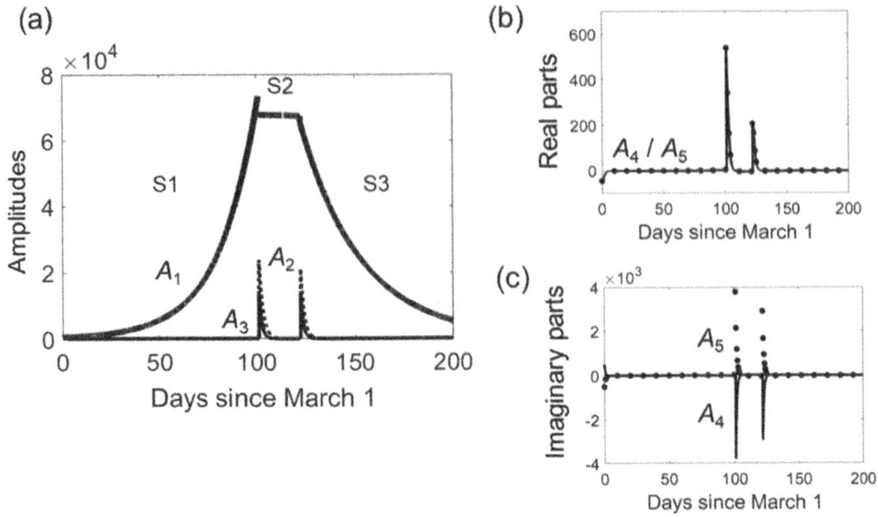

Fig. 8.22 Three-stage amplitude description of the first COVID-19 wave in Pakistan during 2020. Panel (**a**): Amplitudes $A_1(t)$ (thick solid line), $A_2(t)$ (dotted line), and $A_3(t)$ (thin solid line) are shown across the stages S1, S2, and S3. Panels (**b**) and (**c**): The real parts (panel (**b**)) and imaginary parts (panel (**c**)) of $A_4(t)$ (solid line) and $A_5(t)$ (full circles) are shown for stages S1, S2, and S3

amplitudes. In particular, the dynamics in stages 1 and 3 was completely determined by A_1. That is, the contributions of the remaining amplitudes to the dynamics were negligible. As far as the bifurcation stage 2 was concerned, A_1 was larger than all other amplitudes. However, A_1 did not vary much over time. Therefore, during the stage at which the epidemic in Pakistan reached its bifurcation (or turning) point, changes of the state $\mathbf{X}^+(t)$ were determined primarily by the remaining amplitudes A_2 to A_5.

References

1. M. Ferraresi, A coronavirus cautionary tale from Italy: don't do what we did, in *Boston Globe*, http://www.bostonglobe.com/2020/03/13/opinion/coronavirus-cautionary-t
2. R.D. Truog, C. Mitchell, G.Q. Daley, The toughest triage: allocating ventilators in a pandemic. N. Engl. J. Med. **382**, 1973–1975 (2020)
3. V.E. Schaye, J.A. Reich, B.P. Bosworth, et al., Collaborating across private, public, community, and federal hospital systems: lessons learned from the COVID-19 pandemic response in NYC. NEJM Catal. **1** (2020). Article CAT.20.0343
4. D.R.Q. Lemos, S.M. D'angelo, L.A.B.G. Farias, et al., Health system collapse 45 days after the detection of COVID-19 in Ceara, Northesat Brazil: a preliminary analysis. J. Braz. Soc. Trop. Med. **53** (2020). Article e20200354
5. S.J.R. da Silva, L. Pena, Collapse of public health system and the emergence of new variants during the second wave of the COVID-19 pandemic in Brazil. One Health **13** (2021). Article 100287

6. D. Scott, P. Raina, M. Sitlhou, The long road to India's unparalleled pandemic catastrophe. *Vox News*, http://www.vox.com/coronavirus-covid19/22628806/india-covid-19-cases
7. R. Dickman, A SEIR-like model with a time-dependent contagion factor describes the dynamics of the Covid-19 pandemic. medRxiv (2020). Article 20169557
8. T.D. Frank, S. Chiangga, SEIR order parameters and eigenvectors of the three stages of completed COVID-19 epidemics: with an illustration for Thailand January to May 2020. Phys. Biol. **18** (2021). Article 046002
9. K. Rock, S. Brand, J. Moir, M.J. Keeling, Dynamics of infectious diseases. Rep. Prog. Phys. **77** (2014). Article 026602
10. H.W. Hethcote, Three basic epidemiological models, in *Applied Mathematical Ecology*. ed. by S. Levin, T.G. Hallam, L.J. Gross (Springer, Berlin, 1989), pp. 119–144
11. J. Wangping, H. Ke, S. Yang, C. Wenzhe, W. Shengshu, Y. Shanshan, W. Jianwei, K. Fuyin, T. Penggang, L. Jing, L. Miao, H. Yao, Extended SIR prediction of the epidemics trend of COVID-19 in Italy and compared with Hunan, China. Front. Med. **7** (2020). Article 169
12. M.G. Pedersen, M. Meneghini, A simple method to quantify country-specific effects of COVID-19 containment measures. MedRxiv (2020). Article 2020.04.07.20057075
13. D. Fanelli, F. Piazza, Analysis and forecast of COVID-19 spreading in China, Italy, and France. Chaos, Solitons Fractals **134** (2020). Article 109761
14. M.J. Willis, V.H.G. Diaz, O.A. Prado-Rubio, M. von Stosch, Insights. Chaos, Solitons Fractals **138** (2020). Article 109937
15. B. Zareie, A. Roshani, M.A. Mansournia, M.A. Rasouli, G. Moradi, A model. Arch. Iran. Med. **23**, 244–248 (2020)
16. G.D. Barmparis, G.P. Tsironis, Estimating the infection horizon of COVID-19 in eight countries with a data-driven approach. Chaos, Solitons Fractals **135** (2020). Article 109842
17. N. Crokidakis, Covid-19 spread in Rio de Janeiro, Brazil: do the policies of social isolation really well. Chaos, Solitons Fractals **136** (2020). Article 109930
18. L. Pang, S. Liu, X. Zhang, T. Tian, Z. Zhao, Transmission dynamics and control strategies of COVID-19 in Wuhan, China. J. Biol. Syst. **28**, 543–561 (2020)
19. T.D. Frank, COVID-19 order parameters and order parameter time constants of Italy and China: a modeling approach based on synergetics. J. Biol. Syst. **28**, 589–608 (2020)
20. M. Gatto, E. Bertuzzo, L. Mari, S. Miccoli, L. Carraro, R. Casagrandi, A. Rinaldo, Spread and dynamics of the COVID-19 epidemic in Italy: effects of emergency containment measures. Proc. Natl. Acad. Sci. U.S.A. **117**, 10484–10491 (2020)
21. T. Sun, Y. Wang, Modeling COVID-19 epidemic in Heilongjiang province, China. Chaos, Solitons Fractals **138** (2020). Article 109949
22. H. Zhao, Z. Feng, Staggered release policies for COVID-19 control: costs and benefits of relaxing restrictions by age and risk. Math. Biosci. **326** (2020). Article 108405
23. J.F. Oliveira, D.C.P. Jorge, R.V. Veiga, M.S. Rodrigues, M.F. Torquato, N.B. da Silva, R.L. Fiaconne, L.L. Cardim, F.A.C. Pereira, C.P. de Castro, A.S.S. Paiva, A.A.S. Amad, E.A. BF. Lima, D.S. Souza, S.T.R. Pinho, P.I.P. Ramos, R.F.S. Andrade, Mathematical modeling of Covid-19 in 14.8 million individuals in Bahia, Brazil. Nat. Commun. **12** (2021). Article 333
24. M. Serhani, H. Labbardi, Mathematical modeling of COVID-19 spreading with asymptomatic infected and interacting people. J. Appl. Math. Comput. **66**, 1–20 (2021)
25. T.D. Frank, COVID-19 interventions in some European countries induced bifurcations stabilizing low death states against high death states: an eigenvalue analysis based on the order parameter concept of synergetics. Chaos, Solitons Fractals **140** (2020). Article 110194
26. T.D. Frank, Emergence and subsiding of the first-wave COVID-19 pandemic in Pakistan 2020: an eigenvalue analysis based on synergetics. Proc. Pak. Acad. Sci. B **57**, 1–7 (2020)
27. T.D. Frank, Rise and decay of the COVID-19 epidemics in the USA and the State of New York in the first half of 2020: a nonlinear physics perspective yielding novel insights. BioMed. Res. Int. **2021** (2021). Article 6645688
28. S.M. Garba, J.M.S. Lubuma, B. Tsanou, Modeling the transmission dynamics of the COVID-19 pandemic in South Africa. Math. Biosci. **328** (2020). Article 108441

29. B. Ivorra, M.R. Ferrandez, M. Vela-Perez, A.M. Ramos, Mathematical modeling of the spread of coronavirus disease 2019 (COVID-19) taking into account the undetected infections. the case of China. Commun. Nonlinear Sci. Numer. Simulat. **88** (2020). Article 105303
30. Z. Zhao, X. Li, F. Liu, G. Zhu, C. Ma, L. Wang, Prediction of the COVID-19 spread in Africa countries and implications for prevention and control: a case study in South Africa, Egypt, Algeria, Nigeria, Senegal, and Kenya. Sci. Total Environ. **729** (2020). Article 138959
31. M. Mandal, S. Jana, S.K. Nandi, A. Khatua, S. Adak, T.K. Kar, A model based study on the dynamics of COVID-19: prediction and control. Chaos, Solitons Fractals **136** (2020). Article 109889
32. K.N. Nabi, Forecasting COVID-19 pandemic: a data-driven analysis. Chaos, Solitons Fractals **139** (2020). Article 110046
33. C.N. Ngonghala, E. Iboi, S. Eikenberry, M. Scotch, C.R. MacIntyre, M.H. Bonds, A.B. Gumel, Mathematical assessment of the impact of non-pharmaceutical interventions on curtailing the 2019 novel Coronavirus. Math. Biosci. **325** (2020). Article 108364
34. N. Crokidakis, Modeling the early evolution of the Covid-19 in Brazil: results from a susceptible-infectious-quarantined-recovered (SIQR) model. Int. J. Mod. Phys. C **31** (2020). Article 2050135
35. Statistisches Bundesamt (Germany), Bevoelkerung 2020, in *Statistisches Bundesamt (Germany)*, https://www.destatis.de/DE/Themen/Gesellschaft-Umwelt/Bevoelkerung/Bevoelker
36. T. Frank, *Determinism and Self-organization of Human Perception and Performance* (Springer, Berlin, 2019)
37. M.C. Mackey, L. Glass, Oscillations and chaos in physiological control systems. Science **197**, 287–289 (1977)
38. R. Friedrich, S. Siegert, J. Peinke, St. Lück, M. Seifert, M. Lindemann, J. Raethjen, G. Deuschl, G. Pfister, Extracting model equations from experimental data. Phys. Lett. A **271**, 217–222 (2000)
39. J.L. Perez-Velazquez, L.G. Dominguez, V. Nenadovic, R.A. Wennberg, Experimental observation of increased fluctuations in an order parameter before epochs of extended brain synchronization. J. Biol. Phys. **37**, 141–152 (2011)
40. L. Zhang, Q. Wang, G. Baier, Spontaneous transitions to focal-onset epileptic seizures: a dynamical study. Chaos **30** (2020). Article 103114
41. J.L. Perez-Velazquez, The biophysical basis of will-less behaviors. Front. Integr. Neurosci. **6** (2012). Article 98
42. Y. Suzuki, A. Nakamura, M. Milosevic, et al., Postural instability via a loss of intermittent control in elderly and patients with Parkinson's diseases. Chaos **30** (2020). Article 113140
43. C. Foley, M.C. Mackey, Dynamic hematological disease: a review. J. Math. Biol. **58**, 285–322 (2009)
44. M.C. Mackey, Periodic hematological disorders: quintessential examples of dynamical diseases. Chaos **30** (1920). Article 063123
45. C. Zhuge, M.C. Mackey, J. Lei, Origins of oscillation patterns in cyclical thrombocytopenia. J. Theor. Biol. **462**, 432–445 (2019)
46. S. Avissar, G. Schreiber, The involvement of guanine nucleotide binding proteins in the pathogenesis and treatment of affective disorders. Biol. Psychiatry **31**, 435–459 (1992)
47. M.B. Bonsall, S.M.A. Wallace-Hadrill, J.R. Geddes, G.M. Goodwin, E.A. Holmes, Nonlinear time-series approaches in characterizing mood stability and mood instability in bipolar disorder. Proc. R. Soc. B **279**, 916–924 (2012)
48. T.D. Frank, A limit cycle model for cycling mood variations of bipolar disorder patients derived from cellular biochemical reaction equations. Commun. Nonlinear Sci. Numer. Simul. **18**, 2107–2119 (2013)
49. T.D. Frank, From systems biology to systems theory of bipolar disorder, in *Systems Theory: Perspectives, Applications and Developments*, ed. by F. Miranda (Nova Publ., New York, 2014), pp. 17–36
50. A. Steinacher, K.A. Wright, Relating the bipolar spectrum to dysregulation of behavioral activation: a perspective from dynamical modelling. PLoS One **8** (2013). Article e63345

51. F. Tretter, P. Gebicke-Haerter, U. an der Heiden, H.W. Mewes, C.W. Turck, Affective disorders as complex dynamic diseases: a perspective from systems biology. Pharmacopsychiatry **44**(Suppl 1), S2–S8 (2011)
52. L. Ciompi, The key role of emotions in the schizophrenia puzzle. Schizophr. Bull. **4**, 318–322 (2015)
53. J. Belair, F. Nekka, J. Milton, Introduction to focus issue: dynamical disease: a translation approach. Chaos **31** (2021). Article 060401
54. S. Alfonso, A.L. Jenner, M. Craig, Translational approaches to treating dynamical diseases in silico clinical trials. Chaos **30** (2020). Article 123128
55. L.R. Dickman, Y. Kuang, Analysis of tumor-immune dynamics in a delayed dendritic cell therapy model. Chaos **30** (2020). Article 113108
56. M. Kromer, A. Khaledi-Nasab, P.A. Tass, Impact of number of stimulation sites on long-lasting desynchronization effect of coordinated reset stimulation. Chaos **30** (2020). Article 083134
57. H. Haken, G. Schiepek, *Synergetik in der Psychologie* (Hogrefe, Gottingen, 2006). (in German)
58. G. Schiepek, I. Tominschek, S. Karch et al., A controlled single case study with repeated fMRI measurements during the treatment of a patient with obsessive-compulsive disorder: testing the nonlinear dynamics approach to psychotherapy. World J. Biol. Psychiatry **10**, 658–668 (2009)
59. G. Schiepek, G. Strunk, The identification of critical fluctuations and phase transitions in short term and coarse-grained time series: a method for real time monitoring of human change processes. Biol. Cybern. **102**, 197–207 (2010)
60. G. Schiepek, I. Tominschek, S. Heinzel, et al., Discontinuous patterns of brain activation in the psychotherapy process of obsessive-compulsive disorder: converging results from repeated fMRI and daily self-reports. PLoS ONE **8** (2013). Article e71863
61. G. Schiepek, W. Aichhorn, M. Gruber, G. Strunk, E. Bachler, B. Aas, Real-time monitoring of psychotherapeutic processes: concept and compliance. Front. Psychol. **7** (2016). Article 604
62. Tageschau, Ende der Corona-Grenzen, http://www.tageschau.de/inland/reisen-grenzen-deutschland-101.htm
63. COVID-19 tracker, Timeline data from Johns Hopkins Center for Systems Science and Engineering. *COVID-19 tracker*, https://vac-lshtml.shinyapps.io/ncov_tracker
64. WHO, Coronavirus disease 2019 WHO Thailand Situation Report, 21 February 2020, in *Coronavirus Disease 2019 WHO Thailand Situation Report, 21 February 2020*, https://www.who.int/thailand/emergencies/novel-coronavirus-2019
65. Thailand Department of Disease Control, *The Coronavirus Disease 2019 Situation: Thailand Situation Update on 25 May 2020*, https://dds.moph.go.th/viralpneumonia/eng/situation.php
66. Thailand Department of Disease Control, *The Coronavirus Disease 2019 Situation: Thailand Situation Update on 1 September 2020*, https://dds.moph.go.th/viralpneumonia/eng/situation.php
67. C. Distante, P. Piscitelli, A. Miani, Covid-19 outbreak progression in Italian regions: approaching the peak by the end of March in Northern Italy and first week of April in Southern Italy. Environ. Res. Publ. Health **17** (2020). Article 3025
68. A. Bouchnita, A. Jebrane, A hybrid multi-scale model of COVID-19 transmission dynamics to assess the potential of non-pharamceutical interventions. Chaos, Solitons Fractals **138** (2020). Article 109941
69. F. Ndairou, I. Area, J.J. Nieto, D.F.M. Torres. Mathematical modeling of COVID-19 transmission dynamics with a case study of Wuhan. Chaos, Solitons Fractals **135** (220). Aticle 109846
70. H.B. Taboe, K.V. Salako, J.M. Tison, C.N. Ngonghala, R.G. Kakai, Predicting COVID-19 spread in the face of control measures in West Africa. Math. Biosci. **328** (2020). Article 108431

71. S. Ullah, M.A. Khan, Modeling the impact of non-pharmaceutical interventions on the dynamics of novel coronavirus with optimal control analysis with a case study. Chaos, Solitons Fractals **139** (2020). Article 110075
72. T.D. Frank, J. Smucker, Characterizing stages of COVID-19 epidemics: a nonlinear physics perspective based on amplitude equations. Eur. Phys. J. Spec. Top. (in press)
73. H.W. Hethcote, The mathematics of infectious diseases. SIAM Rev. **42**, 599–653 (2000)

Chapter 9
Models of Virus Dynamics

This chapter is about mechanistic, biophysical and biochemical processes of viral infectious diseases in the human body as seen from a nonlinear physics perspective. These processes can be described by a variety of nonlinear physics models. They apply to virus infections in general. Before reviewing some of the benchmark models, some facts and hypotheses about the SARS coronavirus 2 (SARS-CoV-2) that causes COVID-19 are presented.

9.1 Coronaviruses

9.1.1 Classification

Coronaviruses are a family of viruses that cause several diseases in humans. Some of the viruses of the coronavirus family have been named with respect to the diseases that they cause. Figure 9.1 provides an overview over some members of the family of coronavirus.

As it is shown in Fig. 9.1, there are four types of coronaviruses that cause only mild to moderate respiratory diseases: 229E, NL63, OC43, and KHU1 [1, 2]. Furthermore, there are three types of coronaviruses that can lead to death: SARS-CoV, MERS-CoV, and SARS-CoV-2 [1–3]. The severe acute respiratory syndrome coronavirus is a coronavirus that causes the viral respiratory disease called severe acute respiratory syndrome. The phrase severe acute respiratory syndrome is abbreviated as SARS [4]. Accordingly, the severe acute respiratory syndrome coronavirus has been abbreviated SARS-CoV. The virus was first detected in 2002 in China. Subsequently, infections with SARS-CoV in Hong Kong and Vietnam were reported. The SARS-CoV infections spread out to other countries as well [5, 6]. However, the epidemic stopped by July 2003 [5]. The SARS-CoV 2002–2003 epidemic claimed

T. D. Frank, *COVID-19 Epidemiology and Virus Dynamics*,
Understanding Complex Systems, https://doi.org/10.1007/978-3-030-97178-6_9

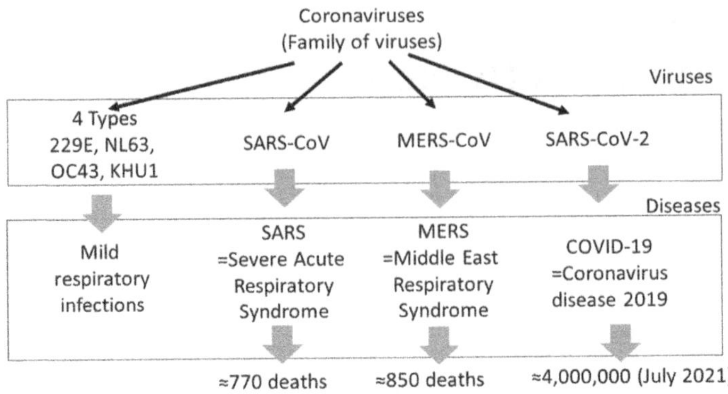

Fig. 9.1 Overview over some of the coronaviruses

about 770 deaths worldwide [5]. Another type of a coronavirus causes the viral respiratory disease called Middle East respiratory syndrome (abbreviated MERS). The virus has been named accordingly as Middle East respiratory syndrome coronavirus (MERS-CoV) [7]. The outbreak of the virus disease took place in 2012 primarily in countries of the Middle East, Africa, and South Asia. Eighty percent of all cases occurred in Saudi Arabia [7]. About 850 people died worldwide due to the disease [8]. Finally, SARS coronavirus 2 (SARS-CoV-2) causes the disease nowadays called coronavirus disease 2019 (COVID-19) [9]. According to an early report by the World Health Organization, first cases of the disease were reported in December 2019 from Wuhan city, China [10] (see also Sects. 1.1 and 5.8). The virus spread out globally during the year 2020. By 2021 the COVID-19 pandemic was still ongoing. As stated in Sect. 1.1, by July 2021 the worldwide COVID-19 death toll reached the mark of 4,000,000 lives.

9.1.2 Possible SARS-CoV-2 Target Cells

When a virus enters a human or animal body it typically invades certain cells. The infected cells subsequently produce the virus and release the produced virus particles. These newly produced particles invade other cells such that a circular causal loop is created. Viruses typically affect certain types of cells. That is, different viruses affect different types of cells. In what follows the type of cells that are affected by a virus will be called target cells. In the context of COVID-19, the question arises which cells does the SARS coronavirus 2 affect? Related to that question, we may ask where does SARS-CoV-2 go in the human body?

SARS-CoV-2 just as the 2002 emerging SARS-CoV seems to enter cells by means of a specific cell receptor called ACE2 [11–18]. Consequently, the virus travels in the human body via the airway or in the blood stream to parts of the body that exhibit cells with ACE2 receptors. Such parts featuring ACE2 receptor cells are the human nose, lung, and brain [11, 18, 19]. COVID-19 has caused in some patients temporary loss of taste and smell or an altered sense of taste and smell [20]. SARS-CoV-2 entering the human brain or the olfactory bulb may explain such experiences [11]. SARS-CoV-2 entering the human lung causes respiratory conditions that may lead to respiratory failure and death [3, 14, 21, 22] and multiple organ failure and death [3, 14, 15].

9.1.3 Target Cells in SARS-CoV-2 Infections of the Human Lung

Respiratory failure is a major cause of COVID-19 associated deaths [3, 14, 21, 22]. The human lung consists of a plenitude of branches (bronchia) that end in small cavities or "bags": the alveoli. In these alveoli gas exchange takes place. Oxygen carried into the lung when breathing in enters alveoli, passes through the walls of the alveoli and reaches the blood stream. In opposite direction, carbon dioxide from the blood stream passes through the alveolar walls and enters the alveoli. It leaves the lung and human body when breathing out. The walls of alveoli consists of epithelial cells of type 1 and 2, which are also called pneumocytes cells of type 1 and 2. These epithelial cells play an important role for the aforementioned gas exchange. The state-of-the-art hypothesis is that SARS-CoV-2 invades epithelial respiratory cells, in general [23–26]. That is, SARS-CoV-2 invades upper respiratory tract epithelial cells (e.g., in the nose) as well as lower respiratory tract epithelial cells (e.g., in the lung). In particular, it has been hypothesized that SARS-CoV-2 invades pneumocytes (i.e., epithelial cells in the alveoli of the lung) [14, 15, 27, 28]. The pneumocytes target cell hypothesis is supported by studies that showed that the 2002 SARS coronavirus affects pneumocytes [12, 29]. It is supported by studies that showed that epithelial cells exhibit ACE2 receptors [18, 19]. Furthermore, the hypothesis is supported by direct findings that in COVID-19 patients epithelial cells of the airway are affected by the disease [30]. Finally, the pneumocytes target cell hypothesis is supported by pathological studies on deceased COVID-19 patients in which affected (i.e., damaged) pneumocytes were found [21, 22] and SARS-CoV-2 particles in pneumocytes cells could be observed [2].

9.2 Models Overview

Models of the dynamics of viruses in the human body apply to various infectious diseases such as influenza, HIV/AIDs, hepatitis, and COVID-19. In what follows, the virus dynamics within an individual will be described by means of a discrete description in terms of set of n variables $X_1, \ldots, X_n$. The variables describe the virus concentration and the numbers of cells of different types (e.g., the number of target cells). That is, the disease state (or health state) of an individual is described in terms of a finite set of variables $X_1, \ldots, X_n$ and the corresponding state vector $\mathbf{X} = (X_1, \ldots, X_n)$. The virus dynamics models that will be considered in what follows assume the general form of Eq. (2.1), which for sake of conveniency is repeated as

$$\frac{\mathrm{d}}{\mathrm{d}t}\mathbf{X} = \mathbf{N}(\mathbf{X}) . \tag{9.1}$$

The key entities that allow for the description of the disease state of an individual are the non-infected target cells (T), the infected target cells (I), and the virus particles (V). The variables T, I, and V stand for these entities. In addition, they describe quantitatively how many items are in the respective categories. That is, T and I describe the number of non-infected and infected target cells, respectively. Likewise, V describes the number of virus particles. Frequently, the units are densities or concentrations. For example, cell counts are typically given in cells per microliter. The amount of virus is typically measured in particles per milliliter (e.g., RNA copies per milliliter [31]). For sake of simplicity, T and I will be simply referred to as target cells and infected cells.

The entities T, I, and V have counterparts in epidemiological models. Accordingly, target cells and infected cells can be regarded as counterparts to susceptible individuals and exposed, non-infectious individuals, respectively. Virus particles can be considered as counterparts to infectious individuals. In the following sections, three benchmark virus dynamics models will be considered: TIV, TV, and TIIV. They can be regarded as counterparts to SIR and SEIR epidemiological models (TIV and TV) or can be derived in analogy to those models (TIIV). The relationship between the virus dynamics models and epidemiological models is summarized in Table 9.1.

Acute Infections Versus Long-Term Infections

In the context of epidemiological models, in Sect. 3.1 a distinction was made between epidemics on short and long time scales. For example, for epidemic waves on short time scales, typically, demographic terms can be neglected, whereas when epidemics on the time scale of generations are considered such terms should be taken into account. A similar distinction can be made for virus infections.

Table 9.1 Some basic virus dynamics models and their counterparts in epidemiology

Virus dynamics models		Epidemiological models
TIV	⇔	SEIR without R
TV	⇔	SIR without R
TIIV as generalization of TIV	⇔	SEIR as generalization of SIR

As such, virus dynamics in the body of individuals is affected by the production of target cells and the natural cell death (i.e., death that is not induced by the virus). Consequently, virus dynamics models may take such processes into account. However, for acute infections that take place on relative short periods (e.g., period less than 30 days) cell reproduction processes and non-disease related cell death processes (i.e., natural death processes) frequently play a negligible role and for this reason are often neglected in the respective models [32, 33]. This implies that acute infections only exhibit virus-free stationary states (which does not necessarily imply that the individual under consideration survives the disease). Furthermore, such models typically describe situations in which the infection period ends and the virus dies out when most of the target cells in the affected region of the body have been turned into infected cells and have died out themselves. Since there is no noteworthy production of target cells over the time scale under consideration, when target cells and infected cells disappear, then the virus disappears as well. Such models are called target cell-limited models (see Sect. 9.3.2). In fact, the TIV, TV, and TIIV models in the absence of cell production terms and cell natural death terms of cells belong to the class of target cell-limited models. Influenza and COVID-19 are two examples of acute virus infections.

Long-term infections or persistent infections take place over longer periods (e.g., several years) or stay with individuals over their lifetimes. Consequently, cell death and cell production plays an important role and needs to be taken into account in virus dynamics models addressing long term infections. Long term infections may exhibit stationary states with non-vanishing viral load and the corresponding models typically account for that possibility. An example of long term infections are untreated HIV infections [31, 34–36].

9.3 TIV Model

9.3.1 Model Formulation

The TIV model describes infections in individuals in terms of the three aforementioned variables T, I, and V, that is, in terms of the numbers (or concentrations) of target cells and infected cells and the number (or concentration) of virus particles. V is also called the viral load. Figure 9.2 shows a flow chart of the mechanistic processes described by the model.

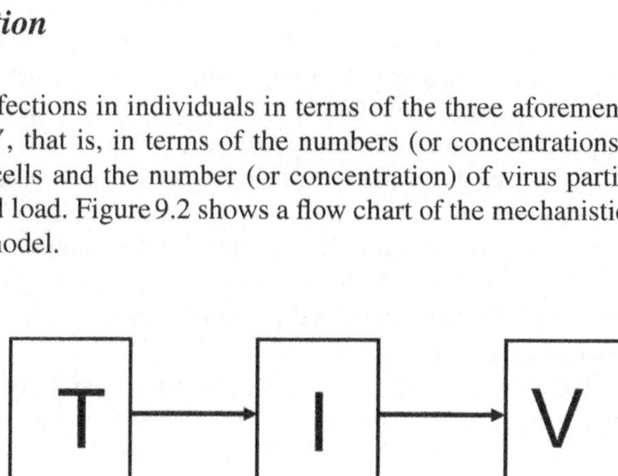

Fig. 9.2 Flow chart of the TIV model

Accordingly, one or several virus particles infect a target cell. The target cell turns into an infected cell and begins to produce and release virus particles [16, 37]. The released virus particles increase the viral load (i.e., virus concentration). In particular, the released virus particles infect other target cells (see also Sect. 1.3). However, infected cells differ from target cells not only by the fact that they produce virus. They also tend to have a shortened lifetime. Due to the altered biochemical processes, frequently, they die relatively quickly as compared to the non-infected target cells. The evolution equations for T, I, and V of the TIV model for acute infections read [32, 33, 35, 38, 39]

$$\frac{\mathrm{d}}{\mathrm{d}t}T = -\beta V T \ , \ \frac{\mathrm{d}}{\mathrm{d}t}I = \beta V T - k_1 I \ , \ \frac{\mathrm{d}}{\mathrm{d}t}V = pI - k_2 V \ . \tag{9.2}$$

Equation (9.2) assumes the general from described by Eq. (9.1). The state vector reads $\mathbf{X} = (T, I, V)$ and we have $n = 3$. In Eq. (9.2) the parameter β denotes the infectivity rate, while p denotes the production rate of virus particles. The parameters k_1 and k_2 denote decay rates in the sense that they describe an exponential decrease of I and V for $\beta = 0$ and $p = 0$, respectively. As such, they denote the death rate of infected cells (k_1) and the clearance rate of the virus (k_2).

9.3.2 *Target Cell-Limited Models*

The TIV model can be used to demonstrate the aforementioned idea of acute infections as target cell-limited infections [32, 33, 35, 38]. Accordingly, the virus infects most of the target cells at the affected sites (e.g., sites in the human lung). The thus infected target cells produce new virus. The virus concentration increases towards a peak value. At the same time, the infected cells die out and the turnover from target cells to infected cells becomes small since there is only a small portion of target cells left and there is no noteworthy production of new target cells. The decay of infected cells implies that the production of new virus particles is reduced and eventually stops. The virus dies out. A key feature of this scenario is that the affected parts of the human body come with a limited number of target cells and do not produce new target cells with a sufficiently fast rate.

As far as the depletion of target cells is concerned, for example, it has been hypothesized that in the case of severe and fatal COVID-19 most of the target cells (i.e., pneumocytes) in affected areas of the human lung get lost [23], which is consistent with the aforementioned pathological studies on deceased COVID-19 patients that found damaged pneumocytes [21, 22] and SARS-CoV-2 particles in pneumocytes [2].

Figure 9.3 illustrates this basic mechanism underlying target cell-limited models by means of a simulation of the TIV model (9.2). Panel (a) of Fig. 9.3 shows T and I as functions of time. During the initial stage of the infection, the number of infected cells I sharply increases at the cost of the target cells. That is, when I increases the number of target cells T rapidly decreases. For the selected parameters, T decreases to an almost negligibly small level. The sharp increase of I in the TIV model is due to the βVT term in the evolution equation of I listed in Eq. (9.2), while the rapid decay of T is due to the counterpart term $-\beta VT$ in the evolution equation of T listed in Eq. (9.2). As soon as T has decayed towards the low level plateau, the number of infected cells decays at a rate k_1 due to the $-k_1 I$ term in evolution equation of I. Panel (b) shows the viral load. While I initially increases (here during the first two days after the infection), the viral load increases as well. The production of V is described mathematically by the pI term in the evolution equation of V. When the number of infected cells begins to decline, the viral load may continue to increase for some period of time. Eventually, V reaches its maximum value at a somewhat later time point as compared to I. In particular, when the number of infected cells has reached a negligibly low level, the evolution of the viral load is entirely determined by the decay term $-k_2 V$ in Eq. (9.2). That is, V decays exponentially at a rate k_2.

As such, the TIV models allows for two scenarios according to which a non-negligible viral load can be present in the presence or absence of infected cells and the decay of the viral load is determined by either k_1 or k_2 (or both). Let us dwell on this issue. For sake of simplicity, let us assume that the target cells have been completely depleted such that Eq. (9.2) becomes $\mathrm{d}I/\mathrm{d}t = -k_1 I, \mathrm{d}V/\mathrm{d}t = pI - k_2 V$. If $k_1 > k_2$ holds (as for the simulation shown in Fig. 9.3) then the decay of infected cells takes place on a faster time scale then the clearance of virus (i.e., the decay of viral load). Moreover, as long as there exists a substantial pool of infected cells virus particles are produced, such that viral load may increase or decay at a rate slower than k_2. Once I has decayed to a negligibly small level, the dynamics of V is given by $\mathrm{d}V/\mathrm{d}t = -k_2 V$. In summary, under the conditions that (i) $k_1 > k_2$ holds, (ii) target cells have been wiped out in a certain region of the body due to the infection ($T \approx 0$), and (iii) the virus dynamics can appropriately be described by the TIV model, then the viral load $V(t)$ can be relatively high for a certain period even in the absence

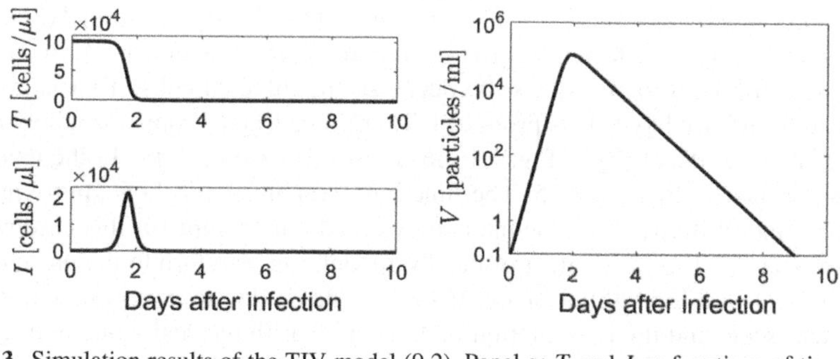

Fig. 9.3 Simulation results of the TIV model (9.2). Panel **a**: T and I as functions of time. Panel **b**: $V(t)$ shown on a logarithmic scale. Parameters and initial conditions: $T(0) = 10,000$ cells/μl, $I(0) = 0$, $V(0) = 0.1$ particles/ml, $\beta = 10^{-4}$/(d $\times$ particles/ml), $k_1 = 10$/d, $k_2 = 2.0$/d, $p = 20$ (particles/ml) $\times$ (μl/cells) /d

of infected cells and the period of decline of the disease as measured in terms of the viral load $V(t)$ is characterized by the rate constant k_2. In contrast, as will be demonstrated below by numerical simulations, if $k_1 < k_2$ holds then the speed of the decay of viral load may be determined by k_1 rather than by k_2 and the viral load decays in the presence of infected cells. As far as the decay rate of V in both scenarios is concerned, roughly speaking, the smaller value of the two parameters k_1 and k_2 determines the dynamics of the disease decline.

As will be shown in Sect. 9.3.3 and has been anticipated in Table 9.1, the TIV model is mathematically equivalent to the SEIR model. Therefore, the phenomenon of a target cell-limited infection can be compared with an epidemic wave without intervention as described by the SEIR model. As discussed in Sect. 8.1, the epidemic wave in the no intervention scenario begins to subside when the number of susceptibles has reached a sufficiently low number such that the disease-free fixed point becomes stable. The same principle holds for the TIV model. The decline stage of the infection begins when the number of target cells T decreases to a sufficiently low level such that the virus-free fixed point becomes stable. More explicitly, when T decreases towards T_{crit} such that eigenvalue $\lambda_{\max}$ of the virus-free fixed point $V_{st} = I_{st} = 0$ becomes zero, then the system is at its bifurcation point. Since $\mathrm{d}T/\mathrm{d}t < 0$ holds for $V > 0$ and $T > 0$, in the next moment of time, the system exhibits a negative maximal eigenvalue $\lambda_{\max}$. The virus-free fixed point is stable. The infection enters the decline stage.

Note that the viral load is typically shown on a logarithmic scale, while cell counts (or concentrations) are typically presented on linear scales. Therefore, when comparing panels (a) and (b) of Fig. 9.3 it seems that V decays relatively slowly as compared to I. This impression is deceiving. In fact, if T is negligibly small, then I decays exponentially at a rate of k_1. Likewise, if I is negligibly small, then V decays exponentially at a rate of k_2. In particular, as argued above, for $k_1 > k_2$ the infected cells decay faster than the viral load. Panel (a) of Fig. 9.4 shows all variables T, I, and V for the simulation presented in Fig. 9.3 in linear scales. In this case $k_1 > k_2$ holds and k_1 is larger than k_2 by a factor 5. As mentioned in the context of Fig. 9.3 and can be seen in panel (a) of Fig. 9.4, $I(t)$ reaches its peak value $I_{\max}$ earlier in time than $V(t)$. Importantly, since k_1 is considerably larger than k_2, the function $I(t)$ decays from its peak value $I_{\max}$ faster to zero as compared to the function $V(t)$. In contrast, for $k_2 > k_1$ the viral load follows the decay of the infected cells. That is, as long as I is substantially large, V is produced. Therefore, I and V approximately decay at the same rate. Panel (b) of Fig. 9.4 illustrates this case $k_2 > k_1$. In the example, k_2 is larger than k_1 by a factor 5. The functions $I(t)$ and $V(t)$ show approximately the same kind of decay. A mathematically worked out example of this case will be given in Sect. 9.5.1 in the context of the TV model. Let us return to the presentation issue of $V(t)$ in a logarithmic scale. If $k_2 > k_1$ holds due to the presentation of I on a linear scale and the presentation of V on a logarithmic scale, one may get the incorrect impression that V decays slowly as compared to I.

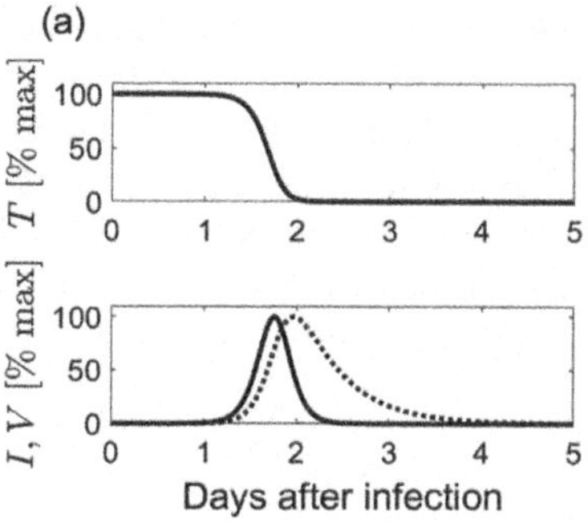

Fig. 9.4 Impact of decay rates k_1 and k_2 on the I, V dynamics. Panel **a**: Solutions T (top subpanel), I (solid line, bottom subpanel), and $V(t)$ (dotted line, bottom subpanel) of Eq. (9.2) are shown for the simulation with $k_1 > k_2$ presented in Fig. 9.3. V is presented on a linear scale. Parameters and initial conditions as for the simulation shown in Fig. 9.3. Panel **b**: As in panel (**a**) but for $k_2 > k_1$. All parameters and initial conditions are kept the same except for k_1 and k_2, which are given by $k_1 = 2.0$/d, $k_2 = 10$/d

9.3.3 Equivalence of TIV and SEIR Models

In order to compare the TIV model (9.2) with the SEIR model defined by Eq. (3.43), let us write the evolution equations of both models next to each other like

$$\frac{\mathrm{d}}{\mathrm{d}t}T = -\beta VT\,,\ \frac{\mathrm{d}}{\mathrm{d}t}I = \beta VT - k_1 I\,,\ \frac{\mathrm{d}}{\mathrm{d}t}V = pI - k_2 V\,,$$
$$\frac{\mathrm{d}}{\mathrm{d}t}S = -\frac{\beta}{N}IS\,,\ \frac{\mathrm{d}}{\mathrm{d}t}E = \frac{\beta}{N}IS - \alpha E\,,\ \frac{\mathrm{d}}{\mathrm{d}t}I = \alpha E - \gamma I\,. \tag{9.3}$$

In Eq. (9.3) for the purpose of the model comparison the evolution equation of R of the SEIR model is ignored. In view of Eq. (9.3), the variables T, I, V can be assigned to the variables S, E, I like $T \leftrightarrow S$, $I \leftrightarrow E$, and $V \leftrightarrow I$. Moreover, as far as the parameters are concerned, the following assignments can be made: $\beta \leftrightarrow \beta/N$ and $k_2 \leftrightarrow \gamma$. The models differ in terms of the transitions $E \to I$ and the virus production stage $I \to V$ involving the species I. With respect to the SEIR model, transitions from E to I are described by the term $-\alpha E$ in the evolution equation of E and the term αE in the evolution equation of I. Both terms are of the same magnitude. With respect to the TIV model, there are two terms with independent parameters k_1 and p, respectively. In the evolution equation of I the decay term reads $-k_1 I$, whereas in the evolution equation for V the production term reads pI. This implies that the SEIR model is a three-parametric model with parameters β/N, α and γ. In contrast, the TIV model is a four-parametric model with parameters β, k_1, p, and k_2.

As mentioned in Sect. 9.2, viral load and cells are frequently quantified in terms of concentrations. However, the respective concentrations are not necessarily measured in the same units (e.g., see Ref. [31]). Frequently, cells are measured in cells per microliter (which is equivalent to cells per cubic millimeter), whereas viral load is measured in virus particles (or copies) per milliliter (where 1 RNA copy may reflect 1 virus particle [40]). The units microliter and milliliter differ by a factor 1000. Moreover, cells and virus particles may or may not be regarded as entities of the same category. In contrast, the SEIR model features the variables S, E, I that all refer to population sizes. They are measured in the same units. In order to map the TIV model to the SEIR model, the viral load may be converted into a cell equivalent viral load V' using

$$V' = \frac{k_1}{p} V \, . \tag{9.4}$$

From the evolution equation of V listed in Eq. (9.2) it follows that the terms pI and $k_2 V$ exhibit the same units. Since k_2 and k_1 exhibit the same units it then follows that V' as defined by Eq. (9.4) exhibits the same units as T and I. Substituting Eq. (9.4) in form of $V = V'p/k_1$ into Eq. (9.2), the TIV model becomes

$$\frac{\mathrm{d}}{\mathrm{d}t} T = -\beta' V' T \, , \quad \frac{\mathrm{d}}{\mathrm{d}t} I = \beta' V' T - k_1 I \, , \quad \frac{\mathrm{d}}{\mathrm{d}t} V' = k_1 I - k_2 V' \tag{9.5}$$

with $\beta' = \beta p / k_1$. The TIV model in form of Eq. (9.5) exhibits three variables T, I, V' measured in the same units. Importantly, only three independent parameters β', k_1, and k_2 occur in Eq. (9.5). Consequently, using the mappings $T \leftrightarrow S$, $I \leftrightarrow E$, $V' \leftrightarrow I$, $\beta' \leftrightarrow \beta/N$, $\alpha \leftrightarrow k_1$, and $\gamma \leftrightarrow k_2$, the TIV model (9.5) turns into the SEIR model (3.43) when ignoring the dynamics of the recovered individuals R. In other words, the two models are mathematically equivalent. This also implies that the original TIV model defined by Eq. (9.2) is mathematically equivalent to the SEIR model (3.43) provided that the variable transformation (9.4) is used. Finally, note that while Eq. (9.5) involves only three parameters β', k_1, and k_2, the overall model involves four parameters: β', k_1, k_2, and p. The model solutions V' are related to viral load concentrations V and observed viral load data by means of Eq. (9.4) that features the virus production rate parameter p.

9.4 Viral Load Patterns, Infection Order Parameters, and Order Parameter Amplitudes

The rise and decay of viral load forms a temporal pattern that exhibits some characteristic features [33]. Figure 9.5 shows the typical shape of a viral load pattern and illustrates four pattern features. First, the pattern shows an initial phase of viral load increase, on the one hand, and a final phase of viral load decay, on the other hand. As discussed in Sect. 8.5 in the context of epidemiological models, the initial stage of an epidemic is determined by the largest eigenvalue $\lambda_{\max}$ when assuming that the system under consideration exhibits only one positive real-valued eigenvalue. As pointed out in Chap. 2, the principles of nonlinear physics and, in particular,

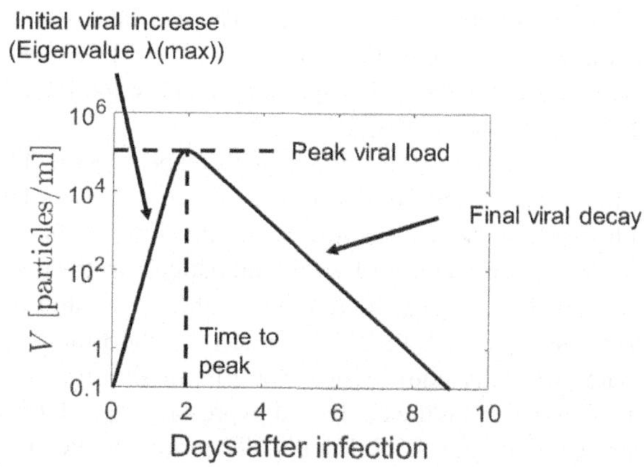

Fig. 9.5 Illustration of some key features of viral load patterns

the amplitude equation description, holds for all kind of systems. In particular, the principles apply to systems described by Eq. (9.1). Therefore, the considerations presented in Sect. 8.5 also hold for virus dynamical models of the form (9.1) such as the TIV, TV, and TIIV models. The state vector $\mathbf{X}$ can be expressed in terms of a superposition $\mathbf{X} = \sum_{k=1}^{n} A_k \mathbf{v}_k$. At the beginning of the infection, that is, in the initial stage, the disease (or health) state of an individual (or the disease/health state of the affected body sites of the individual) is close to the virus-free fixed point $\mathbf{X}_{st}$, which is unstable. As discussed in Chap. 2, the amplitudes A_j related to eigenvalues with positive real parts dominate the initial dynamics away from an unstable fixed point. For sake of simplicity, let us consider the situation of a single positive (real-valued) eigenvalue, which is the case that holds for the TIV, TV, and TIIV models (see Chap. 10). Let $k(\max)$ denote the index of this largest, real-valued eigenvalue. Let $\mathbf{v}_{k(\max)}$ denote the associated eigenvector and $v_{k(\max),V}$ denote the coefficient in the direction of the V-axis in the state space spanned by $X_1, \ldots, X_n$ (which includes the variable V). Then, for an infection starting at time $t = 0$, it follows that during the initial stage of the infection

$$\mathbf{X}(t) \approx A_{k(\max)}(t)\mathbf{v}_k(\max) \;\Rightarrow\; V(t) \approx v_{k(\max),V} A_{k(\max)}(t) \tag{9.6}$$

and $A_{k(\max)}(t) = A_{k(\max)}(0) \exp\{\lambda_{k(\max)} t\}$ such that

$$V(t) \approx v_{k(\max),V} A_{k(\max)}(0) \exp\{\lambda_{k(\max)}(t)\} \tag{9.7}$$

or

$$V(t) \approx b \exp\{\lambda_{k(\max)} t\} \tag{9.8}$$

with $b = v_{k(\max),V} A_{k(\max)}(0)$. Consequently, the eigenvalue $\lambda_{k(\max)}$ describes the rate of increase of V. In the logarithmic scale of V it corresponds to the slope of the graph $V(t)$. The vector $\mathbf{v}_k$ is the order parameter of the infection. The amplitude $A_{k(\max)}(t)$

is the order parameter amplitude. The special case of the TIV model will be worked out in Sect. 10.1 (see also Refs. [39, 41–43]). Equation (9.8) will be derived more explicitly for the more comprehensive TIIV model in Sect. 10.4.4 and the simplified TV model in Sect. 10.6.2.

As discussed in Sect. 9.3.2, with respect to the TIV model for $k_1 > k_2$ the final viral decay is determined by the rate constant k_2. That is, when V is presented in a logarithmic scale, then the declining slope of the graph V equals $-k_2$. Two more characteristic features of viral load patterns as the one shown in Fig. 9.5 are the peak viral load V_{max} (i.e., the maximal value of V) and the time it takes for the disease dynamics to reach the peak value (i.e., the time to peak). Since the TV model is mathematically equivalent to the SIR model, for the TV model the value V_{max} can be obtained from the analytical expression I_{max} defined by Eq. (3.25), which will be discussed in Sect. 9.5.2. Finally, depending on the virus dynamics model at hand, the time to peak may be given in terms of approximate analytical expressions [33] or needs to be determined by means of numerical simulations.

9.5 TV Model

9.5.1 Model Derivation

The TV model is a simplified version of the TIV model [33]. It is assumed that a time-scale separation holds such that $I(t)$ and $V(t)$ can be regarded as slow and fast dynamics, respectively. Consequently, one of the two variables can be eliminated and the three-variable TIV model reduces to a two-variable model. More precisely, the main ideas are the following. If I in the TIV model (9.2) is constant over time, then $V(t)$ approaches the fixed point

$$\frac{\mathrm{d}}{\mathrm{d}t}V = pI - k_2 V = -k_2\left(V - \frac{p}{k_2}I\right) \Rightarrow V \to \frac{p}{k_2}I\,. \tag{9.9}$$

The approach towards the fixed point exhibits the time constant $\tau = 1/k_2$. If k_2 is large, then τ is small. Let us assume that k_2 is sufficiently large such that $I(t)$ varies only slowly on the time scale defined by τ. Then, $V(t)$ is approximately given by the fixed point value if I would be constant like

$$V(t) = \frac{p}{k_2}I(t)\,. \tag{9.10}$$

Eliminate I in TIV model (9.2) by substituting Eq. (9.10) into the evolution equation of I in Eq. (9.2), we obtain

$$\frac{\mathrm{d}}{\mathrm{d}t}T = -\beta VT\,,\ \frac{\mathrm{d}}{\mathrm{d}t}V = r\beta VT - k_1 V \tag{9.11}$$

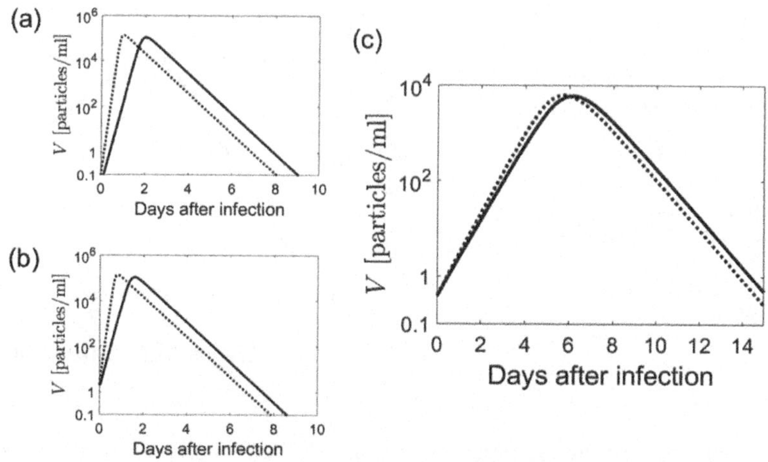

Fig. 9.6 The accuracy of the TV model approximation of the TIV model depends on the choice of suitable initial conditions and the k_1, k_2 relationship. Panels **a**, **b**, and **c** show solutions $V(t)$ obtained from the TIV model (9.2) (solid lines) and the TV model (9.11) (dotted lines) for various initial conditions and parameter values of k_1 and k_2. See text for details

with $r = p/k_2$. The model has been referred to as TV model [33]. The TV model exhibits the fixed points (i.e., stationary states) $\mathbf{X}_{st} = (T_{st}, V_{st})$ with $T_{st} \geq 0$ and $V_{st} = 0$. From the evolution equation of $V(t)$ presented in Eq. (9.11) it follows that the fixed point $\mathbf{X}_{st} = (T_{st}, 0)$ is unstable for $r\beta T_{st} > k_1$ and neutrally stable for $r\beta T_{st} < k_1$.

Figure 9.6 shows simulations of the TIV and TV model for the same parameter sets to compare their solutions. Panel (a) uses $k_2 > k_1$ where k_2 is five times larger than k_1. Obviously, the solutions of the TIV and TV model differ. They differ for two reasons. First, the initial conditions $V(0)$ and $I(0)$ used in the simulation were not consistent with Eq. (9.10). Second, the time scale separation was not sufficiently large. In order to fix the first issue, in a second simulation $k_2/k_1 = 5$ was used again but $V(0)$ and $I(0)$ were selected such that they satisfied Eq. (9.10). Panel (b) shows the simulation results. As expected, the solutions of the TV and TIV model initially are close together due to the consistent initial values. However, the solutions diverge because the two variables I and V cannot be regarded as slow and fast variables that evolve on clearly different time scales. In a third simulation the parameter k_2 was increased to obtain a ratio of $k_2/k_1 = 25$. The initial values were the same as for the second simulation (i.e., consistent with Eq. (9.10)). The simulation results are shown in panel (c). The solutions of the TV model can be regarded as an approximation of the solution of the TIV model with moderate accuracy.

As stated in the context of Eq. (9.8) and Fig. 9.5, the initial slope of $V(t)$ is determined by λ_{max}. Therefore, the simulations in panels (a) and (b) reveal that the eigenvalues λ_{max} of the TV and TIV models for the parameter ratio $k_2/k_1 = 5$ differ considerably. In contrast, when increasing the ratio to $k_2/k_1 = 25$, the difference

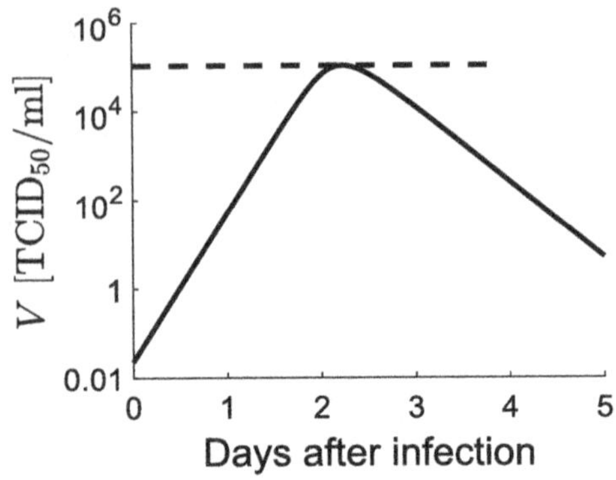

Fig. 9.7 TV model solution describing the viral load pattern of participant 1 of an influenza research study. $V(t)$ describes the solution of the TV model (9.11) for $T(0) = 4 \cdot 10^8$ cells, $I(0) = 0$, $V(0) = 0.023$ TCID$_{50}$/ml, $\beta = 2.67 \cdot 10^{-5}$ /(d $\times$ (TCID$_{50}$/ml)), $r\beta = 3.32 \cdot 10^8$/(d $\times$ cell), $k_1 = 5.59$/d

becomes smaller and the maximal eigenvalues of the two models become approximately the same.

The TV model was applied in Ref. [33] to describe acute influenza infections of twelve volunteers that participated in two different research studies on influenza infections. The nasal sections of the volunteers were infected with a particular influenza virus on day 0. The viral concentration found in nasal swabs was measured on several days after the infection. To this end, viral load was measured in units of 50% tissue culture infectious dose (TCID$_{50}$) per milliliter. The TV model (9.11) was fitted to the observed data. Figure 9.7 shows the model solution for participant 1 of the first research study mentioned in Ref. [33]. The dashed line in Fig. 9.7 indicates the peak viral load as computed from an analytical expression that will be derived in Sect. 9.5.2 below.

9.5.2 *Equivalence of TV and SIR Models*

The TV model (9.11) can be compared with the SIR model (3.16). To this end, it is convenient to consider the two-variable version of the SIR model (3.22) for which R is ignored or given by $R = N - S - I$. In order to ease the presentation, the respective TV and SIR model equations (Eq. (9.11) and Eq. (3.22)) are listed below next to each other like

$$\begin{aligned} \frac{\mathrm{d}}{\mathrm{d}t} T &= -\beta V T \, , \quad \frac{\mathrm{d}}{\mathrm{d}t} V = r\beta V T - k_1 V \, , \\ \frac{\mathrm{d}}{\mathrm{d}t} S &= -\frac{\beta}{N} I S \, , \quad \frac{\mathrm{d}}{\mathrm{d}t} I = \frac{\beta}{N} I S - \gamma I \, . \end{aligned} \tag{9.12}$$

Consequently, the variables T and V may be assigned to the variables S and I like $T \leftrightarrow S$ and $V \leftrightarrow I$. Moreover, the TV model parameter k_1 can be regarded as the counterpart of SIR parameter γ: $k_1 \leftrightarrow \gamma$. The difference between the two models can be found in the transition mechanisms $S \to I$ of the SIR model and the virus production process $T \to V$ of the TV model. With respect to the SIR model, transitions from S to I are described by means of the terms $-\beta IS/N$ and $\beta IS/N$ that occur in the evolution equations for S and I, respectively, and are of the same magnitude. In contrast, the TV model features two terms with independent parameters β and r. In the evolution equations for T and V the terms $-\beta VT$ and $r\beta VT$ occur. The parameter r typically differs from 1 (e.g., see the interpretation of r as $r = p/k_2$ in the context of the derivation of the TV model from the TIV model). Consequently, the SIR model is a two-parametric model with parameters β/N and γ, whereas the TV model is a three-parametric model with parameters β, r, and k_1.

Nevertheless, the approach used in Sect. 9.3.3 can be applied to the TV model. In doing so, it can be shown that the TV model is equivalent to the SIR model. Let us dwell on this issue. First of all, the TV model involves two variables that are not necessarily measured in the same units. Therefore, it would be desirable to replace V by a variable that measures virus concentration in cell equivalents. Second, in the derivation of the TV model, the variable I was eliminated using Eq. (9.10). However, Eq. (9.10) can alternatively be used to eliminate V. In doing so, a model that involves the variables T and I that reflect cells is obtained. More precisely, given the TV model in the form of Eq. (9.11), let us introduce the infected cell variable I like

$$I = \frac{V}{r} \,. \tag{9.13}$$

Note that as such Eq. (9.13) simply corresponds to Eq. (9.10). However, Eq. (9.13) should be regarded as an equation independent of the derivation of the TV model from the TIV model. To reiterate, the departure point is the TV model defined by Eq. (9.11) featuring the parameters β, r, and k_1. Subsequently, the variable I is defined by Eq. (9.13) and may be interpreted as viral load measured in cell units or alternatively interpreted as the concentration (or number) of infected cells within a TIV modeling framework. Substituting Eq. (9.13) into Eq. (9.11), the TV model becomes

$$\frac{\mathrm{d}}{\mathrm{d}t} T = -\beta' IT \,, \quad \frac{\mathrm{d}}{\mathrm{d}t} I = \beta' IT - k_1 I \tag{9.14}$$

with $\beta' = r\beta$. Equation (9.14) involves only two independent parameters β' and k_1. The TV model in the form (9.14) is equivalent to the SIR model (3.22) when replacing the cell variables T and I by the human population variables S and I, respectively, and, likewise, when replacing β' and k_1 by β/N and γ, respectively. As such the TV model described by Eqs. (9.13) and (9.14) corresponds still to a three-parametric model with parameter β', k_1, and r because in order to relate the solutions $I(t)$ to viral load counts $V(t)$ Eq. (9.13) is used that features the third parameter r.

For the SIR model an analytical expression for I_{max} is given by Eq. (3.25). Making the aforementioned replacements, from Eq. (3.25) it follows that I_{max} (i.e., the peak viral load measured in cell units) of the TV model (9.14) is given by

$$I_{max} = I_0 + T_0 \left(1 - \frac{1}{\xi}[1 + \ln(\xi)] \right) \tag{9.15}$$

with $\xi = \beta' T_0 / k_1 = r\beta T_0 / k_1$, where $I_0 = I(t_0)$ and $T_0 = T(t_0)$ are the initial values of T and I at time t_0. Using Eq. (9.13) again, from Eq. (9.15) it follows that the peak viral load V_{max} can be computed from

$$V_{max} = V_0 + r T_0 \left(1 - \frac{1}{\xi}[1 + \ln(\xi)] \right) \tag{9.16}$$

(see also Ref. [33] for an alternative derivation of Eq. (9.16)). As an example, V_{max} was computed from Eq. (9.16) for the viral load trajectory of patient 1 shown in Fig. 9.7. The value V_{max} thus obtained is indicated in Fig. 9.7 by means of a dashed line and is identical with the peak value of the numerical solution $V(t)$ of the TV model.

9.6 TIIV Model

The TIIV model is a generalization of the TIV model. The motivation for that generalization is the general observation that being infected does not imply to be infectious. In epidemiology, this idea motivates to generalize the SIR model to the SEIR model by introducing a latent class: the compartment of exposed individuals who are not infectious. As far as the virus dynamics in individuals is concerned, models may be constructed that account for a transient phase or transient period during which infected cells do not release virus. This phase is sometime referred to as eclipse phase or latent infection period [44] (although strictly speaking the two are not the same and the eclipse phase is part of the latent infection period). The additional class that refers to the latent stage or latent period of infected cells (i.e., the stage or period in which they do not release virus particles) turns the three-variable TIV model into a four-variable model: the TIIV model [35, 38, 45]. The rationale to introduce an additional variable is to obtain a biologically more accurate model of virus dynamics [32, 45–47].

For example, Sedmak and Grossberg [48] infected chicken embryo cell cultures with the influenza virus A/Hong Kong/t/68 (H3N2). They found that the infected cells did not produce virus for about 6 h [48]. In a similar experimental study, Madin-Darby canine kidney (MDCK) cell cultures were infected by the equine influenza virus A/Equi/2 (H3N8). It was found that infected cells did not produce virus for about 4.5 h [49]. The TIIV model has been used as a tool for determining the rate constant of $I_1 \rightarrow I_2$ transitions from infected, not yet virus-producing cells to virus-producing cells. For example, Beauchemin et al. [46] found in a TIIV model-based study on MDCK cell cultures that were infected by the influenza virus A/Albany/1/98 (H3N2) that the rate constant was about 0.31/hour (or 7.44/d), which can be interpreted in term of a mean duration of the latent infection period of 3.2 h. Baccam et al. [32] and Canini and Carrat [50] determined TIIV model rate constants for influenza A infections in human volunteers and found values of about 4.0/d and 2.8/d, respectively. Moreover, using the TIIV model as an analytical tool, Patel et al. [51] examined the virus dynamics of the respiratory syncytial virus (RSV) in human volunteers and found a rate constant for infected cells to turn into virus-producing cells of about 0.07/hour or 1.7/d. In summary, the two advantages of the TIIV model with respect to the TIV model is that it increases the biological realism of the virus dynamics description at hand [32, 45] and allows to obtain insights into the $I_1 \rightarrow I_2$ transition dynamics that turns infected, not yet virus-producing cells into virus-producing cells.

The variables of the TIIV model are the number of target cells (T), the number of infected, not yet virus-producing cells (I_1), the number of virus producing infected cells (I_2), and the viral load (V). The model itself reads

$$\frac{\mathrm{d}}{\mathrm{d}t}T = -\beta V T\ , \quad \frac{\mathrm{d}}{\mathrm{d}t}I_1 = \beta V T - k_1 I_1\ ,$$
$$\frac{\mathrm{d}}{\mathrm{d}t}I_2 = k_1 I_1 - k_2 I_2\ , \quad \frac{\mathrm{d}}{\mathrm{d}t}V = p I_2 - k_3 V. \tag{9.17}$$

The rate parameter k_1 describes the transition speed of the $I_1 \rightarrow I_2$ transitions of cells. The terms $-k_1 I_1$ and $k_1 I_1$ that describe in Eq. (9.17) this transition match each other in magnitude. In contrast, just as for the TIV model, for the TIIV model the terms that describe the virus production at the cost of I_2 cells, namely, pI_2 and $-k_2 I_2$, in general, are not of the same magnitude.

The phrases not yet virus-producing and virus-producing that are used above to characterize cells of type I_1 and I_2, respectively, should be understood in the context of Eq. (9.17). That is, I_1 cells do not release virus, while I_2 cells do so (see the pI_2 term). Virus particles may be assembled in I_1 cells and in this sense may be produced. The key point is that I_1 cells are in a transient stage in which they do not release virus as it is described by Eq. (9.17). For sake of simplicity, in the above and in what follows the phrases not yet virus-producing and virus-producing are used (instead of the phrases not virus-releasing and virus-releasing) to indicate that in the evolution equation of V there is no pI_1 term but there is a pI_2 term.

Introducing the variable

$$V' = \frac{k_2}{p} V \tag{9.18}$$

that measures viral load in cell equivalents, Eq. (9.17) becomes

$$\frac{\mathrm{d}}{\mathrm{d}t}T = -\beta' V' T\ , \quad \frac{\mathrm{d}}{\mathrm{d}t}I_1 = \beta' V' T - k_1 I_1\ ,$$
$$\frac{\mathrm{d}}{\mathrm{d}t}I_2 = k_1 I_1 - k_2 I_2\ , \quad \frac{\mathrm{d}}{\mathrm{d}t}V' = k_2 I_2 - k_3 V' \tag{9.19}$$

with $\beta' = p\beta/k_2$. Equations (9.18) and (9.19) involve the parameters β', k_1, k_2, and p and provide an alternative description of the TIIV model (9.17) with the advantage

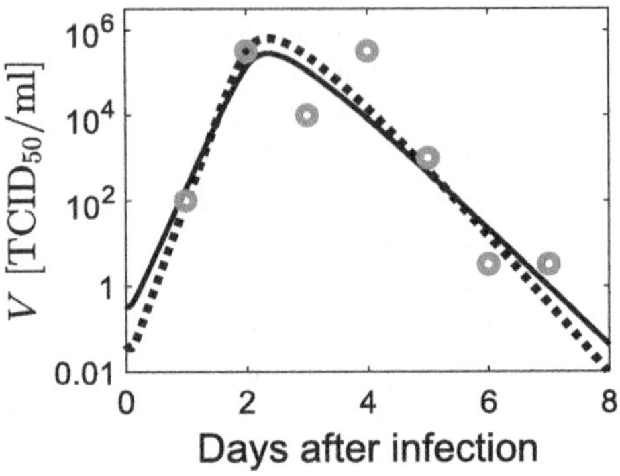

Fig. 9.8 Viral load data from volunteer 1 (gray circles) are shown together with the TIV model (solid line) and TIIV model (dotted line) solutions describing the viral load pattern of volunteer 1. $V(t)$ (solid line) describes the solution of the TIV model (9.2) for $T(0) = 4 \cdot 10^8$ cells, $I(0) = 0$, $V(0) = 0.35$ $TCID_{50}$/ml, $\beta = 3.4 \cdot 10^{-5}$ /(d × ($TCID_{50}$/ml)), $k_1 = 3.4$/d, $k_2 = 3.3$/d, and $p = 7.9 \cdot 10^{-3}$ ($TCID_{50}$/ml) /(d × cell). $V(t)$ (dotted line) describes the solution of the TIIV model (9.17) for $T(0) = 4 \cdot 10^8$ cells, $I(0) = 0$, $V(0) = 0.043$ $TCID_{50}$/ml, $\beta = 4.9 \cdot 10^{-5}$ /(d × ($TCID_{50}$/ml)), $k_1 = 3.9$/d, $k_2 = 4.2$/d, $k_3 = 4.3$/d, and $p = 2.8 \cdot 10^{-2}$ ($TCID_{50}$/ml) /(d × cell)

that the variables T, I_1, I_2, and V' are all cell-like variables measured in the same units.

The TIIV model (9.17) has been extensively used in the literature. For example, the TIIV model was applied in Ref. [32] to describe acute influenza infections in a sample of six volunteers. Just as in Ref. [33], the nasal sections of the volunteers were infected with a particular influenza virus. For the first week after the infection, daily $TCID_{50}$ measurements were made to determine the time course of the viral load. The TIV model (9.2) was fitted to the observed data. The data (gray circles) and the TIV model solution (solid line) are shown in Fig. 9.8 for volunteer 1. The data were also fitted to the TIIV model (9.17) (see the dashed line). The solutions of both models exhibit the general temporal pattern depicted in Fig. 9.5 and, accordingly, interpret the observed data in terms of such a pattern.

9.7 Beyond Acute Virus Infections

Virus dynamics in individuals that evolves over longer periods is affected by the natural death of target cells, on the one hand, and the production of target cells, on the other hand. In order to take such processes into account, virus dynamics models can be supplemented with the respective terms. For example, the TIV model (9.2) for long term infections becomes [31, 37]

$$\frac{d}{dt}T = B - \mu T - \beta V T \ , \ \frac{d}{dt}I = \beta V T - k_1 I \ , \ \frac{d}{dt}V = pI - k_2 V \ . \tag{9.20}$$

Just as in the case of epidemiological models, in Eq. (9.20) the parameter B describes the constant production rate of target cells, whereas μ describes the natural death rate of target cells. In contrast, k_1 describes the total death rate that includes the natural death rate μ of (non-infected) target cells. That is, k_1 takes into account that due to the virus infection the death rate is increased from its baseline level μ of the natural death rate to a higher level $k_1 > \mu$. In the absence of an infection (i.e., for $I = V = 0$), the concentration of target cells approaches the non-vanishing stationary value $T_{st} = B/\mu$. Likewise, under appropriate circumstances the model exhibits an asymptotically stationary fixed point with a non-vanishing viral load $V_{st} > 0$ and $T_{st} > 0$, $I_{st} > 0$ [31]. This fixed point reflects a chronic virus infection. The model (9.20) and similar models taking target cell death and target cell production into account have been frequently used to discuss viral load dynamics of HIV and the disease progression of AIDS (e.g., see Refs. [31, 34–36]).

9.8 Modeling Studies of SARS-CoV-2 Dynamics in COVID-19 Patients

The study of the viral load dynamics of SARS-CoV-2 in the human body from a nonlinear physics perspective is an active field of research. Only a few studies will be mentioned in what follows. In Ref. [41, 52, 53] the TIV model (9.2) was applied to describe the time course of the SARS-CoV-2 load in the upper and lower tracts in a sample of COVID-19 patients. Panel (a) of Fig. 9.9 shows the viral load measured in the lower tract of patient 1 discussed in Ref. [41, 53, 54]. The graph was obtained in Ref. [41]. The viral load reflects SARS-CoV-2 concentrations in the human lung of that patient. The best-fit model solution $V(t)$ obtained from the TIV model (9.2) describing the time course of the infection is shown as well. The studies [41, 53, 54] and the application of the TIV model to COVID-19 patient data will be discussed in Sect. 10.2.1. In Ref. [52, 53] TIV model solutions were compared to solutions of the more comprehensive TIIV model (9.17) with respect to the COVID-19 patients from the aforementioned sample. The TIIV model produced qualitatively the same kind of fits and viral load patterns as the TIV model [52, 53]. Panel (b) of Fig. 9.9 shows the best-fit solution of the TIIV model for the patient 1 data as re-obtained in Ref. [55]. When comparing the TIV and TIIV modeling approach, the TIIV modeling approach comes with the advantage that it allows to study the role of latently infected cells in the disease progression of COVID-19 patients [55]. The TIIV model and its application to COVID-19 patient data will be discussed in Sect. 10.5.

The TIIV model (9.17) was also applied to viral load data of COVID-19 patients in a study by Goncalves et al. [56]. However, in this study the evolution equation of V as shown in Eq. (9.17) was modified by adding a $-\beta T V$. Consequently, an evolution equation of the form $\mathrm{d}V/\mathrm{d}t = pI - k_3 V - \beta V T$ was considered. As such the additional term may be motivated by chemical reaction equations. For example, the synthesis reaction in which molecules A and B produce a molecule C like

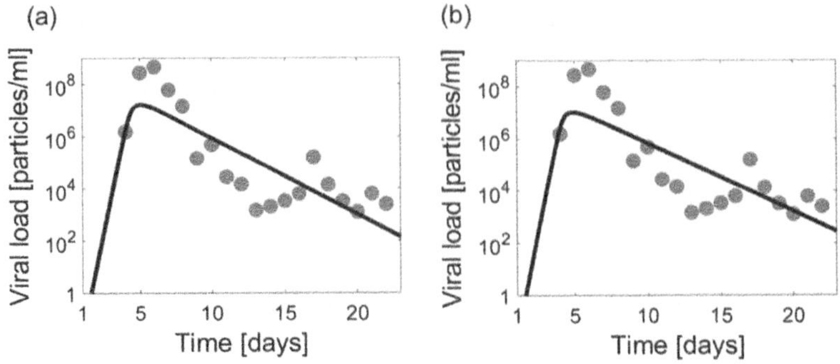

Fig. 9.9 SARS-CoV-2 viral load trajectory (gray circles) of patient 1 and its description in terms of TIV model (panel (**a**)) and TIIV model (panel (**b**)) solutions $V(t)$ (solid lines)

$A + B \rightarrow C$ can be described with the chemical reaction equations $\mathrm{d}a/\mathrm{d}t = -kab$, $\mathrm{d}b/\mathrm{d}t = -kab$, and $\mathrm{d}c/\mathrm{d}t = kab$, where a, b, and c denote the concentrations of A,B, and C [57]. By analogy, if T and V are considered as two variables measured in the same units then the chemical reaction $T + V \rightarrow I$ leads to the evolution equations $\mathrm{d}T/\mathrm{d}t = -\beta TV$, $\mathrm{d}I/\mathrm{d}t = \beta TV$, and $\mathrm{d}V/\mathrm{d}t = -\beta TV$. However, as mentioned in the context of the TIV, TV, and TIIV models, cell concentrations and viral load are frequently measured in different units. Therefore, writing the βTV term in the evolution equation of T and the evolution equation of V typically leads to a contradiction. A possible solution is to make the explicit assumption that a virus particle can be considered as an equivalent to a cell such that V and the cell variables are measured in the same units. An alternative solution is to consider the TIIV model in terms of Eq. (9.19). Having said that the suggested term $-\beta VT$ in Ref. [56] describes a decay of the viral load due to the fact that a virus particle or a few virus particles invade a non-infected target cell and turn it into an infected cell. However, the concentration of virus typically decays to a negligible amount due to the infection of cells. The main cause of a decay of the viral load is the clearance of the virus [46] (as described by the $-k_V V$ terms in the TIV, TV, and TIIV models). Consequently, in good approximation the chemical reaction scheme describing the production of infected cells does not read $T + V \rightarrow I$ but $T + V \rightarrow I + V$, which is a reaction equation of a catalytic reaction. In this case, from the chemical reaction scheme $T + V \rightarrow I + V$ the following kinetic reaction equations are obtained: $\mathrm{d}T/\mathrm{d}t = -\beta TV$, $\mathrm{d}I/\mathrm{d}t = \beta TV$, and $\mathrm{d}V/\mathrm{d}t = 0$. This approach leads to the standard TIIV model presented in Eq. (9.17).

A generalized TIIV model was considered by Neant et al. [58] and Czuppon et al. [59] to describe the dynamics of SARS-CoV-2 concentrations in COVID-19 patients. The model takes into account that not all virus particles may be infectious. When virus particles are produced inside infected cells mutations may occur or virus particles may be assembled in an non-functional way such that the produced particles do not possess the ability to infect other cells. Let V_I and V_{NI} denote the concentrations (or numbers) of the infectious and non-infectious virus particles, respectively. Then, the

TIIV model (9.17) turns into a TIIVV model and reads [58]

$$\frac{d}{dt}T = -\beta T V_I \, , \; \frac{d}{dt}I_1 = \beta T V_I - k_1 I_1 \, , \; \frac{d}{dt}I_2 = k_1 I_1 - k_2 I_2 \, ,$$
$$\frac{d}{dt}V_I = \eta p I_2 - k_3 V_I \, , \; \frac{d}{dt}V_{NI} = (1-\eta) p I_2 - k_3 V_{NI} \, . \tag{9.21}$$

In Eq. (9.21) the parameter $\eta \in [0, 1]$ describes the fraction of produced infectious virus particles. Note that in Ref. [59] a $-\beta T V_I$ term was added in the evolution equation of V_I. As argued above and can be seen in the literature [35, 58], this term, however, is frequently not considered and may lead to modeling inconsistencies.

References

1. Virginia Department of Health. Epidemiological fact sheet: Coronaviruses, accessed 2021: https://www.vdh.virginia.gov/epidemiology/epidemiology--fact--sheets/coronaviruses/
2. N. Zhu, D. Zhang, W. Wang, X. Li, B. Yang et al., A novel coronavirus from patients with pneumonia in China, 2019. New England Journal of Medicine **382**, 727–733 (2020)
3. Y. Jin, H. Yang, W. Ji, W. Wu, S. Chen, W. Zhang, G. Duan, Virology, epidemiology, pathogenesis, and control of COVID-19. Viruses **12**:article 372 (2020)
4. World Health Organization. Health topics: severe acute respiratory syndrome (SARS). *World Health Organization*, https://www.who.int/health-topics/severe-acute-respiratory-syndrom, Accessed 2021
5. Centers for Disease Control and Prevention, USA. CDC SARS Response timeline. *Centers for Disease Control and Prevention, USA*, https://www.cdc.gov/about/history/sars/timeline.htm, Accessed 2021
6. J. Wallinga, P. Teunis, Different epidemic curves for severe acute respiratory syndrome reveal similar impacts of control measures. Am. J. Epidemiol. **160**, 509–516 (2004)
7. World Health Organization. Fact sheet: Middle East respiratory syndrome coronavirus (MERS-CoV). *World Health Organization*, https://www.who.int/news-room/fact-sheets/detail/middle-east-respiratory-syndrome-coronavirus-(mers-cov), Accessed 2021
8. World Health Organization. Health topics: Middle East respiratory syndrome coronavirus (MERS-CoV). *World Health Organization*, https://www.who.int/health-topics/middle-east-respiratory-syndrome-coronavirus-mers, Accessed 2021
9. World Health Organization. Naming the coronavirus disease (COVID-19) and the virus that causes it. *Naming the coronavirus disease (COVID-19) and the virus that causes it*, https://www.who.int/emergencies/diseases/novel-coronavirus-2019
10. World Health Organization. Novel coronavirus (2019-nCoV) Situation Report-1, 21. Jan. 2020. *Novel coronavirus (2019-nCoV) Situation Report-1, 21. Jan. 2020* https://www.who.int/emergencies/diseases/novel-coronavirus-2019/situation-reports, 2020
11. A.M. Baig, A. Khaleeq, U. Ali, H. Syeda, Evidence of the COVID-19 virus targeting the CNS: tissue distribution, host-virus interaction, and proposed neurotropic mechanisms. ACS Chem. Neurosci. **11**, 995–998 (2020)
12. B.L. Haagmans, T. Kuiken, B.E. Martina, R.A.M. Fouchier, G.F. Rimmelzwaan, G. van Amerongen, D. van Riel, T. de Jong, S. Itamura, K.H. Chan, M. Tashiro, A.D.M.E. Osterhaus, Pegylated interferon-alpha protects type 1 pneumocytes against SARS coronavirus infection in macaques. Nat. Med. **10**, 290–293 (2004)
13. M. Kielian, Enhancing host cell infection by SARS-CoV-2. Science **370**, 765–766 (2020)
14. F.N. Novikov, V.S. Stroylov, I.V. Svitanko, V.E. Nebolsin, Molecular basis of COVID-19 pathogenesis. Russ. Chem. Rev. **89**, 858–878 (2020)

15. A. Saha, B. Saha, Novel coronavirus SARS-CoV-2 (Covid-19) dynamics inside the human body. Rev. Med. Virol. **30**:article e2140 (2020)
16. M. Scudellari, How the coronavirus infects our cells. Nature **595**, 640–644 (2021)
17. J. Shang, Y. Wan, C. Luo, G. Ye, Q. Geng, A. Auerbach, F. Li, Cell entry mechanisms of SARS-CoV-2. Proc. Natl. Acad. Sci. USA **117**, 11727–11734 (2020)
18. H. Zhang, J.M. Penninger, Y. Li, N. Zhong, A.S. Slutsky, Angiotensin-converting enzyme 2 (ACE2) as a SARS-CoV-2 receptor: molecular mechanisms and potential therapheutic target. Intensive Care Med. **46**, 586–590 (2020)
19. X. Zou, K. Chen, J. Zou, J. Han, J. Hao, Z. Han, Single cell RNA-seq data analysis on the receptor ACE2 expression reveals the potential risk of different human organs vulnerable to 2019-nCoV infection. Front. Med. **14**, 185–192 (2020)
20. L. Mao, H. Jin, M. Wang, Y. Hu, S. Chen, Q. He, J. Chang, C. Hong, Y. Zhou, D. Wang, X. Miao, Y. Li, Neurologic manifestations of hospitalized patients with coronavirus disease 2019 in Wuhan, China. JAMA Neurol. **77**, 683–690 (2020)
21. R.B. Martines, J.M. Ritter, E. Matkovic, J. Gary, B.C. Bollweg, H. Bullock, C.S. Goldsmith et al., Pathology and pathogenesis of SARS-CoV-2 associated with fatal coronavirus disease, United States. Emerg. Infect. Dis. **26**, 2005–2015 (2020)
22. Z. Xu, L. Shi, Y. Wang, J. Zhang, L. Huang, C. Zhang et al., Pathological findings of COVID-19 associated with acute respiratory distress syndrome. The Lancet Respiratory Med. **8**, 420–422 (2020)
23. Mason. Pathogenesis of COVID-19 from a cell biology perspective. Eur. Respir. J. **55**:article 2000607 (2020)
24. Mason. Thoughts on the alveolar phase of COVID-19. Am. J. Physiol. Lung Cell Mol. Physiol. **319**, L115–L120 (2020)
25. E. Mochan, T.J. Sego, L. Gaona, E. Rial, G.B. Ermentrout, Compartmental model suggests importance of innate immune response to COVID-19 infection in rhesus macaques. Bull. Math. Biol. **83**, 79 (2021)
26. J.M. Sallenave, L. Guillot, Innate immune signaling and preteolytic pathways in the resolution or exacerbation of SARS-CoV-2 in COVID-19: key therapeutic targets? Front. Immunol. **11**:article 1229 (2020)
27. F. Zahedipour, S.A. Hosseini, T. Sathyapalan, M. Majeed, T. Jamialahmadi, K. Al-Rasadi, M. Banach, A. Sahebkar, Potential effects of curcumin in the treatment of COVID-19 infection. Phytother. Res. **34**, 2911–2920 (2020)
28. L. Zhou, Z. Niu, X. Jiang, Z. Zhang, Y. Zheng et al., SARS-CoV-2 Tarets by the pscRNA profiling of ACE2, TMPRSS2 and furin proteases. iScience **23**:article 101744 (2020)
29. O.Y. Cheung, P. Graziano, M.W. Smith, Acute lung injury, in *Practical Pulmonary Pathology: A Diagnostic Approach*. ed. by K.O. Leslie, M.R. Wick (Elsevier, New York, 2018), pp. 125-146.e3
30. T. Chua, S. Lukassen, S. Trum, B.P. Hennig, D. Wnedisch, F. Pott, O. Debnath, L. Thuermann, F. Kurth, M.T. Voelker et al., COVID-19 severity correlates with airway epithelium-immune cell interactions identified by single-cell analysis. Nat. Biotechnol. **38**, 970–979 (2020)
31. M.A. Stafford, L. Corey, Y. Cao, E.S. Daar, D.D. Ho, A.S. Perelson, Modeling plasma virus concentration during HIV primary infection. J. Theor. Biol. **203**, 285–301 (2000)
32. P. Baccam, C. Beauchemin, C.A. Macken, F.G. Hayden, A.S. Perelson, Kinetics of influenza A virus infection in humans. J. Virol. **80**, 7590–7599 (2006)
33. C. Hadjichrysanthou, E. Cauet, E. Lawrence, C. Vegvari, F. de Wolf, R.M. Anderson, Understanding the within-host dynamics of influenza a virus: from theory to clinical implications. J. R. Soc. Interface **13**:article 20160289 (2016)
34. S. Bonhoeffer, R.M. May, G.M. Shaw, M.A. Nowak, Virus dynamics and drug therapy. Proc. Natl. Acad. Sci. USA **94**, 6971–6976 (1997)
35. A.S. Perelson, R. Ke, Mechanistic modeling of SARS-CoV-2 and other infectious diseases and the effect of therapeutics. Clin. Pharmacol. Therapeutics **109**, 829–840 (2021)
36. C. Zitzmann, L. Kaderali, Mathematical analysis of viral replication dynamics and antiviral treatment strategy: from basic models to age-based multi-scale modeling. Front. Microbiol. **9**:article 01546 (2018)

37. M.A. Nowak, R.M. May, *Viral Dynamics: Mathematical Principles of Immunology and Virology* (Oxford University Press, New York, 2000)
38. A. Boianelli, V.K. Nguyen, T. Ebensen, K. Schulze, E. Wilk, N. Sharma, S. Stegemann-Koniszewski, D. Bruder, F.R. Toapanta, C.A. Guzman, M. Meyer-Hermann, E.A. Hernandez-Vargas, Modeling influenza virus infection: a roadmap for influenza research. Viruses **7**, 5274–5304 (2015)
39. H. Ikeda, R.J. de Boer, K. Sato, S. Morita, N. Misawa, Y. Koyanagi, K. Aihara, S. Iwami. Improving the estimation of the death rate of infected cells from time course data during acute phase of virus infections: application to acute HIV-1 infection in a humanized mouse model. Theor. Biol. Med. Model. **11**:article 22 (2014)
40. R. Sender et al., Proc. Natl. Acad. Sci. USA **118**:article 2024815118 (2021)
41. T.D. Frank, SARS-Coronavirus-2 nonlinear dynamics in patients: three-dimensional state and amplitude state description. J. Phys. Soc. Jpn. **90**:article 073802 (2021)
42. M.A. Nowak, A.L. Lloyd, G.M. Vasquez, T.A. Wiltrout, L.M. Wahl, N. Bischofberger, J. Williams, A. Kinter, A.S. Fauci, V.M. Hirsch, J.D. Lifson, Viral dynamics of primary viremia and antiretrovial therapy in Simian immunodeficiency virus infection. J. Virol. **71**, 7518–7525 (1997)
43. A.M. Smith, F.R. Adler, A.S. Perelson, An accurate two-phase approximative solution to acute viral infection model. J. Math. Biol. **60**, 711–726 (2010)
44. B.P. Holder, P. Simon, L.E. Liao, Y. Abed, X. Bouhy, C.A.A. Beauchemin, G. Boivin, Assessing the in vitro fitness of an oseltamivir-resistant seasonal a/h1n1 influenza train using a mathematical model. PLoS One **6**:article e14767 (2011)
45. K. Best, J. Guedj, V. Madelain, X. de Lamballerie, S.Y. Lim, C.E. Osuna, J.B. Whitney, A.S. Perelson, Zika plasma viral dynamics in nonhuman primates provides insights into earlz infection and antiviral strategies. PNAS **114**, 8847–8852 (2017)
46. C.A.A. Beauchemin, J.J. McSharry, G.L. Drusano, J.T. Nguyen, G.T. Went, R.M. Ribeiro, A.S. Perelson, Modeling amantadine treatment of influenza A virus in vitro. J. Theor. Biol. **254**, 439–451 (2008)
47. A.S. Perelson, D.E. Kirschner, R. De Boer, Dynamics of HIV infection of CD4+ T cells. Math. Biosci. **114**, 81–125 (2013)
48. J.J. Sedmak, S.E. Grossberg, Interferon bioassay: reduction in yield of myxovirus neuroaminidases. J. Gen. Virol. **21**, 1–7 (1973)
49. E. Möhler, D. Flockerzi, H. Sann, U. Reichl, Mathematical model of influenza A virus production in large-scale microcarrier culture. Biotechnol. Bioeng. **90**, 48–58 (2005)
50. L. Canini, F. Carrat, Population modeling of influenza a/h1n1 virus kinetics and symptom dynamics. J. Virol. **85**, 2764–2770 (2011)
51. K. Patel et al., Respiratory syncytial virus a dynamics and the effect of lumicitabine, a nucleoside viral replication inhibitor, in experimentally infect.ed humans. J. Antimicrob Chemother. **74**, 442–452 (2019)
52. E.N. Hernandez-Vargas, J.X. Velasco-Hernandez, In-host mathematical modelling of COVID-19 in humans. Ann. Rev. Control **50**, 448–456 (2020)
53. S. Wang, Y. Pan, Q. Wang, H. Miao, A.N. Brown, L. Rong, Modeling the viral dynamics of SARS-CoV-2 infection. Math. Biosci. **328**:article 108438 (2020)
54. M.M. Böhmer, U. Buchholz, V.M. Corman, M. Hoch, K. Katz et al., Investigation of a COVID-19 outbreak in Germany resulting from a single travel-associated primary case: a case series. The Lancet Infecti. Dis. **20**, 920–928 (2020)
55. T.D. Frank, SARS-coronavirus-2 infections: biological instabilities characterized by order parameters, Phys. Biol. **19**:article 036001 (2022)
56. A. Goncalves, Y. Bertrand, R. Ke, E. Comets, X. de Lamballerie, D. Malvy, Pizzorno, O. Terrier, M.R. Calatrava, F. Mentre, P. Smith, A.S. Perelson, J. Guedj, Timing of antiviral treatment initiation is critical to reduce SARS-CoV-2 viral load. CPT: Pharmacometrics Syst. Pharmacol. **9**, 509–514 (2020)
57. T.D. Frank, Unifying mass-action kinetics and Newtonian mechanics by means of Nambu brackts. J. Biol. Phys. **37**, 375–385 (2011)

58. N. Neant et al., Modeling SARS-CoV-2 viral kinetics and association with mortality in hospitalized patients from the French COVID cohort. PNAS **118**, e2017962118 (2021)
59. P. Czuppon, F. Debarre, A. Goncalves, O. Tenaillon, A.S. Perelson, J. Guedj, F. Blanquart, Success of prophylactic antiviral therapy for SARS-CoV-2: predicted critical efficiacies and impact of different drug-specific mechanisms of action. PLoS Comput. Biol. **17**, e1008752 (2021)

Chapter 10
Virus Dynamics in Humans: Unstable Directions and Order Parameters

This chapter applies the nonlinear physics perspective reviewed in Chap. 2 to the virus dynamics in humans. Accordingly, it is shown that the disease state of an infected individual evolves in a certain way: it evolves along an unstable direction or order parameter that defines how the affected cells and the virus concentration change relative to each other over time. The corresponding unstable amplitude or order parameter amplitude describes the exponential increase or decay of all variables in the initial stage of the infection. These key characteristics of virus infections are worked out for the virus dynamics models introduced in the previous chapter. The characteristics are also demonstrated for SARS-CoV-2 infections observed in a sample of COVID-19 patients.

10.1 Analysis of the TIV Model

10.1.1 3D Approach: Original Model

The TIV model is defined by Eq. (9.2). Equation (9.2) is repeated as

$$\frac{\mathrm{d}}{\mathrm{d}t}T = -\beta V T,\ \frac{\mathrm{d}}{\mathrm{d}t}I = \beta V T - k_1 I,\ \frac{\mathrm{d}}{\mathrm{d}t}V = pI - k_2 V \tag{10.1}$$

with $k_1, k_2, \beta, p > 0$. Due to the analogy between the TIV and SEIR model (see Sect. 9.3.3), the analysis of the TIV model can be conducted in the same way as the analysis of the SEIR model presented in Sect. 5.7.1. The following analysis can be found, for example, in Ref. [1]. The state vector of the TIV model reads $\mathbf{X} = (T, I, V)$. The virus-free fixed point (i.e., stationary state) is given by $\mathbf{X}_{st} = (T_{st}, 0, 0)$, where

T. D. Frank, *COVID-19 Epidemiology and Virus Dynamics*,
Understanding Complex Systems, https://doi.org/10.1007/978-3-030-97178-6_10

$T_{st} > 0$ denotes the target cell concentration characteristic for healthy adults. Let δ denote the relative state of target cell concentrations defined by $T = T_{st} + \delta$. The relative state vector reads $\mathbf{u} = \mathbf{X} - \mathbf{X}_{st} = (\delta, I, V)$. Assuming $\mathbf{u}$ is small in magnitude, Eq. (10.1) is approximately described by the linearized model

$$\frac{\mathrm{d}}{\mathrm{d}t}\mathbf{u} = L\,\mathbf{u}\ ,\ L = \begin{pmatrix} 0 & 0 & -\beta T_{st} \\ 0 & -k_1 & \beta T_{st} \\ 0 & p & -k_2 \end{pmatrix} . \tag{10.2}$$

The linearization matrix L exhibits the eigenvalue $\lambda_1 = 0$. The two remaining eigenvalues are defined by the characteristic equation [1–3]

$$(\lambda + k_1)(\lambda + k_2) = p\beta T_{st}\ , \tag{10.3}$$

which yields [1, 2]

$$\lambda_{2,3} = -\frac{k_1 + k_2}{2} \pm \sqrt{\frac{(k_1 + k_2)^2}{4} + \beta_w - \beta_{w,crit}}\ ,$$
$$\beta_w = pT_{st}\beta\ ,\ \beta_{w,crit} = k_1 k_2 \tag{10.4}$$

with $\lambda_{2,3} \in \mathbb{R}$. In Eq. (10.4) β_w denotes the weighted infectivity parameter and $\beta_{w,crit}$ is its critical value. Equation (10.4) assumes the form of Eq. (5.27). From the general discussion of Eq. (5.27) presented in Sect. 5.4.3 it follows that $\lambda_3 < 0$ holds for any parameters values and

$$\begin{aligned} &\beta_w > \beta_{w,crit} \Rightarrow \lambda_2 > 0 \Rightarrow \mathbf{X}_{st} \text{ saddle (i.e., unstable)}\ , \\ &\beta_w = \beta_{w,crit} \Rightarrow \lambda_2 = 0 \Rightarrow \text{system at bifurcation point}\ , \\ &\beta_w < \beta_{w,crit} \Rightarrow \lambda_2 < 0 \Rightarrow \mathbf{X}_{st} \text{ neutrally stable node}\ . \end{aligned} \tag{10.5}$$

Note that if $\lambda_2 < 0$ (i.e., $\beta_w < \beta_{w,crit}$) holds for $\mathbf{X}_{st} = (T_{st}, 0, 0)$, then $\lambda_2 < 0$ also holds for any stationary state $\mathbf{X}_{st} = (Y, 0, 0)$ with $Y < T_{st}$. Consequently, for $\lambda_2 < 0$ perturbations that move the state $\mathbf{X}$ out of $\mathbf{X}_{st} = (T_{st}, 0, 0)$ along the T-axis do not decay in magnitude over time, which makes that $\mathbf{X}_{st}$ is neutrally stable rather than asymptotically stable.

Let us derive next the amplitude equations of the TIV model. To this end, the simplified picture will be taken in which all variables of the TIV model are assumed to be measured in the same units. This implies that 1 virus particle is considered as equivalent to 1 cell. Let $\mathbf{v}_j = (v_{j,T}, v_{j,I}, v_{j,V})^T$ denote the eigenvectors of L with $j = 1, 2, 3$. For $\lambda_1 = 0$ the eigenvector reads $\mathbf{v}_1 = (1, 0, 0)$. For the remaining eigenvalues the eigenvectors can be determined in analogy to the procedure discussed in Sect. 5.7.1. The result reads [1]

$$\mathbf{v}_j = \frac{1}{Z_j}\begin{pmatrix} -\beta T_{st}(\lambda_j + k_1) \\ \beta T_{st}\lambda_j \\ \lambda_j(\lambda_j + k_1) \end{pmatrix} \tag{10.6}$$

for $j = 2, 3$ with $Z_j = \sqrt{(\beta T_{st})^2[(\lambda_j + k_1)^2 + \lambda_j^2] + \lambda_j^2(\lambda_j + k_1)^2}$. These eigenvectors correspond to the SEIR model eigenvectors (5.71) when replacing βT_{st} by β. The TIV state $\mathbf{X}$ can then be expressed in terms of the eigenvectors $\mathbf{v}_j$ and their corresponding amplitudes A_j in the usual way like

$$\mathbf{X} = \mathbf{X}_{st} + \sum_{j=1}^{3} A_j \mathbf{v}_j \,. \tag{10.7}$$

Equation (10.7) describes the mapping from amplitude space (A_1, A_2, A_3) to state space $\mathbf{X} = (T, I, V)$ (as discussed in general in Sect. 2.6.3) that can be cast into the form

$$\mathbf{X} - \mathbf{X}_{st} = \mathbf{u} = M \begin{pmatrix} A_1 \\ A_2 \\ A_3 \end{pmatrix} , \; M = \begin{pmatrix} 1 & v_{2,T} & v_{3,T} \\ 0 & v_{2,I} & v_{3,I} \\ 0 & v_{2,V} & v_{3,V} \end{pmatrix} . \tag{10.8}$$

The matrix M exhibits the determinant

$$|M| = \beta T_{st} \frac{\lambda_2 \lambda_3 (\lambda_3 - \lambda_2)}{Z_2 Z_3} \tag{10.9}$$

with $|M| \neq 0$ for $\beta_w \neq \beta_{crit}$ (i.e., $\lambda_2 \neq 0$). The inverse mapping from state space $\mathbf{X} = (T, I, V)$ to the amplitude space (A_1, A_2, A_3) is given by the inverse matrix M^{-1} like

$$\begin{pmatrix} A_1 \\ A_2 \\ A_3 \end{pmatrix} = M^{-1}\mathbf{u} = M^{-1}(\mathbf{X} - \mathbf{X}_{st}) \,. \tag{10.10}$$

As discussed in Sect. 5.6, the rows of M^{-1} correspond to the biorthogonal vectors $\mathbf{w}_i$ with $\mathbf{w}_i \mathbf{v}_k = \delta_{ik}$. Since the TIV model matrix M defined by Eq. (10.8) exhibits the same structure as the SEIR model matrix M defined by Eq. (5.73), the biorthogonal vectors $\mathbf{w}_i$ of the TIV model assume the same form as the corresponding vectors $\mathbf{w}_i$ defined by Eqs. (5.74) and (5.75) of the SEIR model. Explicitly, $\mathbf{w}_i$ read

$$\mathbf{w}_1 = \frac{1}{\lambda_2 \lambda_3} \begin{pmatrix} \lambda_2 \lambda_3 \\ (\lambda_2 + k_1)(\lambda_3 + k_1) \\ -k_1 \beta T_{st} \end{pmatrix} \tag{10.11}$$

and

$$\mathbf{w}_2 = \frac{1}{|M|} \begin{pmatrix} 0 \\ v_{3,V} \\ -v_{3,I} \end{pmatrix} , \; \mathbf{w}_3 = \frac{1}{|M|} \begin{pmatrix} 0 \\ -v_{2,V} \\ v_{2,I} \end{pmatrix} . \tag{10.12}$$

The biorthogonal eigenvectors can be used to apply the vector calculation methods (see Sect. 2.9.3) to derive the amplitude equations of the TIV model. Using the relative state $\mathbf{u} = \mathbf{X} - \mathbf{X}_{st}$ the TIV model (10.1) can be cast into the form

$$\frac{\mathrm{d}}{\mathrm{d}t}\begin{pmatrix}\delta\\ I\\ V\end{pmatrix}=\begin{pmatrix}0 & 0 & -\beta T_{st}\\ 0 & -k_1 & \beta T_{st}\\ 0 & p & -k_2\end{pmatrix}\begin{pmatrix}\delta\\ I\\ V\end{pmatrix}+\beta\delta V\begin{pmatrix}-1\\ 1\\ 0\end{pmatrix}$$
$$=L\begin{pmatrix}\delta\\ I\\ V\end{pmatrix}+\beta\delta V\begin{pmatrix}-1\\ 1\\ 0\end{pmatrix} \tag{10.13}$$

(see also the SEIR model counterpart given by Eq. (5.64)). Equation (10.13) assumes the structure of the general relative state equation $\mathrm{d}\mathbf{u}/\mathrm{d}t = L\mathbf{u}+\mathbf{R}$ (see Eq. (2.89)) discussed in Sect. 2.9.3 with the remainder term $\mathbf{R}$ given by

$$\mathbf{R}(\mathbf{u})=\beta\delta V\begin{pmatrix}-1\\ 1\\ 0\end{pmatrix} . \tag{10.14}$$

As previously discussed in Sect. 2.9.3, multiplying Eq. (10.13) by $\mathbf{w}_i$, we obtain the evolution equation for the amplitude A_i in form of $\mathrm{d}A_i/\mathrm{d}t = \lambda_i A_i + G_i$ with

$$G_i=\mathbf{w}_i\mathbf{R}=\beta\delta V\mathbf{w}_i\begin{pmatrix}-1\\ 1\\ 0\end{pmatrix}=C_i\beta\delta V\ ,\ C_i=w_{i,I}-w_{i,T}\ . \tag{10.15}$$

The product δV can be expressed in terms of amplitudes A_1, A_2, A_3 using Eq. (10.8). From $\mathrm{d}A_i/\mathrm{d}t = \lambda_i A_i + G_i$, Eqs. (10.8) and (10.15) it then follows that the amplitude equations of the TIV model read [1]

$$\frac{\mathrm{d}}{\mathrm{d}t}A_i=\lambda_i A_i + C_i p_2(A_1,A_2,A_3)\ ,$$
$$p_2=\beta\left(\sum_{k=1}^{3}v_{k,T}A_k\right)\left(\sum_{k=1}^{3}v_{k,V}A_k\right)\ ,$$
$$C_1=\frac{k_1(\lambda_2+\lambda_3+k_1)}{\lambda_2\lambda_3}\ ,\ C_2=\frac{v_{3,V}}{|M|}\ ,\ C_3=-\frac{v_{2,V}}{|M|} \tag{10.16}$$

for $i = 1, 2, 3$ with $\lambda_1 = 0$ and $\lambda_{2,3}$ defined by Eq. (10.4). For the initial state $\mathbf{X}_0 = (T_{st}, 0, V(0))$ at $t_0 = 0$ the initial amplitudes can be computed from

$$\begin{pmatrix}A_1(0)\\ A_2(0)\\ A_3(0)\end{pmatrix}=M^{-1}(\mathbf{X}(0)-\mathbf{X}_{st})=M^{-1}\begin{pmatrix}0\\ 0\\ V(0)\end{pmatrix}\Rightarrow A_i(0)=w_{i,V}V(0)\ . \tag{10.17}$$

Solving Eq. (10.16) under the initial conditions (10.17) produces the amplitude dynamics that corresponds to the state dynamics defined by Eq. (10.1) for the initial state $\mathbf{X}_0 = (T_{st}, 0, V(0))$.

For $\beta_w > \beta_{w,crit}$ the infectious disease in the human body emerges in a certain way that will be discussed in what follows. The virus-free fixed point $\mathbf{X}_{st} = (T_{st}, 0, 0)$ corresponds to a saddle with an unstable direction $\mathbf{v}_2$ ($\lambda_2 > 0$), a stable direction $\mathbf{v}_3$ ($\lambda_3 < 0$), and a neutrally stable direction $\mathbf{v}_1$ ($\lambda_1 = 0$). Consequently, A_3 initially decays exponentially with a time constant $\tau_3 = 1/|\lambda_3|$, A_2 increase exponentially, A_1 can be regarded as constant relative to A_3 and A_2. As argued in Sect. 5.7.1, there is an intermediate period denoted by T_i characterized by the time constant τ_3 during which $A_3(t)$ decays in magnitude. It is assumed that at T_i the contribution of A_3 to $\mathbf{X}$ as defined by Eq. (10.7) becomes negligibly small. Furthermore, let T_L denote the period for which the linear approximation holds. The case is considered in which τ_3 is sufficiently short such that $T_L > T_i$ holds. Consequently, for $t \in [T_i, T_L]$ Eq. (10.7) reduces to

$$\mathbf{X}(t) \approx \mathbf{X}_{st} + A_1(0)\mathbf{v}_1 + \mathbf{v}_2 A_2(t) \tag{10.18}$$

(see Eq. (5.81)), where it has been taken into account that variations of A_1 can be neglected. Let $V_a(t)$ denote the approximation of the solution $V(t)$ of the TIV model. Then, Eq. (10.18) implies

$$V(t) \approx V_a(t) = v_{1,V} A_1(0) + v_{2,V} A_2(t) = a + b \exp\{\lambda_2 t\} \tag{10.19}$$

with $a = v_{1,V} A_1(0)$ and $b = v_{2,V} A_2(0)$. The exact solution for $V(0)$ is given by $V(0) = \sum_{k=1}^{3} v_{k,V} A_k(0) = a + b + v_{3,V} A_3(0)$. Consequently, the approximation $V_a(t)$ defined by Eq. (10.19) exhibits an initial mismatch with $V_a(0) = a + b \neq V(0)$. However, if the term $b \exp\{\lambda_2 t\}$ increases sufficiently fast with respect to the constant $v_{3,V} A_3(0)$, then the initial mismatch can be neglected after a relative short period.

Alternatively, from the aforementioned considerations about the eigenvalues λ_1, λ_2, and λ_3, it follows that the state dynamics approximately is given by [1]

$$\frac{\mathrm{d}}{\mathrm{d}t}\mathbf{X} \approx \mathbf{v}_2 \frac{\mathrm{d}}{\mathrm{d}t} A_2 \tag{10.20}$$

such that

$$\Delta \mathbf{X} \approx \mathbf{v}_2 \Delta A_2 \tag{10.21}$$

with $\Delta \mathbf{X} = \mathbf{X}(t) - \mathbf{X}(t')$ and $\Delta A_2 = A_2(t) - A_2(t')$ with $t > t'$ for $t', t \in [T_i, T_L]$. Finally, if the initial state $\mathbf{X}$ is located sufficiently close to the fixed point $\mathbf{X}_{st}$ such that all initial amplitudes are relatively small in magnitude, then the fundamental approximation

$$\mathbf{X}(t) \approx \mathbf{X}_{st} + \mathbf{v}_2 A_2(t) \tag{10.22}$$

holds (see Sect. 2.7). All three approximations given by Eqs. (10.18), (10.21), and (10.22) imply that the unstable eigenvector $\mathbf{v}_2$ characterizes how the variables T, I, and V evolve relative to each other during the initial stage of the disease. The vector

$\mathbf{v}_2$ characterizes the emerging disease and corresponds to the order parameter of the disease. The order parameter amplitude A_2 increasing exponentially like

$$A_2(t) = A_2(0)\exp\{\lambda_2 t\} \tag{10.23}$$

and captures explicitly the temporal aspects of the dynamics during this initial stage.

Let us return to the approximations (10.21) and (10.22). From Eq. (10.21) it follows that in the initial disease stage the changes of the subsystems X_i relative to each other over time, where X_i describing the disease (or health) state of an individual, are determined by the order parameter like $\Delta X_i/\Delta X_k \approx v_{2,i}/v_{2,k}$, which reads explicitly

$$\frac{\Delta T}{\Delta I} \approx \frac{v_{2,T}}{v_{2,I}}, \quad \frac{\Delta T}{\Delta V} \approx \frac{v_{2,T}}{v_{2,V}}, \quad \frac{\Delta I}{\Delta V} \approx \frac{v_{2,I}}{v_{2,V}}. \tag{10.24}$$

From Eq. (10.22) it follows that the approximation $V_a(t)$ of $V(t)$ reads

$$V_a(t) = b\exp\{\lambda_2 t\} \tag{10.25}$$

with $b = v_{2,V}A_2(0)$ (i.e., b is defined just as in the case of Eq. (10.19)). Again the approximative solution exhibits an initial mismatch $V_a(0) = b \neq V(0)$. Importantly, Eqs. (10.19) and (10.25) demonstrate that in good approximation the viral load increases exponentially with a rate constant λ_2 of the unstable eigenvector as previously discussed in Sect. 9.4. For the TIV model, the eigenvalue λ_2 corresponds to the maximal eigenvalue λ_{max}. That is, λ_2 determines the initial slope of the pattern $V(t)$ shown in Fig. 9.5.

10.1.2 3D Approach: Scaled Model

Let us consider the scaled TIV model (9.5) that is repeated as

$$\frac{\mathrm{d}}{\mathrm{d}t}T = -\beta' V' T, \quad \frac{\mathrm{d}}{\mathrm{d}t}I = \beta' V' T - k_1 I, \quad \frac{\mathrm{d}}{\mathrm{d}t}V' = k_1 I - k_2 V' \tag{10.26}$$

with $V = pV'/k_1$, $\beta' = \beta p/k_1 > 0$, $k_1, k_2, \beta, p > 0$, and state vector $\mathbf{X} = (T, I, V')$. The amplitude equations can be derived in two ways. First, since the scaled TIV model is equivalent to the SEIR model, the amplitude equations can directly be obtained from the SEIR model amplitude equations. For the sake of conveniency, let β_{SEIR} denote the effective contact rate β of the SEIR model defined by Eq. (3.43). Then, as stated in Sect. 9.3.3, the SEIR model variables S, E, I are replaced by the TIV model variables like $S \to T$, $E \to I$, and $I \to V'$. Moreover, in the equations of the SEIR model the following substitutions have to be made:

$$N = T_{st} \ , \ \alpha = k_1 \ , \ \gamma = k_2 \ , \ \frac{\beta_{SEIR}}{N} = \beta' \ \Rightarrow \ \beta_{SEIR} = \beta' T_{st} \ . \tag{10.27}$$

In particular, substituting Eq. (10.27) into Eq. (5.67), the eigenvalues $\lambda_{2,3}$ of the scaled TIV model read

$$\lambda_{2,3} = -\frac{k_1 + k_2}{2} \pm \sqrt{\frac{(k_1 + k_2)^2}{4} + k_1(\beta' T_{st} - k_2)} \ . \tag{10.28}$$

Furthermore, $\lambda_1 = 0$ holds. Substituting Eq. (10.27) into Eq. (5.71), the eigenvectors $\mathbf{v}_{2,3}$ of the scaled TIV model read

$$\mathbf{v}_i = \frac{1}{Z_i} \begin{pmatrix} -\beta' T_{st}(\lambda_i + k_1) \\ \beta' T_{st} \lambda_i \\ \lambda_i(\lambda_i + k_1) \end{pmatrix} = \begin{pmatrix} v_{i,T} \\ v_{i,I} \\ v_{i,V'} \end{pmatrix} \tag{10.29}$$

with $Z_i = \sqrt{(\beta' T_{st})^2[(\lambda_i + k_1)^2 + \lambda_i^2] + \lambda_i^2(\lambda_i + k_1)^2}$ for $i = 2, 3$. Furthermore, $\mathbf{v}_1 = (1, 0, 0)$ holds. Substituting Eq. (10.27) into Eqs. (5.78) and (5.80), the amplitude equations of the scaled TIV model read

$$\frac{\mathrm{d}}{\mathrm{d}t} A_i = \lambda_i A_i + C_i p_2(A_1, A_2, A_3) \ ,$$
$$p_2 = \beta' \left(\sum_{k=1}^{3} v_{k,T} A_k \right) \left(\sum_{k=2,3} v_{k,V'} A_k \right) , \tag{10.30}$$

with

$$C_1 = k_1 \frac{\lambda_2 + \lambda_3 + k_1}{\lambda_2 \lambda_3} \ , \ C_2 = \frac{v_{3,V'}}{|M|} \ , \ C_3 = -\frac{v_{2,V'}}{|M|} \tag{10.31}$$

with $|M| = v_{2,I} v_{3,V'} - v_{3,I} v_{2,V'}$. As usual, the dynamics of the state $\mathbf{X}(t)$ can then be computed from the amplitude dynamics like $\mathbf{X}(t) = \mathbf{X}_{st} + \sum_{k=1}^{3} A_k(t) \mathbf{v}_k$. Substituting Eq. (10.27) into Eqs. (5.74) and (5.75), the biorthogonal vectors of the scaled TIV model read

$$\mathbf{w}_1 = \frac{1}{\lambda_2 \lambda_3} \begin{pmatrix} \lambda_2 \lambda_3 \\ (\lambda_2 + k_1)(\lambda_3 + k_1) \\ -k_1 \beta' T_{st} \end{pmatrix} \tag{10.32}$$

and

$$\mathbf{w}_2 = \frac{1}{|M|} \begin{pmatrix} 0 \\ v_{3,V'} \\ -v_{3,I} \end{pmatrix} \ , \ \mathbf{w}_3 = \frac{1}{|M|} \begin{pmatrix} 0 \\ -v_{2,V'} \\ v_{2,I} \end{pmatrix} \ . \tag{10.33}$$

In analogy to Eq. (10.17), the initial amplitudes can then be obtained from

$$A_i(0) = w_{i,V'} V'(0) = w_{i,V'} \frac{k_1}{p} V(0) \, . \tag{10.34}$$

The second way to derive Eqs. (10.28) to (10.33) is to acknowledge that the scaled TIV model (10.26) from a mathematical point of view is a special case of the original TIV model (10.1). That is, if the substitutions $V = V'$, $\beta = \beta'$, and $p = k_1$ are made in the original TIV model (10.1), then Eq. (10.1) becomes Eq. (10.26). Likewise, if the substitutions $V = V'$, $\beta = \beta'$, and $p = k_1$ are used in the relevant equations of Sect. 10.1.1 then Eqs. (10.28) to (10.33) can be obtained. In particular, substituting $\beta = \beta'$ and $p = k_1$ into Eq. (10.4), we obtain $\beta_w = k_1 \beta' T_{st}$ such that Eq. (10.4) becomes the eigenvalue equation (10.28). Substituting $V = V'$ and $\beta = \beta'$ into Eq. (10.6) yields the eigenvector equation (10.29). Substituting $V = V'$ and $\beta = \beta'$ into Eq. (10.16) yields the amplitude equation (10.30). Substituting $V = V'$ and $\beta = \beta'$ into Eqs. (10.11) and (10.12) yields the biorthogonal vectors (10.32) and (10.33), respectively.

The scaled model (10.26) is equivalent to the original model (10.1) when using the variable and parameter transformations $V = pV'/k_1$ and $\beta = k_1\beta'/p$, respectively. Therefore, for any parameter set k_1, k_2, p, and β (or β') the models exhibit the same eigenvalues. This can be shown explicitly. When substituting $\beta' = p\beta/k_1$ into the eigenvalue equation (10.28) of the scaled TIV model, then Eq. (10.28) becomes the eigenvalue equation (10.4) of the original TIV model. In other words, the eigenvalue equations (10.4) and (10.28) produce the same eigenvalues for any parameters k_1, k_2, p, β and $\beta' = p\beta/k_1$.

Let us return to the scaled TIV model and the eigenvalue equation (10.28). In analogy to Eq. (10.5), for the scaled TIV model and the virus-free fixed point $\mathbf{X}_{st} = (T_{st}, 0, 0)$ based on Eq. (10.28) the following conclusions can be drawn:

$$\begin{aligned}
\beta' T_{st} > k_2 &\Rightarrow \lambda_2 > 0 \Rightarrow \mathbf{X}_{st} \text{ saddle (i.e., unstable)} \, , \\
\beta' T_{st} = k_2 &\Rightarrow \lambda_2 = 0 \Rightarrow \text{system at bifurcation point} \, , \\
\beta' T_{st} < k_2 &\Rightarrow \lambda_2 < 0 \Rightarrow \mathbf{X}_{st} \text{ neutrally stable node} \, .
\end{aligned} \tag{10.35}$$

10.1.3 2D Approach

In Chap. 6 in the context of epidemiological models it was discussed a stability analysis may be restricted to a subset of the original variables of interest. In the context of epidemiological models, the stability analysis may focus only on the variables describing actually infectious individuals (i.e., infectious individuals who are in the position to infect others) and some auxiliary variables that allow to construct an autonomous linearized model (see Sect. 6.1.1). In particular, in Sect. 6.2.2 an amplitude equation description of the SEIR model has been developed that is based only on the infectious compartment variables E and I. In a similar vain, let us evaluate the dynamics of the TIV model when focusing on the dynamics in the I-V

subspace [4]. To this end, the variables T, I, V are arranged in the state vector $\mathbf{X}$ like $\mathbf{X} = (I, V, T)$. Subsequently, the subspace vector $\mathbf{X}^+ = (I, V)$ and the subspace variable $X^- = T$ are introduced such that $\mathbf{X} = (\mathbf{X}^+, X^-)$. Let us consider the fixed point $\mathbf{X}_{st} = (0, 0, T_{st})$ with $\mathbf{X}^+_{st} = (0, 0)$. From Eq. (10.1) it follows that the dynamics close to the fixed point in the two-dimensional I-V subspace D^+ is determined by the linearized evolution equations

$$\frac{\mathrm{d}}{\mathrm{d}t} I = \beta V T_{st} - k_1 I \ , \ \frac{\mathrm{d}}{\mathrm{d}t} V = pI - k_2 V \ . \tag{10.36}$$

The linearized equations form an autonomous dynamical system. They can be written like

$$\frac{\mathrm{d}}{\mathrm{d}t}\mathbf{X}^+ = L^+\mathbf{X}^+ \ , \ L^+ = \begin{pmatrix} -k_1 & \beta T_{st} \\ p & -k_2 \end{pmatrix} . \tag{10.37}$$

In what follows the simplified perspective will be taken according to which a virus particle can be considered as the equivalent counterpart of a cell such that I and V are measured in the same units. If so, the eigenvalues of the matrix L^+ are given by Eq. (10.4) when changing the indices from 2,3 to 1,2 such that

$$\lambda_{1,2} = -\frac{k_1 + k_2}{2} \pm \sqrt{\frac{(k_1 + k_2)^2}{4} + \beta_w - \beta_{w,crit}} \ , \tag{10.38}$$

where the upper (lower) sign holds for λ_1 (λ_2) such that $\lambda_2 < 0$ holds for any parameter set k_1, k_2, β, p, and T_{st}. Close to the fixed point, the amplitudes evolve like $A_1(t) = A_1(0)\exp\{\lambda_1 t\}$ and $A_2(t) = A_2(0)\exp\{-|\lambda_2|t\}$, which implies that A_2 decays in magnitude in any case, whereas A_1 increases (decreases) in magnitude for $\beta_w > \beta_{w,crit}$ ($\beta_w < \beta_{w,crit}$). The state $\mathbf{X}^+(t)$ evolves like

$$\mathbf{X}^+(t) = \sum_{j=1,2} A_j(t)\mathbf{v}_j \ , \mathbf{v}_j = \frac{1}{Z_j}\begin{pmatrix} \beta T_{st} \\ \lambda_j + k_1 \end{pmatrix} , \tag{10.39}$$

where $\mathbf{v}_j$ denote the eigenvectors $\mathbf{v}_j = (v_{j,I}, v_{j,V})$ for $j = 1, 2$ and $Z_j = \sqrt{(\beta T_{st})^2 + (\lambda_j + k_1)^2}$. The 2D eigenvectors $\mathbf{v}_{1,2}$ point in the I-V space in the same directions as the 3D eigenvectors $\mathbf{v}_{2,3}$ defined by Eq. (10.6). That is, by comparing Eqs. (10.6) and (10.39) it can be seen that $v_{2,I}(3D)/v_{2,V}(3D) = v_{1,I}(2D)/v_{1,V}(2D)$ and $v_{3,I}(3D)/v_{3,V}(3D) = v_{2,I}(2D)/v_{2,V}(2D)$ holds, which indicates that the 2D approach is consistent with the 3D approach.

Let us consider the case $\lambda_1 > 0$ for which A_1 increases exponentially during the initial stage of the infection. When assuming that the linearized model still holds when A_2 has decayed towards a negligible small value, then the evolution of the state $\mathbf{X}^+$ is entirely determined by A_1 such that

$$\Delta\mathbf{X}^+ \approx \mathbf{v}_1 \Delta A_1 \tag{10.40}$$

holds with $\Delta\mathbf{X} = \mathbf{X}(t + \Delta t) - \mathbf{X}(t)$, $\Delta A_1 = A_1(t + \Delta t) - A_1(t)$, and $\Delta t > 0$. The vector $\mathbf{v}_1$ for $\lambda_1 > 0$ corresponds to an unstable eigenvector and describes the IV order parameter of the infectious disease under consideration.

10.1.4 2D Versus 3D Approach

As discussed in Chap. 6 for epidemiological models, the 2D approach to analyze the TIV model has the advantage that it reduces the dimensionality of the problem at hand. In general, analytical solutions can frequently be derived for lower dimensional problems but not for higher dimensional problems. Another advantage of the 2D approach of the TIV model presented in Sect. 10.1.3 is that the superposition $\mathbf{X} = \sum_{k=1}^{2} A_k \mathbf{v}_k$ (see Eq. (10.39)) holds at the bifurcation point $\beta_w = \beta_{w,crit}$ for which $\lambda_1 = 0$ and $\lambda_2 = -(k_1 + k_2)$. In this case the eigenvectors read $\mathbf{v}_1 = (\beta T_{st}, k_1)/Z_1$ and $\mathbf{v}_2 = (\beta T_{st}, -k_2)/Z_1$ and are linearly independent. In contrast, the superposition $\mathbf{X} = \mathbf{X}_{st} + \sum_{k=1}^{3} A_k \mathbf{v}_k$ of the 3D approach (see Eq. (10.7)) involving the 3D eigenvectors $\mathbf{v}_k$ defined by Eq. (10.6) has only been worked out for $\beta_w \neq \beta_{w,crit}$. At the bifurcation point $\beta_w = \beta_{w,crit}$ the linearization matrix (10.2) exhibits in addition to the first eigenvalue $\lambda_1 = 0$ a second eigenvalue $\lambda_2 = 0$. This implies that $|M| = 0$ (see Eq. (10.9)), which means that the three eigenvectors $\mathbf{v}_k$ are no longer linearly independent. In other words, the 3D approach requires that the special case $\beta_w = \beta_{w,crit}$ needs to be worked out separately if this case is explicitly considered from an amplitude space perspective. An example of how such an analysis can be performed is given in Sect. 4.2.4 in the context of the SIR model. As illustrated explicitly in Chap. 6, the disadvantage of the 2D approach is that the set of nonlinear amplitude equations do not constitute an autonomous system. The 2D approach produces a non-autonomous amplitude equation description when the nonlinear terms are taken into account.

10.2 TIV Model and Viral Load in a Sample of COVID-19 Patients

10.2.1 3D Approach: TIV Order Parameters of COVID-19 Patients

In what follows data from eight COVID-19 patients will be considered. The patients are described in Refs. [5, 6]. Using the notation of Ref. [6] they are referred to as patients 1, 2, 3, 4, 7, 8, 10, and 14. The patients developed symptoms during January/February 2020. That time they were located in Germany. For all eight patients the disease was considered to be mild. SARS-CoV-2 viral load was measured in the sputum of the patients. Consequently, the measured viral load reflected virus concentrations in the lungs of the patients. Daily measurements were made for periods up to 25 days after symptoms onset. Wang et al. [7] fitted the viral load trajectories to the TIV model (10.1). In doing so, estimates for the TIV model parameters β, p, k_1, and k_2 occurring in Eq. (10.1) were obtained for each patient. In a subsequent study [1], the model parameters β, p, k_1, and k_2 were used to compute for each patient the unstable eigenvector $\mathbf{v}_2$, the eigenvalues λ_2, λ_3, and amplitude equation coefficients C_1, C_2, C_3 from Eqs. (10.4), (10.6), and (10.15).

In what follows the results presented in Ref. [1] will be reviewed. Figure 10.1 shows the analysis results for patient 1. Panel (a) shows the viral load data (gray circles) [6] of patient 1 and the model solution $V(t)$ (solid line) by solving the TIV model (10.1). The viral load pattern of patient 1 reached a peak at about 5 to 6 days after symptoms onset. Subsequently, the viral load decayed in a more or less monotonic fashion. The period of decline was longer as compare to the initial stage of increasing viral load, which is a general feature of SARS-CoV-2 infections in COVID-19 patients [8, 9]. The model solution $V(t)$ captures the relatively short period of viral load increase and the longer period of disease decline.

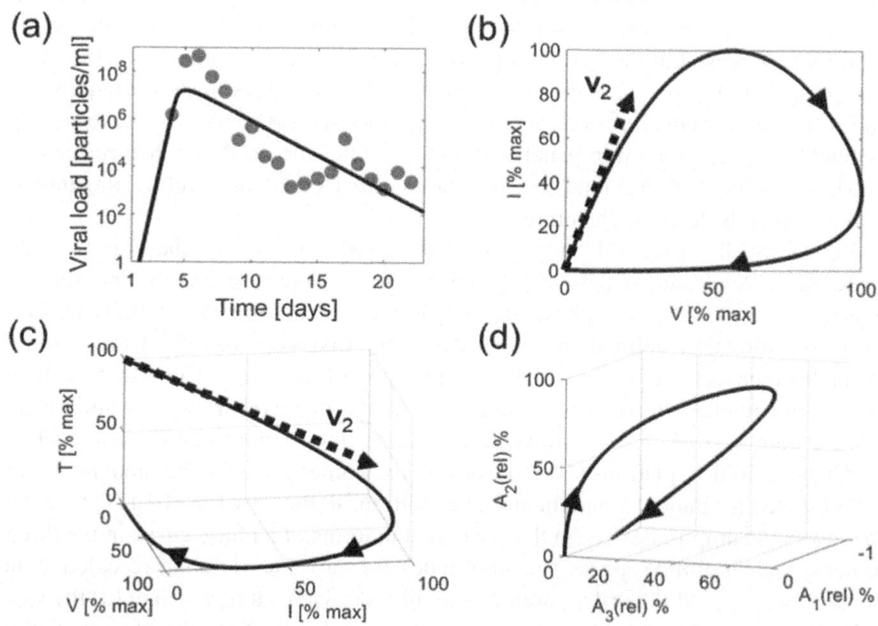

Fig. 10.1 Progression of COVID-19 in patient 1 in state and amplitude space as seen within a TIV modeling framework. Panel **a**: Viral load trajectory $V(t)$ over time of patient 1 (gray circles) solution $V(t)$ (solid line) computed from Eq. (10.1). Panels **b** and **c**: The evolution of the disease state $\mathbf{X}(t)$ is shown as phase curves (solid lines) in the V-I subspace and the full 3D TIV model space. The order parameter $\mathbf{v}_2$ (thick dotted line) is presented as well as computed from Eq. (10.6) and magnified in length for the sake of visibility. Panel **d**: Amplitude dynamics computed from Eq. (10.31). Parameters and initial conditions [7]: $\beta = 9.8 \cdot 10^{-7}/(\text{d} \times \text{(particles/ml)})$, $p = 970$ particles/(d × cell), $k_1 = 2/\text{d}$, $k_2 = 0.67/\text{d}$, $T_{st} = 6 \cdot 10^4$ cells/ml, $V(0) = 10^{-4}$ particles/ml

Panels (b) and (c) show the evolution of the state $\mathbf{X} = (T, I, V)$ in terms of phase curves in the 2D V-I subspace (panel (b)) and full 3D state space (panel (c)). The variables are presented in relative units of their respective maximal values. Note that in panels (b) and (c) the viral load is shown on a linear scale (in contrast to panel (a), where it is shown on a logarithmic scale).

The initial state was given by $T(0) = 100\%T_{\max} \Rightarrow T(0) = T_{\max}$ (with $T_{\max} = T_{st}$), $I(0) = 0\%I_{\max} \Rightarrow I(0) = 0$, and $V(0) = (V(0)/V_{\max})V_{\max}$. For patient 1, $V(0)$ was less than 1% of $V_{\max}$. As discussed in Sect. 9.3.2 in the context of target cell-limited models, according to the TIV model (10.1) the infection initially converted relatively quickly target cells into infected cells such that the number of infected cells I in patient 1 increased. At the same time, the infected cells produced free virus such that the virus concentration V increased as well (see panel (b)). During this initial stage the number of target cells dropped (see panel (c)). The lack of the target cells resulted in a de-acceleration of the increase in infected cells. At some point in time, I reached the maximum $I = 100\%I_{\max}$ (see panel (b)). At that time point, the viral load was approximately at a 50% level of its maximum value (see panel (b)). After $I(t)$ passed in time its maximum value $I_{\max}$, the infected cells were removed at a rate k_1 faster than new infected cells were produced. As a result, the number of infected cells I decayed. Virus was still produced by the remaining infected cells such that the viral load increased until it eventually reached its maximum value $V = 100\%V_{\max}$ (see panel (b)). For patient 1 the peak was at about 5 to 6 days as shown in panel (a) and corresponds to the locations at which V reached the 100% mark in the respective 2D and 3D spaces shown in panels (b) and (c). After the viral load had reached its maximum value, the viral load decayed. In this stage of disease decline, the number of infected cells decreased even further.

Let us dwell on the role of the unstable eigenvector $\mathbf{v}_2$ for the course of the SARS-CoV-2 infection in patient 1. Panel (b) shows the projection of $\mathbf{v}_2$ into the I-V plane. As can be seen, the phase curve $I(V)$ of $\mathbf{X}^{(2)}(t) = (I, V)$ initially evolved along the direction defined by $\mathbf{v}_2$, as expected. Likewise, panel (c) presents the unstable eigenvector $\mathbf{v}_2$ in the full three-dimensional state space. Consistent with the observation made in the two-dimensional I-V plane, panel (c) shows that initially the disease state $\mathbf{X} = (T, I, V)$ followed the unstable eigenvector or order parameter $\mathbf{v}_2$.

The amplitude equations (10.16) were solved numerically for the model parameters obtained for patient 1 and the initial conditions defined by Eq. (10.17) [1]. Panel (d) shows the amplitudes $A_j(t)$ thus obtained in terms of a phase curve in the three-dimensional amplitude space. The amplitudes are shown as variables rescaled to the maximum $A_{2,\max}$ of the order parameter amplitude A_2. That is, in panel (d) the variables $A_j(rel) = 100A_j/A_{2,\max}$ for $j = 1, 2, 3$ are shown. Note that the simulation of Eq. (10.16) shows that the amplitudes A_2 and A_3 were positive during the SARS-CoV-2 infection of patient 1. In contrast, A_1 was negative. For this reason $A_1(rel)$ is shown in negative percentage values. As far as the hypothesized dominant role of the order parameter amplitude A_2 is concerned, in panel (d) it can be seen that initially only A_2 varied over time and increased, while A_1 and A_3 remained approximately constant at a level close to zero percent. This initial increase of A_2 corresponds to the initial evolution of the state $\mathbf{X} = (T(t), I(t), V(t))$ along the eigenvector $\mathbf{v}_2$ as

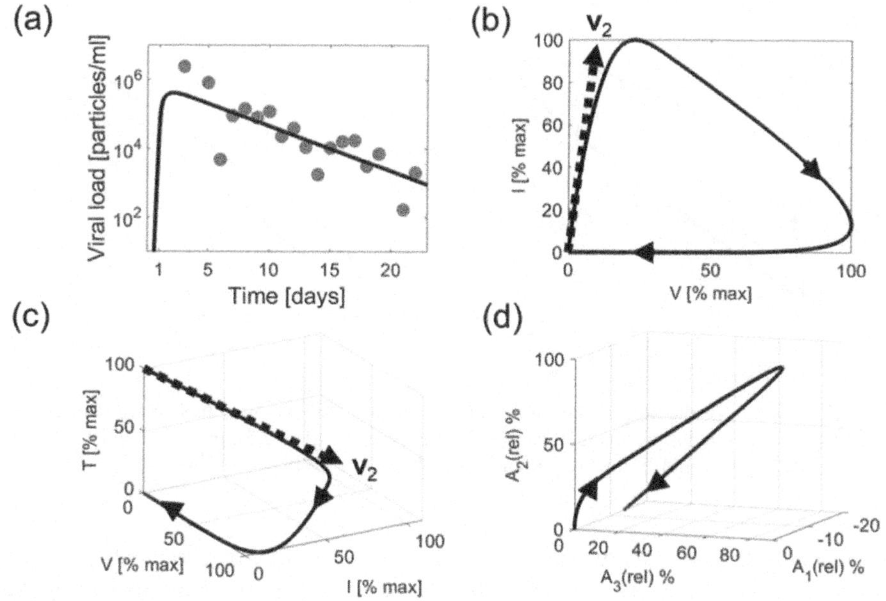

Fig. 10.2 COVID-19 TIV model dynamics of patient 3 in state and amplitude space. Panels as in Fig. 10.1. Parameters and initial conditions [7]: $\beta = 5 \cdot 10^{-4}$/(d $\times$ (particles/ml)), $p = 19$ particles/(d $\times$ cell), $k_1 = 2$/d, $k_2 = 0.3$/d, $T_{st} = 6 \cdot 10^4$ cells/ml, $V(0) = 10^{-4}$ particles/ml

shown in panels (b) and (c). At the end of the period during which A_2 entirely dominated the amplitude dynamics, the amplitude A_3 began to increase. A_1 eventually showed some variations as well and assumed negative values. However, during the whole course of the infection, A_1 remained at a level less than 1% of the maximum value $A_{2,\max}$, see panel (d). In summary, panel (d) illustrates the dominant role of the amplitude A_2 of the unstable eigenvector during the initial disease stage.

For all eight patients the modeling results looked qualitatively similar as in Fig. 10.1. Patient data showed some quantitative differences across patients. For example, the model-based analysis of the data from patient 3 suggest that the viral load increased during the initial stage of the infection more rapidly for patient 3 as compared to patient 1. In this context, Fig. 10.2 presents the sputum viral load data [6] and the modeling results of patient 3. In panels (a), (b), (c), (d) of Fig. 10.2 the same variables are presented as in the corresponding panels of Fig. 10.1. Visual inspection of the panels (a) presented in Figs. 10.1 and 10.2 reveals the aforementioned observation that for patient 3 the initial stage of the infection was characterized by a more rapid increase of viral load. As a result, the trajectory $\mathbf{X}(t) = (T, I, V)$ in the 2D subspace I-V and the 3D state space exhibits a more square-edged shape (see panels (b) and (c) of Fig. 10.2). The phase curve $\mathbf{A}(t) = (A_1, A_2, A_3)$ in amplitude space looks edgy as well (see panel (d)). It resembles the shape of an inverted V. Irrespective of these differences between patients 1 and 3, for patient 3 the initial

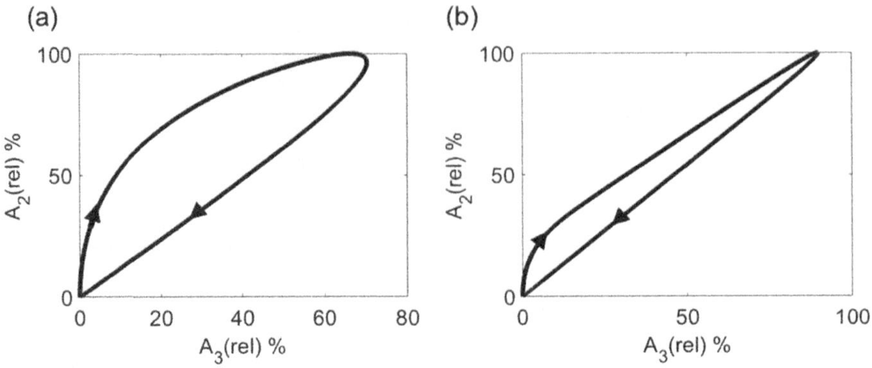

Fig. 10.3 Details of the progression of COVID-19 as seen from an amplitude space perspective. The amplitude dynamics presented in panels (**d**) of Figs. 10.1 and 10.2 for patients 1 and 3 is shown as 2D phase curves $A_2(A_3)$ in panel (**a**) and panel (**b**), respectively

infection dynamics in state space followed closely the order parameter $\mathbf{v}_2$ (see panels (b) and (c)). Likewise, the amplitude dynamics was initially dominated by the evolution of A_2 (see panel (d)). The pair $\mathbf{v}_2$ and A_2 determined the initial stage of the SARS-CoV-2 infection of patient 3.

In order to highlight the dominant role of the order parameter amplitude A_2, Fig. 10.3 presents the amplitude dynamics obtained for patients 1 and 3 in the amplitude subspace A_2-A_3. That is, Fig. 10.3 presents projections of the three-dimensional plots of panels (d) shown in Figs. 10.1 and 10.2 into the respective A_2-A_3 planes. Panels (a) for patient 1 and (b) for patient 3 visualize that in the initial stage of the disease A_2 increased, while A_3 remained almost constant at a zero level. This is consistent with the theoretical consideration according to which the fixed point of interest $\mathbf{X}_{st} = (T_{st}, 0, 0)$ from which the infectious disease emerges corresponds to a saddle with an unstable direction in $\mathbf{v}_2$ ($\lambda_2 > 0$), a stable direction in $\mathbf{v}_3$ ($\lambda_3 < 0$), and a neutrally stable direction in $\mathbf{v}_1$ ($\lambda_1 = 0$).

Let us present the analysis of the remaining patients 2,4,7,8,10,14 in a somewhat condensed form. Figure 10.4 shows the viral load trajectories observed (gray circles) [6] and computed from Eq. (10.1) (solid lines). Panels (a) to (f) correspond to patients 2 to 14. For all patients $T_{st} = 6 \cdot 10^4$ cells/ml and $V(0) = 10^{-4}$ particles/ml was used [7]. As mentioned at the beginning of this section (i.e., Sect. 10.2.1), parameters (here listed as referring to a 1 ml volume) for patient 2 ($\beta = 18 \cdot 10^{-7}$/(d × particle), $p = 625$ particles/(d × cell), $k_1 = 2$/d, $k_2 = 0.65$/d), patient 4 ($\beta = 0.00065$/(d × particle), $p = 2200$ particles/(d × cell), $k_1 = 2$/d, $k_2 = 1$/d), patient 7 ($\beta = 19 \cdot 10^{-7}$/(d × particle), $p = 430$ particles/(d × cell), $k_1 = 2$/d, $k_2 = 0.6$/d), patient 8 ($\beta = 0.000091$/(d × particle), $p = 5800$ particles/(d × cell), $k_1 = 2$/d, $k_2 = 0.6$/d), patient 10 ($\beta = 9.7 \cdot 10^{-7}$/(d × particle), $p = 940$ particles/(d × cell), $k_1 = 2$/d, $k_2 = 0.5$/d), and patient 14 ($\beta = 0.000047$/(d × particle), $p = 120$ particles/(d × cell), $k_1 = 2$/d, $k_2 = 2$/d) were taken from Ref. [7]. All patients showed the same

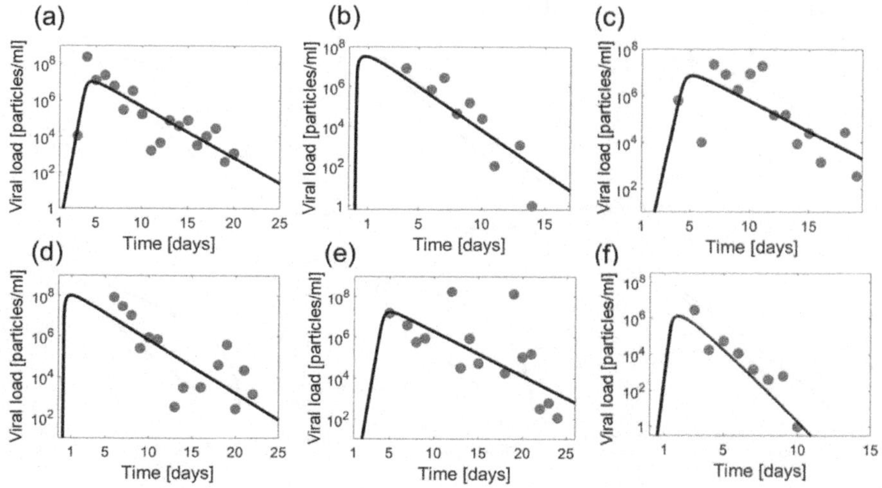

Fig. 10.4 Viral load dynamics observed (gray circles) and computed (solid lines) for the six patients 2 (**a**), 4 (**b**), 7 (**c**), 8 (**d**), 10 (**e**), and 14 (**f**). See text for parameters and initial conditions

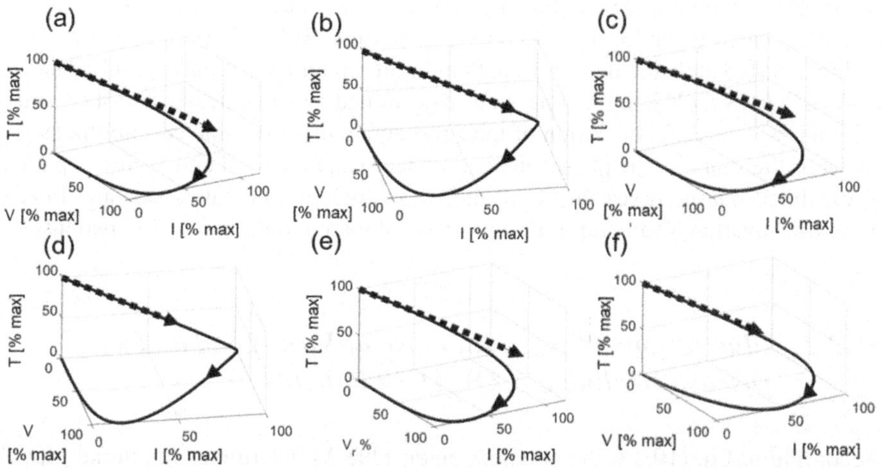

Fig. 10.5 Phase curves (solid lines) computed from Eq. (10.1) for the six patients 2 (**a**), 4 (**b**), 7 (**c**), 8 (**d**), 10 (**e**), and 14 (**f**). The patient-specific order parameters $\mathbf{v}_2$ (thick dotted lines) were computed from Eq. (10.6)

characteristic temporal pattern composed of a relatively short (or fast) period of increase and a relatively long (or slow) period of decay of viral load.

Figure 10.5 shows the corresponding phase curves for the six patients as computed from Eq. (10.1). The patient-specific order parameters $\mathbf{v}_2$ are shown as well. Figure 10.6 presents the amplitude descriptions of the SARS-CoV-2 infections of the

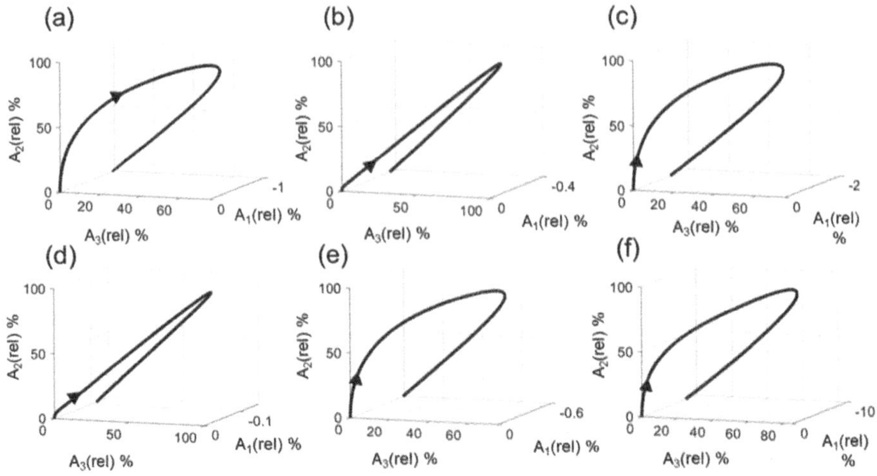

Fig. 10.6 Amplitude dynamics computed from Eq. (10.31) for the six patients 2 (**a**), 4 (**b**), 7 (**c**), 8 (**d**), 10 (**e**), and 14 (**f**)

six patients as computed from Eq. (10.16). Note that in Figs. 10.5 and 10.6 the same kind of rescaled variables have been used as in Figs. 10.1 and 10.2.

Figures 10.5 and 10.6 demonstrate that for all remaining patients the initial stages of their SARS-CoV-2 infections were determined by their respective TIV order parameters $\mathbf{v}_2$ and the evolution of their respective order parameter amplitudes A_2. As far as the shape of the phase curves is concerned, patients 4 and 8 (just as patient 3) exhibited phase curves that were more square-edged in state space and looked more like inverted Vs in amplitude space as compared to the remaining patients.

10.2.2 Illustrations of λ_{max} Increase of Viral Load and k_2 Disease Decline in COVID-19 Patients

According to Eq. (10.25) the maximal eigenvalue λ_2 determines the initial growth rate of the viral load $V(t)$. When presenting $V(t)$ in a logarithmic scale, then λ_{max} corresponds to the slope of the function $V(t)$. Figure 10.7 shows again the viral load trajectories of patients 1 and 3 (as presented in panels (a) of Figs. 10.16 to 10.19). The approximation (10.25) is shown as well as dotted gray line. As expected, λ_{max} describes the slope of the viral load increase.

Let us address the decline period of the disease. As can be seen in panels (c) of Figs. 10.1 and 10.2, the disease state of patients 1 and 3 in the final stages of their diseases evolved along the V-axis. Likewise, Fig. 10.5 shows that for all of the remaining patients except for patient 14 the disease dynamics evolved along the V-axis during the final stages of their diseases. As discussed in Sect. 9.3.2, this features

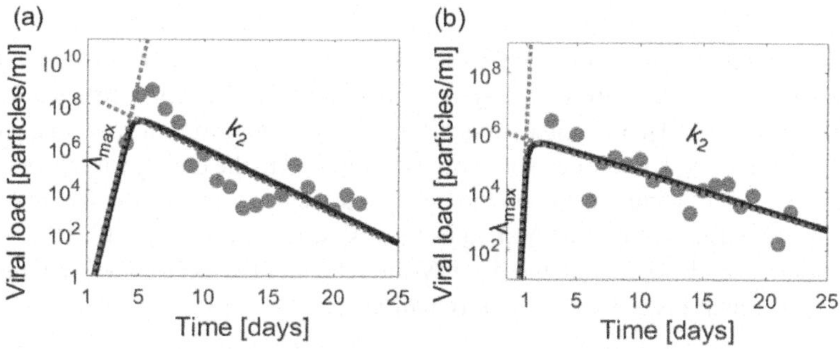

Fig. 10.7 The phase of initial viral load increase is determined by λ_{max} (i.e., λ_2), whereas the decay of viral load is determined by k_2. Panels **a** and **b** refer to patients 1 and 3, respectively

follows from the TIV model equation (10.1) for $k_1 < k_2$. If $k_1 < k_2$ then I decays faster relative to V such that the final stage is determined by the decay of V and the phase curve of $\mathbf{X}(t)$ evolves along the V-axis. In fact, for all patients except for patient 14 the relation $k_2 < k_1$ was satisfied. For patient 14 the parameters were of the same magnitude (i.e., $k_1 = k_2$), which explains why the disease state $\mathbf{X}$ did not evolve in the final stage along the V-axis. More explicitly, if k_2 determines the decline of viral load then $V(t)$ can be approximated by $V_a(t)$ defined by

$$V_a = V_{max} \exp\{-k_2(t - t_p)\} , \tag{10.41}$$

where V_{max} denotes the maximal (or peak) viral load (as obtained from the model fit $V(t)$) and t_p denotes the time point at which $V(t)$ becomes maximal. The approximation (10.41) is plotted in Fig. 10.7 for patients 1 and 3 in panels (a) and (b), respectively. For patients 1 and 3, the approximative function $V_a(t)$ fits the exact solution $V(t)$ in good approximation. In summary, the viral load trajectories for patients 1 and 3 can approximately be described by $V(t) = b\exp\{\lambda_{max}t\}$ (see Eq. (10.25)) in the acute phase (i.e., initial stage of viral load increase) and $V_a = V_{max}\exp\{-k_2(t - t_p)$ (see Eq. (10.41) in the stage of viral load decrease. Similar considerations on two-phase approximations of viral load patterns can be found, for example, in Ref. [10].

The model-based analysis of viral load data of COVID-19 patients presented above and the model-based analysis that will be presented below is subjected to limitations. In general, fitting a model with a relatively small number of data points results in situations for which several parameter sets can be found that fit the data with the same kind of accuracy. While theoretical considerations suggest that the increase of viral load in the initial stage of a SARS-CoV-2 infection increases with a rate given by λ_{max}, the viral load decay in the stage of disease decline may not be determined entirely by k_2 (see Sect. 9.3.2). The analysis presented in Fig. 10.7 demonstrate how a k_2 disease decline could look like in COVID-19 patients. Figure 10.7 does not provide any evidence that such a disease decline is a characteristic feature of the disease progression of COVID-19.

As a second note, it should be pointed out that for all patients but one patient the estimated parameters k_1 and k_2 were such that $k_2 < k_1$ holds, which leads to the k_2 decay that was discussed above and is illustrated in Figs. 10.1 (panel (c)), 10.2 (panel (c)), 10.5, and 10.7. That is, the estimated parameters suggest that for those patients with $k_2 < k_1$ the viral dynamics was slow as compared to the cell dynamics or at least evolved on a similar time scale as the cell dynamics. However, this finding is just opposite to the assumption that motivates the TV model. As discussed in Sect. 9.5.1, the TV model can be derived from the TIV model under the assumption that the viral dynamics is much faster than the cell dynamics (i.e., $k_2 \gg k_1$).

10.2.3 2D Approach: Initiation of Disease Decline by Self-induced Bifurcations

In this section, the 2D approach discussed in Sect. 10.1.3 will be applied to describe the time course of SARS coronavirus 2 load in COVID-19 patients. In particular, the role of the maximal eigenvalue will be discussed. In the context of the 2D approach, the maximal eigenvalue is given by λ_1, see Eq. (10.38). According to the nonlinear physics perspective of SARS-CoV-2 infections, the virus-free state of an infected individual is unstable at the beginning of an infection. With respect to the 2D perspective of the TIV model the circumstances at the time of infection are given by $\beta_w = pT_{st}\beta > \beta_{w,crit} = k_1 k_2$ and $\lambda_1 > 0$. Note that as such any state $\mathbf{X} = (\mathbf{X}^+, X^-)$ with $X^- = T \geq 0$ and $\mathbf{X}^+ = (I, V) = (0, 0)$ is a fixed point of the TIV model. During the course of the disease, the number of target cells $T(t)$ decays over time from the starting value $T(0) = T_{st}$. Consequently, the state $\mathbf{X} = (0, 0, T(t))$ can be considered as a sliding fixed point, whose stability changes when the number of target cells T falls below a critical value such that $pT\beta < k_1 k_2$. When the fixed point becomes stable, the infection subsides in the sense that the viral load V begins to decay. Therefore, the eigenvalue λ_1 may be considered as an (implicitly) time-dependent variable. If so, the function $\lambda_1(t)$ can capture the switch from an unstable to a stable virus-free fixed point. In order to conduct a time-resolved eigenvalue analysis, in Eq. (10.38) the parameter T_{st} is replaced by $T(t)$ such that Eq. (10.38) becomes

$$\lambda_1(t) = -\frac{k_1 + k_2}{2} + \sqrt{\frac{(k_1 + k_2)^2}{4} + pT(t)\beta - k_1 k_2}\,. \tag{10.42}$$

In what follows, the 2D approach and the time-resolved eigenvalue analysis will be applied to the data from the sample of eight patients discussed in Sect. 10.2.1. From the patient-specific TIV model parameters reported in Sect. 10.2.1, for each patient the fixed eigenvalue λ_1 was computed from Eq. (10.38) and the corresponding unstable 2D eigenvector $\mathbf{v}_1$ was computed from Eq. (10.39). Subsequently, the dynamics of the disease state $\mathbf{X}$ was computed from Eq. (10.1) (just as for the 3D approach conducted in Sect. 10.2.1). The subspace variable $\mathbf{X}^+$ was plotted in the

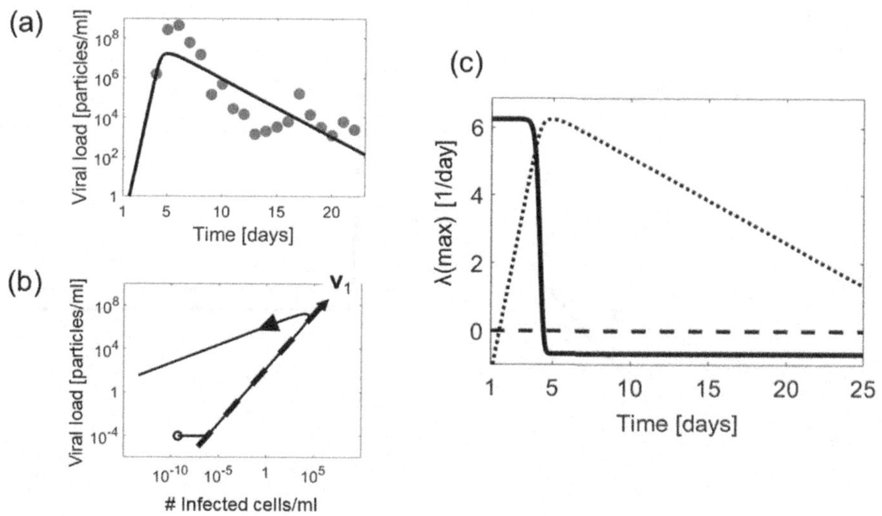

Fig. 10.8 Two-dimensional eigenvalue and eigenvector analysis of the viral load dynamics of patient 1. Panel **a**: Observed viral load trajectory (gray circles) and TIV model solution $V(t)$ (solid line) computed from Eq. (10.1). Panel **b**: Phase curve $V(I)$ (solid line) in logarithmic scales and order parameter $\mathbf{v}_1$ (thick dashed line) computed from Eq. (10.39). The vector $\mathbf{v}_1$ is shown in a magnified scale for the sake of visibility. Panel (**c**): Time-course of the eigenvalue λ_{max} (i.e., λ_1) (solid line) as computed from Eq. (10.38). The viral load graph $V(t)$ shown in panel (**a**) is depicted in panel (**c**) (dotted line) in a rescaled manner such that its peak corresponds to the maximum value of λ_{max}

I-V plane and compared with the 2D eigenvector $\mathbf{v}_1$. Finally, the time-resolved eigenvalue $\lambda_1(t)$ was computed from Eq. (10.42) using the solution $T(t)$ of the TIV model (10.1).

Figure 10.8 shows the results of the eigenvalue analysis for patient 1 [4]. Panel (a) repeats panel (a) of Fig. 10.1, that is, it shows the measured viral load over time (gray circles) and the model fit $V(t)$ (solid line) obtained from the TIV model (10.1). Panel (b) of Fig. 10.8 presents the functions $V(t)$ versus $I(t)$ as computed from Eq. (10.1) as $V(I)$ phase curve (solid thin line). The circle indicates the initial state. The order parameter $\mathbf{v}_1$ is plotted as well (dashed thick line). The disease dynamics $\mathbf{X}^+(t)$ closely followed the order parameter while the viral load increased over time. After the number of infected cells reached its maximum value (and shortly after that the viral load reached its peak value) the dynamics branched off in a different direction. As such, panel (b) of Fig. 10.8 is the projection of panel (c) of Fig. 10.1 into the V-I plane. However, in panel (c) of Fig. 10.1 linear scales and relative variables are used, while in panel (b) of Fig. 10.8 logarithmic scales are used and the variables are not scaled.

Panel (c) of Fig. 10.8 shows λ_{max} (i.e., λ_1) (solid line) as function of time as computed from Eq. (10.42). The viral load graph $V(t)$ as presented in panel (a) (solid black line) is presented in panel (c) as a rescaled function (dotted line) such that V_{max} is at the same height as $\lambda_{max}(0)$. Note that $V(t)$ is presented on the logarithmic

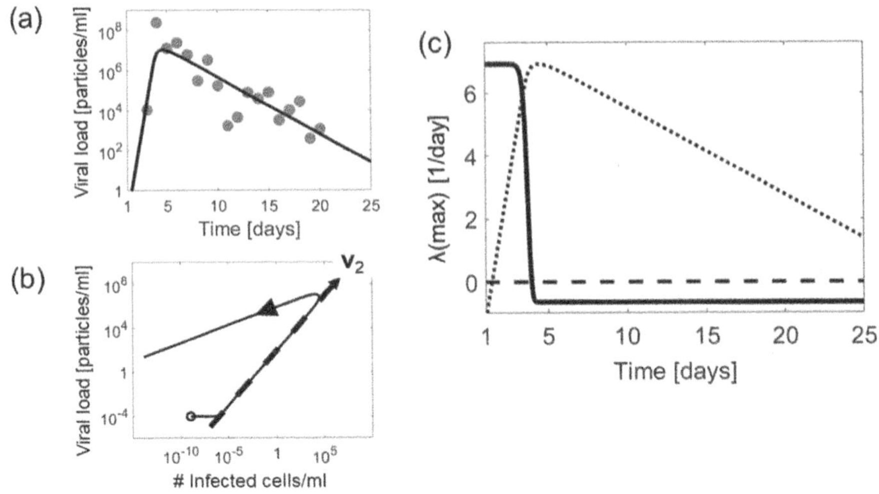

Fig. 10.9 As in Fig. 10.8 but for patient 2

scale (not shown in panel (c)), whereas $\lambda_{max}(t)$ is presented on a linear scale. As can be seen in panel (c), according to the TIV model-based analysis, the eigenvalue λ_{max} of the affected sites of patient 1 dropped from a positive to a negative value at around day 5. Subsequent to this switch of the sign of λ_{max}, the viral load reached its peak and began to decay. Consequently, the course of the disease for patient 1 was characterized by an initial period during which the maximal eigenvalue λ_{max} was positive and the virus-free fixed point unstable. During this initial period, viral load increased and the number of target cells decayed. There was a switching point at around day 5 when the number of target cells in the affected regions of the lung of patient 1 were depleted and reached a sufficiently low value such that λ_{max} turned from a positive to a negative value. The virus-free fixed point became neutrally stable. The viral load decayed.

Figure 10.9 presents the same kind of analysis for patient 2. Panel (a) of Fig. 10.9 corresponds to panel (a) of Fig. 10.4 and shows the viral load data and the model solution $V(t)$. Panel (b) of Fig. 10.9 reveals that during the initial stage with an increasing number of infected cells and an increasing viral load the disease dynamics $\mathbf{X}^+(t)$ followed the order parameter $\mathbf{v}_1$. After the number of infected cells reached its maximum value, the disease state $\mathbf{X}^+$ branched off from the order parameter $\mathbf{v}_1$. Viral load increased slightly at the turning point. Subsequently, the number of infected cells and the viral load decreased. With the help of panel (c), the two stages can be explained using the time-resolved eigenvalue analysis. As can be seen in panel (c), during the initial stage of the disease the eigenvalue λ_{max} (i.e. λ_1) was positive, while in the stage of disease decline the eigenvalue was negative. According to the TIV model, the switch of the eigenvalue was caused by the decrease in the number of target cells in the affected regions of the lung of patient 2.

The viral load trajectories of the remaining six patients 3,4,7,8,10 and 14 can be found in Figs. 10.2 and 10.4. Figures 10.10 and 10.11 present the results of the time-

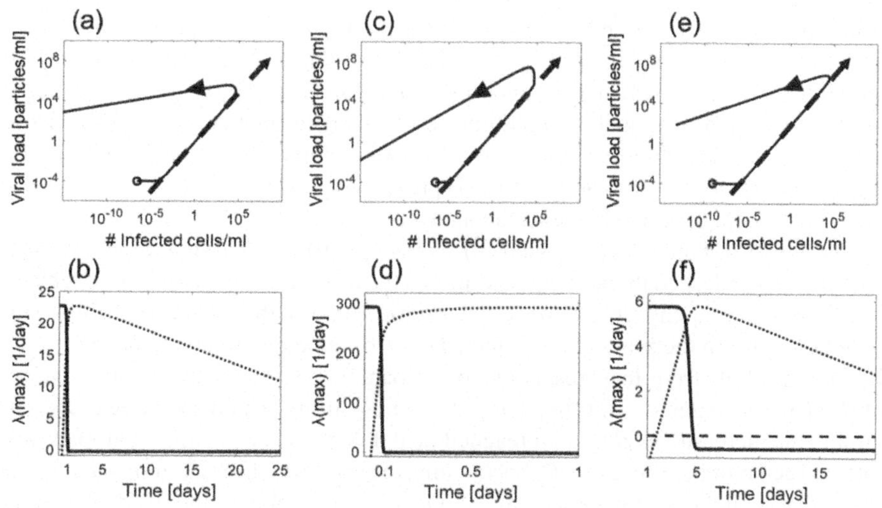

Fig. 10.10 Panels **a**, **c**, and **e** show the disease progression in terms of phase curves $V(I)$ computed from Eq. (10.1) for patients 3, 4, and 7, respectively. The patient-specific order parameters $\mathbf{v}_1$ (thick dashed lines) are shown as well as computed from Eq. (10.39). Panels **b**, **d**, and **f** present the graphs $\lambda_1(t)$ (solid lines) obtained from Eq. (10.38). For comparison purposes, the corresponding viral load trajectories $V(t)$ are shown as well (dotted lines) as functions rescaled to the maximum values of λ_1

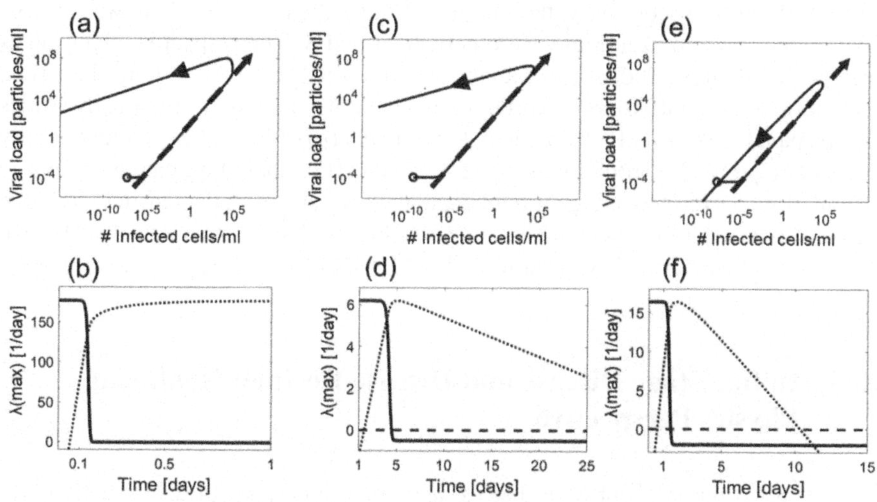

Fig. 10.11 As in Fig. 10.10 but for patients 8 (**a** and **b**), 10 (**c** and **d**), and 14 (**e** and **f**)

fixed and time-resolved analysis methods for the remaining six patients 3,4,7,8,10 and 14 in a condensed form. Figure 10.10 shows the results of the two analysis methods for patients 3, 4, and 7. Panels (a), (c), (e) and (b), (d), (f) present the results of the fixed eigenvalue analysis and time-resolved eigenvalue analysis, respectively. Likewise, Fig. 10.11 reports the results of the two methods for patients 8, 10, and 14. Again, panels (a), (c), (e) and (b), (d), (f) present the results of the fixed eigenvalue analysis and time-resolved eigenvalue analysis, respectively.

As shown in panels (a), (c), and (e) of Figs. 10.10 and 10.11, the 2D IV order parameters $\mathbf{v}_1$ (dashed thick lines) determined the disease progressions of all remaining COVID-19 patients during the initial disease stages of their diseases. Close to the time points when the numbers of infected cells became maximal, the disease states $\mathbf{X}^+$ branched off from their respective order parameters. The diseases progressions of the patients entered a turning point dynamics during which the patient-specific viral loads increased slightly and reached peak loads. Subsequently, viral loads and infected cell numbers decayed. Consequently, the results of the remaining patients are consistent with those of patients 1 and 2. Overall, the results of the fixed eigenvalue and fixed eigenvector analysis concerning the role of the order parameter obtained by means of the 2D approach are consistent with the corresponding results of the 3D approach presented in Sect. 10.2.1.

As shown in panels (b), (d), and (f) of Figs. 10.10 and 10.11, the time-resolved eigenvalue analysis shows that for all but two patients the switch of the leading eigenvalue from a positive to a negative value took place after at least 1 day after symptoms onset. Accordingly, the initial stage of increasing viral load was at least 1 day long. The two exceptions were patients 4 and 8. The respective eigenvalues switched at about 0.10 days (i.e., about two-and-a-half hours) and 0.15 days (i.e., about three-and-a-half hours). In order to make these very early switching dynamics visible, panel (d) of Fig. 10.10 (patient 4) and panel (b) of Fig. 10.11 (patient 8) only show 1 day on the horizontal axis. For all patients (i.e., including patients 4 and 8) the initial stages of increasing viral load and the final stages of decreasing viral load were characterized by a positive and negative eigenvalue λ_2, respectively. The two stages were characterized by unstable and stable virus-free fixed points, respectively.

10.3 Initial-Stage Disease and Disease Decline: Nonlinear Physics Perspective

In Sect. 8.4.2 the concept of a disease as a state emerging in a bifurcation was reviewed. Accordingly, the initial stage of such a disease is characterized by an instability of an appropriately defined healthy state. From a nonlinear physics perspective, infectious diseases belong to the class of diseases emerging from unstable healthy states. The TIV model that has been discussed above and the TIIV model that will be discussed in the following sections are two descriptions of the nonlinear physics of infectious diseases that reveal this feature of infectious diseases. When acknowledging the pivot role of instabilities for the emergence of diseases, in general, and infectious diseases, in particular, then it becomes clear that undoing a disease requires either making the original healthy state stable again or stabilizing a different favorable (i.e., healthy) state [11]. As reviewed briefly in Sect. 8.4.2, this notion that treatment and therapy is a form of bifurcation has been supported by a number of clinical studies.

On the level of infectious diseases spreading in populations (rather than presenting themselves in individuals), a fundamental principle of the mathematical modeling of epidemics and pandemics [12, 13] is in fact that infectious disease outbreaks are instability phenomena. Again, when acknowledging the pivot role of instabilities for the emergence of infectious diseases in population, then the impact of intervention measures on epidemics as it has been discussed from the nonlinear physics perspective in Chap. 8 becomes clear. Accordingly, intervention measures (when implemented successfully) switch the sign of positive (real-valued) eigenvalues of disease-free states of populations such that these eigenvalues become negative. In other words, intervention measures can induced bifurcations. The sign switching phenomenon and, in doing so, evidence for bifurcations induced by interventions was observed for the COVID-19 epidemics in Europe (see Sect. 8.4.4), Thailand (see Sect. 8.4.5), the state of New York (see Sect. 8.7.1), and Pakistan (see Sect. 8.7.2). Returning to the level of infectious diseases in individuals, such bifurcations leading to disease decline in individuals are either self-induced or assisted by medication.

In Sect. 10.2.3, in the context of the TIV model, bifurcations have been addressed that are induced by a dramatic decay of the number of target cells. In this case, the disease decline is initiated by self-induced bifurcations. More sophisticated virus dynamics models have been suggested to take the immune reaction of the human body [14–17] and medication of patients into account [14, 18–20]. The time-resolved eigenvalue analysis presented in Sect. 10.2.3 may be applied to obtain insights into the stabilization of the virus-free fixed point induced by the natural immune reaction and/or medication.

10.4 Analysis of the TIIV Model

10.4.1 Stability Analysis

The TIIV model is defined by Eq. (9.17) which is here repeated as

$$\frac{\mathrm{d}}{\mathrm{d}t}T = -\beta V T \,, \ \frac{\mathrm{d}}{\mathrm{d}t}I_1 = \beta V T - k_1 I_1 \,, \ \frac{\mathrm{d}}{\mathrm{d}t}I_2 = k_1 I_1 - k_2 I_2 \,, \ \frac{\mathrm{d}}{\mathrm{d}t}V = p I_2 - k_3 V \tag{10.43}$$

with $k_1, k_2, k_3, \beta, p > 0$. Accordingly, the disease state (or health state) of an individual is described by the state vector $\mathbf{X} = (T, I_1, I_2, V)$. The model exhibits fixed points of the form $\mathbf{X}_{st} = (Y, 0, 0, 0)$ with $Y \geq 0$ that describe virus-free states. The virus-free fixed point $\mathbf{X}_{st} = (T_{st}, 0, 0, 0)$ denotes the virus-free state of healthy adults, where $T_{st} > 0$ corresponds to their target cell concentrations in the absence of the virus infection under consideration. Just as for the TIV model, let us introduce the relative target concentration δ and the relative state vector $\mathbf{u}$ by $T = T_{st} + \delta$ and $\mathbf{u} = (\delta, I_1, I_2, V)$, respectively. Then, the TIIV model equations (10.43) can be linearized around $\mathbf{X}_{st} = (T_{st}, 0, 0, 0)$ like

$$\frac{\mathrm{d}}{\mathrm{d}t}\mathbf{u} = L\,\mathbf{u}\,, \; L = \begin{pmatrix} 0 & 0 & 0 & -\beta T_{st} \\ 0 & -k_1 & 0 & \beta T_{st} \\ 0 & k_1 & -k_2 & 0 \\ 0 & 0 & p & -k_3 \end{pmatrix} . \tag{10.44}$$

The linearization matrix L exhibits the eigenvalue $\lambda_1 = 0$. The remaining three eigenvalues can be obtained from the determinant of the appropriate 3×3 submatrix of L like

$$\begin{vmatrix} -k_1 - \lambda & 0 & \beta T_{st} \\ k_1 & -k_2 - \lambda & 0 \\ 0 & p & -k_3 - \lambda \end{vmatrix} = 0 . \tag{10.45}$$

Equation (10.45) leads to the cubic characteristic equation

$$(\lambda + k_1)(\lambda + k_2)(\lambda + k_3) = p\beta T_{st} k_1 . \tag{10.46}$$

Before discussing the roots of Eq. (10.46), let us consider the special case in which the TIIV model reduces to the TIV model. To this end, let us assume that the dynamics of the variable I_1 reflecting the non-virus producing infected cells can be regarded as fast relative to all remaining variables. Then, we may put $k_1 I_1 = \beta V T \Rightarrow I_1 = \beta V T / k_1$. Substituting this quasi-stationary state into Eq. (10.43), the TIV model of the form $\mathrm{d}T/\mathrm{d}t = -\beta V T$, $\mathrm{d}I_2/\mathrm{d}t = \beta V T - k_2 I_2$, and $\mathrm{d}V/\mathrm{d}t = p I_2 - k_3 V$ is obtained. In particular, let us assume that $k_1 \gg k_2, k_3$ such that I_1 can be considered as fast variable relative to I_2 and V. If k_1 is of an order of magnitude larger than k_2 and k_3 like $k_1 \propto O(2)$, while $k_2, k_3 \propto O(1)$, then solutions λ of Eq. (10.46) that are of the order of k_2 and k_3 can be obtained by assuming that $\lambda + k_1 \approx k_1$ (because $\lambda \propto O(1) \ll k_2 \propto O(2)$). Consequently, Eq. (10.46) becomes $(\lambda + k_2)(\lambda + k_3) = p\beta T_{st}$, which corresponds to the quadratic eigenvalue equation (10.3) of the TIV model that involves k_1 and k_2 rather than k_2 and k_3.

Let us return to the general case of roots λ defined by Eq. (10.46). Let $\lambda_{\max}$ define the largest, real-valued eigenvalue. A graphical analysis of Eq. (10.46) can be conducted to discuss the roots of Eq. (10.46) and, in particular, $\lambda_{\max}$. To this end, Eq. (10.46) is split into two functions f_1 and f_2 of a variable x, respectively, whose function values must intersect each other like

$$f_1(x) = f_2\,, \; f_1(x) = (x + k_1)(x + k_2)(x + k_3)\,, \; f_2 = p\beta T_{st} k_1 . \tag{10.47}$$

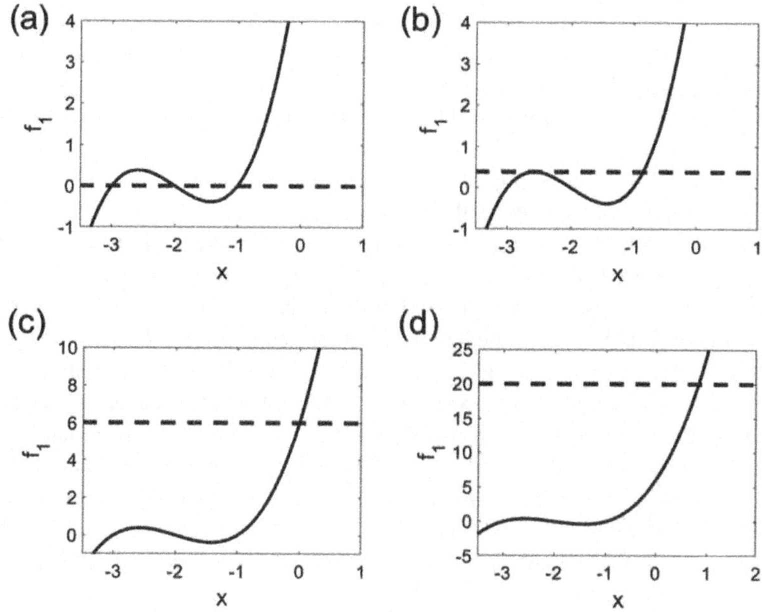

Fig. 10.12 Graphical illustration of real-valued solutions of the cubic equation (10.47) for $k_1 = 1$, $k_2 = 2$, and $k_3 = 3$

The left-hand side function f_1 corresponds to a cubic function with zeros at $-k_1$, $-k_2$, and $-k_3$, while the right-hand side function f_2 corresponds to a constant. Real-valued solutions of Eq. (10.47) correspond to intersection points of f_1 and f_2. Let k_{min} denote the smallest of parameters of k_1, k_2, k_3. Then, for $x > -k_{\text{min}}$ the function f_1 increases monotonically (e.g., see Fig. 10.12).

Figure 10.12 demonstrates four fundamental cases. Case 1 is given by $f_2 = 0$ and illustrated in panel (a). In this case, there are three real-valued eigenvalues $\lambda_i = -k_i < 0$ with $i = 1, 2, 3$. The largest eigenvalue corresponds to $\lambda_{\text{max}} = -k_{\text{min}}$. In general, the function f_2 describes a "horizontal bar". Let us increase this "bar" gradually. In case 2 (see panel (b)), the "bar" is lifted such that it hits the local maximum of f_2. There is a double real-valued eigenvalue $\lambda_2 = \lambda_3$ that corresponds to the location of the local maximum and another real-value eigenvalue λ_1 with $\lambda_1 > \lambda_{2,3}$ but $\lambda_1 < 0$. This eigenvalue correspond to the maximal eigenvalue. Increasing f_2 further, Eq. (10.47) exhibits two complex-valued and one real-valued solution. It can be shown that the complex-valued eigenvalues exhibit negative real parts (see Sect. 10.7). Case 3 (see panel (c)) corresponds to the critical case (i.e., the bifurcation point) at which the maximal eigenvalue becomes zero and a positive eigenvalue emerges. In case 3 the constant f_2 is just as large as the intersection point of f_1 with the vertical axis. That is, $f_2 = f_1(0)$ holds. In this case, Eq. (10.47) reads

$$f_1(0) = k_1 k_2 k_3 = p\beta T_{st} k_1 \Rightarrow k_2 k_3 = p\beta T_{st} \,. \tag{10.48}$$

If increasing f_2 even further, case 4 (see panel (d)) occurs for which there is a single real-valued positive eigenvalue $\lambda_{max} > 0$ and two complex valued eigenvalues with negative real parts. Note that the critical condition at which $\lambda_{max} = 0$ holds corresponds to the critical condition of the TIV model (i.e., $\beta_w = \beta_{w,crit}$ when replacing k_1 and k_2 by k_2 and k_3).

In summary, from the graphical analysis it follows that there are two main cases. In the first case, there is one real-valued eigenvalue and a pair of complex-conjugated eigenvalues. In this case, the real-valued eigenvalue corresponds to λ_{max} and can be positive or negative. The complex-valued pair of eigenvalues exhibits a negative real part. The second case was not addressed in the discussion above. Let us assume that the local maximum of $f_1(x_{max})$ exceeds the interception value $f_1(0)$. In this case as long as $f_2 < f_1(x_{max})$ all eigenvalues are real: $\lambda_{2,3,4} \in \mathbb{R}$. The maximal eigenvalue can be positive or negative. However, the remaining two eigenvalue (as can be seen from the graphical constructions in Fig. 10.10) are negative. Consequently, in the second main case, we have either $\lambda_{max} > 0$ or $\lambda_{max} \leq 0$ but the two remaining real-valued eigenvalues are negative. Irrespective which of the two main cases applies, the critical case for which $\lambda_{max} = 0$ holds is given by Eq. (10.48) such that (in analogy to the TIV model) the weighted infectivity parameter β_w and its critical value $\beta_{w,crit}$ can be defined like [21]

$$\beta_w = p\beta T_{st} \ , \ \beta_{w,crit} = k_2 k_3 \ . \tag{10.49}$$

For $\beta_w > \beta_{w,crit}$ we have $\lambda_{max} > 0$, while for $\beta_w < \beta_{w,crit}$ we have $\lambda_{max} < 0$. Consequently, the virus-free fixed point is unstable for $\beta_w > \beta_{w,crit}$ and describes a saddle with a single unstable direction. In addition to the unstable direction, the saddle either exhibits two stable directions or a plane with a stable focus [22]. Furthermore, there is a neutrally stable direction (namely, the axis given by T) as indicated by the eigenvalue $\lambda_1 = 0$.

Figure 10.13 illustrates solutions of the TIIV model for $\beta_w > \beta_{w,crit}$ (panel (a)) and $\beta_w < \beta_{w,crit}$ (panel (b)). In the former case ($\beta_w > \beta_{w,crit}$), the model describes a patient with an unstable virus-free fixed point. The viral load plotted in a linear scale shows a wave-like pattern. For the selected parameters, during the course of the disease the target cells decay almost to zero. In the latter case ($\beta_w < \beta_{w,crit}$), the virus-free fixed point is neutrally stable, V decays from its initial value. Only a small portion of target cells get infected during the time it takes to clear the virus out of the body.

10.4.2 TIIV Model Amplitude Equations

As mentioned in Sect. 9.6, the TIIV model involves variables that are typically not measured in the same units: the cell variables T, I_1, I_2, on the one hand, and the viral load V, on the other hand. The rescaled model presented in Sect. 9.6 may be used to solve this issue. In what follows, the simplified perspective will be used

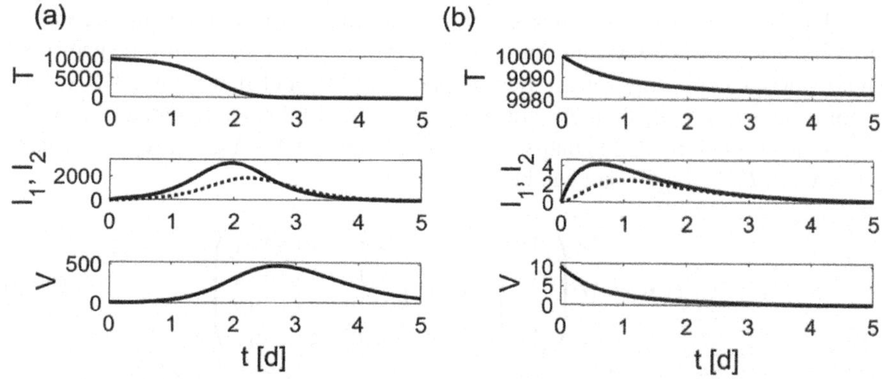

Fig. 10.13 Panel **a**: Wave-solution of the TIIV model (10.43) reflecting exposure to a virus that leads to an initial increase of viral load. T (top subpanel), I_1 and I_2 (middle subpanel with I_2 given by the dotted line), and V (bottom subpanel) are shown as functions of time. Parameters and initial conditions: $T(0) = T_{st} = 10,000$, $V(0) = 10$, $I(0) = 0$, $\beta = 0.01$/d, $k_1 = 2.0$/d, $k_2 = 3.0$/d, $k_3 = 1.5$/d, and $p = 0.5$, which leads to $\beta_{w,crit} = 4.5$ and $\beta_{crit} = \beta_{w,crit}/(pT_{st}) = 0.0009$/d. Panel **b**: Solution of the TIIV model for the case in which a virus invades an area of the body of an individual that exhibits a stable virus-free state. Parameters and initial conditions as for the simulation shown in panel (**a**) but with $\beta = 0.0002$/d, which implies $\beta < \beta_{crit} \Rightarrow \beta_w < \beta_{w,crit}$

according to which one virus particle will be considered as equivalent to 1 cell. In order to derive the amplitude equations of the TIIV model (10.43) the standard procedure (e.g., as used for the TIV model) outlined in Chap. 2 can be applied. First, the eigenvectors $\mathbf{v}_j = (v_{j,1}, v_{j,2}, v_{j,3}, v_{j,4})$ with $j = 1, 2, 3, 4$ of L defined by Eq. (10.44) are determined. The eigenvector $\mathbf{v}_1$ related to $\lambda_1 = 0$ reads $\mathbf{v}_1 = (1, 0, 0, 0)$. The remaining eigenvectors for $j = 2, 3, 4$ read [21]

$$\mathbf{v}_j = \frac{1}{Z_j} \begin{pmatrix} -\beta T_{st} p(\lambda_j + k_1) \\ \beta T_{st} p \lambda_j \\ \lambda_j(\lambda_j + k_1)(\lambda_j + k_3) \\ p\lambda_j(\lambda_j + k_1) \end{pmatrix}, \tag{10.50}$$

as can be shown by a detailed calculation. In Eq. (10.50) the parameter Z_j is a normalization factor that has to be chosen carefully. If λ_j is complex, then the corresponding eigenvector is complex as well. In this case, the complex-valued vector is normalized such that the dot product (scalar product) yields $\sqrt{\mathbf{v}_j \mathbf{v}_j^*} = 1$, where $\mathbf{v}_j^*$ is the complex-conjugate of $\mathbf{v}_j$. It is assumed that the eigenvectors are linearly independent. Having obtained the eigenvectors, the amplitudes A_1, A_2, A_3, A_4 of the four-dimensional amplitude space are introduced by means of the superposition

$$\mathbf{X}(t) = \mathbf{X}_{st} + \sum_{j=1}^{4} A_j(t)\mathbf{v}_j \, . \tag{10.51}$$

Note that the eigenvectors are dimensionless. This implies that if, for example, the cell variables and viral load are measured in cell/ml and RNA copies/ml, respectively, where 1 RNA copy may reflect 1 virus particle [23], and if we put 1 RNA copy as equivalent to a cell, then A_j are measured in cell/ml (or alternatively in RNA copies/ml). Equation (10.51) maps the amplitude space (A_1, A_2, A_3, A_4) to the state space $\mathbf{X} = (T, I_1, I_2, V)$ and can be written like

$$\mathbf{X} - \mathbf{X}_{st} = \mathbf{u} = M \begin{pmatrix} A_1 \\ A_2 \\ A_3 \\ A_4 \end{pmatrix}, \ M = \begin{pmatrix} 1 & v_{2,1} & v_{3,1} & v_{4,1} \\ 0 & v_{2,2} & v_{3,2} & v_{4,2} \\ 0 & v_{2,3} & v_{3,3} & v_{4,3} \\ 0 & v_{2,4} & v_{3,4} & v_{4,4} \end{pmatrix}, \tag{10.52}$$

as discussed in Chap. 2. The inverse M^{-1} of M yields the mapping

$$\begin{pmatrix} A_1 \\ A_2 \\ A_3 \\ A_4 \end{pmatrix} = M^{-1}\mathbf{u} = M^{-1}(\mathbf{X} - \mathbf{X}_{st}) \tag{10.53}$$

from state space to amplitude space. The matrix M is composed of the biorthogonal vectors $\mathbf{w}_i$ like

$$M^{-1} = \begin{pmatrix} \mathbf{w}_1 \\ \mathbf{w}_2 \\ \mathbf{w}_3 \\ \mathbf{w}_4 \end{pmatrix} \tag{10.54}$$

(see Sects. 2.6.2 and 5.6) for which $\mathbf{w}_j \mathbf{v}_k = \delta_{jk}$ holds. The amplitude equations may be derived using the vector calculation method (see Sect. 2.9.3). To this end, from Eq. (10.43) the evolution equation of the relative state of the form $d\mathbf{u}/dt = L(\mathbf{X}_{st})\mathbf{u} + \mathbf{R}(\mathbf{u})$ may be derived. In analogy to Eq. (10.13), the result reads

$$\frac{\mathrm{d}}{\mathrm{d}t} \begin{pmatrix} \delta \\ I_1 \\ I_2 \\ V \end{pmatrix} = L \begin{pmatrix} \delta \\ I_1 \\ I_2 \\ V \end{pmatrix} + \beta\delta V \begin{pmatrix} -1 \\ 1 \\ 0 \\ 0 \end{pmatrix}. \tag{10.55}$$

Multiplying Eq. (10.55) by $\mathbf{w}_i$ yields

$$\frac{\mathrm{d}}{\mathrm{d}t} A_j = \lambda_j A_j + G_j(A_1, A_2, A_3, A_4), \ G_i = \beta\delta V \mathbf{w}_i \begin{pmatrix} -1 \\ 1 \\ 0 \\ 0 \end{pmatrix} = \beta\delta V(w_{i,2} - w_{i,1}) \tag{10.56}$$

for $j = 1, 2, 3, 4$. Replacing δ and V by means of the amplitudes, leads to the closed set of amplitude equations [21]

$$\frac{\mathrm{d}}{\mathrm{d}t}A_i = \lambda_i A_i + C_i p_2(A_1, A_2, A_3, A_4) ,$$
$$p_2 = \beta\left(\sum_{k=1}^{4} v_{k,1}A_k\right)\left(\sum_{k=2}^{4} v_{k,4}A_k\right) ,$$
$$C_i = w_{i,2} - w_{i,1} \tag{10.57}$$

with $\lambda_1 = 0$ and λ_i for $i = 2, 3, 4$ given by the solutions of Eq. (10.46). The initial conditions for the amplitude equation model (10.57) are derived from the initial state $\mathbf{X}(0) = (T_{st}, 0, 0, V(0))$ at $t = 0$ with an initial viral load $V(0)$. From Eq. (10.53) it follows that

$$\begin{pmatrix} A_1(0) \\ A_2(0) \\ A_3(0) \\ A_4(0) \end{pmatrix} = M^{-1}(\mathbf{X}(0) - \mathbf{X}_{st}) = M^{-1}\begin{pmatrix} 0 \\ 0 \\ 0 \\ V(0) \end{pmatrix} \Rightarrow A_i(0) = w_{i,4}V(0) . \tag{10.58}$$

In summary, solving Eq. (10.57) for the initial conditions (10.58) yields the amplitude dynamics that corresponds to the state dynamics defined by Eq. (10.43) for the initial conditions $\mathbf{X}(0) = (T_{st}, 0, 0, V(0))$.

As discussed above, for $\beta_w > \beta_{w,crit}$ the virus-free fixed point $\mathbf{X}(0) = (T_{st}, 0, 0, V(0))$ is unstable and the virus can multiply in the human body. The disease emerges in the human body from the virus-free fixed point $\mathbf{X}_{st} = (T_{st}, 0, 0, 0)$ that corresponds to a saddle characterized by one unstable direction $\mathbf{v}_{k(\max)}$ related to the real-valued eigenvalue $\lambda_{max} > 0$, a neutrally stable direction $\mathbf{v}_1$ (related to $\lambda_1 = 0$), and two stable directions $\mathbf{v}_m$ and $\mathbf{v}_l$ with two real-valued negative eigenvalues or a two-dimensional plane spanned by two complex conjugated eigenvectors $\mathbf{v}_m = \mathbf{v}_l^*$ with complex eigenvalues exhibiting a negative real part. The amplitude equation model (10.58) describes the dynamics away from this saddle point as seen in the four-dimensional amplitude space.

10.4.3 The TIIV Unstable Eigenvector and Order Parameter

Let us consider the classical scenario of dynamical systems that satisfy $\mathrm{d}\mathbf{X}/\mathrm{d}t = \mathbf{N}(\mathbf{X})$ (see Eq. (2.1)) and evolve close to an unstable fixed point as discussed in Chap. 2. Accordingly, in the context of the TIIV model, a virus infection begins in the human body as a disease state close to a saddle point characterized by the eigenvalues $\lambda_1 = 0$, $\lambda_{k(\max)} = \lambda_{\max} > 0$, and $\lambda_m, \lambda_l < 0$ or $\mathbb{R}\{\lambda_m\} = \mathbb{R}\{\lambda_l\} < 0$, where m, l describe the indices of the two remaining eigenvalues other than λ_1 and $\lambda_{\max}$. Without loss of generality, let λ_4 denote the maximal eigenvalue such that $m = 2$ and $l = 3$. The virus-free fixed point $\mathbf{X}_{st} = (T_{st}, 0, 0, 0)$ is defined in amplitude space by $\mathbf{A} = (A_1, A_2, A_3, A_4) = (0, 0, 0, 0)$. This implies that in the aforementioned classical scenario the initial amplitudes $A_j(0)$ correspond to small perturbations. In this case, $A_4(t)$ increases exponentially in magnitude like $A_4(t) = A_4(0)\exp\{\lambda_4 t\}$ with an e-folding time $\tau = 1/\lambda_4$ and a doubling time of the amplitude $T_2 = \ln(2)\tau$. A_2 and A_3 decay in the amount ($\lambda_{2,3} < 0$ or $\mathbb{R}\{\lambda_2\} = \mathbb{R}\{\lambda_3\} < 0$). A_1 can be regarded as constant relative to A_2, A_3, A_4 because the linearized evolution equation of A_1 is characterized by $\lambda_1 = 0$ and reads $\mathrm{d}A_1/\mathrm{d}t = 0$. In summary, after an intermediate period T_i during which A_2 and A_3 have decayed in magnitude to negligibly small values, the initial dynamics of the disease state is given by

$$\mathbf{X}(t) = \mathbf{X}_{st} + A_1(0)\mathbf{v}_1 + A_4(t)\mathbf{v}_4 \; . \tag{10.59}$$

Alternatively, the approximation $\mathrm{d}\mathbf{X}/\mathrm{d}t \approx \mathbf{v}_4 \mathrm{d}A_4/\mathrm{d}t$ may be considered, which leads to

$$\frac{\mathrm{d}}{\mathrm{d}t}\mathbf{X} \approx \mathbf{v}_4 \frac{\mathrm{d}}{\mathrm{d}t} A_4 \;\Rightarrow\; \Delta\mathbf{X} \approx \mathbf{v}_4 \Delta A_4 \tag{10.60}$$

with $\Delta\mathbf{X} = \mathbf{X}(t + \Delta t) - \mathbf{X}(t)$, $\Delta A_4 = A_4(t + \Delta t) - A_4(t)$, and $\Delta t > 0$. Equations (10.59) and (10.60) are the counterparts to Eqs. (10.18) and (10.21), respectively, that hold for the TIV model. Accordingly, the unstable eigenvector or order parameter $\mathbf{v}_4$ and the corresponding amplitude A_4 determine the emergence of the disease in the human body. From Eq. (10.60) it follows that during the initial stage of the infectious disease changes of the state variables relative to each other are determined by $\Delta X_j/\Delta X_k \approx v_{4,j}/v_{4,k}$, For example, cell concentrations change relative to the viral load V like

$$\frac{\Delta T}{\Delta V} \approx \frac{v_{4,1}}{v_{4,4}} \, , \; \frac{\Delta I_1}{\Delta V} \approx \frac{v_{4,2}}{v_{4,4}} \, , \; \frac{\Delta I_2}{\Delta V} \approx \frac{v_{4,3}}{v_{4,4}} \; . \tag{10.61}$$

In particular, the order parameter $\mathbf{v}_4$ determines how the number of virus-producing infected cells I_2 and the number of infected, not yet virus-producing cells I_1 change relative to each other:

$$\frac{\Delta I_2}{\Delta I_1} \approx \frac{v_{4,3}}{v_{4,2}} \;\Rightarrow\; \theta = \arctan\left(\frac{v_{4,3}}{v_{4,2}}\right) \; . \tag{10.62}$$

The angle θ defined by Eq. (10.62) describes the direction of the order parameter $\mathbf{v}_4$ in the I_1-I_2 plane with respect to the I_1 axis. In particular, the angle θ determines the qualitative relationship between changes in these two cell populations like

$$\begin{aligned} \theta > 45^\circ \;&\Rightarrow\; \frac{\Delta I_2}{\Delta I_1} > 1 \, , \\ \theta < 45^\circ \;&\Rightarrow\; \frac{\Delta I_2}{\Delta I_1} < 1 \; . \end{aligned} \tag{10.63}$$

This means that if $\theta > 45^\circ$ holds, then during the initial disease stage the number of virus-producing cells (I_2) increases faster than the number of cells in the latent period (I_1). In contrast, if $\theta < 45°$ holds, then the number of cells in the latent period (I_1) increases faster than the number of the virus-producing cells (I_2).

10.4.4 *Dominant Role of the TIIV Order Parameter During Initial Infection and Disease Decline*

Let us demonstrate the role of $\mathbf{v}_4$ for the two cases presented in Fig. 10.13. Figure 10.14 refers to the first case $\beta_w > \beta_{w,crit}$ and $\lambda_{max} > 0$ shown in panel (a) of Fig. 10.13. For illustration purposes panel (a) of Fig. 10.13 is repeated in Fig. 10.14 as panel (a). The phase curve in the three-dimensional I_1-I_2-V subspace is shown in panel (b) of Fig. 10.14. The order parameter $\mathbf{v}_4$ is plotted (in a magnified scale) as well. As expected, the disease state evolves initially along $\mathbf{v}_4$.

The second case $\beta_w > \beta_{w,crit}$ and $\lambda_{max} < 0$ is shown in Fig. 10.15 and refers to the simulation results shown in panel (b) of Fig. 10.13. Panel (b) of Fig. 10.13 is repeated as panel (a) of Fig. 10.15. Panel (b) of Fig. 10.15 shows $\mathbf{X}(t)$ as phase curve in the I_1-I_2-V subspace together with the eigenvector $\mathbf{v}_4$. The eigenvector (or remnant order parameter) $\mathbf{v}_4$ of $\lambda_{\max}$ describes a stable direction. Importantly, for this simulation the eigenvalues λ_2 and λ_3 exhibit negative real parts that are relatively larger in the amount. Consequently, $\mathbf{v}_4$ describes a direction characterized by a slowly decaying dynamics. While initially the disease state does not evolve along $\mathbf{v}_4$ (see panel (b) of Fig. 10.15), as soon as the amplitudes A_2 and A_3 have decayed to zero at rates $|\lambda_2|$ and $|\lambda_3|$, respectively, the dynamics of the disease decline in the I_1-I_2-V subspace is entirely determined by the direction $\mathbf{v}_4$ and the temporal aspects of the decline are determined by the exponential decay $A_4(t) = A_4(0)\exp\{\lambda_4 t\}$ of the amplitude A_4

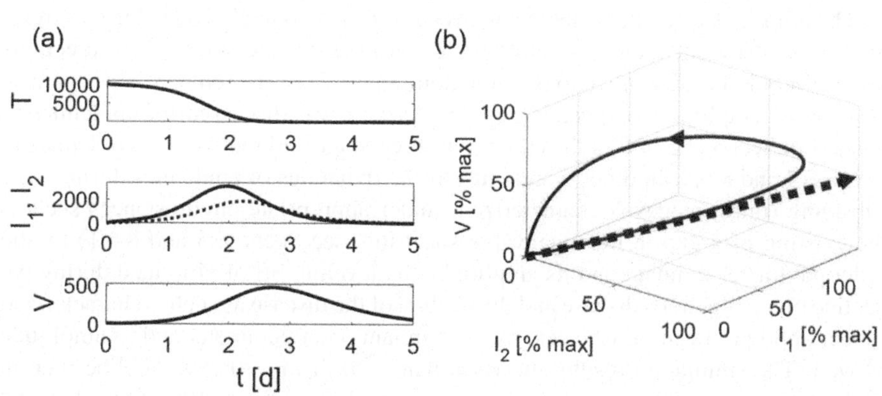

Fig. 10.14 Eigenvector analysis of an infection of an unstable area. Panel **a** repeats the simulation results presented in panel (**a**) of Fig. 10.13. Panel **b** presents the corresponding phase curve (solid line) and the TIIV model order parameter (dotted line). As it can be seen, the phase curve follows during the initial stage of the disease the order parameter. See Fig. 10.13 for parameters and initial conditions

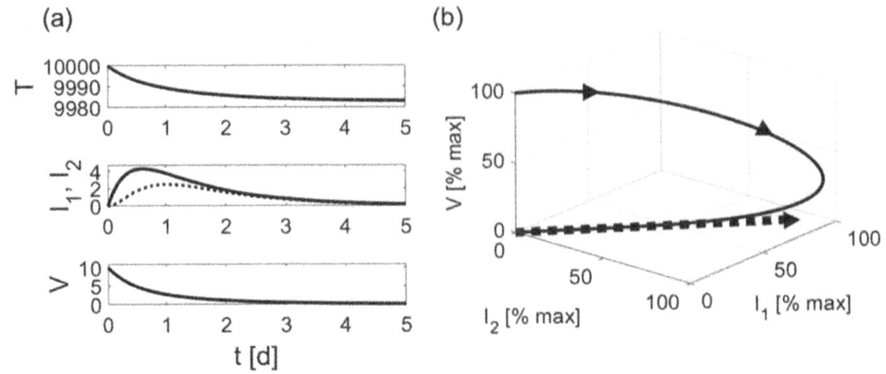

Fig. 10.15 Eigenvector analysis of an infection of a stable area. Panel **a** repeats the simulation results presented in panel (**b**) of Fig. 10.13. Panel **b** presents the corresponding phase curve (solid line) and the remnant of the order parameter (dotted line). As it can be seen, the phase curve follows the order parameter remnant during the final disease decline. See Fig. 10.13 for parameters and initial conditions

of the maximal eigenvalue. Equation (10.59) holds and implies that the viral load decays at a rate $|\lambda_4|$ like

$$V(t) = b\exp\{-|\lambda_4|t\} \tag{10.64}$$

with $b = v_{4,4}A_4(0)$. Note that as far as the dynamics in the I_1-I_2-V subspace is concerned, the amplitude A_1 does not play a role because the term $A_1\mathbf{v}_1$ in the superposition given by Eq. (10.51) does not make any contribution to the variables of that subspace. This is also the reason why the $\mathbf{v}_1 A_1(0)$ does not occur in (10.64) although it occurs in Eq. (10.59).

The dominant role of the stable eigenvector of the maximal eigenvalue (i.e., remnant order parameter) and its amplitude during the decline stage of an infectious disease should be seen in analogy to the dominant role of the remnant order parameter and its amplitude in subsiding epidemics as it was discussed for epidemics in general in Sect. 6.1.3 and in the context of three-stage and multi-stage epidemics in Sects. 8.5 and 8.6.1. In other words, just as the dynamics of epidemics during their subsiding (third) stages is characterized (under appropriate circumstances such as the existing of a gap in the eigenvalue spectrum, see Sects. 8.5 and 8.6.1) by the order parameter remnant and its amplitude, the leveling off of viral load during the decline of an infectious disease and the decline of the disease as such is characterized (under appropriate circumstances) by the remnant order parameter and its amplitude as well. The simulation results shown in panel (b) of Fig. 10.15 should be seen in analogy to the simulation results presented in panel (d) of Fig. 8.18 and the observed subsiding dynamics of the first-wave epidemics in Thailand (panel (b) of Fig. 8.14), state of New York (see Fig. 8.18), and Pakistan (see Fig. 8.21).

In the context of the TIV model and self-induced bifurcations leading to disease decline, these considerations have been worked out in Sect. 10.3. While in the simulation shown in Fig. 10.15 the bifurcation parameter β_w has been put to a value smaller than the critical value such that $\lambda_{max} < 0$ holds, as argued in Sect. 10.3, the eigenvalue λ_{max} may be considered as an (implicitly) time-dependent quantity. During the course of an infectious disease λ_{max} may change due to a disease-related decrease in the number of target cells from a positive to a negative value. If this is the case and if, in addition, the time scales of the amplitudes of the virus-cell system are separated (i.e., there is a gap between λ_{max} and the real-part of the remaining non-zero eigenvalues), then the scenario demonstrated in Fig. 10.15 applies.

10.5 TIIV Model and Viral Load in a Sample of COVID-19 Patients

In the study by Wang et al. [7] the data of the sample of eight COVID-19 patients described in Sect. 10.2.1 was interpreted in the context of the TIIV model (10.43). Wang et al. fitted for each patient the TIIV model parameters β, p, k_1, k_2, and k_3 of Eq. (10.43) to the sputum viral load data. In a subsequent study [21], based on the estimates obtained by Wang et al., for each patient the unstable eigenvector $\mathbf{v}_4$, the eigenvalues λ_2, λ_3, λ_4, and the amplitude equation coefficients C_1, C_2, C_3, C_4 were computed from Eqs. (10.46), (10.50), and (10.57), respectively. To this end, the cubic equation (10.46) was solved numerically. Subsequently, the state space TIIV model (10.43) and the amplitude space TIIV model (10.57) were solved numerically for each patient.

10.5.1 Main Results Illustrated for Four Patients

Figure 10.16 shows the viral load measurements [6] of patient 1 as well as the TIIV modeling results. Panel (a) shows the course of the infection of patient 1 in terms of the measured viral load (gray circles) as shown in previous panels (e.g., panel (a) of Fig. 10.1). It also shows the best-fit solution $V(t)$ of Eq. (10.43). Just as in the context of the TIV model, the solution $V(t)$ of the TIIV model captures the qualitative features of the viral load pattern of patient 1 by showing a relatively quickly increasing viral load dynamics and a relatively slowly leveling off dynamics.

Panels (b) and (c) show the phase curves (T, I_1, I_2, V) computed from Eq. (10.43) in two subspaces: the two-dimensional I_1-I_2 subspace (panel (b)) and the three-dimensional I_1-I_2-V subspace (panel (c)). All variables are shown as percentage values of their respective maximum values X_{max} like $X_r = 100\,X/X_{max}$. As can be seen in panel (b), at the beginning of the infection target cells were turned into latently infected cells (I_1) and infected virus-producing cells (I_2), that is, I_1 and I_2 increased. At a certain point in time the number of cells I_1 in the latent period reached a maximum value ($I_{1,r} = 100\%$). Subsequently, I_1 decreased over time,

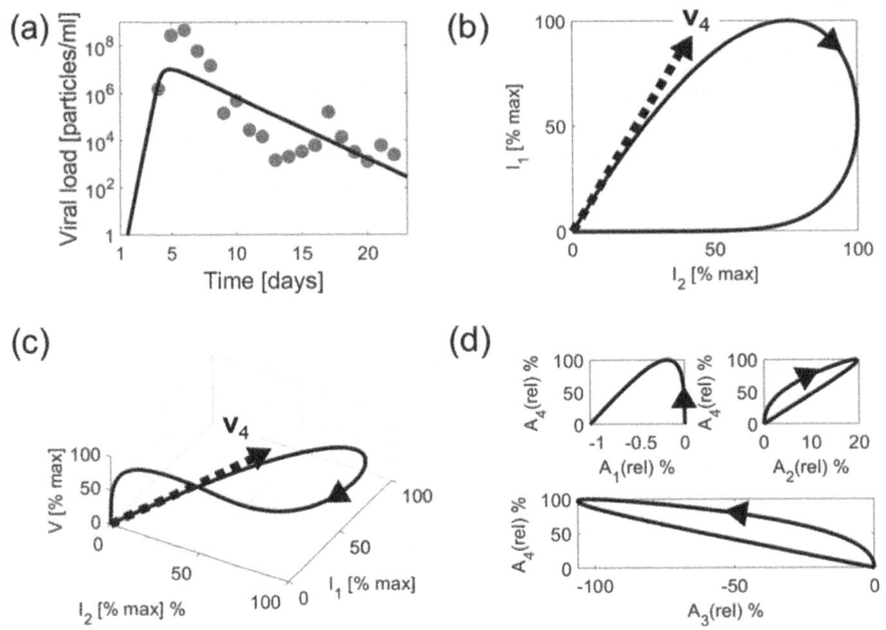

Fig. 10.16 Progression of COVID-19 in patient 1 in state and amplitude space as seen within a TIIV modeling framework. Panel **a**: Viral load trajectory $V(t)$ over time of patient 1 (gray circles) solution $V(t)$ (solid line) computed from Eq. (10.43). Panels **b** and **c**: The evolution of the disease state $\mathbf{X}(t)$ is shown as phase curves (solid lines) in the I_1-I_2 and (I_1, I_2, V) subspaces. The order parameter $\mathbf{v}_4$ (thick dotted line) is also shown (magnified in length) as obtained from Eq. (10.50). Panel **d**: Amplitude dynamics computed from Eq. (10.57). The three panels show A_4 versus A_1, A_2, and A_3, respectively. Parameters and initial conditions [7]: $\beta = 1.9 \cdot 10^{-6}$/(d $\times$ (particles/ml)), $p =$ 590 particles/(d $\times$ cell), $k_1 = 33$/d, $k_2 = 2$/d, $k_3 = 0.6$/d, $T(0) = T_{st} = 6 \cdot 10^4$ cells/ml, $I_1(0) = I_2(0) = 0$, $V(0) = 10^{-4}$ particles/ml

while I_2 (i.e., the number of virus-producing cells) still increased until I_2 assumed its maximum value ($I_{2,r} = 100\%$). Subsequently, both I_1 and I_2 decreased towards zero. Panel (c) presents the dynamics of the two types of infected cells together with the dynamics of the viral load. The two-dimensional phase curve shown in panel (b) corresponds to the projection of the three-dimensional phase curve shown in panel (c) onto the I_1-I_2 plane. Note that the top view presented in panel (c) may create a somewhat deceiving impression about the dynamics of $V(t)$. Therefore, panel (c) should be viewed together with panel (a). As shown in panel (a) of Fig. 10.16 the viral load $V(t)$ increased monotonically until it reached its maximum value $V_{\max}$ and, subsequently, decreased. Panel (c) demonstrates that the viral load V continuously increased while the infected cells of type I_1 and I_2 increased and, subsequently, decreased. At the time point t_p, when the viral load reached its peak value ($V_{\max}$ in panel (a) or $V_r = 100\%$ in panel (c)), the concentrations of the two types of infected cells were decaying. Finally, the number of target cells decayed during the whole course of the infection (graph not shown).

Table 10.1 Results of the TIIV model-based analysis of SARS-CoV-2 infections in a sample of eight patients ("P." stands for patient)

P.	λ_2 [1/d]	λ_3 [1/d]	λ_4 [1/d] (λ_{max})	τ [d]	τ [h]	T_2 [h]	V_{max} [copies/ml]	$A_{4,max}$ [copies/ml]	θ[°]
1	-30.4	-11.5	6.3	0.16	3.8	2.7	$1.0 \cdot 10^7$	$5.6 \cdot 10^6$	76
2	$-14.9 - 6.7i$	$-14.9 + 6.7i$	7.2	0.14	3.3	2.3	$7.9 \cdot 10^6$	$4.0 \cdot 10^6$	4
3	-70.9	-35.1	21.7	0.05	1.1	0.8	$3.9 \cdot 10^5$	$4.7 \cdot 10^5$	1
4	$-6.5 + 7.8i$	$-6.5 - 7.8i$	7.0	0.14	3.4	2.4	$9.4 \cdot 10^6$	$4.0 \cdot 10^6$	13
7	$-4.7 + 5.9i$	$-4.7 - 5.9i$	5.6	0.18	4.3	3.0	$7.6 \cdot 10^6$	$3.1 \cdot 10^6$	0.5
8	$-65.5 + 8.8i$	$-65.5 - 8.8i$	31.4	0.03	0.8	0.5	$7.2 \cdot 10^7$	$3.3 \cdot 10^7$	49
10	$-6.1 - 8.7i$	$-6.1 + 8.7i$	9.1	0.11	2.7	1.8	$6.9 \cdot 10^6$	$2.8 \cdot 10^6$	3
14	-73.0	-28.1	17.1	0.06	1.4	1.0	$1.4 \cdot 10^6$	$7.2 \cdot 10^5$	6

Importantly, the unstable eigenvector $\mathbf{v}_4$ is plotted in panels (b) and (c). As can be seen in both panels, the disease state of the SARS-CoV-2 infection of patient 1 evolved during the initial stage of the disease closely along the direction specified by $\mathbf{v}_4$. That is, at the beginning of the infection the disease state $\mathbf{X}(t)$ satisfied a saddle dynamics characterized by the direction $\mathbf{v}_4$ and the amplitude $A_4(t)$. After the initial period, the dynamics branched off from the $\mathbf{v}_4$ direction. Panels (b) and (c) demonstrate that the SARS-CoV-2 initial stage infection dynamics in patient 1 followed a disease order parameter.

The eigenvalues λ_2, λ_3, λ_4 of patient 1 are reported in Table 10.1 and are discussed in more detail in Sect. 10.5.2. At this stage, it should only be mentioned that the saddle dynamics of patient 1 exhibited three real-valued eigenvalues λ_2, λ_3, and λ_4 (in addition to $\lambda_1 = 0$). Consequently, the four amplitudes $A_1, \ldots, A_4$ were real-valued. Panel (d) presents the amplitude dynamics $\mathbf{A}(t) = (A_1, A_2, A_3, A_4)$ in terms of several phase curves. Since the order parameter amplitude A_4 is supposed to play the dominate role, the amplitude dynamics is illustrated in terms of phase curves A_4 versus A_1, A_4 versus A_2, and A_4 versus A_3. All amplitudes $A_1, \ldots, A_4$ were rescaled by the maximum value $A_{4,max}$ of A_4 like $A_j(rel) = 100 A_j / A_{4,max}$.

Note that in general the sign of an amplitude depends on the selected sign of its corresponding eigenvector. Without loss of generality, any eigenvector may be multiplied by -1 in order to obtain an eigenvector again. Such a multiplication by -1 implies that the corresponding amplitude is multiplied by -1 as well. That is, the substitution $\mathbf{v}_k \to -\mathbf{v}_k$ implies $A_k \to -A_k$. Throughout this book the signs of some of the eigenvectors $\mathbf{v}_k$ have been chosen to obtain for the sake of presentation purposes amplitude functions A_k that are positive,

The amplitude A_1 of patient 1 (and in fact for all patients) was found to be negative. This is consistent with the fact that A_1 corresponds to the amplitude of the eigenvector $\mathbf{v}_1 = (1, 0, 0, 0)$ and, consequently, primarily describes the decay of target cells. In line with the comment from above, an eigenvector $\mathbf{v}_1 = (-1, 0, 0, 0)$ would have produced a positive amplitude A_1 rather than a negative one. Regardless of the choice of the sign of $\mathbf{v}_1$, the amplitude A_1 captures the decay of the target cells. For patient 1 (and in fact for all patients) A_4 was positive. A_2 and A_3 assumed positive or negative values. In all two-dimensional plots, the amplitude A_4 was plotted on the vertical axis. Consequently, if A_4 increases while its to-be-compared amplitude A_j with $j = 1, 2, 3$ varies only little, then a vertical graph in the A_j-A_4 plane under consideration occurs. The two-dimensional phase curves shown in panel (d) demonstrate that in the very beginning of the infection of patient 1, A_4 increased while variations of the remaining amplitudes A_j with $j = 1, 2, 3$ were negligibly small. That is, it was found that all three phase curves $A_4(A_1)$, $A_4(A_2)$, and $A_4(A_3)$ of the initial stage SARS-CoV-2 infection of patient 1 were given by vertical lines close to the unstable virus-free fixed point $\mathbf{A}_{st} = (0, 0, 0, 0)$. This initial increase of A_4 that was not accompanied by any substantial dynamics of one of the remaining amplitudes, reflects again that the disease state $\mathbf{X}$ of the SARS-CoV-2 infection of patient 1 initially evolved along the order parameter $\mathbf{v}_4$ (as can be seen in panels (b) and (c)).

When taking a more quantitative view, it was found that the amplitude A_2 over the whole course of the SARS-CoV-2 infection attained a maximum value of about 20% of $A_{4,max}$ (see the top-right subpanel in panel (d)). The amplitude A_3 attained a maximum value of about 106% of $A_{4,max}$ (see the bottom subpanel in panel (d)). That is, while A_3 did not make an initial contribution to the evolution of the disease state $\mathbf{X}(t)$, A_3 played an essential role over the whole course of the disease.

As mentioned above, fixed points of the TIIV model assume the form $\mathbf{X} = (Y, 0, 0, 0)$ with $Y = T(\infty) \geq 0$, which implies

$$\mathbf{X}_{st} = (Y, 0, 0, 0) \Leftrightarrow \mathbf{A}_{st} = (A_{1,st}, 0, 0, 0) \tag{10.65}$$

(for a counterpart on the level of epidemiological models see Eq. (4.55)). In particular, from Eq. (10.51) it follows that

$$T(\infty) = T_{st} + A_{1,st} \Rightarrow T_{st} - T(\infty) = -A_{1,st} \ , \tag{10.66}$$

where T_{st} correspond to the concentration of target cells of healthy adults before the infection takes place. Consequently, $A_{1,st}$ measures the decay of target cells during the infection (see also Eq. (4.56) for a counterpart on the epidemiological level). Returning to Fig. 10.16, the drop of A_1 to a stationary value $A_{1,st}$ of about 2% of $A_{4,\max}$ as shown in panel (d) of Fig. 10.16 (see the top left subpanel) reflects that eventually the disease state of patient 1 approached a neutrally stable fixed point $\mathbf{X}_{st} = (Y, 0, 0, 0)$ (i.e., the coronavirus disease 2019 of patient 1 eventually declined). The target cell concentration in the affected areas dropped during the course of the disease by $|A_{1,st}|$ (i.e., 2% of $A_{4,\max}$).

Figure 10.17 presents data and modeling results for patient 2. As in Fig. 10.16, in Fig. 10.17 panels (a), (b), (c), and (d) present the viral load dynamics (panel (a)), the state space dynamics (panels (b) and (c)), and the amplitude dynamics (panel (d)) of the coronavirus disease 2019 of patient 2 as seen from a TIIV modeling perspective. In particular, in panels (b) and (c) the unstable eigenvector $\mathbf{v}_4$ of the virus-free state

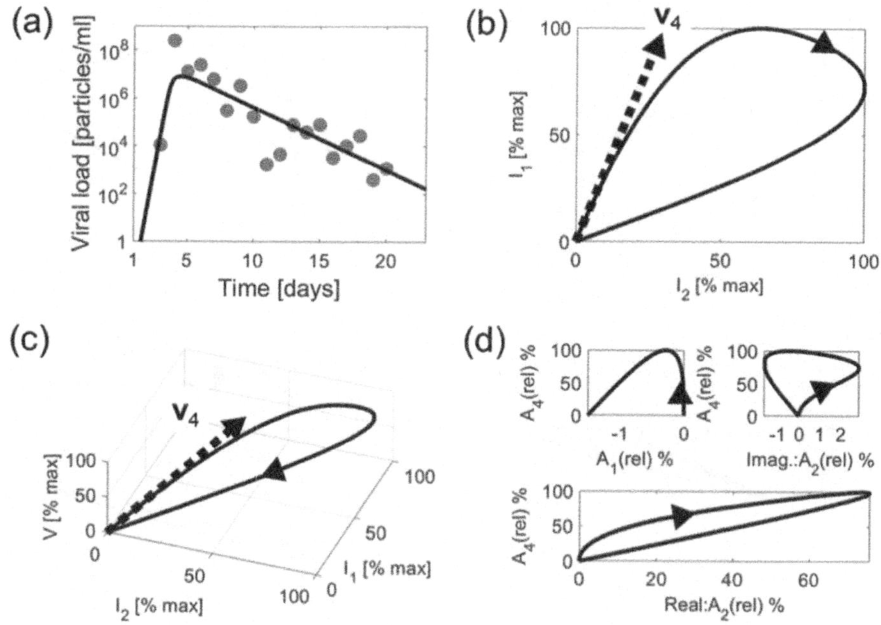

Fig. 10.17 COVID-19 TIIV model dynamics of patient 2 in state and amplitude space. Panels as in Fig. 10.16. Parameters and initial conditions [7]: $\beta = 3.6 \cdot 10^{-6}$/(d $\times$ (particles/ml)), $p = 1.5 \cdot 10^4$ particles/(d $\times$ cell), $k_1 = 0.6$/d, $k_2 = 2$/d, $k_3 = 20$/d, $T(0) = T_{st} = 6 \cdot 10^4$ cells/ml, $I_1(0) = I_2(0) = 0$, $V(0) = 10^{-4}$ particles/ml

of the affected regions in the lung of patient 2 is shown. As can be seen in panels (b) and (c), as expected, the initial dynamics of COVID-19 in patient 2 followed closely the unstable eigenvector $\mathbf{v}_4$.

The unstable virus-free state of patient 2 was given in terms of a saddle point involving two complex-conjugated eigenvalues with negative real parts [22] in addition to the eigenvalues $\lambda_1 = 0$ and $\lambda_4 = \lambda_{\max} > 0$ (see Table 10.1 for details). Consequently, the eigenvectors $\mathbf{v}_2$ and $\mathbf{v}_3$ and their corresponding amplitudes A_2 and A_3 satisfied $\mathbf{v}_3 = \mathbf{v}_2^*$, $A_3 = A_2^*$, and

$$A_2\mathbf{v}_2 + A_3\mathbf{v}_3 = 2\mathrm{Re}(A_2\mathbf{v}_2) \tag{10.67}$$

As a result, the real and imaginary parts of A_2 (or alternatively A_3) can be considered as the two independent quantities occurring in the superposition $A_2\mathbf{v}_2 + A_3\mathbf{v}_3$. For this reason, in panel (d) the dynamics of the amplitudes A_1, A_2, A_3, A_4 is shown in terms of the phase curves A_4 versus A_1, A_4 versus the imaginary part of A_2, and A_4 versus the real part of A_2.

Visual inspection of the three phase curves reveals that close to the disease onset only A_4 increased in magnitude while variations in A_1, Imag(A_2), and Re(A_2) were negligibly small. Due to this phenomenon the initial SARS-CoV-2 infection of patient

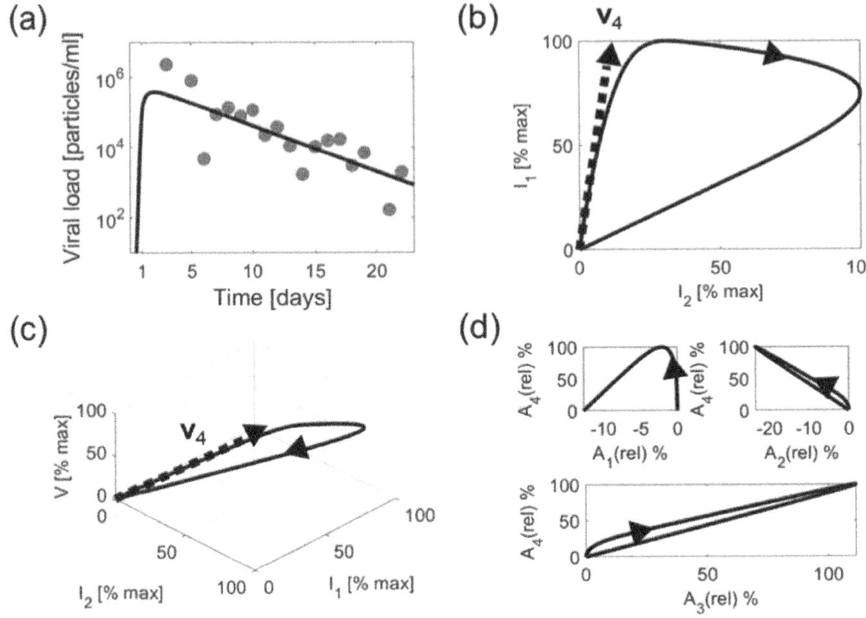

Fig. 10.18 COVID-19 disease of patient 3 described in state and amplitude space of the TIIV model (10.43). Panels as in Fig. 10.16. Parameters and initial conditions [7]: $\beta = 6 \cdot 10^{-4}$/(d $\times$ (particles/ml)), $p = 5000$ particles/(d $\times$ cell), $k_1 = 0.3$/d, $k_2 = 2$/d, $k_3 = 82$/d, $T(0) = T_{st} = 6 \cdot 10^4$ cells/ml, $I_1(0) = I_2(0) = 0$, $V(0) = 10^{-3}$ particles/ml

2 was characterized by vertical lines – just as for patient 1. Panels (b), (c), and (d) taken together illustrate that the TIIV model order parameter $\mathbf{v}_4$ and its order parameter amplitude A_4 determined entirely the SARS-CoV-2 initial stage infection in the affected regions of patient 2.

Panel (d) of Fig. 10.17 may be used to conduct a more quantitative discussion of the SARS-CoV-2 infection of patient 2. Accordingly, the imaginary part of the amplitude A_2 (or A_3) played a minor role during the course of the infection of patient 2. It only reached a maximum value of 2% of $A_{4,max}$ (see top-right subpanel). In contrast, the real part of A_2 (or A_3) attained a maximum value of about 76% of $A_{4,max}$ (see bottom subpanel) and contributed substantially to the progression of COVID-19 in patient 2 when considering the whole 25 days time span until the viral load decreased to non-detectable levels.

Figures 10.18 and 10.19 present data and the results of the model-based analyses for patients 3 and 4. Panels (a), (b), (c), and (d) show the respective viral load dynamics, the state space dynamics, and amplitude dynamics. The order parameters $\mathbf{v}_4$ of the two patients are shown in panels (b) and (c) of the respective figures. As can be seen, the SARS-CoV-2 infections for both patients followed in their initial stages their respective order parameters and subsequently branched off. The eigenvalues λ_2, λ_3, λ_4 of patients 3 and 4 will be discussed in Sect. 10.5.2 and are reported

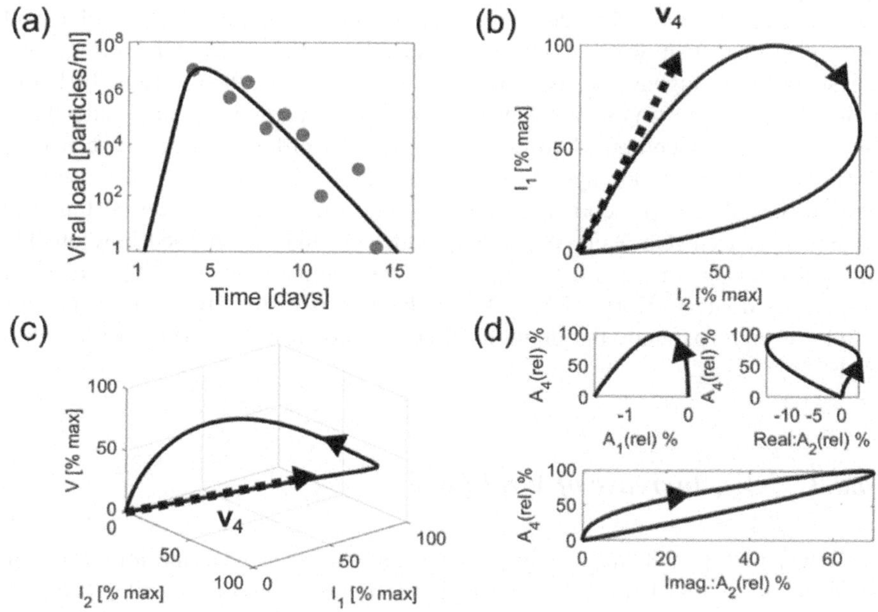

Fig. 10.19 COVID-19 disease of patient 4 described in state and amplitude space of the TIIV model (10.43). Panels as in Fig. 10.16. Parameters and initial conditions [7]: $\beta = 5.1 \cdot 10^{-6}$/(d × (particles/ml)), $p = 1200$ particles/(d × cell), $k_1 = k_2 = k_3 = 2$/d, $T(0) = T_{st} = 6 \cdot 10^4$ cells/ml, $I_1(0) = I_2(0) = 0$, $V(0) = 10^{-4}$ particles/ml

in Table 10.1. The SARS-CoV-2 infection of the affected lung regions of patient 3 was characterized by real-valued eigenvalues. In contrast, the infection of the affected lower respiratory tract regions of patient 4 exhibited a pair of complex-conjugated eigenvalues. Accordingly, for patient 3 panel (d) shows the dynamics of A_1, A_2, A_3, A_4 by means of two-variable phase curves, whereas for patient 4 panel (d) shows the dynamics of A_1, $\mathrm{Re}(A_2)$, $\mathrm{Imag}(A_2)$, and A_4. Irrespective of this difference, panels (d) in Figs. 10.18 and 10.19 illustrates that for patients 3 and 4 initially only A_4 increased substantially over time, whereas the remaining amplitude variables stayed almost constant. That is, the initial multiplication of the virus in the bodies of patients 3 and 4 was characterized by amplitude phase curves that started off as vertical lines.

10.5.2 Eigenvalues, Doubling Times, and Peak Viral Loads

The results obtained for the remaining four patients were qualitatively similar to those shown in Figs. 10.16, 10.17, 10.18 and 10.19 [21]. Table 10.1 reports the eigenvalues λ_2, λ_3, λ_4 for all eight patients. The virus-free fixed points of three patients exhibited only real-valued eigenvalues. The fixed points of the remaining five patients were characterized by a maximal real-valued eigenvalue and two eigenvalues that were complex-conjugated to each other. On average, the largest (positive) eigenvalue $\lambda_{\max}$ was M = 13.2/d (SD = 9.4/d). Table 10.1 also shows the e-folding times τ of the dominant (order parameter) amplitude A_4 in days and hours as computed from $\tau = 1/\lambda_{\max}$. Accordingly, on average it took M=0.11 days (SD = 0.06 days) or M = 2.6 h (SD = 1.3 h) for A_4 to increase by a factor e (i.e., 2.71). The corresponding doubling times in hours can be found in Table 10.1 as well. On average, the amplitudes A_4 that determined how quickly the SARS coronavirus 2 infected the lower respiratory tracts (i.e., the lungs) of the patients doubled every 1.8 h (SD = 0.9 h), that is every 110 minutes. For the sample of eight patients, the doubling times varied from 0.5 h (i.e., 30 minutes) to 3 h.

10.5.3 $\lambda_{\max}$ *Increase of Viral Load*

According to Eq. (9.8) the maximal eigenvalue λ_4 determines the initial growth rate of the viral load $V(t)$. When presenting $V(t)$ in a logarithmic scale, then $\lambda_{\max}$ corresponds to the slope of the function $V(t)$. Figure 10.20 shows again the viral load trajectories of patients 1 to 4 (as presented in panels (a) of Figs. 10.16 to 10.19). The approximation (9.8) is shown as well as dotted gray line. As expected, $\lambda_{\max}$ as listed in Table 10.1 corresponds to the slope of the viral load increase.

10.5.4 *Peak Viral Load* $V_{\max}$ *Determined by Order Parameter Amplitude* $A_{4,\max}$

Table 10.1 reports the model peak viral loads $V_{\max}$ of the patients. For a given patient $V_{\max}$ was defined as the maximal value of the TIIV model solutions $V(t)$ obtained for that patient. The patient specific values $A_{4,max}$ are reported as well. Figure 10.21 shows two scatter plots of $V_{\max}$ versus $A_{4,max}$. Patient 8 showed a $V_{\max}$ value that was by a factor 7 to 180 larger than the $V_{\max}$ values of all remaining patients. In panel (a) of Fig. 10.21 the $(V_{\max}, A_{4,\max})$ data point of patient 8 appears as an isolated point in the scatter plot. For this reason the data from patient 8 was excluded [24]. Panel (b) of Fig. 10.21 shows the scatter plot thus obtained. A correlation analysis based on the data of the remaining seven patients as shown in panel (b) revealed a positive linear correlation between $V_{\max}$ and $A_{4,max}$ with $R^2 = 0.93$. Accordingly, the maximal value $A_{\max}$ of the order parameter amplitude determined the peak viral load $V_{\max}$ in the sample of COVID-19 patients under consideration.

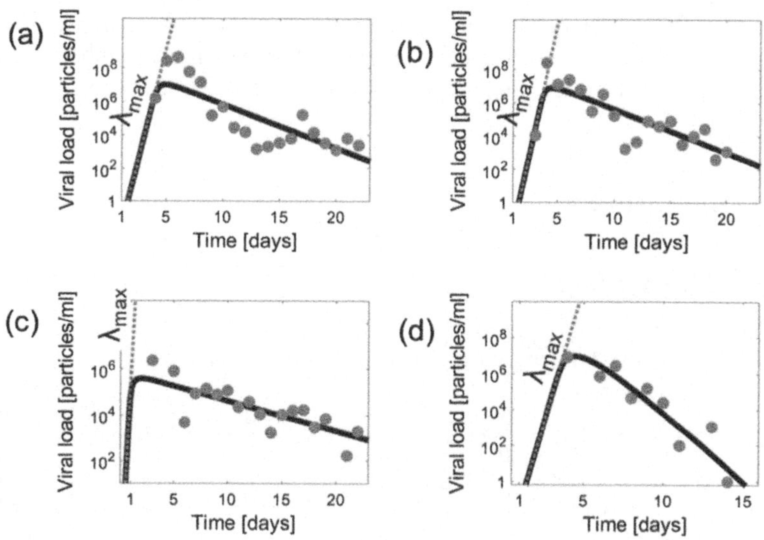

Fig. 10.20 The eigenvalue $\lambda_{\max}$ (i.e., λ_4) determines for patients 1 to 4 entirely the rate of the initial viral load increase. Gray circles and solid black lines show data and TIIV model solutions $V(t)$, respectively, as presented in panels (**a**) of Figs. 10.16 to 10.19. Dotted gray lines show the approximation defined by Eq. (9.8). Panels **a**, **b**, **c**, and **d** refer to patients 1,2,3, and 4, respectively

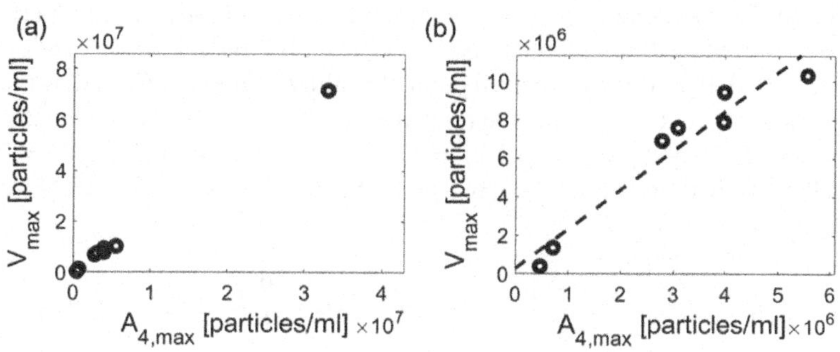

Fig. 10.21 Correlation between $V_{\max}$ and $A_{4,max}$ illustrated by means of scatter plots. Panels **a** and **b** present data from all patients (panel (**a**)) and when patient 8 is excluded (panel (**b**)), respectively. In panel **b** the regression line (dashed line) is shown as well

10.5.5 Latent Stage Determined by Order Parameter

In Table 10.1 the order parameter angles θ are reported, which according to Eqs. (10.62) and (10.63) determine quantitatively and qualitatively the relationship between variations in the number of latently infected cells ΔI_1 and the number of virus-producing infected cells ΔI_2. Angles larger than 45° were found for patients 1 and 8. The disease progression of the remaining patients was characterized by angles smaller than 45°. Consequently, during the initial COVID-19 stage of patients 1 and 8 the number of virus-producing cells increased faster than the number of latently infected cells. For the remaining patients, the opposite was true.

10.6 Other Models

10.6.1 TIIVV Model

The TIIVV model is defined by Eq. (9.21) and here repeated as

$$\begin{aligned} &\frac{\mathrm{d}}{\mathrm{d}t}T = -\beta T V_I\ , \ \frac{\mathrm{d}}{\mathrm{d}t}I_1 = \beta T V_I - k_1 I_1\ , \ \frac{\mathrm{d}}{\mathrm{d}t}I_2 = k_1 I_1 - k_2 I_2\ , \\ &\frac{\mathrm{d}}{\mathrm{d}t}V_I = \eta p I_2 - k_3 V_I\ , \ \frac{\mathrm{d}}{\mathrm{d}t}V_{NI} = (1-\eta) p I_2 - k_3 V_{NI}\ , \end{aligned} \tag{10.68}$$

where V_I and V_{NI} denote infectious and non-infectious virus particles (see Sect. 9.8). The dynamics of the non-infectious virus particles V_{NI} does not affect the stability of the virus-free fixed point. Consequently, the stability properties of the virus-free fixed point of the TIIVV model are like those of the TIIV model when considering the coefficient ηp in the evolution equation of V_I as counterpart to the parameter p in the evolution equation of V of the TIIV model. More explicitly, let $\mathbf{X}_{st} = (T_{st}, 0, 0, 0, 0)$ denote the fixed point of healthy adults not affected by the virus under consideration. Let δ denote the relative state of target cells defined in the usual way by $T = T_{st} + \delta$. Then, using $T V_I = (T_{st} + \delta) V_I \approx T_{st} V_I$ and $\mathrm{d}T/\mathrm{d}t = \mathrm{d}\delta/\mathrm{d}t$, for small perturbations out of the fixed point $\mathbf{X}_{st}$ the TIIVV model (10.68) can be linearized like

$$\begin{aligned} &\frac{\mathrm{d}}{\mathrm{d}t}\delta = -\beta T_{st} V_I\ , \ \frac{\mathrm{d}}{\mathrm{d}t}I_1 = \beta T_{st} V_I - k_1 I_1\ , \ \frac{\mathrm{d}}{\mathrm{d}t}I_2 = k_1 I_1 - k_2 I_2\ , \\ &\frac{\mathrm{d}}{\mathrm{d}t}V_I = \eta p I_2 - k_3 V_I\ , \ \frac{\mathrm{d}}{\mathrm{d}t}V_{NI} = (1-\eta) p I_2 - k_3 V_{NI}\ . \end{aligned} \tag{10.69}$$

Consequently, the relative state vector $\mathbf{u} = (\delta, I_1, I_2, V_I, V_{NI})$ in the case of small perturbations evolves like

$$\frac{\mathrm{d}}{\mathrm{d}t}\mathbf{u} = L\,\mathbf{u}\ ,\ L = \begin{pmatrix} 0 & 0 & 0 & -\beta T_{st} & 0 \\ 0 & -k_1 & 0 & \beta T_{st} & 0 \\ 0 & k_1 & -k_2 & 0 & 0 \\ 0 & 0 & \eta p & -k_3 & 0 \\ 0 & 0 & (1-\eta)p & 0 & -k_4 \end{pmatrix}. \tag{10.70}$$

The linearization matrix exhibits five eigenvalues. Two eigenvalues are given by $\lambda_1 = 0$ with eigenvector $\mathbf{v} = (1, 0, 0, 0, 0)$ and $\lambda_5 = -k_4$ with eigenvector $\mathbf{v} =$

$(0, 0, 0, 0, 1)$. The remaining three eigenvalues can be computed from a 3×3 matrix similar to the 3×3 submatrix of the matrix L defined in Eq. (10.44) of the TIIV model. In particular, the three remaining eigenvalues can be computed from the determinant (10.45) when replacing p by ηp. From Eq. (10.46) it then follows that the eigenvalues are given by the cubic equation

$$(\lambda + k_1)(\lambda + k_2)(\lambda + k_3) = \eta p \beta T_{st} k_1 \ . \tag{10.71}$$

Consequently, for the TIIVV model the bifurcation parameter β_w and its critical value $\beta_{w,crit}$ read

$$\beta_w = \eta p \beta T_{st} \ , \ \beta_{w,crit} = k_2 k_3 \ . \tag{10.72}$$

For $\beta_w > \beta_{w,crit}$ ($\beta_w < \beta_{w,crit}$) the virus-free fixed point $\mathbf{X}_{st} = (T_{st}, 0, 0, 0, 0)$ of the TIIVV is unstable (neutrally stable).

10.6.2 TV Model

The scaled TV model is defined by

$$V = rI \tag{10.73}$$

(see Eq. (9.13)) and

$$\frac{\mathrm{d}}{\mathrm{d}t} T = -\beta' I T \ , \ \frac{\mathrm{d}}{\mathrm{d}t} I = \beta' I T - k_1 I \ . \tag{10.74}$$

(see Eq. (9.14)) with $\beta' = r\beta$. As discussed in Sect. 9.5.2, the scaled TV model (10.74) is equivalent to the SIR model (3.22). Exploiting this equivalence, the amplitude space description of the scaled TV model can be conveniently obtained. For sake of clarity, let β_{SIR} denote the effective contact rate β occurring in the SIR model (3.22). Then, when S and I in the SIR model (3.22) are replaced by T and I (where I denotes the equivalent of the number of virus particles when measured in cells) and the substitutions

$$\gamma = k_1 \ , \ \frac{\beta_{SIR}}{N} = \beta' \tag{10.75}$$

are made, then the SIR model (3.22) becomes the scaled TV model (10.74). In order to derive the amplitude equations of the scaled TV model (10.74) for the virus-free fixed point with $T = T_{st}$ and $V_{st} = 0 \Rightarrow I_{st} = 0$, the additional substitutions

$$N = T_{st} \Rightarrow \beta_{SIR} = \beta' T_{st} \tag{10.76}$$

can be used. Let us summarize the results. The state vector $\mathbf{X} = (T, I)$ of the scaled TV model reads

$$\begin{pmatrix} T \\ I \end{pmatrix} = \begin{pmatrix} T_{st} \\ 0 \end{pmatrix} + \sum_{i=1}^{2} A_i \mathbf{v}_i \tag{10.77}$$

with the eigenvector $\mathbf{v}_1 = (1, 0)$ and

$$\mathbf{v}_2 = \frac{1}{\sqrt{1+g^2}} \begin{pmatrix} -1 \\ g \end{pmatrix} \tag{10.78}$$

with

$$g = \frac{\beta' T_{st} - k_1}{\beta' T_{st}} , \tag{10.79}$$

which can be shown by using the substitutions listed in Eqs. (10.75) and (10.76) in combination with Eqs. (4.19) and (4.20) that are referring to the SIR model. For $\beta' T_{st} > k_1$ the virus-free fixed point is unstable (see Sect. 9.5.1). In this case $\mathbf{v}_2$ denotes the unstable eigenvector of the TV model, that is, the TV order parameter.

The amplitudes A_1 and A_2 of the scaled TV model that occur in Eq. (10.77) satisfy the evolution equations

$$\begin{aligned} \frac{\mathrm{d}}{\mathrm{d}t} A_1 &= \frac{1-g}{\sqrt{1+g^2}} A_2 p(A_1, A_2) , \\ \frac{\mathrm{d}}{\mathrm{d}t} A_2 &= \lambda_2 A_2 + A_2 p , \\ p &= \beta' \left(A_1 - \frac{A_2}{1+g^2} \right) , \end{aligned} \tag{10.80}$$

which can be shown by substituting the relations listed in (10.75) and (10.76) into Eqs. (4.38), (4.39), and (4.42). The eigenvalue λ_2 that occurs in the amplitude equation description of the scaled TV model in Eq. (10.80) reads

$$\lambda_2 = \beta' T_{st} - k_1 \tag{10.81}$$

(see again Eq. (4.19)). The initial amplitudes $A_1(0)$ and $A_2(0)$ can be computed from the initial states $T(0) = T_{st}$ and $I(0) = V(0)/r$ by inverting Eq. (10.77), which leads to

$$A_1(0) = \frac{I(0)}{g} , \quad A_2(0) = \frac{\sqrt{1+g^2}}{g} I(0) , \tag{10.82}$$

which can be shown by putting $S_0 - N = T(0) - T_{st} = 0$ in Eqs. (4.50). For an acute infection for which $\lambda_2 > 0$ holds it follows that during the initial stage A_2 dominates over A_1 (see Sect. 4.2.6) such that $\mathbf{X} \approx \mathbf{X}_{st} + \mathbf{v}_2 A_2$ (see Eq. (4.60)), which implies $V(t)$ can be approximated by $V_a(t)$ defined by

$$V_a(t) = rI(t) = rv_{2,I} A_2(t) = b \exp\{\lambda_2 t\} \tag{10.83}$$

with $b = rv_{2,I} A_2(0)$. Accordingly, λ_2 corresponds to $\lambda_{\max}$ shown in Fig. 9.5.

10.7 Complex-Valued Eigenvalues λ of the TIIV Model and Analytical Expressions for λ

Let $\lambda = x + iy$ denote a complex-valued solution of Eq. (10.46), where x and y denote the real and imaginary parts of λ. Substituting this ansatz into Eq. (10.46), a complex-valued equation is obtained. When solving the imaginary part of the equation, the relation

$$y^2 = (x + k_2)(x + k_3) + (x + k_1)(2x + k_2 + k_3) \tag{10.84}$$

is obtained. Substituting this expression for y^2 into the real part of the aforementioned complex-valued equation, yields a cubic equation for x that reads

$$(x + \gamma_{12})(x + \gamma_{13})(x + \gamma_{23}) = -p\beta T_{st} k_1 \tag{10.85}$$

with $\gamma_{12} = k_1 + k_2$, $\gamma_{13} = k_1 + k_3$, $\gamma_{23} = k_2 + k_3$, which implies $\gamma_{ij} > 0$. Since the right-hand side of Eq. (10.85) is negative for any parameters p, T_{st}, and k_1 larger than zero, it follows that Eq. (10.85) does not exhibit a solution for $x \geq 0$. Consequently, any complex-valued eigenvalue λ of the virus-free fixed point of the TIIV model as defined by Eq. (10.46) exhibits a negative real part.

In general, roots of cubic equations such as Eq. (10.46) may be determined using analytical rather than numerical approaches. In particular, analytical expressions for the solutions of Eq. (10.46) can be found in Ref. [10].

References

1. T. D. Frank. SARS-Coronavirus-2 nonlinear dynamics in patients: three-dimensional state and amplitude state description. J. Phys. Soc. Jpn. **90**:article 073802 (2021)
2. H. Ikeda, R. J. de Boer, K. Sato, S. Morita, N. Misawa, Y. Koyanagi, K. Aihara, and S. Iwami. Improving the estimation of the death rate of infected cells from time course data during acute phase of virus infections: application to acute HIV-1 infection in a humanized mouse model. Theor. Biol. Med. Model. **11**:article 22 (2014)

3. M.A. Nowak, Lloyd, G.M. Vasquez, T.A. Wiltrout, L.M. Wahl, N. Bischofberger, J. Williams, A. Kinter, A.S. Fauci, V.M. Hirsch, J.D. Lifson. Viral dynamics of primary viremia and antiretrovial therapy in Simian immunodeficiency virus infection. J. Virol. **71**, 7518–7525 (1997)
4. T.D. Frank, Eigenvalue analysis of SARS-CoV-2 viral load data: illustration for eight COVID-19 patients. Int. J. Data Sci. Anal. (in press)
5. M.M. Böhmer, U. Buchholz, V.M. Corman, M. Hoch, K. Katz et al., Investigation of a COVID-19 outbreak in Germany resulting from a single travel-associated primary case: a case series. The Lancet Infect. Dis. **20**, 920–928 (2020)
6. R. Wölfel, V.M. Corman, W. Guggemos, M. Seilmaier, S. Zange et al., Virological assessment of hospitalized patients with covid-2019. Nature **581**, 465–469 (2020)
7. S. Wang, Y. Pan, Q. Wang, H. Miao, A.N. Brown, L. Rong, Modeling the viral dynamics of SARS-CoV-2 infection. Math. Biosci. **328:**article 108438 (2020)
8. Y. Pan, D. Zhang, P. Yang, L.M. Poon, Q. Wang, Viral load of SARS-CoV-2 in clinical samples. Lancet **20**, 411–412 (2020)
9. K.A. Walsh, K. Jordan, B. Clyne, D. Rohde, L. Drummond et al., SARS-CoV-2 detection, viral load and infectivity over the course of an infection. J. Infect. **81**, 357–371 (2020)
10. A.M. Smith, F.R. Adler, A.S. Perelson, An accurate two-phase approximative solution to acute viral infection model. J. Math. Biol. **60**, 711–726 (2010)
11. T. Frank, *Determinism and Self-organization of Human Perception and Performance* (Springer, Berlin, 2019)
12. O. Diekmann, J.A.P. Heesterbeek, *Mathematical Epidemiology of Infectious Diseases* (Wiley, Chichester, 2000)
13. P. van den Driessche, J. Watmough, Reproduction numbers and sub-threshold endemic equilibria for compartment models of disease transmission. Math. Biosci. **180**, 29–48 (2002)
14. A. Balti, D. Zugaj, F. Fenneteau, P.O. Tremblay, F. Nekka, Dynamical systems analysis as an additional tool to inform treatment outcomes: the case study of a quantitative systems pharamcology model of immuno-oncology. Chaos **31**:article 023124 (2021)
15. K.A. Pawelek, G.T. Huynh, M. Quinlivan, A. Cullinane, L. Rong, A.J. Perelson. Modeling within-host dynamics of influenza virus infection including immune response. PLoS Comput. Biol. **8**:article e1002588 (2012)
16. R.A. Saenz, M. Quinlivan, D. Elton, S. MacRae, A.S. Blunden, J.A. Mumford et al., Dynamics of influenza virus infection and pathology. J. Virol. **84**, 3974–3983 (2010)
17. N. Neant et al., Modeling SARS-CoV-2 viral kinetics and association with mortality in hospitalized patients from the French COVID cohort. PNAS **118**, e2017962118 (2021)
18. P. Czuppon, F. Debarre, A. Goncalves, O. Tenaillon, A.S. Perelson, J. Guedj, F. Blanquart, Success of prophylactic antiviral therapy for SARS-CoV-2: predicted critical efficiacies and impact of different drug-specific mechanisms of action. PLoS Comput. Biol. **17**, e1008752 (2021)
19. A. Goncalves, Y. Bertrand, R. Ke, E. Comets, X. de Lamballerie, D. Malvy, Pizzorno, O. Terrier, M.R. Calatrava, F. Mentre, P. Smith, A.S. Perelson, J. Guedj. Timing of antiviral treatment initiation is critical to reduce SARS-CoV-2 viral load. CPT: Pharmacometrics Syst. Pharmacol. **9**, 509–514 (2020)
20. A.S. Perelson, R. Ke, Mechanistic modeling of SARS-CoV-2 and other infectious diseases and the effect of therapeutics. Clin. Pharmacol. Therapeutics **109**, 829–840 (2021)
21. T.D. Frank. SARS-coronavirus-2 infections: biological instabilities characterized by order parameters. Phys. Biol. **19**:article 036001 (2022)
22. D. Cai, K.I. Nishikawa, B. Lembege, Magnetotail field topology in a three-dimensional gobal particle simulation. Plasma Phys. Control. Fusion **48**, B123–B135 (2006)
23. R. Sender et al., The total number and mass of SARS-CoV-2 virions. Proc. Natl. Acad. Sci. USA **118**:article 2024815118 (2021)
24. F.J. Gravetter, L.B. Wallnau, *Statistics for the Behavioral Sciences*, 10th edn. (Cengage, Boston, 2017)

Index

T. D. Frank, *COVID-19 Epidemiology and Virus Dynamics*,
Understanding Complex Systems, https://doi.org/10.1007/978-3-030-97178-6

The manufacturer's authorised representative in the EU is Springer Nature Customer Service Centre GmbH, Europaplatz 3, 69115 Heidelberg, Germany. If you have any concerns regarding our products, please contact ProductSafety@springernature.com

Printed and bound by CPI Group (UK) Ltd, Croydon, CR0 4YY
28/09/2026
02224088-0018